T0249574

Physics and Engineering
of Radiation Detection

Physics and Engineering
of Radiation Detection

Syed Naeem Ahmed

Queen's University, Kingston, Ontario

AMSTERDAM • BOSTON • HEIDELBERG • LONDON
NEW YORK • OXFORD • PARIS • SAN DIEGO
SAN FRANCISCO • SINGAPORE • SYDNEY • TOKYO

Academic Press is an imprint of Elsevier

Academic Press is an imprint of Elsevier

525 B Street, Suite 1900, San Diego, CA 92101-4495, USA
84 Theobald's Road, London WC1X 8RR, UK
Radarweg 29, PO Box 211, 1000 AE Amsterdam, The Netherlands
30 Corporate Drive, Suite 400, Burlington, MA 01803, USA

First edition 2007

British Library Cataloguing in Publication Data
A catalogue record for this book is available from the British Library

Library of Congress Cataloging-in-Publication Data
A catalog record for this book is availabe from the Library of Congress

For information on all Academic Press publications
visit our web site at books.elsevier.com

Printed and bound by CPI Group (UK) Ltd, Croydon, CR0 4YY

Transferred to Digital Print 2011

ISBN–13: 978-0-12-045581-2

ISBN–10: 0-12-045581-1

Dedicated to:
my parents whose loving memories still guide me through difficult times; my wife, Rahat, without whose utmost love, support, and encouragements it would not have been possible to write this book; and to my children Hiba, Shozab, and Shanib, who have taught me things about life I could never have learned myself.

Contents

2   Interaction of Radiation with Matter   65

Preface

Contrary to common perception, radiation has enormous potential of benefiting mankind. For example, use of radiation in medical diagnostics in the form of CT and nuclear scans has enabled the physicians to perform diagnoses that would not have been possible otherwise. Another example is the use of radiation to destroy cancerous growths, a process generally known as *radiation therapy*. It is true that radiation can induce harm as well but a close examination reveals that its benefits certainly outweigh its potential hazards. This realization has led to rapid advancements in theory and applications of radiation interactions and its measurements. New types of radiation detectors and sources are being constantly developed in laboratories around the world. Also, a number of annually held international conferences are devoted just to the developments of radiation measuring devices.

During my career as a physicist, working primarily on radiation detectors, I always felt a need for a book that would not only discuss the technological aspects of the field but would also give a thorough account of the underlying physical principles. The scarcity of such books led me to think about writing one myself. However, those who understand the field would appreciate that writing such a book is not an easy task due mainly to the fast paced developments in the related technologies. The strategy that I therefore adopted was to concentrate on theories, methodologies, and technologies that are of fundamental value in terms of understanding the conceptual basis of the radiation devices.

The sole purpose of the book is not to introduce the reader to the working principles of different types of radiation detectors. It has been designed and written such that it encompasses all aspects of design, development, and effective use of the detection devices. Therefore chapters on statistics, data analysis, software for data analysis, dosimetry, and spectroscopy have also been included. It can be used as a text for related courses in physics, nuclear engineering, physical chemistry, and medical physics. It can also be used as a reference by professionals and students working in the related fields.

Most of the courses related to radiation measurements start with an introduction to different types of radiation and their sources. I have adopted the same strategy. The first chapter introduces the reader to various types of radiation and their sources. It also includes sections on radioactivity and its measurements. Chapter 2 deals with the mechanisms by which radiation interacts with matter. Those who want to understand the working principles of radiation detectors, must go through this chapter as thoroughly as possible. The next chapter introduces the reader to the principles of gas filled detectors, such as proportional counters. Gas filled detectors are the earliest built radiation detectors and are still extensively used in different fields. The important concepts, such as electron-ion pair generation, recombination, drift and diffusion of charges, avalanche creation, and breakdown have been thoroughly discussed with necessary mathematics. Some specific types of gas filled detectors have also been discussed. Chapter 4 deals with liquid filled detectors. Liquid filled detectors have recently gained considerable popularity. This chapter gives the reader an overview of the charge production and transport processes in liquids and how different types of liquid filled detectors are built. Solid state detectors are perhaps the most widely used detectors today. Chapter 5 deals with different types of solid state detectors, such as semiconductor detectors, diamond detectors, and thermoluminescent detectors. A major problem with solid state detectors is their vulnerability to radiation. Radiation damage mechanisms are therefore thoroughly discussed in this chapter. Scintillation detectors and photodetectors are the topics of chapter 6. This chapter not only introduces the reader to the basic scintillation mechanisms but also discusses important properties of the commonly used scintillators. For subsequent photodetection, the transfer to scintillation photons are a major issue. The topic of light guides has therefore been given due attention. Detection of these photons is the next step, which can be accomplished with different types of detectors. Two such devices, that is photomultiplier tubes and avalanche photodiode detectors have been thoroughly discussed in this chapter. Chapter 7 deals with position sensitive detection and imaging. The basic principles of position sensitive and imaging devices as well as related techniques have been discussed here. The reader is also introduced to a number of position sensitive and imaging devices. Signal processing is the heart of today's electronic radiation detectors. A major portion of the manpower and capital is therefore invested in designing and building electronics for detection devices. This chapter exposes the reader to the basic electronic circuitry used in radiation detectors. Different types of preamplifiers, shapers, filters, discriminators, and analog-to-digital converters have been discussed here. The issues of electronics noise have also been given due attention. Chapter 9 gives a detailed discussion of the statistics and data analysis techniques. The topics related to probability, error propagation, correlation, regression, time series analysis, and counting statistics have been discussed in detail. Chapter 10 gives an overview of different data analysis software packages that are freely and commercially available. This chapter is not intended to be a manual of these software packages. It introduce the reader to their capabilities with regard to analyzing data that has been acquired through radiation detection devices. Dosimetry is the topic of chapter 11. Since dosimetry plays a central role in assuring health and safety of individuals exposed to radiation, this chapter gives a detailed account of the subject. The harmful effects of radiation and how to guard against them have also been discussed in this chapter. Chapter 12 introduces the reader to the topics related to radiation spectroscopy. Different spectroscopic techniques related to different types of particle detectors have been introduced here. Also included are topics of mass spectroscopy and time spectroscopy. Chapter 13 deals with the topic of data acquisition. The major data acquisition standards of NIM, CAMAC, VME, FASTBUS, and PCI have been introduced.

Writing this book has been a long and tedious process. The highly demanding work at the Sudbury Neutrino Observatory including on-call periods did not leave much choice other than cutting down on family time. Coming back from work and then immediately start working on the book till midnight every day is not a very family oriented approach. However my wife, Rahat, not only didn't complain but actually kept encouraging me all the way through. Without her support, it would not have been possible at all to write this book. My children Hiba, Shozab, and Shanib also deserve appreciation for their understanding.

A number of academics and friends who evaluated the original book proposal and gave valuable suggestions deserve special thanks. I thank Andy Klein, Bashar Issa, David Hamby, Edward Waller, John Antolak, Nicholas Hangiandreou, Nikolaj Pavel, Paul Jennesen, Robert LeClair, Steven Biegalski, Sukesh Aghara, and Timothy DeVol for their support and highly professional suggestions.

Thanks are also due to the academics who evaluated the first draft of the complete manuscript. Their valuable suggestions helped me modify the contents and reshape various sections. I am highly indebted to Andrea Kritcher, David Bradley, James Chu, Leslie Braby, Steven Biegalski, and Todd Allen for sparing time from their busy schedules to do thorough reviews of approximately 750 pages and providing me with highly valuable suggestions for improvement.

I would like to thank Susan Rabiner to have faith in me to complete the project and for taking care of everything else so that I could concentrate on writing the book. I also thank Jeremy Hayhurst for showing remarkable professionalism from the beginning till the end of the manuscript production. Thanks are also due to Derek Coleman, whose highly efficient and professional approach in finalizing the manuscript has amazed me.

Chapter 1

Properties and Sources of Radiation

Mass and energy are the two entities that make up our Universe. At the most basic level, these two entities represent a single reality that sometimes shows itself as mass and sometimes as energy. They are intricately related to each other through Einstein's famous mass-energy relation, $E = mc^2$. Just like matter, energy is also capable of moving from one point in space to another through particles or waves. These carriers of energy always originate from some source and continue their travel in space until they get absorbed or annihilated by some material. The term "radiation" is used to describe this transportation of mass and energy through space.

Since the realization of its potentials, radiation has played a central role in our technological developments in a variety of fields. For example, we all enjoy the benefits of radiation in medical diagnostics and treatment. On the other hand, world has also witnessed the hazards of radiation in the form of atomic explosions and radiation exposure.

Whether we think of radiation as a hazard or a blessing, its study is of paramount importance for our survival and development. If we looked carefully at the benefits and harms brought in by the use or misuse of radiation, we would reach the conclusion that its advantages clearly outweigh its disadvantages. Radiation has unlimited potential and its proper use can open up doors to great developments for the mankind.

This chapter will introduce the reader to different types of radiation, their properties, and their sources. The mechanisms through which the particles interact with matter will be discussed in detail in the next chapter.

1.1   Types of Radiation

Radiation can be categorized in different ways, such as ionizing and non-ionizing, particles and waves, hazardous and non-hazardous etc. However none of these categorizations draw solid boundaries between properties of individual particles comprising the radiation, rather they show the bulk behavior of particle beams. For example, it will not be correct to assert that an electron always ionizes atoms with which it interacts by arguing that it belongs to the category of ionizing particles. All we can say is that if a large number of electrons interact with a large number of atoms, the predominant mode of interaction will lead to the ionization of atoms.

Sometimes radiation is characterized on the basis of its wave and particle properties. However, as we will explore in the next section, this characterization is somewhat vague and can be a cause of confusion. The reason is that, according to modern physics, one can associate a wavelength to every particle whether it carries a mass or not. This implies that a particle having mass can act as a wave and take part in the formation of interference and diffraction patterns. On the other hand, light, which can be well described by its wave character, comprises of photons, which are particles having no rest mass. Hence we can conclude that one should not characterize the radiation based on its particle and wave properties.

Let us have a look at the third category we mentioned above: hazardous and non-hazardous. There are particles that pass through our bodies in large numbers every second (such as neutrinos from the Sun) but do not cause any observable damage. Still, there is a possibility that some of these particles would cause mutations in our body cells, which *could* ultimately lead to cancer[1]. On the other hand there are particles, such as neutrons, that are known to be extremely hazardous to the body but no one can ever be absolutely certain that a neutron entering our body would in deed cause any harm. In fact, due to background radiation around us, our bodies get small doses of neutrons all the time and still majority of us do not experience any adverse effects.

Based on the above arguments it is safe to say that the categorization of radiation found in the literature on these basis should not be considered to represent individual particles. What this really means is that if we have a very large number of a certain kind of particles, there is a high probability that most of them would behave in the manner characteristic of their categorization. Long exposure from a highly intense beam of neutrons would most definitely cause skin burns and most probably cancer but it would be wrong to assume that a single neutron would do the same.

The words probability and chance were mentioned in the preceding paragraphs. What does particle interaction have to do with chance? Well, the theoretical foundations of particle interaction is *quantum mechanics*, which quantifies the variables related to particle motion, such as momentum, energy, and position, in probabilistic terms. We talk about the *probability* of a particle being present at a specific place at a specific time. Nothing is absolute in quantum mechanics. We will learn more about this when we study the concept of *cross section* in the next chapter.

1.2   Waves or Particles?

If we think about light without any prior knowledge, we would assume it to be composed of waves that are continuously emitted from a source (such as a light bulb). In fact, this was the dominant perception amongst scientists until the start of the 20th century. During those days a major problem of theoretical physics had started boggling the minds of the physicists. They had found it impossible to explain the dependence of energy radiated by a black body (a heated cavity) on the wavelength of emitted radiation if they considered light to have continuous wave characteristics. This mystery was solved by Max Planck who developed a theory in which light waves were not continuous but quantized and propagated in small wave packets. This wave packet was later called a *photon*. This theory and the corresponding mathematical model were extremely successful in explaining the black body spectrum. The concept was further confirmed by Einstein when he explained

---

[1]This is a purely hypothetical situation and no data exists that could verify this assertion. The argument is based on the interaction mechanisms of particles.

the photoelectric effect, an effect in which a photon having the right amount of energy knocks off a bound electron from the atom.

Max Planck proposed that the electromagnetic energy is emitted and absorbed in the form of discrete bundles. The energy carried by such a bundle (that is, a photon) is proportional to the frequency of the radiation.

$$E = h\nu \tag{1.2.1}$$

Here $h = 6.626 \times 10^{-34} Js$ is the Planck's constant that was initially determined by Max Planck to solve the black body spectrum anomaly. It is now considered to be a universal constant. Planck's constant has a very important position in quantum mechanics. The frequency ν and wavelength λ of electromagnetic radiation are related to its velocity of propagation in vacuum c by $c = \nu\lambda$. If radiation is traveling through another medium, its velocity should be calculated from

$$v = n\nu\lambda, \tag{1.2.2}$$

where n is the refractive index of the medium. The refractive index of most materials has a nonlinear dependence on the radiation frequency.

These experiments and the consequent theoretical models confirmed that sometimes radiation behaves as particles and not continuous waves. On the other hand there were effects like interference, which could only be explained if light was considered to have continuous wave characteristics.

To add to this confusion, de Broglie in 1920 introduced the idea that sometimes particles (such as electrons) behave like waves. He proposed that one could associate a wavelength to any particle having momentum p through the relation

$$\lambda = \frac{h}{p}. \tag{1.2.3}$$

For a particle moving close to the speed of light (the so called relativistic particle) and rest mass m_0 (mass of the particle when it is not moving), the above equation can be written as

$$\lambda = \frac{h}{m_0 v}\sqrt{1 - \frac{v^2}{c^2}}. \tag{1.2.4}$$

For slow moving particles with $v << c$, the de Broglie relation reduces to

$$\lambda = \frac{h}{mv}. \tag{1.2.5}$$

De Broglie's theory was experimentally confirmed at Bell Labs where electrons diffraction patterns consistent with the wave picture were observed. Based on these experiments and their theoretical explanations, it is now believed that all the entities in the Universe simultaneously possess localized (particle-like) and distributed (wave-like) properties. In simple terms, particles can behave as waves and waves can behave as particles[2]. This principle, known as wave-particle duality, has played a central role in the development of quantum physics.

---

[2]A more correct statement would be: *all particles, regardless of being massless or massive, carry both wave and particle properties.*

Example:
Compare the de Broglie wavelengths of a proton and an alpha particle moving with the same speed. Assume the velocity to be much smaller than the velocity of light. Consider the mass of the α-particle to be about 4 times the mass of the proton.

Solution:
Since the velocity is much slower than the speed of light, we can use the approximation 1.2.5, which for a proton and an α-particle becomes

$$\lambda_p = \frac{h}{m_p v}$$

$$\text{and} \quad \lambda_\alpha = \frac{h}{m_\alpha v}.$$

Dividing the first equation with the second gives

$$\frac{\lambda_p}{\lambda_\alpha} = \frac{m_\alpha}{m_p}.$$

An α-particle comprises of two protons and two neutrons. Since the mass of a proton is approximately equal to the mass of a neutron, we can use the approximation $m_\alpha \approx 4m_p$ in the above equation, which then gives

$$\lambda_p \approx 4\lambda_\alpha.$$

This shows that the de Broglie wavelength of a proton is around 4 times larger than that of an α-particle moving with the same velocity.

1.3   Radioactivity and Radioactive Decay

Radioactivity is the process through which nuclei spontaneously emit subatomic particles. It was discovered by a French scientist Henri Bacquerel in 1896 when he found out that an element, uranium, emitted something invisible that fogged his photographic plates. The term *radioactivity* was suggested by Marie Curie about 4 years later. Originally three types of radiation were discovered:

▶ α-rays (helium nuclei with only 2 protons and 2 neutrons)

▶ β-rays (electrons)

▶ γ-rays (photons)

Later on it was found that other particles such as neutrons, protons, positrons and neutrinos are also emitted by some decaying nuclei. The underlying mechanisms responsible for emission of different particles are different. To add to this complication, a particle can be emitted as a result of different modes of decay. For example, a common decay mode resulting in the emission of neutrons is *spontaneous fission*. During this process a heavy nucleus spontaneously splits into two *lighter* nuclides called *fission fragements* and emits several neutrons. But this is not the only mode of neutron decay. Another particle just mentioned was neutrino. Neutrino is an extremely light and low interacting chargeless particle that was discovered in 1952. It solved the mystery of the variable electron energy in beta decays: the electron, being very light as compared to the other heavy decay product in beta decays, was supposed to carry the same energy in each decay. However, it was observed that the emitted electrons had a whole spectrum of energy with a cut off characteristic to the decaying atom. It took several decades for scientists to discover that some of the energy is actually taken away by a very light particle called neutrino. Now we know that this particle in beta decays is actually anti-neutrino[3].

When nuclei emit subatomic particles, their configuration, state, and even identity may change. For example, when a nucleus emits an alpha particle, the new nucleus has 2 protons and 2 neutrons less than the original one. Except for γ-decay, in which the nucleus retains its identity, all other decays transform the nucleus into a totally different one. This process is called radioactive decay .

There are a number of naturally occurring and man-made radioactive elements that decay at different rates. Although the underlying mechanism of these decays is fairly complicated, their gross outcome can be easily predicted by considering the conservation of electrical charge. Before we write general decay equations, let us first have a look at some examples.

$$
\begin{array}{llll}
\text{Alpha decay:} & {}^{222}_{88}Ra & \rightarrow & {}^{218}_{86}Rn & + & \alpha \\[2mm]
\text{Beta decay:} & {}^{131}_{53}I & \rightarrow & {}^{131}_{54}Xe & + & e & + & \bar{\nu} \\[2mm]
\text{Gamma decay:} & {}^{152}_{86}Dy^* & \rightarrow & {}^{152}_{86}Dy & + & \gamma \\[2mm]
\text{Spontaneous fission:} & {}^{256}_{100}Fm & \rightarrow & {}^{140}_{54}Xe & + & {}^{112}_{46}Pd & + & 4n
\end{array}
$$

Here ${}^{n+p}_{p}X$ represents an element X with p protons and n neutrons. X^* represents an atom in an excited state.

The term beta decay as used in the above example is sometimes conventionally used to represent only emission of electrons. However, there are actually three kinds of beta decays: electron decay, electron capture, and positron decay. The first two involve electrons while the third involves the emission of the *anti-electron* or positron, which we will represent by e^+ in this book. A positron has all the properties of an electron with the exception of electrical charge, which is positive in this case. In this book the symbol e will be used to represent either the electron or the unit electrical charge.

Electron capture occurs when a nucleus captures one of the electrons orbiting around it and as a result goes into an excited state. It quickly returns to the ground state by emitting a photon and a neutrino. The positron emission is very similar to the electron decay with the exception that during this process instead of an electron a positron is emitted. Let us have a look at examples of these two processes.

---

[3]Anti-neutrino is the antiparticle of neutrino. The presence of neutrino in β-decays was suggested by Wolfgang Pauli. It was named *neutrino* by Enrico Fermi.

$$\text{Electron capture:} \quad {}^{81}_{36}Kr + e \quad \rightarrow \quad {}^{81}_{35}Br \quad + \quad \gamma \quad + \quad \nu$$

$$\text{Positron decay:} \quad {}^{40}_{19}K \quad \rightarrow \quad {}^{40}_{18}Ar \quad + \quad e^+ \quad + \quad \nu$$

The electron captured in the first of these reactions actually transforms a proton into a neutron, that is

$$p + e \rightarrow n + \nu.$$

That is why the daughter nuclide has one proton less and one neutron more than the parent nucleus. Electron capture transforms the nuclide into a different element. In a similar fashion, the positron emission is the result of transformation of a proton into a neutron, that is

$$p \rightarrow n + e^+ + \nu.$$

This implies that in case of positron emission the daughter nuclide has one proton less and one neutron more than the parent nuclide. This is also apparent from the positron emission example of potassium-40 given earlier. It is interesting to note that in this reaction the mass of the proton is less than the combined mass of the neutron, positron, and the neutrino[4]. This means that the reaction is possible only when enough energy is available to the proton. That is why there is a threshold energy of 1.022 MeV needed for positron emission. Below this energy the nuclide can decay by electron capture, though.

Note that the electron capture reaction above shows that a photon is also emitted during the process. This photon can be an x-ray or a γ-ray photon. The x-ray photon is emitted as one of the electrons in the higher orbitals fills the gap left by the electron captured by the nucleus. Since in most cases a K-shell electron is captured by the nucleus, the orbital is quickly filled in by another electron from one of the higher energy states. The difference in the energy is released in the form of an x-ray photon. It can also happen that the nucleus being in an excited state after capturing an electron emits one or more γ-rays. The electron capture reaction then should actually be written as a two step process, that is

$$\begin{aligned} {}^{81}_{36}Kr + e &\rightarrow {}^{81}_{35}Br^* + \nu \\ {}^{81}_{35}Br^* &\rightarrow {}^{81}_{35}Br + \gamma \end{aligned}$$

In general, the subsequent γ-decay is not specific to electron capture. It can occur in a nucleus that has already undergone any other type of decay that has left it in an excited state. It is a natural way by which the nuclei regain their stability. Sometimes it takes a number of γ-decays for a nucleus to eventually reach a stable state.

Although γ-emission is the most common mode of de-excitation after a decay, it is not the only one. Another possible process is the so called *internal conversion*. In this process, the excess energy is transferred to an orbital electron. If the supplied energy is greater than the binding energy of this electron, the electron gets expelled from the orbital with a kinetic energy equal to the difference of the atom's excess energy and its binding energy.

---

[4] As of the time of writing this book, the neutrino mass is still unknown. However it has been confirmed that the mass is very small, probably 100,000 times less than the mass of an electron. We will also learn later that there are actually three *or more* types of neutrinos.

The process of internal conversion can occur to electrons in any electronic orbit. If an electron from one of the inner shells is expelled, it leaves behind a vacancy that could be filled up by an electron in one of the higher shells. If that happens, the excess energy is emitted in the form of an x-ray photon. This photon can either escape the atom or can knock off another electron from the atom. The knocked-off electron is known as *Auger electron*. The process of internal conversion followed by emission of an Auger electron is graphically depicted in Fig.1.3.1. Auger electron emission is not specific to the decay process. It can happen whenever an electron from one of the inner electronic orbitals leaves the atom. An example of such a process is the *photoelectric effect*, which we will study in some detail in the next Chapter.

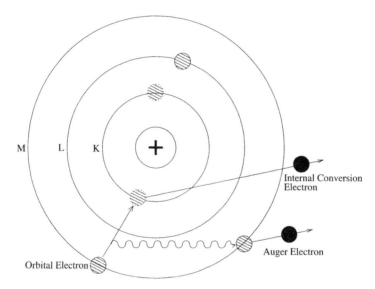

Figure 1.3.1: Depiction of internal conversion leading to the emission of an Auger electron. The internal conversion electron from the K-shell creates a vacancy that must be filled in order for the atom to regain stability. Another electron from the M-shell fills this gap but releases some energy in the process (shown as a photon). This photon is shown to have knocked off another electron from the M-shell. The end result is the emission of an internal conversion electron and an Auger electron.

At this point the reader might be wondering why the radioisotopes emitting neutrons and protons have not been mentioned so far. These decays are in fact possible. However, such isotopes are not found in nature except for the ones that undergo spontaneous fission. On the other hand, one can produce such a radionuclide by bombarding a suitable material with high energy particles, something that can be done at a particle accelerator. The basic idea behind such a process is to depart enough energy to a nucleus abundant in protons and neutrons such that it becomes unstable. This instability forces it to eventually decay by emitting a proton or a neutron. There are also some β-decays, which are followed by proton emission. Such a process is generally known as *beta-delayed proton emission*. An example of the

decay by proton emission is the decay of Indium-109, which decays with a half life
[5] of only 103 μs.

$$_{53}^{109}I \rightarrow \, _{52}^{108}Te + p$$

This extremely small half life is actually typical of all of the known proton emitters
with the exception of a few that have half lives in the range of a few hundred
milli-seconds. Such short half lives severely limit the usefulness of such materials as
proton sources.

Just like proton emitters, it is also possible to produce isotopes that decay by
neutron emission. They suffer from the same extremely short half lives typical of
proton emitters, rendering them useless as neutron sources in normal laboratory
environments. Exception to this are a few isotopes, which decay by spontaneous
fission and in the process also emit neutrons. The most common example of this is
californium-252, which with a half live of 2.65 years is commonly used in laboratories
as a source of neutrons.

It is now worthwhile to write general equations for different types of radioactive
decays. For a nucleus X with p protons, n neutrons and e electrons, which transforms
into another nucleus Y, the general decay equations can be written as follows.

$$\text{Alpha decay:} \qquad _{p}^{n+p}X \qquad \rightarrow \qquad _{p-2}^{n+p-4}Y \; + \; \alpha$$

$$\text{Electron decay:} \qquad _{p}^{n+p}X \qquad \rightarrow \qquad _{p+1}^{n+p}Y \; + \; e \; + \; \bar{\nu}$$

$$\text{Electron capture:} \quad _{p}^{n+p}X + e \quad \rightarrow \qquad _{p-1}^{n+p}Y \; + \; \gamma \; + \; \nu$$

$$\text{Positron decay:} \qquad _{p}^{n+p}X \qquad \rightarrow \qquad _{p-1}^{n+p}Y \; + \; e^{+} \; + \; \nu$$

$$\text{Gamma decay:} \qquad _{p}^{n+p}X^{*} \qquad \rightarrow \qquad _{p}^{n+p}X \; + \; \gamma$$

$$\text{Proton emission:} \qquad _{p}^{n+p}X \qquad \rightarrow \qquad _{p-1}^{n+p-1}Y \; + \; p$$

Here we have deliberately avoided using the term *proton decay* to describe the decay
by proton emission. The reason is that proton decay is explicitly used for the
decay of the proton itself, a process that is expected to occur in nature, albeit
with an extremely low probability. Dedicated detectors have been built around the
world to observe such a phenomenon. The reader is encouraged to verify that in all
these reactions the total electrical charge is always conserved. This conservation of
electrical charge is one of the fundamental laws of nature.

The general equation for spontaneous fission process can not be written as there
are a number of modes in which a nucleus may fission. That is, there is generally a
whole spectrum of nuclides in which a decaying nucleus may split. Also the number
of neutrons emitted is variable and depends on the particular mode of decay.

It should be remembered that during radioactive decays the emitted particles
originate from the nuclei. For example during the process of electron decay, a neutron
inside the nucleus decays into a proton, an electron and an anti-neutrino. The proton
stays inside the nucleus while electrons and anti-neutrino are emitted out. Similarly,
during gamma decay, the photon is emitted from the nucleus and is not the result

---

[5]Half life is the time taken by half of the sample atoms to decay. The term will be formally defined later
in the Chapter.

of electronic transitions between atomic levels. The particles that are not emitted from the nucleus can be byproducts of a decay but should not be confused with the direct decay products.

Example:
Write down equations for alpha decay of Thorium-232 and electron decay of Sodium-24.

Solution:
Using the general decay equations defined above and the periodic table of elements we find the required decay equations.

$$^{232}_{90}Th \rightarrow {}^{228}_{88}Ra + \alpha$$
$$^{24}_{11}Na \rightarrow {}^{24}_{12}Mg + e + \bar{\nu}$$

1.3.A Decay Energy or Q-Value

Decay energy is a term used to quantify the energy released during the decay process. It can be used to determine whether a certain decay mode for a nucleus is possible or not. To understand this, let us suppose a nucleus X goes through a decay mode that transforms it into a nucleus Y with a subsequent emission of a particle d. This reaction can be written as

$$X \rightarrow Y + d. \tag{1.3.1}$$

According to the law of conservation of energy, the total energy before and after the decay should be equal, that is

$$E_{0,X} + T_X = E_{0,Y} + T_Y + E_{0,d} + T_d, \tag{1.3.2}$$

where E_0 stands for the rest energy and T represents the kinetic energy. The rest energy can be computed from the Einstein relation $E_0 = m_0 c^2$, with m_0 being the rest mass. Since the decaying nucleus X can be assumed to be at rest therefore we can safely use $T_X = 0$ in the above relation. If we represent the rest masses of X and Y by m_X and m_Y, the above equation would read

$$(m_X - m_Y - m_d)c^2 = T_Y + T_d. \tag{1.3.3}$$

Now, it is evident that the left hand side of this relation must be positive in order for the kinetic energy to be positive and meaningful. In other words, the decay would be possible only if the left hand side is positive valued. Both the left and right hand sides of this relation are termed as the decay energy or the Q-value. That is

$$Q_d = T_Y + T_d \tag{1.3.4}$$
$$\text{or} \quad Q_d = (m_X - m_Y - m_d)c^2. \tag{1.3.5}$$

The first relation above requires the knowledge of the kinetic energy taken away by the decaying nucleus and the emitted particle. These energies are difficult to determine experimentally, though. Therefore, generally one uses the second relation containing the mass terms to determine the decay energy for any decay mode the isotope is expected to go through. If the energy comes out to be negative then the decay is not possible unless energy is supplied through an external agent, such as by bombarding the material with high energy particles. A positive Q-value signifies that the isotope is unstable with respect to that particular mode of decay. Note that if a nucleus has a positive Q-value for one decay mode, it does not guarantee that it can decay through other modes as well (see example below).

Since atomic data tables list isotope masses in $a.m.u.$ therefore one must multiply masses in the above equation by $a.m.u.$ to kg conversion factor. We can also substitute this conversion factor, the value of c, and the conversion factor for Joules to MeV in the above relation to transform it into a more computationally convenient form.

$$
\begin{aligned}
Q_d &= (m_X - m_Y - m_d)8.94 \times 10^{16} \quad J \qquad (m \text{ in } kg) &\text{(1.3.6)}\\
&= (m_X - m_Y - m_d)931.502 \quad MeV \quad (m \text{ in } a.m.u.) &\text{(1.3.7)}
\end{aligned}
$$

Care should be exercised when substituting masses in the above relation. The above relation as it stands is valid for nuclear masses. If one wishes to use atomic masses, the mass of electrons should be properly accounted for as explained later in this section.

The Q-value can be used to determine the kinetic energies of the daughter nucleus and the emitted particle. To demonstrate this, let us substitute $T = \frac{1}{2}mv^2$ in equation 1.3.5. This gives

$$
T_d = Q_d - \frac{1}{2}m_Y v_Y^2, \tag{1.3.8}
$$

where m_Y and v_Y represent the mass and velocity of the daughter nucleus respectively. The velocity of the daughter nucleus can be determined by applying the law of conservation of linear momentum, which for this case gives

$$
m_Y v_Y = m_d v_d. \tag{1.3.9}
$$

Note that here we have assumed that the parent was at rest before the decay. The velocity v_Y from this equation can then be substituted into equation 1.3.A to get

$$
T_d = \left[\frac{m_Y}{m_Y + m_d}\right] Q_d. \tag{1.3.10}
$$

Similary the expression for the kinetic energy of the daughter nucleus is given by

$$
T_Y = \left[\frac{m_d}{m_d + m_Y}\right] Q_d. \tag{1.3.11}
$$

Let us now write the Q-value relations for α and β decays.

$$
\alpha\text{-decay:} \quad Q_\alpha = (m_X - m_Y - m_\alpha)c^2
$$

$$
\beta\text{-decay:} \quad Q_\beta = (m_X - m_Y - m_\beta)c^2
$$

Note that the above relation is valid for nuclear masses only. For atomic masses, the following equations should be used

$$\alpha\text{-decay:} \quad Q_\alpha = (M_X - M_Y - M_\alpha)\,c^2$$

$$\beta\text{-decay:} \quad Q_\beta = (M_X - M_Y)\,c^2$$

Here M stands for atomic mass, which means that M_α is the mass of the helium atom and not the helium nucleus as in case of equation 1.3.A.

Example:
Determine if actinium-225 can decay through α as well as β modes.

Solution:
The α decay reaction for actinium-225 can be written as

$$^{225}_{89}Ac \rightarrow ^{221}_{87}Fr + ^4_2He.$$

The Q-value, in terms of *atomic masses*, for this reaction is

$$
\begin{aligned}
Q_\alpha &= (MAc - M_{Fr} - M_\alpha)\,931.502 \\
&= (225.023229 - 221.014254 - 4.002603)\,931.502 \\
&= 5.93\ MeV.
\end{aligned}
$$

If actinium went through β decay, the decay equation would be written as

$$^{225}_{89}Ac \rightarrow ^{225}_{90}Th + e,$$

with a Q-value, in terms of *atomic masses*, given by

$$
\begin{aligned}
Q_\beta &= (M_{Ac} - M_{Th})\,931.502 \\
&= (225.023229 - 225.023951)\,931.502 \\
&= -0.67\ MeV.
\end{aligned}
$$

Since the Q-value is positive for α decay therefore we can say with confidence that actinium-225 can emit α particles. On the other hand a negative Q-value for β decay indicates that this isotope can not decay through electron emission.

1.3.B   The Decay Equation

Radioactive decay is a random process and has been observed to follow Poisson distribution (see chapter on statistics). What this essentially means is that the rate of decay of radioactive nuclei in a large sample depends only on the number of decaying nuclei in the sample. Mathematically this can be written as

$$\frac{dN}{dt} \quad \propto \quad -N$$

$$\text{or} \quad \frac{dN}{dt} = -\lambda_d N. \tag{1.3.12}$$

Here dN represents the number of radioactive nuclei in the sample in the time window dt. λ_d is a proportionality constant generally referred to in the literature as *decay constant.* In this book the subscript d in λ_d will be used to distinguish it from the wavelength symbol λ that was introduced earlier in the chapter. Conventionally both of these quantities are represented by the same symbol λ. The negative sign signifies the fact that the number of nuclei in the sample decrease with time. This equation, when solved for the number N of the radioactive atoms present in the sample at time t, gives

$$N = N_0 e^{-\lambda_d t}, \tag{1.3.13}$$

where N_0 represents the number of radioactive atoms in the sample at $t = 0$.

Equation 1.3.12 can be used to determine the decay constant of a radionuclide, provided we can somehow measure the amount of *decayed* radionuclide in the sample. This can be fairly accurately accomplished by a technique known as mass spectroscopy (the details can be found in the chapter on spectroscopy). If the mass of the isotope in the sample is known, the number of atoms can be estimated from

$$N = \frac{N_A}{w_n} m_n, \tag{1.3.14}$$

where $N_A = 6.02 \times 10^{23}$ is the Avagadro number, w_n is the atomic weight of the radionuclide and m_n is its mass as determined by mass spectroscopy.

Although this technique gives quite accurate results but it requires sophisticated equipments that are not always available in general laboratories. Fortunately there is a straightforward experimental method that works almost equally well for nuclides that do not have very long half lives. In such a method the rate of decay of the sample is measured through a particle detector capable of counting individual particles emitted by the radionuclide. The rate of decay A, also called the activity, is defined as

$$A = -\frac{dN}{dt} = \lambda_d N. \tag{1.3.15}$$

Using this definition of activity, equation 1.3.13 can also be written as

$$A = A_0 e^{-\lambda_d t}, \tag{1.3.16}$$

where $A_0 = \lambda_d N_0$ is the initial activity of the sample.

Since every detection system has some intrinsic efficiency ϵ with which it can detect particles therefore the measured activity C would be lower than the actual activity by the factor ϵ.

$$\begin{aligned} C &= \epsilon A \\ &= \epsilon \left[-\frac{dN}{dt} \right] \\ &= \epsilon \lambda_d N \end{aligned} \tag{1.3.17}$$

The detection efficiency of a good detection system should not depend on the count rate (as it would imply nonlinear detector response and consequent uncertainty in determining the actual activity from the observed data). and hence the above equation can be used to determine the count rate at $t = 0$.

$$C_0 = \epsilon \lambda_d N_0 \tag{1.3.18}$$

The above two equation can be substituted in equation 1.3.13 to give

$$C = C_0 e^{-\lambda_d t}. \tag{1.3.19}$$

What this equation essentially implies is that the experimental determination of the decay constant λ is independent of the efficiency of the detection system, although the counts observed in the experiment will always be less than the actual decays. To see how the experimental values are used to determine the decay constant, let us rewrite equations 1.3.16 and 1.3.19 as

$$\ln(A) = -\lambda t + \ln(A_0) \quad \text{and} \tag{1.3.20}$$
$$\ln(C) = -\lambda t + \ln(C_0). \tag{1.3.21}$$

Hence if we plot C versus t on a semilogarithmic graph, we should get a straight line with a slope equal to $-\lambda$. Figure 1.3.2 depicts the result of such an experiment. The predicted activity has also been plotted on the same graph using equation 1.3.20. The difference between the two lines depends on the efficiency, resolution, and accuracy of the detector.

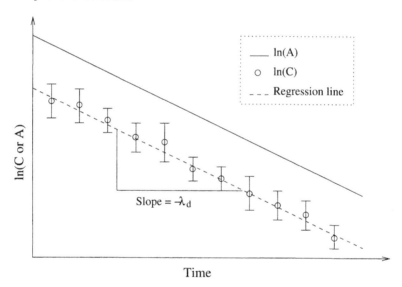

Figure 1.3.2: Experimental determination of decay constant.

Equation 1.3.13 can be used to estimate the average time a nucleus would take before it decays. This quantity is generally referred to as the "lifetime" or "mean life" and denoted by τ or T. In this book it will be denoted by the symbol τ. The mean life can be calculated by using

$$\tau = \frac{1}{\lambda_d}. \tag{1.3.22}$$

Another parameter, which is extensively quoted and used, is the *half life*. It is defined as the time required by half of the nuclei in a sample to decay. It is given by

$$T_{1/2} = 0.693\tau = \frac{\ln(2)}{\lambda_d}. \tag{1.3.23}$$

Since mean and half lives depend on the decay constant, therefore the experimental procedure to determine the decay constant can be used to find these quantities as well. In fact, whenever a new radionuclide is discovered its half life is one of the first quantities that are experimentally determined. The half life of a radionuclide can range from a micro second to million of years. However this experimental method to determine the half life does not work very well for nuclides having long half lives. The reason is quite simple: for such a nuclide the disintegration rate is so low that the counts difference between two points in time will be insignificantly small. As we saw earlier in this section, for such radionuclides other techniques such as mass spectroscopy are generally employed.

Example:
Derive the equations for mean and half lives of a radioactive sample.

Solution:
To derive the equation for mean life we take the weighted mean of the decay time t

$$\tau = \frac{\int_0^\infty t dN}{\int_0^\infty dN}$$

Using $N = N_0 e^{-\lambda_d t}$, the integral in the denominator becomes

$$
\begin{aligned}
\int_0^\infty dN &= -\lambda_d N_0 \int_0^\infty e^{-\lambda_d t} dt \\
&= N_0 \left| e^{-\lambda_d t} \right|_0^\infty \\
&= -N_0.
\end{aligned}
$$

The integral in the numerator can be solved through integration by parts as follows.

$$
\begin{aligned}
\int_0^\infty t dN &= -\lambda_d N_0 \int_0^\infty t e^{-\lambda_d t} dt \\
&= -\lambda_d N_0 \left[\left| -\frac{t e^{\lambda_d t}}{\lambda_d} \right|_0^\infty + \frac{1}{\lambda_d} \int_0^\infty e^{-\lambda_d t} dt \right]
\end{aligned}
$$

The first term on the right side vanishes for $t = 0$ and at $t \rightarrow \infty$ (a function vanishes at infinity if its derivative vanishes at infinity). Therefore the integral becomes

$$
\begin{aligned}
\int_0^\infty t dN &= -N_0 \int_0^\infty e^{-\lambda_d t} dt \\
&= \frac{N_0}{\lambda_d} \left| e^{-\lambda_d t} \right|_0^\infty \\
&= -\frac{N_0}{\lambda_d}.
\end{aligned}
$$

Hence the mean life is

$$
\begin{aligned}
\tau &= \frac{-N_0/\lambda_d}{-N_0} \\
&= \frac{1}{\lambda_d}
\end{aligned}
$$

Since half life represents the time taken by half of the atoms in a sample to decay, we can simply replace N by $N_0/2$ in equation 1.3.13 to get

$$\frac{1}{2} = e^{-\lambda_d T_{1/2}}$$
$$e^{\lambda_d T_{1/2}} = 2$$
$$T_{1/2} = \frac{\ln(2)}{\lambda_d}$$
$$= \ln(2)\tau = 0.693\tau.$$

Example:
The half life of a radioactive sample is found to be 45 days. How long would it take for 2 moles of this material to decay into 0.5 mole.

Solution:
Since $T_{1/2} = 45$ days, therefore

$$\lambda_d = \frac{\ln(2)}{T_{1/2}}$$
$$= \frac{\ln(2)}{45} = 15.4 \times 10^{-3} day^{-1}.$$

Since mole M is proportional to the number of atoms in the material, therefore equation 1.3.13 can also be written in terms of number of moles as follows.

$$M = M_0 e^{-\lambda_d t}$$

Rearrangement of this equation gives

$$t = \frac{1}{\lambda_d} \ln\left(\frac{M_0}{M}\right).$$

Hence the time it will take for 1.5 moles of this material to decay is

$$t = \frac{1}{15.4 \times 10^{-3}} \ln\left(\frac{2.0}{0.5}\right)$$
$$\approx 90 \text{ days.}$$

1.3.C   Composite Radionuclides

A problem often encountered in radioactivity measurements is that of determining the activity of individual elements in a composite material. A composite material is the one that contains more than one radioisotope at the same time. Most of the radioactive materials found in nature are composite.

Let us suppose we have a sample that contains two isotopes having very different half lives. Intuitively thinking, we can say that the semilogarithmic plot of activity versus time in such a case will deviate from a straight line of single isotopes. The best way to understand this is by assuming that the composite material has one *effective* decay constant. But this decay constant will have time dependence since as time passes the sample runs out of the short lived isotope. Hence equations 1.3.20 and 1.3.21 will not be linear any more.

Figure 1.3.3 shows the activity plot of a composite radioactive material. Since we know that each individual isotope should in fact yield a straight line therefore we can extrapolate the linear portion of the graph backwards to get the straight line for the isotope with longer half life. We can do this because the linear portion shows that the shorter lived isotope has fully decayed and the sample now essentially contains only one radioactive isotope. Then the straight line for the other isotope can be determined by subtracting the total activity from the activity of the long lived component.

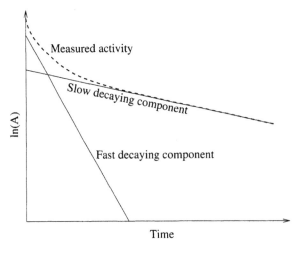

Figure 1.3.3: Experimental determination of decay constants of two nuclides in a composite decaying material.

Example:
The following table gives the measured activity in counts of a composite radioactive sample with respect to time. Assuming that the sample contains two radioactive isotopes, compute their decay constants and half lives.

t (min)	0	30	60	90	120	150	180	210	240	270	300
A (cts/min)	2163	902	455	298	225	183	162	145	133	120	110

Solution:
Following the procedure outlined in this section, we plot the activity as a function of time on a semilogarithmic graph (see Fig. 1.3.4). It is apparent from the plot that after $t = 120$ minutes $\ln(A)$ varies linearly with time. Using least square fitting algorithm we fit a straight line through points between $t = 150$ and $t = 300$ minutes. The equation is found to be

$$\ln(A) = -3.28 \times 10^{-3}t + 5.68.$$

This straight line represents the activity of the long lived component in the sample. Its slope gives the decay constant of the long lived component, which can then be used to obtain the half life. Hence we get

$$\lambda_1 = 3.28 \times 10^{-3} \text{ min}^{-1}$$
$$\Rightarrow T_{1/2,1} = \frac{0.693}{\lambda_1} = \frac{0.693}{3.28 \times 10^{-3}}$$
$$= 211.3 \text{ days.}$$

To obtain the decay constant of the short lived component, we extrapolate the straight line obtained for the long lived component up to $t = 0$ and then subtract it from the observed data (see Fig.1.3.4). The straight line thus obtained is given by

$$\ln(A) = -3.55 \times 10^{-2}t + 7.53.$$

The slope of this line gives the decay constant of the second isotope, which can then be used to determine its half life. Hence we have

$$\lambda_2 = 3.55 \times 10^{-2} \text{ min}^{-1}$$
$$\Rightarrow T_{1/2,2} = \frac{0.693}{\lambda_1} = \frac{0.693}{3.55 \times 10^{-2}}$$
$$= 19.5 \text{ days.}$$

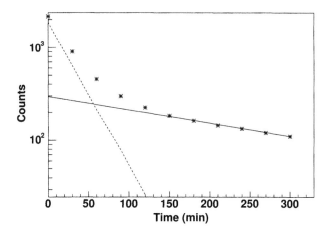

Figure 1.3.4: Determination of decay parameters of two nuclides from observed *effective* activity. The actual data are represented by (*). The solid and dashed lines represent the long lived and short lived components respectively.

1.3.D   Radioactive Chain

We saw earlier that when a radionuclide decays, it may change into another element or another isotope. This new *daughter* radionuclide may as well be unstable and radioactive. The decay mode and half life of the daughter may also be different from the *parent*. Let us see how our radioactive decay equations can be modified for such a situation.

I will start with a sample composed of a parent and a daughter radionuclide. There will be two processes happening at the same time: production of daughter (or decay of parent) and decay of daughter. The net rate of decay of the daughter will then be the difference of these two rates, that is

$$\frac{dN_D}{dt} = \lambda_{dP} N_P - \lambda_{dD} N_D, \tag{1.3.24}$$

where subscripts P and D represent parent and daughter respectively. Using $N_P = N_{0P} e^{-\lambda_{dP} t}$, this equation can be written as

$$\frac{dN_D}{dt} + \lambda_{dD} N_D - \lambda_{dP} N_{0P} e^{-\lambda_{dP} t} = 0. \tag{1.3.25}$$

Solution of this first order linear differential equation is

$$N_D = \frac{\lambda_{dP}}{\lambda_{dD} - \lambda_{dP}} N_{0P} \left(e^{-\lambda_{dP} t} - e^{-\lambda_{dD} t} \right) + N_{0D} e^{-\lambda_{dD} t}. \tag{1.3.26}$$

Here N_{0P} and N_{0D} are the initial number of parent and daughter nuclides respectively. In terms of activity $A (= \lambda N)$, the above solution can be written as

$$A_D = \frac{\lambda_{dD}}{\lambda_{dD} - \lambda_{dP}} A_{0P} \left(e^{-\lambda_{dP} t} - e^{-\lambda_{dD} t} \right) + A_{0D} e^{-\lambda_{dD} t}. \tag{1.3.27}$$

Equations 1.3.26 and 1.3.27 have decay as well as growth components, as one would expect. It is apparent from this equation that the way a particular material decays depends on the half lives (or decay constants) of both the parent and the daughter nuclides. Let us now use equation 1.3.27 to see how the activity of a freshly prepared radioactive sample would change with time. In such a material, the initial concentration and activity of daughter nuclide will be zero $N_{0D} = 0, A_{0D} = 0$. This condition reduces equation 1.3.26 to

$$A_D = \frac{\lambda_{dD}}{\lambda_{dD} - \lambda_{dP}} A_{0P} \left(e^{-\lambda_{dP} t} - e^{-\lambda_{dD} t} \right). \tag{1.3.28}$$

The first term in parenthesis on the right side of this equation signifies the buildup of daughter due to decay of parent while the second term represents the decay of daughter. This implies that the activity of the daughter increases with time and, after reaching a maximum, ultimately decreases (see figure 1.3.5). This point of maximum daughter activity t_D^{max} can be easily determined by requiring

$$\frac{dA_D}{dt} = 0.$$

Applying this condition to equation 1.3.28 gives

$$t_D^{max} = \frac{\ln(\lambda_{dD}/\lambda_{dP})}{\lambda_{dD} - \lambda_{dP}}. \tag{1.3.29}$$

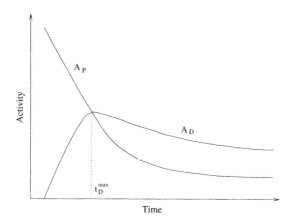

Figure 1.3.5: Typical parent and daughter nuclide activities.

Example:
Derive the relation for the time behavior of buildup of a stable nuclide from a radioactive element.

Solution:
Assuming the initial concentration of daughter to be zero ($N_{0D} = 0$), equation 1.3.26 can be used to determine the concentration of daughter nuclide at time t.

$$N_D = \frac{\lambda_{dP}}{\lambda_{dD} - \lambda_{dP}} N_{0P} \left(e^{-\lambda_{dP}t} - e^{-\lambda_{dD}t} \right)$$

Since the daughter nuclide is stable therefore we can substitute $\lambda_{dD} = 0$ in the above equation to get the required relation.

$$N_D = N_{0P} \left(1 - e^{-\lambda_{dP}t} \right)$$

In most cases the radioactive decay process does not stop at the decay of the daughter nuclide as depicted by equation 1.3.26. Instead the nuclides continue to decay into other unstable nuclides until a stable state is reached. Assuming that the initial concentrations of all the nuclides except for the parent is zero, equation 1.3.28 can be generalized for a material that undergoes several decays. The generalization was first done by Bateman in 1910 (8). The Bateman equation for the concentration of ith radionuclide is

$$N_i(t) = \lambda_{d1}\lambda_{d2}.....\lambda_{d(i-1)} N_{01} \sum_{j=1}^{i} \frac{e^{-\lambda_{dj}t}}{\prod_{k=1, k \neq j} (\lambda_{dk} - \lambda_{dj})}, \qquad (1.3.30)$$

provided $N_{0i} = 0$ for $i > 1$. In terms of activity the Bateman equation can be written as

$$A_i(t) = \lambda_{d2}.....\lambda_{di} A_{01} \sum_{j=1}^{i} \frac{e^{-\lambda_{dj}t}}{\prod_{k=1, k \neq j} (\lambda_{dk} - \lambda_{dj})}. \qquad (1.3.31)$$

Example:
A 50 μCi radioactive sample of pure $^{222}_{88}Rn$ goes through the following series of decays.

$^{222}_{86}Rn$ $(T_{1/2} = 3.82$ days $) \rightarrow ^{218}_{84}Po$ $(T_{1/2} = 3.05$ min $) \rightarrow ^{214}_{82}Pb$ $(T_{1/2} = 26.8$ min $) \rightarrow ^{214}_{83}Bi$ $(T_{1/2} = 19.7$ min $)$

Compute the activity of its decay products after 3 hours.

Solution:

Activity of $^{222}_{86}Rn$
The decay constant of $^{222}_{86}Rn$ can be calculated from its half life as follows.

$$\begin{aligned} \lambda_{d1} &= \frac{\ln(2)}{T_{1/2}} = \frac{0.693}{3.82 \times 24 \times 60} \\ &= 1.26 \times 10^{-4} \ \text{min}^{-1} \end{aligned}$$

Since we have a pure sample of radon-222 therefore its activity after 3 hours can be calculated from equation 1.3.16.

$$\begin{aligned} A_1 &= A_{01}e^{-\lambda_{d1}t} \\ &= 50\left[e^{-1.26\times 10^{-4}\times 3\times 60}\right] \\ &= 48.88 \ \mu Ci \end{aligned}$$

Activity of $^{218}_{84}Po$
The decay constant of $^{218}_{84}Po$ is

$$\begin{aligned} \lambda_{d2} &= \frac{\ln(2)}{T_{1/2}} = \frac{0.693}{3.05} \\ &= 0.227 \ \text{min}^{-1}. \end{aligned}$$

Since polonium-218 is the first daughter down the radioactive chain of radon-222, we use $i = 2$ in Bateman equation 1.3.31 to get

$$\begin{aligned} A_2 &= \lambda_{d2}A_{01}\left[\frac{e^{-\lambda_{d1}t}}{(\lambda_{d2} - \lambda_{d1})} + \frac{e^{-\lambda_{d2}t}}{(\lambda_{d1} - \lambda_{d2})}\right] \\ &= 0.227 \times 50\left[\frac{e^{-1.26\times 10^{-4}\times 3\times 60}}{(0.227 - 1.26 \times 10^{-4})} + \frac{e^{-0.227\times 3\times 60}}{(1.26 \times 10^{-4} - 0.227)}\right] \\ &= 48.9 \ \mu Ci \end{aligned}$$

A point worth noting here is that the second term in the parenthesis on the right side of the above equation is negligible as compared to the first term and could have safely been omitted from the calculations.

Activity of $^{214}_{82}Pb$

The decay constant of $^{214}_{82}Pb$ is

$$\lambda_{d3} = \frac{\ln(2)}{T_{1/2}} = \frac{0.693}{26.8}$$
$$= 0.0258 \quad \text{min}^{-1}.$$

To calculate the activity of this isotope of lead after 3 hours we use $i = 3$ in Bateman equation 1.3.31.

$$A_3 = \lambda_{d2}\lambda_{d3}A_{01}\left[\frac{e^{-\lambda_{d1}t}}{(\lambda_{d2} - \lambda_{d1})(\lambda_{d3} - \lambda_{d1})} + \frac{e^{-\lambda_{d2}t}}{(\lambda_{d1} - \lambda_{d2})(\lambda_{d3} - \lambda_{d2})}\right] +$$
$$\lambda_{d2}\lambda_{d3}A_{01}\left[\frac{e^{-\lambda_{d3}t}}{(\lambda_{d1} - \lambda_{d3})(\lambda_{d2} - \lambda_{d3})}\right]$$

Due to high decay constants of $^{218}_{84}Po$ and $^{214}_{82}Pb$ the second and third terms on right hand side can be neglected. Hence

$$A_3 \approx \lambda_{d2}\lambda_{d3}A_{01}\left[\frac{e^{-\lambda_{d1}t}}{(\lambda_{d2} - \lambda_{d1})(\lambda_{d3} - \lambda_{d1})}\right]$$
$$= 0.227 \times 0.0258 \times 50 \left[\frac{e^{-1.26\times10^{-4}\times3\times60}}{(0.227 - 1.26 \times 10^{-4})(0.0258 - 1.26 \times 10^{-4})}\right]$$
$$= 49.14 \quad \mu Ci.$$

Activity of $^{214}_{83}Bi$

The decay constant of $^{214}_{83}Bi$ is

$$\lambda_{d4} = \frac{\ln(2)}{T_{1/2}} = \frac{0.693}{19.7}$$
$$= 0.0352 \quad \text{min}^{-1}.$$

The Bateman's equation for $i = 4$ will also contain negligible exponential terms as we saw in the previous case. Hence we can approximate the activity of $^{214}_{83}Bi$ after 3 hours by

$$A_4 \approx \lambda_{d2}\lambda_{d3}\lambda_{d4}A_{01}\left[\frac{e^{-\lambda_{d1}t}}{(\lambda_{d2} - \lambda_{d1})(\lambda_{d3} - \lambda_{d1})(\lambda_{d4} - \lambda_{d1})}\right]$$
$$= 49.32 \quad \mu Ci.$$

1.3.E   Decay Equilibrium

Depending on the difference between the decay constants of parent and daughter nuclides it is possible that after some time their activities reach a state of equilibrium. Essentially there are three different scenarios leading to different long term states of the radioactive material. These are termed as *secular equilibrium, transient equilibrium*, and *no equilibrium* states. For the discussion in this section we will assume a radioactive material that has a parent and a daughter only. However the assertions will also be valid for materials that go through a number of decays.

E.1   Secular Equilibrium

If the activity of parent becomes equal to that of the daughter, the two nuclides are said to be in secular equilibrium. This happens if the half life of parent is much greater than that of the daughter, i.e.

$$T_{1/2}^P \gg T_{1/2}^D \quad \text{or} \quad \lambda_{dP} \ll \lambda_{dD}$$

Let us see if we can derive the condition of equal activity from equation 1.3.28. It is apparent that if $\lambda_{dP} > \lambda_{dD}$ then as $t \to \infty$

$$e^{-\lambda_{dD}t} \gg e^{-\lambda_{dP}t},$$

and hence we can neglect the second term on right side of equation 1.3.28. The daughter activity in this case is given approximately by

$$A_D \simeq \frac{\lambda_{dD}}{\lambda_{dD} - \lambda_{dP}} A_{0P} e^{-\lambda_{dP}t} \qquad (1.3.32)$$

$$= \frac{\lambda_{dD}}{\lambda_{dD} - \lambda_{dP}} A_P$$

Or

$$\frac{A_P}{A_D} \simeq 1 - \frac{\lambda_{dP}}{\lambda_{dD}}. \qquad (1.3.33)$$

As $\lambda_{dD} \gg \lambda_{dP}$ we can neglect second term on right hand side of this equation. Hence

$$A_P \simeq A_D$$

This shows that if the half life of parent is far greater than that of the daughter then the material eventually reaches a state of secular equilibrium in which the activities of parent and daughter are almost equal. The behavior of such a material with respect to time is depicted in Fig.1.3.6.

An example of a material that reaches secular equilibrium is $^{237}_{93}Np$. Neptunium-237 decays into protactinium-233 through α-decay with a half life of about 2.14×10^6 years. Protactinium-233 undergoes β-decay with a half life of about 27 days.

Example:
How long would it take for protactanium-233 to reach secular equilibrium with its parent neptunium-237?

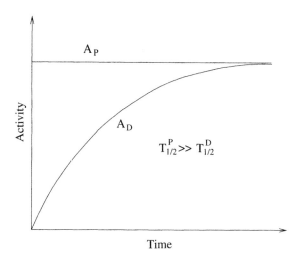

Figure 1.3.6: Activities of parent and daughter nuclides as a function of time for a material that eventually reaches the state of secular equilibrium. The half life of parent in such a material is so large that it can be considered stable.

Solution:

Since the half life of the parent is much larger than that of the daughter, therefore we can safely assume that at the state of secular equilibrium the activity of the daughter will be nearly equal to the initial activity of the parent, that is.

$$A_D \simeq A_{0P}.$$

Substitution of this equality in equation 1.3.32 gives

$$\frac{\lambda_{dD}}{\lambda_{dD} - \lambda_{dP}} e^{-\lambda_{dP}t} \simeq 1$$

$$\Rightarrow t \simeq \frac{1}{\lambda_{dP}} \ln\left(\frac{\lambda_{dD}}{\lambda_{dD} - \lambda_{dP}}\right).$$

The decay constants of the two materials are

$$\lambda_{dP} = \frac{\ln(2)}{T_{1/2}^P}$$

$$= \frac{\ln(2)}{2.14 \times 10^6 \times 365} = 8.87 \times 10^{-10} \text{ day}^{-1}$$

$$\lambda_{dD} = \frac{\ln(2)}{T_{1/2}^D}$$

$$= \frac{\ln(2)}{27} = 2.57 \times 10^{-2} \text{ day}^{-1}$$

Hence the required time is given by

$$t \simeq \frac{1}{8.87 \times 10^{-10}} \ln\left(\frac{2.57 \times 10^{-2}}{2.57 \times 10^{-2} - 8.87 \times 10^{-10}}\right).$$

$$= 38.9 \text{ days}$$

Interestingly enough, this is exactly the mean life of protactanium-233 and it shows that after about one mean life of the daughter its activity becomes approximately equal to that of the parent.

E.2   Transient Equilibrium

The parent and daughter nuclides can also exist in a transient state of equilibrium in which their activities are not equal but differ by a constant fraction. This happens when the half life of the parent is only slightly higher than that of the daughter, i.e.

$$T_{1/2}^P > T_{1/2}^D \quad \text{or} \quad \lambda_{dP} < \lambda_{dD}.$$

The approximate activity 1.3.33 derived earlier is valid in this situation as well.

$$\frac{A_P}{A_D} \simeq 1 - \frac{\lambda_{dP}}{\lambda_{dD}}$$

However now we can not neglect the second term on the right side as we did in the case of secular equilibrium. In this case the equation depicts that the ratio of parent to daughter activity is a constant determined by the ratio of parent to daughter decay constant. Figure 1.3.5 shows the typical behavior of such a material. A common example of transient equilibrium decay is the decay of Pb_{00}^{212} into Bi_{00}^{212}.

E.3   No Equilibrium

If the half life of parent is less than the half life of daughter, i.e.

$$T_{1/2}^P < T_{1/2}^D \text{ or } \lambda_{dP} > \lambda_{dD},$$

then the activity due to parent nuclide will diminish quickly as it decays into the daughter. Consequently the net activity will be solely determined by the activity of the daughter. Figure 1.3.7 depicts this behavior graphically.

1.3.F   Branching Ratio

In the preceding sections we did not make any assumption with regard to whether there was a single mode or multiple modes of decay of the nuclides. In fact the majority of the nuclides actually decay through a number of modes simultaneously with different decay constants. Branching ratio is a term that is used to characterize the probability of decay through a mode with respect to all other modes. For example if a nuclide decays through α and γ modes with branching ratios of 0.8 and 0.2, it would imply that α-particle is emitted in 80% of decays while photons are emitted in 20% of decays. The total decay constant $\lambda_{d,t}$ of such a nuclide having N decay modes is obtained by simply adding the individual decay constants.

$$\lambda_{d,t} = \sum_{i=1}^{n} \lambda_{d,i} \qquad (1.3.34)$$

Here $\lambda_{d,i}$ represents the decay constant of the ith mode for a material that decays through a total of n modes. The total decay constant can be used to determine the

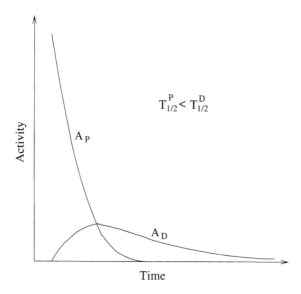

$$T^P_{1/2} < T^D_{1/2}$$

A_P

A_D

Figure 1.3.7: Activities of parent and daughter nuclides as a function of time for a material that never reaches the state of equilibrium. The parent in such a material is shorter lived than the daughter.

effective activity and other related quantities. The expressions for the effective half and mean lives can be obtained by substituting $T_{1/2,i} = 0.693/\lambda_{d,i}$ and $\tau_i = 1/\lambda_{d,i}$ for the ith decay mode in the above equation. This gives

$$\frac{1}{T_{1/2.e}} = \sum_{i=1}^{n} \frac{1}{T_{1/2,i}} \qquad (1.3.35)$$

$$\text{and} \quad \frac{1}{\tau_e} = \sum_{i=1}^{n} \frac{1}{\tau_i}, \qquad (1.3.36)$$

where $T_{1/2,e}$ and τ_e represent the effective half and mean lives respectively.

1.3.G   Units of Radioactivity

Since the most natural way to measure activity of a material is to see how many disintegrations per unit time it is going through, therefore the units of activity are defined in terms of disintegrations per second. For example 1 Becquerel corresponds to 1 disintegration per second and 1 Curie is equivalent to 3.7×10^{10} disintegrations per second. Curie is a much bigger unit than Becquerel and is therefore more commonly used. However for most practical sources used in laboratories, Curie is too big. Therefore its subunits of milli-Curie and micro-Curie are more commonly found in literature. The subunits of Curie and interconversion factors of Curie and Becquerel are given below.

$$1Ci \quad = \quad 3.7 \times 10^{10} \text{ disintegrations/sec}$$

$$1mCi \quad = \quad 10^{-3}Ci = 3.7 \times 10^{7} \text{ disintegrations/sec}$$

$$1\mu Ci \quad = \quad 10^{-6}Ci = 3.7 \times 10^{4} \text{ disintegrations/sec}$$

$$1Bq \quad = \quad 1 \text{ disintegration/sec}$$

$$1Bq \quad = \quad 2.703 \times 10^{-11}Ci$$

1.4   Activation

It is possible to induce radioactivity into materials by letting them interact with radiation. The process is known as activation and is extensively used to produce radioactive particle sources and activation detectors. The radiation emitted by the activated material is generally referred to as *residual radiation*. Most of the activated materials emit γ and β particles but, as we will see later, it is possible to activate materials that emit heavier particles.

To activate the material, it must be irradiated. As soon as the irradiation starts, the material starts decaying. This means that both the processes, irradiation and decay, are happening at the same time. The rate of decay would, of course, depend on the half life of the activated material. Let R_{act} be the rate of activation in the sample. The rate of change in the number of *activated* atoms N in the material is then given by

$$\frac{dN}{dt} = R_{act} - \lambda_d N, \tag{1.4.1}$$

where the second term on the right hand side represents the rate of decay. λ_d is the decay constant of the activated material. Integration of the above equation yields

$$N(t) = \frac{R_{act}}{\lambda_d}\left(1 - e^{-\lambda_d t}\right), \tag{1.4.2}$$

where we have used the boundary condition: at $t = 0$, $N = 0$. We can use the above equation to compute the activity A of the material at any time t. For that we multiply both sides of the equation by λ_d and recall that $\lambda_d N \equiv A$. Hence we have

$$A = R_{act}\left(1 - e^{-\lambda_d t}\right). \tag{1.4.3}$$

Note that the above equation is valid for as long as irradiation is in process at a *constant* rate. In activation detectors, a thin foil of an activation material is placed in the radiation field for a time longer compared to the half life of the activated material. The foil is then removed and placed in a setup to detect the decaying particles. The count of decaying particles is used to determine the activation rate and thus the radiation field.

The activation rate R_{act} in the above equations depends on the radiation flux[6] as well as the activation cross section of the material. In general, it has energy

---

[6]Radiation or particle flux represents the number of particles passing through a unit area per unit time. We will learn more about this quantity in the next chapter.

dependence due to the energy dependence of the activation cross section. However, to get an estimate, one can use a cross section averaged over the whole energy spectrum of the incident radiation. In that case, the *average* activation rate is given by

$$R_{act} = V\Phi\sigma_{act}, \qquad (1.4.4)$$

where V is the total volume of the sample, Φ is the radiation flux, and σ_{act} is the spectrum-averaged activation cross section.

The behavior of equation 1.4.3 is graphically depicted in Fig.1.4.1. Since the decay rate depends on the number of activated atoms, the number of atoms available for decay increases with time. The exponential increase in the activity eventually reaches an asymptotic value equal to the activation rate.

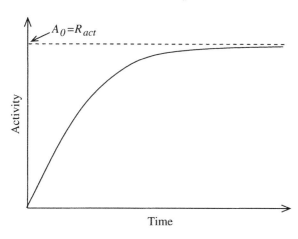

Figure 1.4.1: Buildup of activity in a sample undergoing activation through constant irradiation. The activity eventually reaches an asymptotic value approximately equal to the activation rate R_{act}.

 ## 1.5   Sources of Radiation

Radiation sources can be broadly divided into two categories: natural and man-made.

1.5.A   Natural Sources

There are three types of natural sources of radiation: cosmic, terrestrial, and internal. Exposure from most of these sources is very minimal and therefore does not cause any measurable damage to our bodies. However, as we will see later in this section, there are some potentially hazardous materials, such as radon in our surroundings, which in deed are a cause of concern since they are capable to delivering high integrated doses.

A.1   Cosmic Radiation Sources

The outer space is filled with radiation that comes from a variety of sources such as burning (for example, our Sun) and exploding (for example, Supernovae) stars. These bodies produce immense amounts of radiation, some of which reach earth.

Fortunately the earth's atmosphere acts as a shield to the worst of these radiations, such as ultraviolet rays from the Sun are blocked by the ozone layer. However not all of the harmful radiation is blocked and some reach the surface of earth causing skin burns and cancer in people who remain exposed to sun light for extended periods of time. The situation is even worse in places where the ozone layer has depleted due to some reason.

On top of these localized sources of radiation there is also a background radiation of low energy photons. This radiation is thought to be the remnant of the so called big bang that created this universe. It is known as cosmic microwave background radiation since the photon spectrum peaks in the microwave region of the electromagnetic spectrum. Although these photons reach the earth's surface but due to their low energies, they are not deemed harmful.

Apart from photons, there are other particles as well that are constantly being produced in the outer space. Most of them, however, never reach the earth either due to magnetic deflection or the earth's upper protective atmosphere. Some of the particles, like muons, electrons, and neutrinos, are produced when other cosmic particles interact with atoms in the upper atmosphere. Shower of these particles reach earth's surface time but due to their low energies and low interaction probabilities, they do not pose any significant health hazard.

Muons and neutrinos directly produced by luminous objects in space also manage to reach earth due to their low interaction capabilities but are not considered hazardous to health due to their extremely low interaction cross sections.

A.2   Terrestrial Radiation Sources

This type of radiation is present in small quantities all around us and is more or less inescapable. Our surroundings, the water we drink, the air we breathe in, and the food we consume, all are contaminated with minute quantities of radiation emitting isotopes. Although these isotopes, in general, are extremely hazardous, they are not supposed to cause any appreciable harm to our bodies except when the are present in higher than normal concentrations.

The main source of terrestrial radiation is the element uranium and its decay products such as thorium, radium, and radon. Although the overall natural concentration of these radioactive materials is within the tolerable range of humans, some parts of the world have been identified where higher levels of uranium and thorium in surface soil have increased the radiation to dangerous levels. Unfortunately man has also contributed to this dilemma by carrying out nuclear explosions and by dumping nuclear waste.

The two isotopes of radon, ^{222}Rn and ^{220}Rn, and their daughter products are the most commonly found hazardous radioactive elements in our surroundings. The main cause of concern with respect to these α-emitting isotopes is their inhalation or digestion, in which case the short range α-particles continue to cause damage to internal organs that can lead to cancer.

A.3   Internal Radiation Sources

Our bodies contain some traces of radioactive elements that expose our tissues to continuous low level radiation. This internal radiation primarily comes from

Potassium-40 and Carbon-40 isotopes. However the absorbed dose and the damage to tissues due to this radiation is minimal.

1.5.B   Man-Made Sources

Right after the discovery of radiation and realization of its potentials, scientists started working on developing sources that can be used to produce radiation in controlled laboratory environments. These sources are made for specific purposes and generally give off one type of radiation. Common examples of such sources are

- ▶ medical x-ray machines,

- ▶ airport x-ray scanners,

- ▶ nuclear medicines,

- ▶ particle accelerators, and

- ▶ lasers.

Out of all these sources, the ones used in medical diagnostics and therapy expose the public to the most significant amounts of radiation. For example a single chest x-ray exposes the patient to about 20 *mrem* of radiation, which is a significant fraction of about 360 *mrem* of total radiation exposure to general public due to all types of radiation. Repeated x-rays of patients are therefore discouraged unless there is absolute medical necessity.

There are also some consumer products that give off radiation although they have been made for some other purpose. Examples of such sources are

- ▶ television,

- ▶ smoke detectors, and

- ▶ building materials.

As we saw in the section on radioactivity, there are a large number of naturally occurring and man-made isotopes that emit different kinds of radiation. Depending on their half lives, types of radiation they emit, and their energies, some of these radioisotopes have found applications in a variety of fields. In the next section, when we discuss individual particle properties, we will also look at some of the commonly used radiation sources made from such isotopes.

Before we go to the next section, let us have a general look at some radioisotopes that are made in laboratories for specific purposes. A variety of methods are used to produce radioisotopes. For example a common method is the bombardment of stable elements with other particles (such as neutrons or protons), a process that desta-bilizes the nuclides resulting in its decay with emission of other particles. Another method is to extract the unstable isotopes from the spent fuel of nuclear reactors. Some of the frequently produced radioactive elements are listed in Table.1.5.1.

Table 1.5.1: Common Radioactive Isotopes of Elements

Element	Common Isotopes (Decay Mode)	Common Use
Cobalt	$Co_{27}^{60}(\beta)$	Surgical instrument sterilization
Technetium	$Tc_{43}^{99}(\beta)$	Medical diagnostics
Iodine	$I_{53}^{123}(\beta, EC), I_{53}^{129}(\beta), I_{53}^{131}(\beta)$	Medical diagnostics
Xenon	$Xe_{54}^{133}(\beta)$	Medical diagnostics
Caesium	$Cs_{55}^{137}(\beta)$	Treat cancers
Iridium	$Ir_{77}^{192}(\beta)$	Integrity check of welds and parts
Polonium	$Po_{84}^{210}(\alpha)$	Static charge reduction in photographic films
Thorium	$Th_{90}^{229}(\alpha)$	Extend life of fluorescent lights
Plutonium	$Pu_{94}^{238}(\alpha)$	α-particle source
Americium	$Am_{95}^{241}(\alpha)$	Smoke detectors

 ## 1.6 General Properties and Sources of Particles and Waves

The interaction of particles with atoms is governed by quantum mechanical processes that depend on the properties of both incident particles and target atoms. These properties include mass, electrical charge, and energy. In most of the cases, the target atoms can be considered to be at rest with respect to the incident particles, which greatly simplifies the mathematical manipulations to predict their properties after the interaction has taken place.

Strictly speaking it is not absolutely necessary to know the internal structure (if any) of a particle to understand its gross properties (except for properties, such as magnetic moment of a neutron, which are not needed in usual radiation detection). However to make our discussion complete, let's have a look at our present knowledge about the elementary particles. On the fundamental level there are only a few particles, which in different combinations form heavier and stable particles. For example both proton and neutron are composed of 3 fundamental particles called quarks. The nature of these quarks determines whether it is a proton or a neutron. There is an extremely successful model in particle physics, called the Standard Model, which tells us that there are 6 types of quarks and another breed of elementary particles called leptons. These are the particles that are not known to have any internal structure and are therefore regarded as elementary particles. There are 6 leptons in Standard Model: electron, tau, muon, and their respective neutrinos. Although these particles are not known to be composed of other particles, but there are theoretical models and empirical evidences that suggest that they might have internal structures.

As a comprehensive discussion on elementary particles is out of the scope of this book therefore we will end this discussion here and refer the curious readers to introductory texts on particle physics (17; 16) for detailed explanations.

In order to understand the interaction mechanisms of different particles it is necessary to first familiarize ourselves with their properties. In the following we will survey some of the important particles with respect to the field of radiation detection and discuss their properties. The way these particles interact with matter will be discussed in the next chapter. We will also discuss some of the most important man-made sources that emit these particles.

1.6.A Photons

A photon represents one quantum of electromagnetic energy and is treated as a fundamental particle in the Standard Model of particle physics. In this model the photon is assumed to have no rest mass (although it is never at rest!). When the photon is traveling in a medium, it slows down due to interaction with the medium and acquires an effective mass. In vacuum, however, it is considered to be massless.

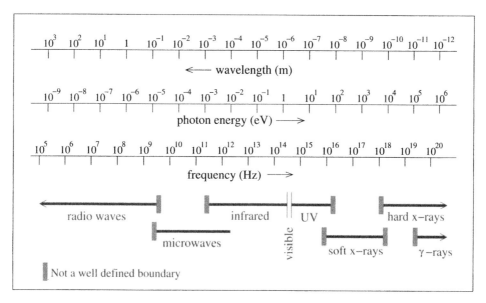

Figure 1.6.1: Electromagnetic spectrum.

Photons do not correspond to only visible light. In fact, light spans a very narrow region of their full energy spectrum (see Fig.1.6.1). Also, what we see as visible light, is not really considered to be composed of individual photons, rather it is a superposition of a number of photons. Photons are involved in all types of electromagnetic interactions. The whole wireless communications on which we are so dependent nowadays is possible by photons largely in the microwave and radio-frequency regions. There are many ways to produce photons that are not in the visible light region of the spectrum. For example whenever a charged particle moves (such as a changing current in a wire), it creates electromagnetic waves around it that propagate in space. These waves are considered to be excitations in the underlying electromagnetic field and a quantum of these excitations is called a photon.

The energy carried by a photon can be absorbed in a number of ways by other particles with which it interacts. Also, like other particles, a photon can scatter off from other particles and even experience gravitational pull. We'll discuss these interactions in detail in the next chapter.

In terms of radiation exposure and biological damage we are generally concerned with high energy photons, such as γ-rays, x-rays, and ultraviolet rays. Having high energies, these photons can penetrate deeper and cause more damage than the low energy photons such as visible light.

Basic Properties of Photons

Rest mass	=	Zero
Electrical charge	=	Zero
Energy	=	$h\nu = \frac{hc}{\lambda}$
Momentum	=	$\frac{h}{\lambda}$
Examples	:	visible light, x-rays, γ-rays

An important property of photons is that they carry momentum even though they have no rest mass. The momentum p_γ of a photon with energy E, frequency ν, and wavelength λ is given by,

$$p_\gamma = \frac{E}{c} = \frac{h\nu}{c} = \frac{h}{\lambda} \tag{1.6.1}$$

A.1 Sources of Photons

Photons play very important roles not only in physics but also in engineering, medical diagnostics, and treatment. For example laser light is used to correct vision, a process called laser surgery of the eye. In medical diagnostics, x-rays are used to make images of internal organs of the body. In the following we will look at some of the most important of sources of photons that have found wide applications in various fields.

X-ray Machine

Since x-rays are high energy photons and can cause considerable damage to tissues, they are produced and employed in controlled laboratory environments. Production of x-rays is a relatively simple process in which a high Z target (i.e. an element having large number of protons, such as tungsten or molybdenum) is bombarded by high velocity electrons (see Fig.1.6.2). This results in the production of two types of x-rays: Bremsstrahlung and characteristic x-rays.

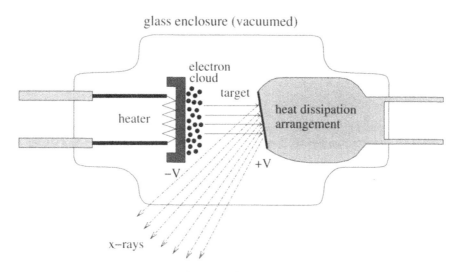

Figure 1.6.2: Sketch of a typical x-ray tube.

Bremsstrahlung (a German word for *braking radiation*) refers to the radiation emitted by charged particles when they decelerate in a medium (see Fig.1.6.3). In case of x-rays, the high energy electrons decelerate quickly in the target material and hence emit Bremsstrahlung. The emitted x-ray photons have continuous energy spectrum (see Fig.1.6.4) since there are no quantized energy transitions involved in this process. Bremsstrahlung are the x-rays that are usually employed to produce images of internal objects (such as internal body organs in medical diagnostics).

The electrons incident on the target may also attain sufficient energies to knock off electrons from the internal atomic shells of the target atoms leaving them in unstable states. To regain atomic stability, the electrons from higher energy levels quickly fill these gaps. Since the energy of these electrons is higher than the energy needed to stay in the new orbits, the excess energy is emitted in the form of x-ray photons (see Fig.1.6.3). Having energies characteristic to the difference in the atomic energy levels, these photons are said to constitute characteristic x-rays (see Fig.1.6.4). The energy of characteristic x-rays does not depend on the energy or intensity of the incident electron beam because the emitted photons always have the energy characteristic to the difference in the corresponding atomic energy levels. Since different elements may have different atomic energy levels, therefore the energy of the emitted characteristic x-ray photons can be fairly accurately predicted.

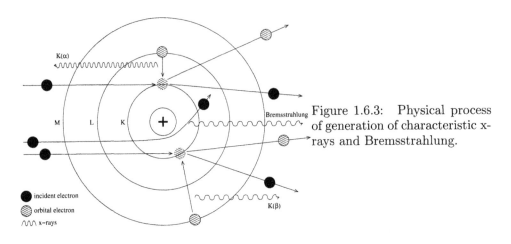

Figure 1.6.3: Physical process of generation of characteristic x-rays and Bremsstrahlung.

X-ray tube spectra generally have more than one characteristic x-ray peaks since there are a number of electronic transitions possible following a vacancy created in one of the inner electronic shells. If an electron from the innermost K-shell is knocked-off, the vacancy can in principle be filled by any of the electrons in the outer shells. If an electron from the L-shell jumps in to fill the vacancy, a photon with an energy of $E_\gamma = E_L - E_K$ is emitted. A large number of such photons would appear as a prominent peak in the spectrum. Such a peak is generally referred to as K_α peak. K_β peak is the result of transitions of M-shell electrons to the K-shell giving off photons with an energy of $E_\gamma = E_M - E_K$ (see Fig.1.6.3).

In an x-ray tube the target (anode) is kept very close (typically 1-3 cm) to the source of electrons (cathode). A high electric potential between cathode and anode accelerates the electrons to high velocities. The maximum kinetic energy in electron volts attained by these electrons is equal to the electric potential (in volts) applied between the two electrodes. For example an x-ray machine working at a potential of 30 kV can accelerate electrons up to a kinetic energy of 30 keV.

X-ray machines are extremely inefficient in the sense that 99% of their energy is converted into heat and only 1% is used to generate x-rays.

Example:
Calculate the maximum velocity attained by an electron in an electric potential of 40 kV.

Solution:
An electron in an electric field of 40 kV can attain a maximum kinetic energy of

$$
\begin{aligned}
T_{max} &= \left(40 \times 10^3\right)\left(1.602 \times 10^{-19}\right) \\
&= 6.408 \times 10^{-15} \quad J \\
&= 40 \quad keV.
\end{aligned}
$$

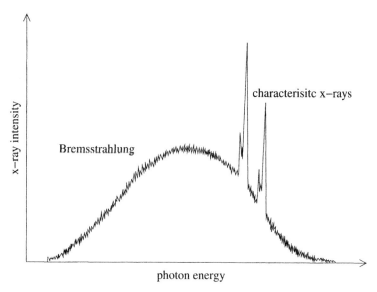

Figure 1.6.4:   A typical x-ray tube spectrum showing Bremsstrahlung continuum and peaks corresponding to characteristic x-rays.

Assuming that the electron velocity is non-relativistic (i.e. $v_e \ll c$), we can write its maximum kinetic energy as

$$T_{max} = \frac{1}{2} m_e v_{max}^2,$$

where v_{max} represents the maximum kinetic energy of the electron. Therefore the required velocity is

$$
\begin{aligned}
v_{max} &= \left[\frac{2T_{max}}{m_e} \right]^{1/2} \\
&= \left[\frac{(2)\left(6.408 \times 10^{-15}\right)}{9.1 \times 10^{-31}} \right]^{1/2} \\
&= 1.2 \times 10^8 \ ms^{-1}.
\end{aligned}
$$

Synchrotron Radiation

X-ray tubes are not the only means of producing x-rays. In high energy facilities where particles are accelerated in curved paths at relativistic velocities using magnetic fields, highly intense beams of photons, called synchrotron radiation, are naturally produced. We saw earlier that when electrons decelerate in a medium they give rise to conventional x-rays called Bremsstrahlung. On the other hand the synchrotron radiation is produced when charged particles are accelerated in *curved paths.* Although conceptually they represent the same physical phenomenon, but they can be distinguished by noting that Bremsstrahlung is a product of tangential acceleration while synchrotron radiation is produced by centripetal acceleration of charged particles.

The spectrum of synchrotron radiation is continuous and extends over a broad energy range from infrared to hard x-rays. In general the spectral distribution is smooth with a maximum near the so-called *critical wavelength* (see Fig.1.6.5). Critical wavelength divides the energy carried by the synchrotron radiation into two halves.

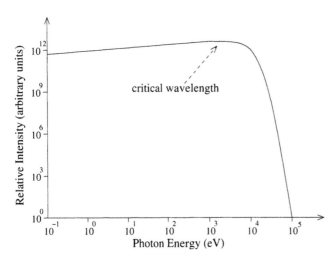

Figure 1.6.5: Typical synchrotron radiation spectrum. Critical wavelength divides the total delivered energy into two halves.

The x-rays produced as synchrotron radiation are extremely intense, highly collimated, and polarized (most of the wave vectors oscillate in the same plane) in contrast to conventional x-rays, which have very low intensities, are very difficult to collimate, and are completely unpolarized. However the production of synchrotron radiation is far more expensive than conventional x-rays and therefore dedicated synchrotron facilities have been developed around the world where beam time is made available to researchers.

LASER

LASER (Light Amplification by Stimulated Emission of Radiation) is generated by exploiting a quantum mechanical phenomenon called stimulated emission of photons. Stimulated emission is an optical amplification process in which the photon population is increased by allowing the incident photons to interact with atoms or molecules in excited states. An excited atoms, when struck by an incident photon of some frequency and phase, emits another photon of the same frequency and phase to relax to the ground state. The initial photon is not destroyed in the process and goes on to create more photons. The result is an intense, highly collimated, and coherent beam of light. In essence, the trick of producing laser is to somehow increase the population of atoms or molecules in the excited state and maintain it through external means. If more atoms or molecules are in excited state than in ground state, the system is said to have reached *population inversion*. Laser light is emitted for as long as this population inversion is maintained.

We saw earlier in the section on radioactive decay that the rate of spontaneous emission is proportional to the number of nuclei in the sample. In the case of stimulated emission this rate is proportional to the product of the number of atoms or molecules of the lasing medium and the radiation density $\rho(\nu)$ of the incident photons.

$$\frac{\partial N}{\partial t} = -B_{21}\rho(\nu)N \tag{1.6.2}$$

Here B_{21} is a constant known as Einstein's B coefficient and depends on the type of atoms. Fig.1.6.6 shows the principle of operation of a typical gas laser. To make stimulated emission possible, energy must be provided from some external source. This so called *pump* can be a simple light source. A semi-transparent mirror at the exiting end and an opaque mirror at the other end of the laser cavity reflects enough light to maintain the population inversion. A focusing lens at the other end is used for further collimation of the laser light.

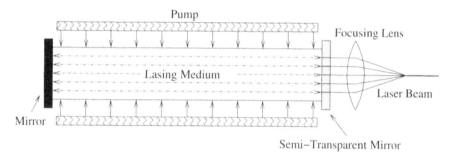

Figure 1.6.6: Principle of lasing action in a gas.

Lasers can be produced either as continuous waves or in the form of short duration pulses by a variety of materials in different states. Following are some of the common types of lasers.

Gas Lasers

The active medium of such a laser is in gaseous atomic, molecular, ionic, or metallic vapor form.

▶ **Atoms:** A very common example of such a laser is He-Ne laser. It emits different wavelengths of laser, such as 632.8 nm, 1152 nm, and 543.5 nm.

▶ **Molecules:** These lasers are produced by molecules of gases such as carbon dioxide and nitrogen. Simple organic molecules, such as CF_4, can also be employed for lasing action. However since these molecules have very narrow energy widths, they must be pumped with another laser, such as CO_2. The wavelengths of these lasers fall in the far infrared region of the spectrum. Lasers can also be produced by molecules in short-lived excited states. These so called Excimer (excited dimer) lasers are generally produced by bonding a noble gas atom (such as argon, krypton, or xenon) with a halogen atom (chlorine, flourine, bromine, iodine). Such molecules can only be produced in a short-lived excited state with a lifetime of less than 10 ns. These lasers are also referred to as Exiplex (excited complex).

▶ **Ions:** The lasing medium of such a laser is generally an ionized noble gas such as Ar^+ or Kr^+. Since most of the wavelengths of these low power lasers fall within the visible region of spectrum, they are extensively used in entertainment industry to produce laser light shows.

▶ **Metal Vapors:** Lasing action can also be produced in metal vapors such as those of gold, silver, and copper. These lasers are very efficient and capable of delivering high pulsed power.

Liquid Lasers

Using liquid as a lasing material is best demonstrated by the so called dye lasers. These lasers use a combination of liquid organic dyes to produce laser light with wavelengths that can be tuned over a specific region of the spectrum. The dyes used in such systems are fluorescent materials. To force the population inversion another light source is used, which may simply be a lamp or another laser. Nd-YAG laser is commonly used to pump dye lasers. Other Excimer or continuous wave lasers are also extensively used for the purpose. Both pulsed and continuous mode of operation are available in dye lasers. A commonly used dye is Rhodamine-6G, which produces a spectrum of light centered at 590 nm. When pumped with a continuous wave laser, it can produce a power of about 1 kWh. The main advantage of dye lasers is that they can be selectively tuned to output laser in a specific wavelengths range. Another advantage of using dyes as lasing media is the very broad range of laser wavelengths that can be produced.

Solid State Lasers

The solid lasing media can be conveniently categorized according to their electrical conduction properties.

▶ **Insulators:** The first lasing material successfully used to generate laser light was Ruby crystal (Al_2O_3 with Cr^{+3} as low level impurity). Since then a large number of electrically insulated solids have been identified as efficient lasing media. Common examples of such materials are Nd-YAG and Ti-Sapphire.

▶ **Semiconductors:** Semiconductors have been found to be extremely cost effective in producing low power laser light. Because of their small size and ease in production these lasers are widely used in consumer products, such as laser printers and CD writers.

New Developments

During the past two decades a lot of research has been carried out to develop highly intense and powerful lasers. Most of these research activities have focused on producing lasers in the ultraviolet and x-ray regions of the electromagnetic spectrum. Production of coherent light in these regions was first successfully demonstrated in the mid 1980s. Since then a large number of these so called x-ray lasers have been developed in different laboratories around the world. A more recent advancement has been the development of the so called free electron lasers. These highly powerful lasers (kW range) produce low wavelength coherent light in brief bursts and are generally tunable to a precise wavelength within their range of operation.

Lasers have found countless applications in many diverse fields. From precision heavy metal cutting to the delicate eye surgery, from CD burners to range finders, lasers are now an essential part of our everyday lives.

Most of the lasers currently in use have spectra that fall within or around the visible light spectrum. A lot of work is now being directed towards production of new laser materials and apparatus that could produce lower wavelength lasers. These lasers would deliver more power to the target in shorter periods of time. Some success has already been achieved in developing these so called, x-ray lasers (their wavelengths lie within the range of conventional x-rays).

Although very useful, laser can be extremely hazardous specially to skin and eye. It can cause localized burning leading to permanent tissue damage and even blindness. Since the possibility and degree of harm depends on its wavelength (or energy), intensity and time of exposure, therefore the regulatory commissions have classified lasers in different classes: Class-I laser is known to be safe and would not cause any damage to eye even after hours of direct exposure. Class-IV laser, on the other hand, is extremely dangerous and can cause irreversible damage such as permanent blindness. The lasers between these two classes are neither absolutely safe nor extremely dangerous and the workers are allowed to directly work with them provided they use appropriate eye protection equipment.

Radioactive Sources of Photons

There are a large number of radioactive elements that emit γ-rays. These γ-rays are often accompanied by α- and β-particles. Besides naturally occurring sources it is possible to produce these isotopes in laboratory as well. This is normally done by bombarding a source material by neutrons. The nuclei, as a result, go into unstable states and try to get rid of these extra neutrons. In the process they also release energy in the form of γ-rays. The two most commonly used radioactive sources of γ-rays are iridium-192 ($^{192}_{77}Ir$) and cobalt-60 ($^{60}_{27}Co$).

The easiest way to produce cobalt-60 is by bombarding cobalt-59 with slow neutrons as represented by the following reaction.

$$^{59}_{27}Co + n \rightarrow\ ^{60}_{27}Co + \gamma \quad (7.492\ MeV) \tag{1.6.3}$$

For this reaction we need slow neutrons. Californium-252 is an isotope that is commonly used as a source of neutrons. $^{252}_{98}Cf$ is produced in Nuclear Reactors and has a half life of approximately 2.64 years. It can produce neutrons through a number of fission modes, such as

$$^{252}_{98}Cf \rightarrow\ ^{94}_{38}Sr + \ ^{154}_{60}Nd + 4n. \tag{1.6.4}$$

However the neutrons produced in this way have higher kinetic energies than needed for them to be optimally captured by cobalt-59. Therefore some kind of *moderator*, such as water, is used to slow down these neutrons before they reach the cobalt atoms. The resultant cobalt-60 isotope is radioactive and gives off 2 energetic γ-rays

Table 1.6.1: Common γ emitters and their half lives.

Element	Isotope	Energy	$T_{1/2}$
Sodium	Na_{11}^{24}	1.368MeV, 2.754MeV	14.959 hours
Manganese	Mn_{25}^{54}	834.838keV	312.3 days
Cobalt	Co_{27}^{60}	1.173MeV, 1.332MeV	5.271 years
Strontium	Sr_{38}^{85}	514.005keV	64.84 days
Yttrium	Y_{39}^{88}	898.036keV, 1.836MeV	106.65 days
Niobium	Nb_{41}^{94}	765.803keV	2.03×10^4 years
Cadmium	Cd_{48}^{109}	88.034keV	462.6 days
Cesium	Cd_{55}^{137}	661.657keV	30.07 years
Lead	Ph_{82}^{210}	46.539keV	22.3 years
Americium	Am_{95}^{241}	26.345keV, 59.541keV	432.2 years

with a half life of around 5.27 years (see Fig.1.6.7), as represented by the reaction below.

$$_{27}^{60}Co \rightarrow {_{28}^{60}}Ni + \beta + \bar{\nu} + 2\gamma \quad (1.17\ MeV, 1.33\ MeV) \qquad (1.6.5)$$

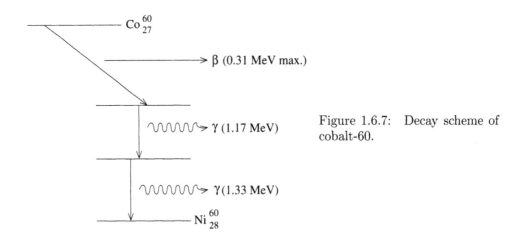

Figure 1.6.7: Decay scheme of cobalt-60.

Table 1.6.1 lists some of the commonly used γ-ray sources and their half lives.

1.6.B   Electrons

According to our understanding so far, electron is one of the fundamental particles of nature. It carries negative electrical charge and has very small mass and radius. Although we talk in terms of radius, none of the experiments so far has been able to associate any particular structure to it. Interestingly enough, even though it appears to have no structure, it seems to be spinning in well defined ways.

Basic Properties of Electrons

Rest mass $\quad\quad=\quad 9.11 \times 10^{-31} kg = 0.511 MeV/c^2$

Electrical charge $\quad=\quad -1.602 \times 10^{-19} C$

Internal structure $\quad:\quad$ Believed to have no internal structure.

Electrons were first discovered by J. J. Thompson in 1897 in Britain, about 6 years after their presence was hypothesized and they were named *electrons* by an Irish physicist, George Stoney. Thompson was able to produce cathode rays (called as such because the rays seemed to be originating from the cathode) by making an electric current pass through a glass bulb containing little amount of air. This produced light of different colors inside the glass bulb and also a faint spot on the wall of the bulb. He managed to change the direction of these rays by applying an electric and a magnetic field across the bulb such that the electric field would move the rays in one direction and the magnetic field would move the rays in the other. With this scheme he was able to find the charge to mass ratio (e/m) of the particles in these rays. Interestingly enough the mass of this particle was found to be hundreds of times smaller than the atom. This was the first discovery of a subatomic particle.

Later on, Ernest Rutherford proposed that the atom actually contained a nucleus where the positive charges were concentrated. Neil Bohr expanded on this idea by assuming that electrons revolved around this nucleus in well defined orbits, a picture that we now know to be correct.

Just like light waves that sometimes behave as particles, electrons also seem to have wave-like properties. For example when one atom comes closer to another, their electron clouds (electrons revolve around the nucleus in a cloud-like orbit such that they appear to be everywhere at the same time) interfere and may form a molecule. Interference of electron waves has also been observed in double slit experiments when electrons were forced to pass through consecutively placed slits.

Our familiar electric current is carried through the metallic wires by electrons that drift at the application of electric potential at the ends of the conductor. But this is not all, by adding or removing electrons from some materials, called semiconductors, scientists have been able to fine tune and control their conduction properties. This has enabled them to construct sophisticated electronic components that now form the backbone of our technological development.

Electrons are also extensively used in medical diagnostics, therapy, material research and a number of other fields.

B.1   Sources of Electrons

Production of electron beams is a relatively simple process and a number of devices have been developed for the purpose. We will discuss here some of the commonly used sources of electrons.

Electron Gun

These are used to produce intense beams of high energy electrons. Two types of electron guns are in common use: thermionic electron gun and field emission electron gun. A third type, photo emission electron gun, is now gaining popularity specially in high energy physics research.

The basic principle of an electron gun is the process in which an electron is provided enough kinetic energy by some external agent to break away from the overall electric field of the material. The three types mentioned above differ in the manner in which a material is stimulated for this ejection. The process is easiest in metals in which almost free electrons are available in abundance. These electrons are so loosely bound that a simple heating of the metal can break them loose. Each metal has a different threshold energy needed to overcome the internal attractive force of the nuclei. This energy is called *work function* and is generally denoted by W.

In a thermionic electron gun, the electrons in a metal are provided energy in the form of heat. Generally tungsten is used in the form of a thin wire as the source due to its low work function ($4.5\ eV$). An electrical current through this thin wire (called *filament*) produces heat and consequently loosely bound electrons leave the metal and accumulate nearby forming the so called *electron cloud*. To make a beam of electrons out of this electron cloud a high electrical potential is applied between two electrodes. The negative electrode (cathode) is placed near the electron cloud while the positive electrode (anode) is placed away from it. Cathode is in the form of a grid so that it could allow the electrons from the filament to pass through it when they experience attractive pull of the anode. The intensity of the electron beam is proportional to the number of electrons emitted by the filament, which in turn depends on the temperature of the filament. Since this temperature is proportional to the current passing through the filament, the current can be used to control the intensity of the electron beam. In practice, as the filament current is raised the electron beam intensity rises until a saturation state is reached in which the intensity remains constant even at higher currents (see Fig.1.6.8). The intensity of the electron beam is generally quoted in Ampere, which is a unit of electrical current and represents the charge passing through a certain point per second.

In a field emission electron gun, the electrons are extracted from a metal using a very high electric field of the order of $10^9\ V/m$. This does not require the source of electrons to be heated but safe application of such a high electrical potential requires a very good vacuum.

It is also possible to liberate electrons from the surface of a metal by illuminating it with photons, a process called photoelectric emission. In order to obtain intense electron beams, this process requires utilization of intense light sources. Currently lasers are being used for this purpose. These photo emission electron tubes are capable of producing very high intensity electron beam pulses.

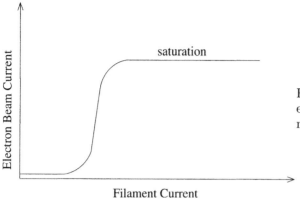

Figure 1.6.8: Dependence of electron beam current on filament current.

Example:
Compute the number of electrons carried in a second by an electron beam of $1.6\ nA$.

Solution:
Electrical current is defined as

$$I = \frac{Q}{t},$$

where Q is the total charge passing in time t. Using this equation we can compute the total charge carried in a second by the beam.

$$
\begin{aligned}
Q &= It = (1.6 \times 10^{-9})(1) \\
&= 1.6 \times 10^{-9}\ C.
\end{aligned}
$$

Since each electron carries a unit charge of 1.6×10^{-19} coulombs, the total number of electrons carried by the beam is

$$
\begin{aligned}
N &= \frac{Q}{1.6 \times 10^{-19}} \\
&= \frac{1.6 \times 10^{-9}}{1.6 \times 10^{-19}} \\
&= 10^{10}\ s^{-1}.
\end{aligned}
$$

Radioactive Sources of Electrons

We saw earlier that cobalt-60 emits β-particles together with γ-rays. Although it can, in principle, be used as a source of electrons but because of the associated high γ-ray background flux, it is not generally used for this purpose. There are a number of other elements as well whose unstable isotopes emit β-particles (see Table.1.6.2) with very low γ-ray backgrounds. Most of these radionuclides are extracted from the spent fuel of nuclear reactors where they are produced as byproducts of the fission reaction.

Table 1.6.2: Common electron emitters and their half lives.

Element	Isotope	Energy (E_{max})	$T_{1/2}$
Sodium	$^{24}_{11}P$	1.393 MeV	14.959 hours
Phosphorus	$^{32}_{15}P$	1.71 MeV	14.262 days
Chromium	$^{51}_{24}Cr$	752.73 keV	27.702 days
Cobalt	$^{60}_{27}Co$	318.13 keV	5.271 years
Copper	$^{64}_{29}Cu$	578.7 keV	12.7 hours
Strontium	$^{90}_{38}Sr$	546.0 keV	28.79 years
Yttrium	$^{90}_{39}Y$	2.28 MeV	64.0 hours
Iodine	$^{125}_{53}I$	150.61 keV	59.408 days
Cesium	$^{137}_{55}Cs$	513.97 keV	30.07 years
Thallium	$^{204}_{81}Th$	763.4 keV	3.78 years

The emission of a β-particle by a radionuclide was described earlier through the reaction

$$^{n+p}_{p}X \rightarrow ^{n+p}_{p+1}Y + e + \bar{\nu}.$$

The energy released in the reaction is taken away by the daughter, the electron, and the neutrino. The daughter nucleus, being much more massive than the other two particles, carries the least amount of kinetic energy. This energy, also called *recoil energy*, is too low to be easily detected[7]. Therefore, for most practical purposes, we can assume that the recoil energy is zero. Most of the energy is distributed between the electron and the neutrino. There is no restriction on either of these particles for the amount of energy they can carry. Although the energy of the neutrino can not be detected by conventional means, it can be estimated from the measured β-particle energy. Fig.1.6.9 shows a typical β-particle energy spectrum. The electrons can carry energy from almost zero up to the *endpoint energy*, which is essentially the decay Q-value.

---

[7]For detection through ionization, the particle must carry energy greater than the ionization threshold of the medium. In this case, the recoil energy is generally lower than the ionization threshold. However, one can utilize other methods of detection, such as scintillation, where the interacting particle excites the medium such that it emits light during de-excitation. Measurement of recoil energy, not necessarily related to β-decay, is not uncommon in particle physics research.

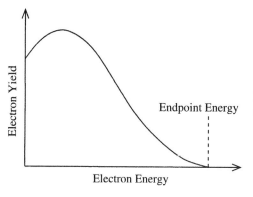

Figure 1.6.9: Typical β-particle energy spectrum in a β-decay.

1.6.C Positrons

A positron is the anti-particle of an electron. It has all the properties of an electron except for the polarity of the electrical charge, which is positive. There a positron can simply be considered an electron having positive unit electrical charge. Whenever an electron and a positron come close to each other, they annihilate each other and produce energy in the form of photons.

$$e + e^+ \rightarrow 2\gamma \tag{1.6.6}$$

Basic Properties of Positrons

Rest mass $= \quad 9.11 \times 10^{-31} \ kg = 0.511 \ MeV/c^2$

Electrical charge $= \quad +1.602 \times 10^{-19} \ C$

Internal structure   :   Believed to have no internal structure.

Positrons have been shown to be extremely effective tools in a variety of fields. Most notably their utility in particle physics research has led to far reaching discoveries made at particle accelerators, such as LEP at CERN. In medical imaging, they are employed in the so called Positron Emission Tomography or simply PET.

C.1 Sources of Positrons

Particle Accelerators

In particle accelerators, positrons are produced through the process of pair production. In this process a photon interacting with the electromagnetic field of a heavy charge creates an electron and a positron.

The positrons thus created are then guided by electric and magnetic fields to form a narrow beam for later collision with other particles. At the Large Electron-Positron Collider (LEP), the positrons are collided with electrons at very high energies, which results in the production of a large number of particles. The particles produced in this way help the physicists to understand the fundamental properties of particles and their interactions.

Radioactive Sources of Positrons

As we described earlier in the Chapter, there are a number of radioactive isotopes of different materials that decay by positron emission. The only problem is that the positrons thus produced have very short lives, i.e. they tend to quickly annihilate with a nearby electron. Table 1.6.3 lists some of the radioisotopes that are commonly used in laboratories.

Table 1.6.3: Common positron emitters and their half lives.

Element	Isotope	$T_{1/2}$
Carbon	$^{11}_{6}C$	20.39 minutes
Nitrogen	$^{13}_{7}N$	9.96 minutes
Oxygen	$^{15}_{8}O$	122.24 seconds
Fluorine	$^{18}_{9}F$	109.77 minutes

1.6.D   Protons

Protons are extremely stable composite particles made up of three quarks. They carry the same amount of electric charge as electrons but in positive polarity. However they are about 1836 times heavier than electrons.

Basic Properties of Protons

Rest mass	$=$	$1.67 \times 10^{-27}\ kg = 938.27\ MeV/c^2$
Electrical charge	$=$	$+1.602 \times 10^{-19}\ C$
Mean life	$>$	10^{25} years
Internal structure	:	Made up of 3 quarks

It was Eernest Rutherford who, in 1911, gave the idea of the atom being composed of positively charged nucleus and separate negative charges. After a series of experiments he reached the conclusion that the nuclei of different elements were always integral multiples of the nucleus of hydrogen atom. He called this basic unit "proton".

Protons have found many useful applications in medicine and research. For example proton beams are used to destroy cancerous tumors. They are also extensively used in high energy physics experiments to explore the fundamental particles and their properties.

D.1 Sources of Protons

The effectiveness of proton beams in several fields, such as radiography, imaging, and fundamental physics research has led to the development of state of the art proton production facilities around the world.

Particle Accelerators

Particle accelerators are one the most important tools of fundamental physics research. The discoveries of different quarks making up protons and neutrons have all been made at particle colliders where particles are first accelerated at very high energies and then made to collide either with other particles or with some other target material. Fermi National Accelerator Laboratory in USA has huge particle accelerators that make up the Tevatron collider. In Tevatron protons and antiprotons are accelerated in opposite directions in a circular ring of about 4 miles in circumference and then made to collide with each other. The resulting shower of millions of particles created as a result of this collision are then tracked and analyzed by very large but extremely sensitive and delicate detection systems.

Cockroft-Walton accelerator is a commonly used device to accelerate protons up to moderate energies (several hundred MeV). In such accelerators, hydrogen ions are generated and pumped into a high electric field, which accelerates the ions in a number of steps. At Fermilab, the protons start their journey from a Cockroft-Walton accelerator and are then further accelerated by passing them through linear and circular accelerators.

Apart from fundamental physics research, particle accelerators are also being extensively used in cancer therapy. Although the success rate for proton therapy is high, its use has been limited largely by the unavailability of dedicated in-house high energy proton accelerators in most of the oncology clinics. For such clinics the other option is to use the facilities of particle accelerators not specifically built for radiation therapy. Many such accelerator laboratories provide beam times for cancer treatment. However, since their uptime is never guaranteed, it becomes difficult to schedule a travel plan for patients and clinicians to the facilities. Also the long distances that sometimes have to be traveled by patients, who are already fatigued by the disease, make the treatment process very uncomfortable. A number of particle accelerators have developed their own cancer therapy stations with trained beam technicians and clinicians. Some of the major hospitals also have small sized accelerators that are mainly used for cancer treatment.

Laser Ion Accelerators

Unlike conventional particle accelerators, laser ion accelerators are relatively small sized devices that are used to accelerate ions at high energies using powerful lasers. This new technique is capable of accelerating ions at very high energies in a very short time and distance. In order to accelerate protons at several MeV, a highly intense laser ($> 10^{19} W/cm^2$) with ultra-short pulses is focused on a solid or gaseous target. This results in a collimated beam of high energy protons, which can be used, for example to destroy cancerous tissues.

A lot of research is currently underway to study the feasibility of these devices for cancer therapy and it is hoped that soon they will be an integral part of oncology clinics and hospitals.

Radioactive Sources of Protons

Table 1.6.4:   Common proton emitters and their half lives.

Element	Isotope	$T_{1/2}$
Indium	$^{109}_{53}I$	103 μs
Cesium	$^{113}_{55}Cs$	17 μs
Thulium	$^{147m}_{69}Tm$	360 μs
Lutetium	$^{151}_{71}Lu$	120 ms
Tantalum	$^{157}_{73}Ta$	300 ms
Rhenium	$^{161}_{75}Re$	370 μs
Bismuth	$^{185m}_{83}Bi$	44 μs

There are no naturally occurring proton emitting isotopes suitable for use in laboratory. However, it is possible to produce such radioisotopes with the help of nuclear reactions, where a target material is bombarded by high energy particles. This process excites the nuclei of the target material and makes them unstable. The excited nuclei decay be proton, neutron, α-particle, and photon emissions. The types of particles emitted actually depend on the target material and the projectile energy. Table 1.6.4 lists a few of the many proton emitting isotopes that have been discovered so far.

1.6.E   Neutrons

Neutrons were the last of the basic atomic constituents to be discovered mainly because of their almost identical mass to protons and no electrical charge. A British scientist, James Chadwick, discovered these particles in 1932. Like protons, they are also composite particles made up of three quarks. However they are not as stable outside of the nucleus as protons and decay within about 15 minutes.

Because neutrons do not carry electrical charge, they can penetrate most materials deeper than other charged particles. Their main interaction mechanism with other particles is through collisions and absorption, releasing other detectable particles. This is how they are generally detected.

Apart from other applications, neutrons are also being extensively used in radiation therapy to destroy cancerous tumor cells in the body. One astonishing application of their use in this area is in destroying metastasis cancer in the body. Scientists are using the high neutron absorption cross section of an element called Boron-10 to do radiation therapy of patients whose cancer cells have spread in the body or who have tumors in inoperable locations. Boron-10 is administered to the cancerous cells and then the body is bombarded by slow neutrons. The neutrons are almost exclusively absorbed by Boron-10 atoms, which as a result release other heavy subatomic particles. Due to their short range, these subatomic particles destroy the tumor cells only in their vicinity thus causing minimal damage to healthy tissues.

Basic Properties of Neutrons

Rest mass $\quad = \quad 1.675 \times 10^{-27} \ kg = 939.55 \ MeV/c^2$

Electrical charge $\quad = \quad$ Zero

Mean life $\qquad$ 14.76 minutes

As discussed earlier, another important application of neutrons is in nuclear power plants where they are used to initiate and sustain chain fission reactions necessary to create heat and thus generate electricity.

E.1 Sources of Neutrons

Neutrons are very valuable particles because of their ability to penetrate deeper in matter as compared to charged particles. Production of free neutrons is therefore of high research significance. In this section we will have a look at the most important of the neutron sources available today.

Spallation Sources

Spallation is a violent reaction in which a target is bombarded by very high energy particles. The incident particle, such as a proton, disintegrates the nucleus through inelastic nuclear reactions. The result is the emission of protons, neutrons, α-particles, and other light and heavy particles. The neutrons produced in such a reaction can be extracted and used in experiments.

A general spallation reaction with a proton as the incident particle can be written as

$$p + ST \rightarrow SF_1 + SF_2 + ... + SF_m + (k)n,$$

where ST is the spallation target and SF represent m spallation fragments. The number k of neutrons produced in this reaction depends on the type of the target and the energy of the incident particle.

The targets used in spallation sources are generally high Z materials such as lead, tungsten, silver, or bismuth. However, it is also possible to generate neutrons by bombarding light elements with high energy charged particles, such as protons. Two examples of such reactions are the production of neutrons by bombarding lithium and beryllium targets by high energy protons.

$$p + {}^7_3Li \quad \rightarrow \quad {}^7_4Be + n$$
$$p + {}^9_4Be \quad \rightarrow \quad {}^9_5B + n$$

The reader should note that the reactions such as these are not strictly classified as spallation reactions since they do not involve the break up of the target nuclei into several constituents. These reactions are more closely related to the nuclear reactions we will discuss shortly.

A big advantage of spallation sources is that they produce neutrons with a wide spectrum of energies ranging from a few eV to several GeV. Another advantage is their ability to generate neutrons continuously or in short pulses. The pulses could be as short as a nanosecond.

Composite Sources

Spallation is not the only means of producing neutrons, that is, it is not absolutely necessary to impinge nuclei with high energy particles to produce neutrons. This can also be accomplished by exciting nuclei such that they emit neutrons during the process of de-excitation. Fortunately for such a process to occur it is not always necessary that the target particle carries a very high energy. In fact, incident particles coming out of radioactive sources are more than sufficient to cause neutron emission from some materials. In this book the sources made with such materials will be termed as *composite sources* as they are made of combining two different materials together.

A composite neutron source consists of a source of incident particles and a target that decays by neutron emission. The incident particle source can either be a radioactive material or a small particle generator. In general, the composite sources are made of a radioactive material acting as the source of incident particles mixed in a target material. A common example is the plutonium-beryllium source, which produces neutrons in the following sequence.

$$^{238}_{94}Pu \quad \rightarrow \quad ^{234}_{92}U + \alpha$$
$$\alpha + {}^9_4Be \quad \rightarrow \quad ^{13}_6C^*$$
$$^{13}_6C^* \quad \rightarrow \quad ^{12}_6C + n$$

The first step of this process involves α-decay of plutonium-238, which emits α-particles of energy around 5.48 MeV with a half life of about 87.4 years. This moderately long half life makes it suitable for long time storage in laboratories. The α-particle impinges on the beryllium-9 target and transforms it into carbon-13 in an unstable state, which ultimately decays into carbon-12 by emitting a neutron.

α-induced reactions are widely used to generate neutrons. Another common example of an $(\alpha - n)$ neutron source is $^{241}Am - Be$. Americium-241 has a half life of 433 years. It decays by α emission with a mean energy of 5.48 MeV. If this isotope is mixed with beryllium-9, the α-particles interact with the beryllium nuclei transforming them into carbon-13 in excited state. The de-excitation of carbon-13 leads to the emission of neutrons. Note the similarity of this process to that of the $^{238}Pu - Be$ source mentioned above.

One can essentially use any α-particle source to make an $(\alpha - n)$ source. In general, the $(\alpha - n)$ reaction can be written as

$$\alpha + {}_{p}^{n+p}X \rightarrow {}_{p+2}^{n+p+3}Y + n. \tag{1.6.7}$$

Just like α-particles, photons can also be used to stimulate nuclei to emit neutrons. A common example of such γ-emitting nuclides is antimony-124. ${}_{51}^{124}Sb$ emits a number of γ-rays, the most probable of which has an energy of around 603 keV. If these photons are then allowed to interact with beryllium nuclei, it may result in the emission of a neutron. Antimony-beryllium neutron sources are commonly used in laboratories. Such sources are sometimes referred to as *photoneutron* sources.

Fusion Sources

Fusion is a reaction in which two light nuclei (hydrogen and its isotopes) are forcibly brought so close together that they form a new heavier nucleus in an excited state. This nucleus releases neutrons and photons to reach its ground state. The fusion reaction can therefore be used to produce neutrons.

To initiate a fusion reaction, a fair amount of energy must be supplied through some external means because the nuclei are repelled by electromagnetic force between the protons. This energy can be provided by several means, such as through charged particle accelerators. The advantage of fusion sources over the spallation sources is that they need relatively lower beam energies to initiate the fusion process. These sources are also more efficient in terms of neutron yield. The fusion process producing neutrons can be written as

$$^{2}D(d, n)^{3}He$$
$$^{3}T(d, n)^{4}He.$$

Here d and D represent Deuterium (an isotope of Hydrogen with one proton and one neutron). T is another isotope of Hydrogen with one proton and 2 neutrons. The nomenclature of the above equations is such that the first term in the brackets represents the incoming particle and the second term the outgoing one. The above reactions can also be written as

$$\begin{align} {}_{1}^{2}D + {}_{1}^{2}D &\rightarrow {}_{2}^{3}He + n \\ {}_{1}^{3}T + {}_{1}^{2}D &\rightarrow {}_{2}^{4}He + n. \end{align}$$

Nuclear Reactors

Nuclear reactors produce neutrons in very large numbers as a result of neutron-induced fission reactions

$$^{235}_{92}U + n \rightarrow FF_1 + FF_2 + (2\text{-}3)n.$$

Here FF represents fission fragments. Although most of these neutrons are used up to induce more fission reactions, still a large number manages to escape the nuclear core.

Radioactive Sources of Neutrons

There are no known naturally occurring isotopes that emit significant number of neutrons. However it is possible to produce such isotopes by bombarding neutron rich isotopes by other particles such as protons. The problem is that these isotopes have very short half lives making them unsuitable for usual laboratory usage.

Table 1.6.5: Common neutron sources and their decay modes.

Source	Isotopes	Reaction Type
Californium	$^{252}_{98}Cf$	Spontaneous Fission
Deuterium-Helium	$^{2}_{1}D$-$^{3}_{2}He$	Nuclear Fusion
Tritium-Helium	$^{3}_{1}T$-$^{4}_{2}He$	Nuclear Fusion
Uranium	$^{235}_{92}U$	Nuclear Fission
Lithium	$^{7}_{3}Li$	Spallation
Beryllium	$^{9}_{4}Be$	Spallation
Plutonium-Beryllium	$^{238}_{94}Pu$-Be	(α, n)
Plutonium-Beryllium	$^{239}_{94}Pu$-Be	(α, n)
Americium-Beryllium	$^{241}_{95}Am$-Be	(α, n)
Americium-Boron	$^{241}_{95}Am$-B	(α, n)
Americium-Fluorine	$^{241}_{95}Am$-F	(α, n)
Americium-Lithium	$^{241}_{95}Am$-Li	(α, n)
Radium-Beryllium	$^{226}_{88}Ra$-Be	(α, n)
Antimony-Beryllium	$^{124}_{51}Sb$-Be	(γ, n)

Perhaps the most extensively used source of neutrons is californium-252. It has two decay modes: α-emission (96.9%) and spontaneous fission (3.1%). The latter produces around 4 neutrons per decay.

$$^{252}_{98}Cf \rightarrow\ ^{94}_{38}Sr +\ ^{154}_{60}Nd + 4n$$

1 mg of californium-252 emits around 2.3×10^9 neutrons per second with an average neutron energy of 2.1 MeV. It also emits a large number of γ-ray photons but the intensity is an order of magnitude lower than that of the neutrons. It therefore does not pose much problem for applications requiring a moderately clean neutron beam. Table 1.6.5 lists some of the common sources of neutrons and their decay modes.

1.6.F Alpha Particles

Alpha particles are essentially helium nuclei with 2 protons and 2 neutrons bound together. The consequence of their high mass and electrical charge is their inability to penetrate as deep as other particles such as protons or electrons. In fact, a typical alpha particle emitted with a kinetic energy of around 5 MeV is not able to penetrate even the outer layer of our skin. On the other hand they interact very strongly with the atoms they encounter on their way, the reason being of course their highly positive charge. Hence α-particle sources pose significantly higher risk than other types of sources of equal strength if the particles are able to reach the internal organs. This can happen, for example if the source is somehow inhaled or digested.

Basic Properties of α-Particles

Rest mass	$=$	$6.644 \times 10^{-27}\ kg = 3.727 \times 10^3\ MeV/c^2$
Electrical charge	$=$	$3.204 \times 10^{-19}\ C$
Mean life	:	Stable
Internal structure	:	Made up of 2 protons and 2 neutrons

The advantage of their extremely low range is that use of gloves in handling an α-particle source is generally sufficient to avoid any significant exposure. It should, however, be noted that because they are readily absorbed by the material, other penetrating particles might be emitted in the process that could harm the body. As noted above, if α-particles somehow enter the body (for example by inhalation of a radionuclide or through an open wound into the blood stream) they can deliver large doses to internal organs leading to irreversible damage.

α-particles can not travel more than a few centimeters in air and readily capture two electrons to become ordinary helium. Because of their low penetration capabilities and very high ionizing power α-particle beams are rarely used in radiation therapy. Even if they are somehow administered near a tumor cell it will be very difficult to control the destruction of cancerous cells while minimizing the damage to the healthy tissues.

α-particles are extremely stable particles having a binding energy of about 28.8 MeV. The protons and neutrons are held together in a very stable configuration by the strong nuclear force.

F.1   Sources of α-Particles

Even with very low penetration capability, α-particles have found many applications in different fields. A lot of research has therefore gone in the direction of designing effective α-particle sources. Fortunately there are a host of naturally occurring radioisotopes that emit α-particles in abundance. A particle accelerator can then be used to increase their energy up to the required energy. Apart from these radioactive sources, other sources with higher α-particle fluxes have also been developed.

Accelerator Based Sources

Just like protons and neutrons, α-particles are also produced during collisions of high energy particles with fixed target materials. At very high incident particle energies the so called spallation reactions tear apart the target nuclei into its constituents. Since the 2 protons plus 2 neutrons configuration of a nucleus is extremely stable therefore such reactions produce α-particles as well together with protons and neutrons.

One example of such a reaction is the collision of high energy neutrons (25-65 MeV) on a cobalt-59 target, which produces one or more α-particles per collision.

Radioactive Sources of α-Particles

There are numerous radionuclides that emit α-particles, some of which are listed in Table.1.6.6.

1.6.G   Fission Fragments

Nuclear fission is a process in which an unstable nucleus splits up into two nuclides and emits neutrons. Two types of fission reactions are possible: spontaneous fission and induced fission. We saw earlier that spontaneous fission occurs in some radionuclides, at least one of which, i.e. californium-252, occurs naturally. These radionuclides are extensively used as neutron sources.

In induced fission, when a fissionable nucleus, such as uranium, captures a neutron it goes into an unstable state and eventually breaks up in two heavy parts. In this process it also emits some neutrons. The two heavier particles are known as fission fragments. The reaction can be written as

$$FM + n \rightarrow FF_1 + FF_2 + (2\text{-}4)n, \tag{1.6.8}$$

where FM represents fissionable material (such as $^{235}_{92}U$), n is the neutron, and FF_1 and FF_2 are the two fission fragments.

Most of these fission fragments are unstable and go through a series of nuclear decays before transforming into stable elements. They generally have unequal masses but in a sample of many fissionable atoms, some of the fragments are always of equal mass (see Fig.1.6.10 for an example of uranium-235 fission fragment spectrum).

Table 1.6.6: Common α emitters, the energies of their most probable emissions, and their half lives (26).

Element	Isotope	Energy (MeV)	$T_{1/2}$
Ameritium	$^{241}_{95}Am$	5.443 (13%), 5.486 (85.5%)	432.2 years
Bismuth	$^{213}_{83}Bi$	5.549 (7.4%), 5.869 (93%)	45.6 minutes
Curium	$^{244}_{96}Cm$	5.763 (23.6%), 5.805 (76.4%)	18.1 years
Californium	$^{252}_{98}Cf$	6.076 (15.7%), 6.118 (82.2%)	2.645 years
Radium	$^{223}_{88}Ra$	5.607 (25.7%), 5.716 (52.6%)	11.435 days
Thorium	$^{229}_{90}Th$	4.845 (56.2%), 4.901 (10.2%)	7340 years
Plutonium	$^{239}_{94}Pu$	5.144 (15.1%), 5.156 (73.3%)	24110 years
Polonium	$^{210}_{84}Po$	5.304 (100%)	138.376 days

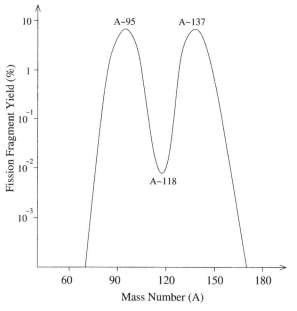

Figure 1.6.10: Typical fission spectrum of uranium-235.

As an example, let us have a look at how uranium-235 fissions. The following equation shows the process for two of the most probable fission fragments: $^{140}_{54}Xe$

and $^{94}_{38}Sr$.

$$^{235}_{92}U + 2n \rightarrow \, ^{236}_{92}U^* \rightarrow \, ^{140}_{54}Xe + \, ^{94}_{38}Sr + 2n \tag{1.6.9}$$

Here $^{236}_{92}U^*$ is an unstable isotope of $^{235}_{92}U$ that eventually breaks up in fragments and releases 2 neutrons. These fission fragments are themselves unstable and go through a series of β-decays till they transmute into stable elements. $^{140}_{54}Xe$ eventually decays into cerium-140 while $^{94}_{38}Sr$ decays into zirconium-94. Because of their high β-yields, these fission fragments and their daughters are regarded as extremely hazardous and special precautions must be taken while handling them.

1.6.H Muons, Neutrinos and other Particles

H.1 Muons

Muon is one of the elementary particles according to the Standard Model of Particle Physics and is classified as a Lepton. Muons carry the same electrical charge (positive or negative) as carried by protons or electrons but they are approximately 9 times lighter than protons. Because of their lower mass, they can penetrate larger distances in the material than protons and are therefore extensively used in probing the magnetic properties, impurities and structural defects of different materials.

Basic Properties of Muons

Rest mass	=	$1.783 \times 10^{-36} \, kg = 105.66 \, MeV/c^2$
Electrical charge	=	$\pm 1.6 \times 10^{-19} \, C$
Mean life	=	$2.19 \times 10^{-6} \, s$
Internal structure	:	Believed to have no internal structure.

With a flux of about 1 muon per square centimeter per second , they are the most abundant of charged particles at sea level that enter the Earth's atmosphere from outer space. The main production mechanism of these muons is the interaction of primary cosmic rays with material in the upper atmosphere. The mean energy of these muons is about 4 GeV. They lose about 2 MeV per g/cm^2 of energy as they pass through the atmosphere which suggests that their original energy is around 6 GeV.

Muons can also be produced in the laboratory by hitting a target (such as graphite) with high energy protons.

H.2 Neutrinos

Neutrinos are also elementary particles belonging to the leptonic category. They come in three distinct *flavors*: electron-neutrino (ν_e), muon-neutrino (ν_μ), and tau-neutrino (ν_τ). *Flavor* is a real term used in Standard Model of particle physics to characterize the elementary particles. Recent experiments have proved that these three types can transform into one another (1). Standard Model does not assign any mass to the neutrinos but recent experiments have suggested that they indeed have masses, albeit very small (a few eV). They do not carry any electrical charge and therefore are extremely difficult to detect.

Basic Properties of Neutrinos		
Rest mass	:	Extremely small (few eV)
Electrical charge	=	Zero
Mean life	>	10^{33} years
Internal structure	:	Believed to have no internal structure.

Neutrinos are extremely abundant in nature and are produced in large numbers in processes like nuclear fusion, stellar collapse, and black-hole collisions. The nuclear fusion reactions inside the core of the Sun produce electron-neutrinos in huge numbers. Some of these electron-neutrinos change into the other two types on their way to the earth and as a result we receive all three types. Their average flux on earth's surface is on the order of 10^6 $cm^{-2}s^{-1}$.

The degree of difficulty in detecting neutrinos can be appreciated by the fact that a neutrino can pass through a thousand earths without making a single interaction. Still, scientists have exploited their abundance to design dedicated neutrino detectors around the world.

H.3   Some Other Particles

All of the particles mentioned above have their counterparts in nature. For example there is an anti-electron called positron having positive electrical charge and all the other properties of an electron. When a positron and an electron come close, they annihilate each other and give out energy in the form of photons. Similarly protons have anti-protons and neutrinos have anti-neutrinos.

Apart from these so called anti-particles there is a host of other particles, some of which exist in nature while others are produced in laboratories by colliding particles at very high energies. For example gluons are the particles that bind the protons and neutrons together inside the nucleus by the *strong nuclear force*. The process of radioactive decay, which we discussed earlier, is the result of the *weak nuclear force* between particles mediated by W and Z particles. During high energy collisions a large number of particles are created. Most of these particles have extremely short mean lives, but with present day technologies, their properties can be studied in detail. This has shed important light on the nature of elementary particles and their underlying physics.

Problems

1. Compute the de Broglie wavelength of an electron moving with a velocity of $2.5 \times 10^8 ms^{-1}$.

2. Compare the de Broglie wavelength of a proton moving at $0.9c$, c being the velocity of light in vacuum, with the wavelength of x-rays $(10^{-10}\ m)$.

3. Scintillators are the materials that produce light when they are exposed to radiation. Most scintillators used to build detectors produce light in blue and green regions of the electromagnetic spectrum with mean wavelengths of 475 nm and 510 nm respectively. Compute the energies and frequencies associated with these photons.

4. What is the minimum de Broglie wavelength of an electron accelerated in a 40 kV x-ray tube.

5. Write down the decay equations for the following radionuclides.
   $^{22}_{11}Na$ (β), $^{60}_{27}Co$ (EC), $^{127}_{55}Cs$ (β), $^{241}_{95}Am$ (α), $^{252}_{98}Cf$ (α)

6. A newly produced radioactive decay sample has an initial activity of 200 MBq. Compute its decay constant, half life, and mean life if after 5 days its activity decreases to 150 MBq.

7. Show that for a massive nucleus undergoing α-decay, the energy of the emitted α particle is approximately given by

$$T_\alpha \approx \left[\frac{A-4}{4}\right] Q_\alpha,$$

   where A is the mass number of the parent nucleus.

8. Compute the Q-values and α-particle energies for plutonium-239, americium-243, and bismuth-214.

9. Verify that Fe-55 can decay by electron capture.

10. Determine if Pb-187 can decay by both α and β emissions.

11. Calculate the rate of α-particle emissions from a 5 mg sample of $^{239}_{94}Pu$.

12. The initial activity of a radioactive sample is found to be 1.5 mCi. Find its decay constant if after 3 years its activity decreases to 1 mC.

13. A radioactive element decays with a half life of 43 days. If initially there are 2.5 moles of this element in a sample, how many of its atoms will be left in the sample after 43 days, 6 months, and 1 year.

14. Estimate the mean life of a radionuclide that has decayed by 75% in 24 hours.

15. The mean life of a radionuclide is 15 days. What fraction of it will remain radioactive after 30 days?

16. Compute the ratio of the number of atoms decayed to the initial number of atoms of a radioactive substance after 4.5 half lives.

17. How many mean lives should pass for a radioactive substance to decay 99.9% of its atoms.

18. The activity of a radioactive sample is found to be 47 Bq. Suppose that the half life of the material is 4 days. Compute its activity 2 days before and 2 days after the measurement was made.

19. The activity of a radioactive material composed of only one type of material is monitored through a detector. The detector is configured in such a way that each output count represents one decay. If the detector count rate decreases from an initial value of 20,000 counts per minute to 3,000 counts per minute after 3 days, estimate the mean life of the element.

20. Estimate the number of thorium-232 atoms present in a sample that also contains 2.5×10^{15} atoms of its daughter radium-228. Assume the two isotopes to be in secular equilibrium.

21. A 1 mCi sample of pure thorium-227 goes through the following series of decays:

$$^{227}_{90}Th(T_{1/2} = 18.5 \text{ days}) \quad \rightarrow \quad ^{223}_{88}Ra(T_{1/2} = 11.4 \text{ days}) \rightarrow$$
$$^{211}_{82}Pb(T_{1/2} = 36.1 \text{ min}) \quad \rightarrow \quad ^{211}_{83}Bi(T_{1/2} = 2.1 \text{ min}) \rightarrow ^{207}_{81}Tl(T_{1/2} = 4.8 \text{ min})$$

Compute the activities of its first four decay products after 10 days.

22. How much uranium-238 is needed to to have an activity of 3.5 μCi. (Take the half life of uranium-238 to be 4.47×10^9 years.)

Bibliography

[1] Ahmad A.Q. et al., **Direct Evidence of Neutrino Flavor Transformation from Neutral-Current Interactions in the Sudbury Neutrino Observatory**, Phys. Rev. Lett. Vol.89, No.1, 011301, 2002.

[2] Als-Nielson, J., McMorrow, D., **Elements of Modern X-ray Physics**, John Wiley & Sons, 2001.

[3] Apparao, K., **Composition of Cosmic Radiation (Library of Anthropology V.1)**, Taylor & Francis, 1975.

[4] Arndt, M. et al., **Wave-Particle Duality of C60**, Nature 401, 1999.

[5] Attix, F.H., **Introduction to Radiological Physics and Radiation Dosimetry**, Wiley, New York, 1986.

[6] Attix, F.H., Roesch, W.C., Tochilin, E., **Radiation Dosimetry**, Academic Press, New York, 1968.

[7] Baldelli, P. et al., **Quasi-Monochromatic X-rays for Diagnostic Radiology**, Institute of Physics Publishing, 2003.

[8] Bateman, H., **The Solution of a System of Differential Equations Occurring in the Theory of Radio-active Transformations**, Proc. Cambridge Phil. Soc., 16, p.423, 1910.

[9] Bateman, H., **Actinide Recycle in LMFBRs as a Waste Management Alternative**, First International Conference on Nuclear Waste Transmutation, Univ. of Texas, 1980.

[10] Bernstein, J., Fishbane, P.M., Gaslorowicz, S.G., **Modern Physics**, Prentice Hall, 2000.

[11] Bragg, L., **The Development of X-Ray Analysis**, Dover, 1993.

[12] Brodsky, A., **Handbook of Radiation Measurement and Protection**, CRC Press, 1978.

[13] Bushong, S.C., **Radiologic Science for Technologists: Physics, Biology, and Protection**, C.V. Mosby, 2001.

[14] Carroll, F.E., **Tunable Monochromatic X-rays: A New Paradigm in Medicine**, American Roentgen Ray Society, 179:483590, 2002.

[15] Catlow, C.R.A., **Applications of Synchrotron Radiation**, Blackie Academic and Professional, 1991.

[16] Cottingham, W.N., & D.A. Greenwood **An Introduction to the Standard Model of Particle Physics**, Cambridge University Press, 1999.

[17] Coughlan, G.D., & J.E. Dodd **The Ideas of Particle Physics: An Introduction for Scientists**, Cambridge University Press, 2003.

[18] Delsanto, P.P., **Radiation and Solid State Physics, Nuclear and High Energy Physics, Mathematical Physics**, CRC, 1998.

[19] Draganic, I.G., **Radiation and Radioactivity on Earth and Beyond**, CRC, 1993.

[20] Duggan, J. and Cloutier, R.J., **Energy Sources for the Future** Proceedings of a Symposium, ORAU, Oak Ridge, 1977.

[21] Eisenbud, M., **Environmental Radioactivity**, Academic Press, New York, 1973.

[22] Eisenbud, M. and Gesell, T., **Environmental Radioactivity from Natural, Industrial, and Military Sources**, Academic Press, 1997.

[23] Ellse, M., **Mechanics & Radioactivity (Nelson Advanced Science)**, Nelson Thornes, 2003.

[24] Firestone, R.B., Baglin, C.M., Chu, S.Y.F., **Table of Isotopes, 8th Edition**, Wiley-Interscience, 1999.

[25] Fischer, R.P. et al., **Generation of Tunable, Monochromatic X-Rays in the Laser Synchrotron Source Experiment**, Proceedings of the 2001 Particle Accelerator Conference, Chicago, IEEE 7803-7191, 2001.

[26] Hagiwara K. et al., Phys. Rev. 66, 010001, 2002.

[27] Helliwell, R., **Reflections on a Decade of Synchrotron X-Radiation Protein Crystallography**, The Rigaku Journal 3(1), 1986.

[28] Helmut, W. **Advanced Radiation Sources and Applications**, Springer, 2005.

[29] Hendee, W.R., **Biomedical Uses of Radiation (2-Volume Set)**, Wiley-VCH, 1999.

[30] IAEA, **Categorization of Radioactive Sources: Safety Standards Series No. Rs-g-1.9 (IAEA Safety Standards)**, IAEA, 2005.

[31] Knight, R.D., **Physics for Scientists and Engineers with Modern Physics: A Strategic Approach**, Addison Wesley Publishing Company, 2003.

[32] Knoll, G.F., **Radiation Detection and Measurement**, John Wiley and Sons, New York, 1999.

[33] L'Annunziata, M.F., **Handbook of Radioactivity Analysis** Academic Press, 1998.

[34] Livingston, H.D., **Marine Radioactivity (Radioactivity in the Environment)**, Pergamon, 2004.

[35] Lowenthal, G., and Alrey, P., **Practical Applications of Radioactivity and Nuclear Radiations**, Cambridge University Press, 2001.

[36] Magill, J., Galy, J., **Radioactivity - Radionuclides - Radiation**, Springer Verlag, 2005.

[37] Maier, R., **Synchrotron Radiation**, CERN Report 91-04, 97-115, 1991.

[38] Marr, G.V., **Handbook on Synchrotron Radiation, Volume 2**, Elsevier, 1987.

[39] Moche D., **Radiation: Benefits/Dangers**, Watts, 1979.

[40] National Council on Radiation Protection and Measurements, **Handbook of Radioactivity Measurements Procedures: With Nuclear Data for Some Biomedically Important Radionuclides** Natl. Council on Radiation, 1985.

[41] Moon, P.B., **Artificial Radioactivity**, Cambridge Univ. Press, 1949.

[42] Mould, R.F., A **Century of X-Rays and Radioactivity in Medicine: With Emphasis on Photographic Records of the Early Years**, Taylor & Francis, 1993.

[43] Pasachoff, N.E., **Marie Curie: And the Science of Radioactivity**, Oxford Univ. Press, 1996.

[44] Podgorsak, E.B., **Radiation Physics for Medical Physicists (Biological and Medical Physics, Biomedical Engineering)**, Springer, 2005.

[45] Pringle, L., **Radiation: Waves and Particles/Benefits and Risks**, Enslow, 1983.

[46] Ritchie, B.G. et al., **Alpha-Decay Properties of 205, 206, 207, 208Fr; identification of 206mFr.**, Physical Review C23:2342, 1981.

[47] Roentgen, W., **Eine neue Art von Strahlen**, Nature 53, 274, 1896.

[48] Romer, A., **Discovery of Radioactivity and Transmutation**, Dover Publications, 1964.

[49] Rothenberg, M.A., **Understanding X-Rays: A Plain English Approach**, Professional Education Systems, 1998.

[50] Selman, J., **The fundamentals of X-ray and radium physics**, C.C. Thomas, 1977.

[51] Shapiro, J., **Radiation Protection: A Guide for Scientists, Regulators and Physicians**, Harvard Univ. Press, 2002.

[52] Stannard, J.N., **Radioactivity and Health: A History**, U.S. Department of Energy, 1989.

[53] Stratern, P., **Curie and Radioactivity**, Random House Inc., 1998.

[54] Suortti, P. and Thomlinson, W., **Medical Applications of Synchrotron Radiation**, Physics in Medicine and Biology, 48:R1-R35, 2003.

[55] Tipler, P.A. and Liewellyn, R., **Modern Physics**, W.H. Freeman, 2002.

[56] Tsoulfanidis, N., **Measurement and Detection of Radiation**, Fuel and Energy Abs., Vol.36, No.4, 1995.

[57] Turner, J.E., **Atoms, Radiation, and Radiation Protection**, McGraw-Hill, 1995.

[58] Tykva, R., **Low-Level Environmental Radioactivity: Sources and Evaluation**, CRC, 1995.

[59] Tykva, R., **Man-Made and Natural Radioactivity in Environmental Pollution and Radiochronology (Environmental Pollution)**, Springer, 2004.

[60] Valkovi, V., **Radioactivity in the Environment : Physicochemical aspects and applications**, Elsevier Science, 2000.

[61] Zganjar, E. et al. **The Investigation of Complex Radioactive Decay Schemes Far from Stability: the Structure of 187Au**. Proceedings of the Fourth International Conference on Nuclei Far from Stability, CERN, Geneva, 1981.

Chapter 2

Interaction of Radiation with Matter

Whenever we want to detect or measure radiation we have to make it interact with some material and then study the resulting change in the system configuration. According to our present understanding, it is not possible to detect radiation or measure its properties without letting it interact with a measuring device. In fact, this can be stated as a universal rule for any kind of measurement. Even our five senses are no exception. For example, our eyes sense photons that strike cells on its retina after being reflected from the objects. Without such interactions we would not be able to see anything[1]. The same is true for radiation detectors, which use some form of radiation interaction to generate a measurable signal. This signal is then used to *reverse engineer* the properties of the radiation. For example, when x-ray photons after passing through an object strike a photographic plate, they *stain*[2] the plate. The photons, which can not pass through the highly absorbing materials in the object, do not reach the photographic plate. The varying intensity, by which the photographic plate is fogged, creates a two dimensional image of the object and in this way the photographic plate acts as a position sensitive detector. Although such a method of radiation detection is still widely used, it is neither very sensitive nor very accurate for measuring properties of radiation, such as its energy and flux. The so called *electronic detectors* provide an alternative and much better means of detecting and measuring radiation. An electronic detector uses a detection medium, such as a gas, to generate an electrical signal when radiation passes through it. This electrical signal can be used to characterize the radiation and its properties. Therefore, a necessary step in building a detector is to understand the interaction mechanisms of radiation in the detection medium. This chapter is devoted to exactly this task. We will look at different types of radiation detectors in the subsequent chapters.

2.1 Some Basic Concepts and Terminologies

Every particle carries some energy with it and is, therefore, capable of exerting force on other particles through processes we call particle interactions. These particle interactions may or may not change the states and properties of the particles involved. The way particles interact with matter depends not only on the types of incident and target particles but also their properties, such as energy and momentum. A lot of work has gone into building theoretical foundations of different interaction

---

[1] In this sense we can say that our eyes are actually photon detectors
[2] The stains become visible after the plate has been developed through appropriate chemicals.

mechanisms, most of which are now well understood through tools like quantum mechanics, quantum electrodynamics, and quantum chromodynamics. However, keeping in view the scope of this book, in this chapter we will not be looking at the theoretical foundations of particle interactions that are based on these theories, rather spend some time on understanding the gross properties that have been empirically and experimentally determined.

Before we go on to look at these mechanisms it is worthwhile to spend some time on understanding some terminologies we will be needing in the following sections.

2.1.A Inverse Square Law

The strength of radiation can be characterized by its flux, which is generally defined as the number of particles passing through a unit area per unit time. Irrespective of the type of source, this flux decreases as one moves away from the source. This decrease in flux depends on the type of source and the type of radiation. For example, laser light, which is highly collimated, does not suffer much degradation in flux with distance. On the other hand, the flux from radioactive sources decreases rapidly as the distance from it increases. The inverse square law, which is based on geometric considerations alone, characterizes this change. It states that the radiation flux is inversely proportional to the square of the distance from the *point* source, that is

$$\Phi \propto \frac{1}{r^2}, \qquad (2.1.1)$$

where r is the distance between the source and the point where flux is to be calculated. This law is the consequence of the isotropic nature of a point source because such a source is expected to radiate equally in all directions. Since the flux is a measure of the amount of radiation passing through an area therefore it should vary according to how the area varies with distance from the source. Now, we know that the surface area around a point is given by $4\pi r^2$, which means that the area varies as r^2. Hence we conclude that the flux, which represents the amount of radiation passing through a *unit area* is proportional to the inverse of the square of the distance or as $1/r^2$.

As stated earlier, this law applies to only point sources. Now, a perfectly point source does not exist in nature and is extremely difficult, if not impossible, to manufacture. However this does not mean that this law can not be applied to practical systems. The reason is that the notion of a *point source* is relative to the distance from the source. For example, a disk source can be be considered a point source if one is at a considerable distance from its center. If this consideration is taken into account then this law can be applied to most sources.

Another very important thing to consider is that the medium through which the radiation travels should neither be scattering nor absorbing. It is understandable that, if there is considerable scattering and absorption, the flux would change more rapidly than r^2. The inverse square law is therefore applied only in vacuum or low pressure gaseous environments, such as air under atmospheric conditions. In liquids and solids the scattering and absorption are not negligible and therefore this law does not hold.

Inverse square law plays an important role in radiation protection, as it sets a minimum distance a radiation worker must retain from a source to minimize the possibility of radiation damage.

Example:
Determine the relative change in flux of γ-rays when the distance from the source is changed from $5\ m$ to $20\ m$ in air.

Solution:
The flux at $5\ m$ from the source is given by

$$\Phi_1 = A\frac{1}{r^2}$$

$$= A\frac{1}{5^2}$$

where A is the proportionality constant. Similarly the flux at $20\ m$ from the source is

$$\Phi_2 = A\frac{1}{20^2}.$$

The relative change in flux is then given by

$$\delta\Phi = \frac{\Phi_1 - \Phi_2}{\Phi_1} \times 100$$

$$= \frac{\frac{1}{5^2} - \frac{1}{20^2}}{\frac{1}{5^2}} \times 100$$

$$= 93.7\%.$$

2.1.B Cross Section

Particle interactions have been very successfully handled by quantum mechanics and there is no reason to believe that these theoretical foundations are not accurate descriptions of natural phenomena. However, these theories have physical and mathematical explanations that sometimes defy our common sense and therefore become difficult to comprehend and visualize. Wave-particle duality that we discussed in the preceding chapter is one such concept. Another important concept is Heisenberg's uncertainty principle, which states that there is always some uncertainty in determining any two properties of a particle simultaneously. For example, according to this principle, it is impossible to find the exact energy of a particle at an exact moment in time or to find its exact momentum at an exact position. Mathematically this concept is represented by

$$\Delta E.\Delta t \geq \frac{\hbar}{2\pi} \tag{2.1.2}$$

$$\Delta p.\Delta x \geq \frac{\hbar}{2\pi}. \tag{2.1.3}$$

Here Δ represents uncertainty in energy E, time t, momentum p, or position x. $\hbar = h/2\pi$ is the so called reduced Planck's constant and is commonly pronounced h-bar or h-cut.

These concepts, together with Planck's explanation of black body spectrum and Einstein's derivation of the photoelectric effect (discussed later in this chapter) led

Schroedinger and Heisenberg to lay the foundations of quantum mechanics. Due to the intrinsic uncertainties in finding particle properties quantum mechanics naturally developed into a probabilistic theory. We talk about the probability of finding a particle at a certain position but never claim to know its exact position with absolute certainty.

So how does quantum mechanics treat particle interactions? That's where the concept of cross section comes in. It tells us how likely it is for a particle to interact with another one in a certain way. Mathematically speaking it has dimensions of area but conceptually it represents probability of interaction.

Cross section is perhaps the most quoted parameter in the fields of particle physics and radiation measurement because it gives a direct measure of what to expect from a certain beam of particles when it interacts with a material. Let us now see how this concept is defined mathematically.

Suppose we have a beam of particles with a flux Φ (number of particles per unit area per unit time) incident on a target. After interacting with the target, some of the particles in the incident beam get scattered. Suppose that we have a detector, which is able to count the average number of particles per unit time (dN) that get scattered per unit solid angle $(d\Omega)$. This average quantity divided by the flux of incident particles is defined as the differential cross section.

$$\frac{d\sigma}{d\Omega}(E, \Omega) = \frac{1}{\Phi}\frac{dN}{d\Omega} \qquad (2.1.4)$$

It is apparent from this equation that the cross section σ has dimension of area. This fact has influenced some authors to define it as a quantity that represents the area to which the incident particle is exposed. The larger this area, the more probable it will be for the incident particle to interact with the target particle. However, it should be noted that this explanation of cross section is not based on any physical principle, rather devised artificially to explain a mathematical identity using its dimensions.

The differential cross section can be integrated to evaluate the total cross section at a certain energy (differential cross section is a function of energy of the incident particles), that is

$$\sigma(E) = \int \frac{d\sigma}{d\Omega}d\Omega. \qquad (2.1.5)$$

The conventional unit of cross section is barn (b) with $1\ b = 10^{-24}\ cm^2$.

2.1.C   Mean Free Path

As particles pass through material, they undergo collisions that may change their direction of motion. The average distance between these collisions is therefore a measure of the probability of a particular interaction. This distance, generally known as the *mean free path*, is inversely proportional to the cross section and the density of the material, that is

$$\lambda_m \propto \frac{1}{\rho_n \sigma}, \qquad (2.1.6)$$

where ρ_n and σ represent the *number* density of the medium and cross section of the particle in that medium. Note that the definition of the mean free path depends on the type of cross section used in the calculation. For example, if scattering cross section is used then the mean free path would correspond to just the scattering process. However in most instances, such as when calculating the shielding required in a radiation environment, one uses total cross section, which gives the total mean free path.

The mean free path has a dependence on the energy distribution of the particles relative to the medium. For particles that can be described by the Maxwellian distribution[3], such as thermal neutrons in a gas under standard conditions, the mean free path can be computed from

$$\lambda_m = \frac{1}{\sqrt{2}\rho_n\sigma}. \tag{2.1.7}$$

In all other cases, the mean free path should be estimated from

$$\lambda_m = \frac{1}{\rho_n\sigma}. \tag{2.1.8}$$

The number density ρ_n in the above relations can be computed for any material from

$$\rho_n = \frac{N_A\rho}{A}, \tag{2.1.9}$$

where A is the atomic weight of the material, N_A is the Avogadro's number, and ρ is the *weight* density of the material. Mean free path is usually quoted in cm, for which we must take A in $g/mole$, N_A in $mole^{-1}$, ρ in g/cm^3, and σ in cm^2.

For a gas, the weight density term in the above relation can be computed from

$$\rho = \frac{PA}{RT}, \tag{2.1.10}$$

where P and T represent respectively the pressure and temperature of the gas and $R = 8.314\ Jmole^{-1}K^{-1}$ is the usual gas constant. Hence for a gas, the mean free path can be computed from the following equations.

$$\lambda_m = \frac{RT}{\sqrt{2}N_AP\sigma} \quad \text{(for Maxwellian distributed energy)} \tag{2.1.11}$$

$$\lambda_m = \frac{RT}{N_AP\sigma} \quad \text{(for non-Maxwellian distributed energy)} \tag{2.1.12}$$

It is apparent from these relations that the mean free path of a particle in a gas can be changed by changing the gas temperature and pressure (see example below).

The mean free path as defined above depends on the density of the medium. This poses a problem not only for reporting the experimental results but also for using the values in computer codes for systems whose density might change with time. Therefore derivative of this term has been defined that does not depend on the density. This is the so called *specific mean free path* and is given by

$$\lambda_p = \frac{A}{N_A\sigma}. \tag{2.1.13}$$

---

[3]The Maxwellian distribution describes how the velocities of molecules in a gas are distributed *at equilibrium* as a function of temperature.

The specific mean free path is generally quoted in units of g/cm^2. In literature one generally finds specific mean free paths for different media, which can then be simply divided by the respective densities to obtain the actual mean free paths.

Example:
Compare the scattering mean free path of moderate energy electrons in helium and argon under standard conditions of temperature and pressure. The scattering cross section of electrons in helium and argon can be taken to be $2.9 \times 10^7 \, b$ and $1.1 \times 10^9 \, b$ respectively.

Solution:
For moderate energy electrons, we can use equation 2.1.11 to compute the mean free path. Substituting $T = 300 \, K$, $P = 1 \, atm = 1.033 \times 10^4 \, kg/m^2$, and the values of N_A and R in this equation yields

$$\lambda_m = \frac{RT}{\sqrt{2} N_A P \sigma}$$

$$= \frac{(8.314)(300)}{\sqrt{2}\,(6.022 \times 10^{23})\,(1.033 \times 10^4)\,\sigma}$$

$$= \frac{2.843 \times 10^{-25}}{\sigma}.$$

The mean free paths in helium and argon are then given by

$$\lambda_{m,hel} = \frac{2.843 \times 10^{-25}}{2.9 \times 10^{-21}}$$

$$= 9.8 \times 10^{-5} \, m$$

$$= 98 \, \mu m$$

$$\text{and} \quad \lambda_{m,arg} = \frac{2.843 \times 10^{-25}}{1.1 \times 10^{-19}}$$

$$= 2.6 \times 10^{-6} \, m$$

$$= 2.6 \, \mu m.$$

2.1.D Radiation Length

We will see later in this chapter that the predominant mode of energy loss for high energy electrons in matter is through a process called *Bremsstrahlung* and for photons it is *pair production*. Both of these processes are intimately connected by underlying physics and hence the overall attenuation of electrons and photons passing through matter can be described by a single quantity called *radiation length*. Radiation length can be defined as the thickness of a material that an electron travels such that it looses $1 - 1/e$ (corresponding to about 63%) of its energy by Bremsstrahlung. Note that here e refers to the exponential factor. For photons the process would be pair production.

Radiation length is an extensively quoted quantity since it relates the physical dimension of the material (such as its depth) to a property of radiation (such as its rate of energy loss). It is almost universally represented by the symbol X_0. The most widely used semi-empirical relation to calculate the radiation length of electrons in any material is given by (54)

$$\frac{1}{X_0} = 4\alpha r_e^2 \frac{N_A}{A} \left[Z^2 \{L_{rad} - f(Z)\} + Z L'_{rad} \right]. \tag{2.1.14}$$

Here $N_A = 6.022 \times 10^{23} mole^{-1}$ is the Avogadro's number, $\alpha = 1/137$ is the electron fine structure constant, and $r_e = 2.8179 \times 10^{-13} cm$ is the classical electron radius. $f(Z)$ is a function, which for elements up to uranium can be calculated from

$$f(Z) = a^2 \left[(1 + a^2)^{-1} + 0.20206 - 0.0369a^2 + 0.0083a^4 - 0.002a^6 \right],$$

with $a = \alpha Z$. X_0 in the above relation is measured in units of g/cm^2. It can be divided by the density of the material to determine the length in cm. Table 2.1.1 lists values and functions for L_{rad} and L'_{rad}.

Table 2.1.1: L_{rad} and L'_{rad} needed to compute radiation length from equation 2.1.14 (19)

Element	Z	L_{rad}	L'_{rad}
H	1	5.31	6.144
He	2	4.79	5.621
Li	3	4.74	5.805
Be	4	4.71	5.924
Others	>4	$\ln(184.15Z^{-1/3})$	$\ln(1194Z^{-2/3})$

The term containing L'_{rad} in equation 2.1.14 can be neglected for heavier elements, in which case the radiation length, up to a good approximation, can be calculated from

$$\frac{1}{X_0} = 4\alpha r_e^2 \frac{N_A}{A} \left[Z^2 \{L_{rad} - f(Z)\} \right]. \tag{2.1.15}$$

Fig.2.1.1 shows the values of radiation length calculated from equations 2.1.14 and 2.1.15 for materials with $Z = 4$ up to $Z = 92$. Also shown are the relative errors assuming equation 2.1.14 gives the correct values.

Another relation, which requires less computations than equation 2.1.14 or even 2.1.15, is

$$X_0 = \frac{716.4A}{Z(Z + 1) \ln(287/\sqrt{Z})}. \tag{2.1.16}$$

Here also, as before, X_0 is in g/cm^2. This relation gives reasonable results for elements with low to moderate atomic numbers. This can be seen in Fig.2.1.2,

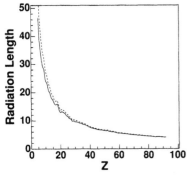

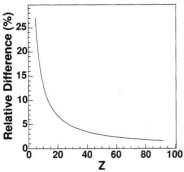

Figure 2.1.1:   Upper plot shows the radiation lengths in g/cm^2 computed from equations 2.1.14(solid line) and 2.1.15(dashed line). The lower plots shows the relative error in the values computed from equation 2.1.15 with respect to the ones computed from equation 2.1.14.

which shows the radiation lengths for different materials computed from equations 2.1.14 and 2.1.15 together with relative error in the latter assuming the former gives the correct results.

For a mixture or compound, the effective radiation length can simply be calculated by taking the weighted mean

$$\frac{1}{X_0} = \sum_i w_i \frac{1}{X_{0i}}, \qquad (2.1.17)$$

where w_i and X_{0i} is the fraction by weight and radiation length of the ith material.

Example:
Estimate the radiation length of high energy electrons in carbon dioxide. CO_2 is commonly used as a fill gas in gas-filled detectors.

Solution:
The effective radiation length can be computed from equation 2.1.17. But we first need to compute the radiation lengths corresponding to the individual elements in CO_2. Since for low Z materials, equation 2.1.16 gives reasonable results, therefore we will use this to determine the individual radiation lengths

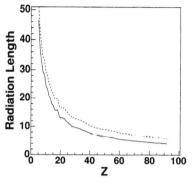

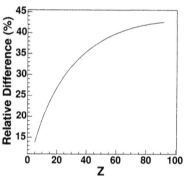

Figure 2.1.2: Upper plot shows the radiation lengths in g/cm^2 computed from equations 2.1.14 (solid line) and 2.1.16 (dashed line). The lower plots shows the relative error in the values computed from equation 2.1.16 with respect to the ones computed from equation 2.1.14.

of carbon and oxygen.

$$X_0^C = \frac{(716.4)(12.01)}{6(6+1)\ln(287/\sqrt{6})}$$
$$= 43.0 \ gcm^{-1}$$
$$X_0^O = \frac{(716.4)(16.0)}{8(8+1)\ln(287/\sqrt{8})}$$
$$= 34.46 \ gcm^{-1}$$

Now we need to calculate the weight fractions of carbon and oxygen in CO_2, which can be done as follows.

$$w_1 = \frac{12.01}{12.01 + 16.0 + 16.0}$$
$$= 0.2730$$
$$w_2 = \frac{16.0}{12.01 + 16.0 + 16.0}$$
$$= 0.3635$$
$$w_3 = w_2.$$

Where the subscript 1 of w refers to carbon while w_2 and w_3 represent the weight fractions of the two oxygen atoms in CO_2.

We can now compute the effective radiation length using equation 2.1.17.

$$
\begin{aligned}
\frac{1}{X_0} &= w_1 \frac{1}{X_0^C} + w_2 \frac{1}{X_0^O} + w_3 \frac{1}{X_0^O} \\
&= \frac{0.2730}{43.0} + \frac{0.3635}{34.46} + \frac{0.3635}{34.46} \\
&= 2.742 \times 10^{-2} \\
\Rightarrow X_0 &= 36.43 \; gcm^{-1}
\end{aligned}
$$

This is our required effective radiation length. As an exercise, let us see how different it is from the one computed using the more accurate relation 2.1.14. The reader is encouraged to carry out the computations to verify that the individual radiation lengths thus calculated are

$$
\begin{aligned}
X_0^C &= 37.37 \; gcm^{-1} \quad \text{and} \\
X_0^O &= 29.37 \; gcm^{-1}.
\end{aligned}
$$

The effective radiation length in CO_2 as calculated from equation 2.1.17 is then given by

$$
\begin{aligned}
\frac{1}{X_0} &= \frac{0.2730}{37.37} + \frac{0.3635}{29.37} + \frac{0.3635}{29.37} \\
&= 3.206 \times 10^{-2} \\
\Rightarrow X_0 &= 31.19 \; gcm^{-1}
\end{aligned}
$$

The percent relative error in the radiation length as computed from equation 2.1.16 as compared to this one is

$$
\begin{aligned}
\epsilon &= \frac{36.43 - 31.19}{36.43} 100 \\
&= 16.8\%.
\end{aligned}
$$

Hence with far less computations, we have gotten a result that is accurate to better than 73%. However, as shown in Fig.2.1.2, the relative error increases as we go higher in Z. Equation 2.1.16 should therefore be used only for low to moderate Z elements.

2.1.E   Conservation Laws

Although the reader should already be aware of the fundamental conservation laws of particle physics but this section has been introduced to serve as a memory refresher. An extensive discussion on the many conservation laws that exist in particle physics is, of course, beyond the scope of this book. We will therefore restrict ourselves to the following three laws that are generally employed to compute kinematic quantities related to the field of radiation detection.

▶ Conservation of energy

▶ Conservation of momentum

► Conservation of electrical charge

Since these laws are directly or indirectly used in computations involved in the detection and measurement of radiation, therefore we will spend some time understanding their basis and learning how they are applied.

E.1 Conservation of Energy

This law dictates that total energy content of a closed system always remains constant irrespective of the time evolution of the system. Here, by the term *closed system*, it is not meant that the system is completely isolated from anything else in the world. Rather it represents a system which is not *influenced* by external forces. Being influenced is certainly a relative term and depends on the accuracy of the result desired. For example, one can not escape the background electromagnetic field, which is present everywhere. But for most practical purposes its strength is too week at the point of collision between two microscopic particles that it can be safely neglected. In this case we can still apply the law of conservation of linear momentum to determine the kinematic quantities related to the collision.

At the fundamental level, the law of conservation of energy ensures that temporal translations do not violate the laws of nature. Since mass and energy are related to each other through Einstein's relation $E = mc^2$, therefore we can also say that it is neither possible to completely destroy mass or energy nor it is possible to create them from nothingness.

Since mass can be converted into energy, therefore with each massive particle an intrinsic energy content can be associated. We call this quantity its *rest energy* to signify the amount of energy it contains when at rest. For particle interactions it has been found that the sum of rest energy and the kinetic energy of a closed system remains constant.

$$E_{rest} + T = constant$$

Here $E_{rest} = m_0 c^2$ is the total rest energy of all particles at any instant having combined rest mass of m_0. T represents the total kinetic energy of all the particles at the same instant in time.

E.2 Conservation of Momentum

Both linear and angular momenta are conserved for any system not subject to external forces. For linear momentum, this law dictates that the total linear momentum of a closed system remains constant. It represents the fact that the laws of physics are the same at all locations in space. This law, together with the law of conservation of energy, is extensively used to determine the kinematic variables related to particle collisions.

Linear momentum is a vector quantity and therefore one must be careful when computing the total momentum of the system. One can not just add the momenta, rather a vector sum must be taken. Mathematically, for the case of collision between particles, this law requires that

$$\sum \vec{p}_i = \sum \vec{p}_f, \tag{2.1.18}$$

where $\vec{p}$ represents the momentum. The subscripts i and f stand for the initial and the final states of the system (such as before and after collision) and the *vector* sum is carried over all moving particles taking part in the collision.

Conservation of angular momentum ensures that there are no preferred directions, that is, the space is isotropic and the laws of physics are the same in all directions.

E.3   Conservation of Electrical Charge

Most of the particles we know hold a property called electrical charge. For example, an electron carries a unit charge of 1.602×10^{-19} coulomb while a proton has the same charge with positive polarity. However there is one fundamental difference between the charge carried by these two particles. The electron does not seem to have any internal structure and is therefore thought to carry the unit charge. On the other hand the proton is composed of three quarks each carrying a fractional charge. The total charge is still the so called *unit charge* but internally the charge is divided between the constituents of the proton. A neutron is also composed of three quarks but in such a way that the fractional charges of these quarks cancel out and the net charge on the neutron appears to be zero. The law of conservation of electrical charge states that the unit charge can neither be created nor destroyed. This law eliminates some of the interaction possibilities. For example, a beta decay **can not** be written as

$$n \rightarrow e + \gamma,$$

since it would imply the creation of electrical charge. Instead the correct reaction is

$$n \rightarrow p + e + \bar{\nu}.$$

2.2   Types of Particle Interactions

Radiation carries energy and whenever it interacts with a detection medium it may deposit some or all of it to the particles in the medium. The result is some form of excitation. This excitation can form a basis of signal formation that can be detected and measured by the processing electronics. Some of these excitations are:

▶ Ionization

▶ Scintillation

▶ Excitation of lattice vibrations

▶ Breakup of Cooper pairs in superconductors

▶ Formation of superheated droplets in superfluids

▶ Excitation of optical states

However not all of these can always be effectively used for particle detection. The most commonly used excitation mechanisms that are used for this purpose are the ionization and scintillation.

An interesting point to note here is that when we talk about the behavior of radiation passing through matter, we are actually referring to the statistical out-comes of interaction of particles with other particles in the material. We saw earlier that at microscopic level, the particle interactions are mainly governed by quantum mechanics and therefore the behavior of radiation can be predicted by statistical quantities such as cross section. Depending on the type of force involved, particles can interact with other particles in a number of ways. For example, the predominant mode of interaction for charged particles and photons at lower energies is electro-magnetic, while neutral particles are mainly affected by short range nuclear forces. In the following sections, we will have a general look at some of the important modes of interaction relevant to the topic of radiation interaction and detection.

2.2.A   Elastic Scattering

Elastic scattering refers to a process in which an incident particle scatters off a target in such a way that the total kinetic energy of the system remains constant. It should be noted that scattering elastically does not imply that there is no energy transfer between incident and target particles. The incident particle can, and in most cases does, loose some of its energy to the target particle. However this energy does not go into any target excitation process, such as, its electronic and nuclear states of the target are not affected.

The best way to mathematically model the elastic scattering process is by con-sidering both the incident and target particles to have no internal structure. Such particles are generally referred to as *point-like* particles.

There are also special types of elastic scattering processes in which the incident particle transfers none or very minimal energy to the target. A common example is the Rayleigh scattering of photons.

2.2.B   Inelastic Scattering

In inelastic scattering process the incident particle excites the atom to a higher electronic or nuclear state, which usually comes back to the ground state by emitting one or more particles. In this type of reaction the kinetic energy is not conserved because some of it goes into the excitation process. However the total energy of the system remains constant.

2.2.C   Annihilation

When a particle interacts with its antiparticle, the result is the annihilation of both and generation of other particles. The most common and oft quoted example of this process is the electron-positron annihilation. Positron is the antiparticle of electron having all the characteristics of an electron except for the unit electrical charge, which in its case is positive. When it comes very close to an electron, the two *annihilate* each other producing other particles. Depending on their energies, different particles can be generated during this process. At low to moderate energies, only photons are produced while at high energies other particles such as Z bosons can be produced.

The annihilation process in itself does not have any threshold energy. That is, the process can also happen even if the electron and positron are at rest. However to produce particles other than photons, the electron and the positron must be allowed to collide with each other at very high energies. This is actually done, for example at Large Electron Positron collider (or LEP) at CERN in Switzerland where particles are accelerated to center of mass energies reaching at $45 - 100 \ GeV$ before collision. The result of the collision is generation of large number of particles, which are then tracked down and identified to get clues about the fundamental particles and their interactions.

At low electron-positron energies, at least two photons are produced in the annihilation process. The reason why only one photon can not be emitted lies in the law of conservation of momentum. According to this law, the total momentum of the emitted photons must be equal to that of the total momentum of electron and positron. Now, if the electron and the positron move in opposite directions with equal kinetic energies or are at rest before annihilation, the net momentum before collision will be zero. This implies that after the annihilation the total momentum must also be zero. We can not have a zero net momentum with only one photon and hence we conclude that at least two photons must be produced. Hence by using a simple argument, we have set a lower limit on the number of photons produced.

Let us now suppose that an electron-positron annihilation process produces only two equal energy photons. If the net momentum before collision was zero then the photons must travel in opposite direction to each other since only then the momentum conservation can be guaranteed. The energy these photons carry can be deduced from the law of conservation of energy, which states that the total energy of the system before and after collision must be equal. The reader may recall that the total energy of a particle is the sum of its kinetic and rest mass energies, such as

$$E = T + m_0c^2, \qquad (2.2.1)$$

where T is the kinetic energy, m_0 is the rest mass, and c is the velocity of light. The total energy before collision can then be written as

$$
\begin{aligned}
E_{total} &= E_e + E_{e+} \\
&= T_e + m_0c^2 + T_{e+} + m_0c^2 \\
&= T_e + T_{e+} + 2m_0c^2,
\end{aligned}
$$

where we have made use of the fact that both the electron and the positron have equal rest masses. The kinetic energies of the two particles will be zero if they were at rest before the annihilation. Hence the total energy before the collision will be

$$E_{total} = 2m_0c^2 = 2 \times 511 \ keV.$$

This shows that to conserve energy, each of the two photons produced must carry an energy of $511 \ keV$. Hence we can conclude that if the conservation of momentum and energy are to be satisfied then at the minimum there must be two photons each carrying $511 \ keV$ of energy and traveling in opposite directions. This concept is also graphically depicted in Fig.2.2.1

An interesting point to note here is that the arguments we gave in the preceding paragraphs do not exclude the possibility of one-photon annihilation process. The assumption of zero net momentum before collision is not always true and therefore it

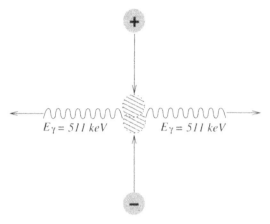

Figure 2.2.1: Depiction of electron-positron annihilation process in which two photons are produced. To ensure conservation of energy and momentum, each of these photons must have an energy of 511 keV and they must travel opposite to each other.

is possible that only a single photon is produced in the process. Similarly production of three or more photons is also possible. However the probability of production of single, three or more photons at low energies is very small and can be safely ignored (39).

The electron-positron annihilation process at low energies is used in a special type of medical imaging called Positron Emission Tomography or PET. In this process images are produced by detecting photons that are emitted as a result of this process. We will discuss this technique in the chapter on imaging.

Since all particles have their inverse counterparts in nature, the annihilation process is in no way limited to electrons and positrons. For example, protons p and anti-protons $\bar{p}$ also get annihilated when they approach each other. In fact, $p\bar{p}$ collisions at the Tevatron collider at Fermilab in USA have given us immense insight into the fundamental particles and their interactions.

2.2.D Bremsstrahlung

Bremsstrahlung is a German word meaning *braking radiation*. It refers to the process in which decelerating charged particles emit electromagnetic radiation. All charged particles can emit this kind of radiation provided they have enough energy. Generally speaking if a charged particle has energy much greater than its rest energy ($E_{rest} \equiv E_0 = m_0c^2$, where m_0 is its rest mass) emits Bremsstrahlung if it encounters resistance while moving through a medium.

In pure Bremsstrahlung process there are no *direct* electronic or nuclear transitions involved. However the radiation emitting particle may excite or ionize atoms and excite nuclei of the medium as it decelerates. These excitations and ionizations may lead to emission of other particles, such as x-rays and γ-rays with characteristic energy peaks in the spectrum. These peaks are generally superimposed on the continuous Bremsstrahlung spectrum and are therefore clearly distinguishable. The most common example of this phenomenon is the emission of x-rays as we saw in the previous chapter. The electrons, as they strike the anode, emit not only characteristic x-rays but also Bremsstrahlung. In fact, as the electron has a very small mass as compared to other charged particles, its Bremsstrahlung is one of the most commonly encountered radiations in laboratories.

If an electron is accelerated through a potential V (as in x-ray machines), the maximum energy it can attain is given by

$$E_{max} = eV, \qquad (2.2.2)$$

where e is the unit electronic charge. The maximum energy of the Bremsstrahlung in the form of photons that this electron can emit will then also be equal to eV, that is

$$E_{brems} \leq E_{max} = eV. \qquad (2.2.3)$$

Since Bremsstrahlung is emitted in the form of photons having energy $E = h\nu = hc/\lambda$ therefore we can write the above equation as

$$h\frac{c}{\lambda} \leq eV$$

$$\Rightarrow \lambda \geq \frac{hc}{eV}. \qquad (2.2.4)$$

Hence we can associate a minimum wavelength λ_{min} with the process below which there will be no Bremsstrahlung photons emitted.

$$\lambda_{min} = \frac{hc}{eV} \qquad (2.2.5)$$

λ_{min} is also called the *cutoff wavelength* for Bremsstrahlung.

Example:
Compute the cutoff Bremsstrahlung wavelength for an electron moving under the influence of a potential of 40 kV.

Solution:
We will use equation 2.2.5 to compute the cutoff wavelength.

$$\begin{aligned}
\lambda_{min} &= \frac{hc}{eV} \\
&= \frac{(6.626 \times 10^{-34})(2.99 \times 10^{8})}{(1.602 \times 10^{-19})(40 \times 10^{3})} \; m \\
&= 30.91 \; fm
\end{aligned}$$

2.2.E   Cherenkov Radiation

We know that velocity of a particle in a medium depends, among other things, also on the nature and density of the medium. The same is true for light particles, or photons. There is no theory in physics that demands that light has constant velocity in all types of media. The special theory of relativity only says that the velocity of light is independent of the frame of reference, not that it is constant everywhere. In water, for example the velocity of light is significantly lower than $c = 2.99 \times 10^{8} \; ms^{-1}$, which is the velocity of light in *vacuum* and is supposed to be constant in *vacuum*.

It is also possible that a high energy charged particle with non-zero rest mass, such as an electron, travels faster than speed of light in that medium. If this happens, the particle emits a special kind of radiation called Cherenkov radiation. The wavelengths of Cherenkov photons lie in and around the visible region of electromagnetic spectrum. In fact, the first Cherenkov radiation was observed by Pavel Cherenkov in 1934 as blue light coming from a bottle of water undergoing bombardment by particles from a radioactive source. This discovery and his subsequent explanation of the process earned him Nobel Prize in Physics in 1958.

Cherenkov radiation has a certain geometric signature: it is emitted in the form of a cone having an angle θ defined by

$$\cos\theta = \frac{1}{\beta n}, \tag{2.2.6}$$

where n is the refractive index of the medium and $\beta = v/c$ with v as the velocity of the particle in the medium.

Since Cherenkov radiation is always emitted in the form of a cone therefore the above equation can be used to determine a value of β (and hence v) below which the particle will not emit any radiation. Since $\cos\theta < 1$ for a cone, therefore using the above relation we can conclude that a necessary condition for the emission of Cherenkov radiation is that

$$\beta > \frac{1}{n}. \tag{2.2.7}$$

Now, since $\beta = v/c$, this condition can be translated into

$$v > \frac{c}{n}. \tag{2.2.8}$$

Here c/n is the velocity of light *in the medium*. This shows that the emission of Cherenkov radiation depends on two factors: the refractive index n of the medium and the velocity v of the particle in that medium. Using this condition one can determine the minimum kinetic energy a particle must possess in order to emit Cherenkov radiation in a medium (see example below).

Example:
Compute the threshold energies an electron and a proton must possess in light water to emit Cherenkov radiation.

Solution:
For both particles the threshold velocity v_{th} can be computed from equation 2.2.8.

$$v > \frac{c}{n} = \frac{2.99 \times 10^8}{1.3}$$
$$\Rightarrow v_{th} = 2.3 \times 10^8 \ ms^{-1}$$

Here we have taken $n = 1.3$ for water and $c = 2.99 \times 10^8 \ ms^{-1}$ is the velocity of light *in vacuum*. Since particles are relativistic, we must use the relativistic kinetic energy relation

$$T = \left[\left(1 - \frac{v^2}{c^2}\right)^{-1/2} - 1\right] m_0 c^2.$$

Here $m_0 c^2$ is simply the rest energy of the particle. For an electron and a proton it is 0.511 MeV and 938 MeV respectively. If we use these numbers then T will also be in MeV irrespective of the units we use for v and c as long as both are in the same units. Hence the above relation can also be written as

$$T_{th} = \left[\left(1 - \frac{(2.3 \times 10^8)^2}{(2.99 \times 10^8)^2} \right)^{-1/2} - 1 \right] \times E_0$$
$$= 0.565 \times E_0,$$

where $E_0 = m_0 c^2$ is the rest energy of the particle. The required threshold kinetic energies of electrons and protons are then given by

$$T_{th}^e = 0.565 \times 0.511$$
$$= 0.289 \; MeV$$
$$T_{th}^p = 0.565 \times 938$$
$$= 539 \; MeV.$$

The above example clearly shows that protons need much higher kinetic energy than an electron to emit Cherenkov radiation. This, of course, is a consequence of the heavier mass of the proton. In general, the heavier the particle the higher kinetic energy it must possess to be able to emit Cherenkov radiation.

2.3   Interaction of Photons with Matter

Since photons are not subject to Coulomb or nuclear forces, their interactions are localized at short distances. This means that although the intensity of a photon beam decreases as it passes through a material and photons are removed from the beam but the energy of individual photons that do not take part in any interaction is not affected.

2.3.A   Interaction Mechanisms

Photons can primarily interact with material in three different ways: photoelectric effect, Compton scattering, and pair production. Other possible interaction mechanisms include Raleigh and Mie processes. These interaction mechanisms have different energy thresholds and regions of high cross-sections for different materials. Whenever a beam of photons of sufficient energy passes through a material, not all of the photons in the beam go through the same types of interaction. With regard to radiation detection, the key then is to look statistically at the process. For example, if we want to know how most of the photons in the beam will interact with the material, we can look at the cross section of all the interactions and find the one that has the highest value at that particular photon energy. Most of the time, the cross sections can either be determined through some semi-empirical relation or extracted from a cross section table that are routinely published by researchers. Fig.2.3.1 shows different photon cross sections as a function of energy for carbon and lead. Another excellent use of these cross sections is the estimation of the attenuation of

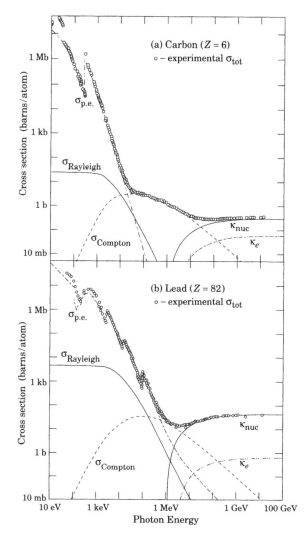

Figure 2.3.1: Photon cross sections as a function of energy for carbon and lead (19).

a photon beam as it passes through a material. Before we go into that discussion, let's have a closer look at the most important photon interaction mechanisms.

A.1 Photoelectric Effect

In the previous chapter we introduced the concept of wave-particle duality, according to which light sometimes behave as particles. Photoelectric effect is one of the processes that confirm this idea. It was originally explained by Einstein and earned him a Nobel Prize. The effect is rather simple: when light shines on a material, electrons can be emitted. The emission of electrons, however, does not depend on the intensity of light, rather on its frequency. If the frequency is lower than a certain value, that depends on the target material, no electrons are emitted. Certainly this can not be explained on the basis of classical wave-like picture of light. Einstein

explained this effect by arguing that light is transferred to the material in packets
called *quanta*, each of which carries an energy equal to

$$E_\gamma = h\nu = \frac{hc}{\lambda},\tag{2.3.1}$$

where ν and λ are the frequency and wavelength of light respectively and c is the
velocity of light in vacuum. Now, since electrons in the material are bonded therefore
to set them free the energy delivered must be greater than their binding energy. For
metals, this energy is called *work function* and generally represented by the symbol
ϕ. Hence, in order for an electron to be emitted from a metal surface, we must have

$$E_\gamma \geq \phi\tag{2.3.2}$$

$$\text{or} \quad \nu \geq \frac{\phi}{h}\tag{2.3.3}$$

$$\text{or} \quad \lambda \leq \frac{hc}{\phi}.\tag{2.3.4}$$

If the photon energy is larger than the work function, the rest of the energy is
carried away by the emitted electron, that is

$$E_e = E_\gamma - \phi.\tag{2.3.5}$$

Proving this theory is quite simple as one can design an experiment that measures
the electron energy as a function of the incident light frequency or wavelength and
see if it follows the above relation.

Since most metals have very low work functions, on the order of a few electron
volts, therefore even very low energy light can set them free. The work function
of metals is approximately half the binding energy of *free* metallic atoms. In other
words, if there were no metallic bonds and the metallic atoms were free, the energy
needed for the photoelectric effect to occur would be twice.

The photoelectric effect can also occur in free atoms. During this process, a photon is completely absorbed by an atom making it unstable. To return to the stable
state, the atom emits an electron from one of its bound atomic shells. Naturally the
process requires that the incident photon has energy greater than or equal to the
binding energy of the most loosely bound electron in the atom. The energy carried
away by the emitted electron can be found by subtracting the binding energy from
the incident photon energy, that is

$$E_e = E_\gamma - E_b,\tag{2.3.6}$$

where E_b is the binding energy of the atom. The *atomic* photoelectric effect is
graphically depicted in Fig.2.3.2

Since the photon completely disappears during the process, the photoelectric
effect can be viewed as the *conversion* of a single photon into a single electron. It
should however be noted that this is just a convenient way of visualizing the process
and does not in any way represents an actual photon to electron conversion. Just
like nuclear reactions we visited in the first chapter, we can also write this reaction
as

$$\gamma + X -> X^+ + e.\tag{2.3.7}$$

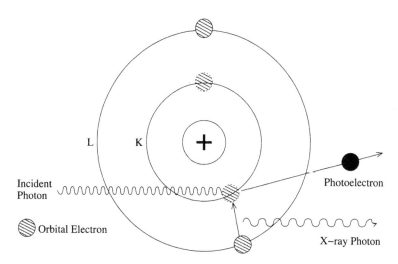

Figure 2.3.2: Depiction of photoelectric effect in a free atom. If the K-shell electron is knocked off by the incident photon, another electron from a higher energy level may fill the gap. This would result in the emission of a photon with an energy equal to the difference of the two energy levels.

Here X is the target atom, which gets an overall positive charge when the electron e is knocked off from one of its shells by the incident gamma ray photon γ.

The cross section for this reaction has a strong Z dependence, that is, the probability of photoelectric effect increases rapidly with atomic number of the target atom. In addition, it also has a strong inverse relationship with energy of the incident photon. Specifically, these dependences can be expressed as

$$\sigma_{pe} \propto \frac{Z^n}{E_\gamma^{3.5}}, \tag{2.3.8}$$

where n lies between 4 and 5. This relation suggests that the probability of photoelectric effect would decrease sharply with higher incident photon energies. Figure 2.3.1 shows the typical dependence of photoelectric cross section on energy of the incident photon in carbon and lead.

Photoelectric effect takes place predominantly in the K atomic shell and therefore generally the cross section related to K-shell interaction is used to estimate the total photoelectric cross section. The K-shell photoelectric cross section is given by

$$\sigma_{pe,K} = \left[\frac{32}{\epsilon^7}\right]^{1/2} \alpha^4 Z^5 \sigma_{Th} \; cm^2/\text{atom}. \tag{2.3.9}$$

Here

$$\epsilon = \frac{E_\gamma}{m_e c^2} \quad \text{and}$$

$$\sigma_{Th} = \frac{8}{3}\pi r_e^2 = 665 \; mbarn.$$

σ_{Th} is called Thompson scattering cross section.

An interesting aspect of photoelectric effect is that if the incident photon has sufficient energy to overcome the binding energy of an inner shell electron, that electron might get ejected leaving a vacancy behind. This vacancy can then be filled by an outer shell electron to stabilize the atom. Such a transition would emit a photon with energy equal to the difference of the two energy levels. These photons are generally in the x-ray region of electromagnetic spectrum and are called fluorescence photons. In experiments with photons where elements with high atomic number (such as lead) are used for shielding, the emission of characteristic x-rays from the shield is not uncommon. This adds to the overall background and care must be taken to eliminate it from the data.

An x-ray photon emitted as a consequence of photoelectric effect can also knock off another orbital electron provided its energy is equal to the binding energy of that electron. This electron is called Auger electron. The process is essentially radiationless because the excess energy of the atom is used and taken away by the Auger electron. Auger process is shown graphically in Fig.2.3.3.

Although here we have shown that the photoelectric effect can lead to the emission of Auger electrons, however this process is not in any way restricted to photoelectric effect. In fact in Auger electron spectroscopy, an electron beam is used to knock off inner shell electrons from target atoms, that eventually leads to the emission of Auger electrons.

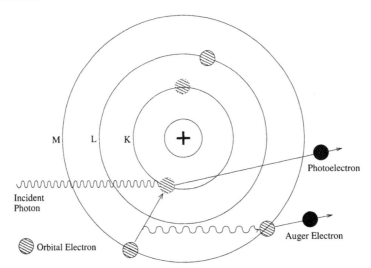

Figure 2.3.3:   Depiction of photoelectric effect leading to the emission of an Auger electron. A K-shell electron is seen to have been knocked off by the incident photon, creating a vacancy that must be filled in order for the atom to be stable again. Another electron from the M-shell fills this gap but releases some energy in the process (shown as a photon) equal to the difference between the two energy level. This photon is shown to have knocked off another electron from the M-shell. The end result is a radiationless electron emission. This electron is called Auger electron.

Example:
Calculate the wavelength below which it would be impossible for photons to ionize hydrogen atoms. The first ionization potential for hydrogen is 13.6 eV.

Solution:
The minimum energy needed to ionize an atom is equal to the binding energy of the most loosely bound electron. Since for hydrogen this energy is 13.6 eV, for photoelectric effect to be possible we must have

$$E_\gamma \geq 13.6 \ eV$$

$$\Rightarrow \frac{hc}{\lambda_{max}} \geq (13.6)\left(1.602 \times 10^{-19}\right) \ J$$

$$\Rightarrow \lambda_{max} \leq \frac{hc}{(13.6)\left(1.602 \times 10^{-19}\right)} \ m$$

$$\leq \frac{\left(6.625 \times 10^{-34}\right)\left(2.99 \times 10^{8}\right)}{(13.6)\left(1.602 \times 10^{-19}\right)}$$

$$\leq 9.09 \times 10^{-8} \ m = 90.9 \ nm$$

Hence a photon beam with a wavelength greater than 90.9 nm will not be able to ionize hydrogen atoms no matter how high its intensity is.

A.2   Compton Scattering

Compton scattering refers to the *inelastic* scattering of photons from free or loosely bound electrons which are at rest. Since the electron is almost free, it may also get scattered as a result of the collision.

Compton scattering was first discovered and studied by Compton in 1923. During an scattering experiment he found out that the wavelength of the scattered light was different from that of the incident light. He successfully explained this phenomenon by considering light to consist of quantized wave packets or photons.

Fig.2.3.4 shows this process for a bound electron. The reader may recall that the binding energies of low Z elements are on the order of a few hundred eV, while the γ-ray sources used in laboratories have energies in the range of hundreds of keV. Therefore the bound electron can be considered *almost free* and *at rest* with respect to incident photons. In general, for orbital electrons, the Compton effect is more probable than photoelectric effect if the energy of the incident photon is higher than the binding energy of the innermost electron in the target atom.

Simple energy and linear momentum conservation laws can be used to derive the relation between wavelengths of incident and scattered photons (see example below),

$$\lambda = \lambda_0 + \frac{h}{m_0 c}[1 - \cos\theta]. \tag{2.3.10}$$

Here λ_0 and λ represent wavelengths of incident and scattered photons respectively. m_0 is the rest energy of electron and θ is the angle between incident and scattered photons (see Fig.2.3.4).

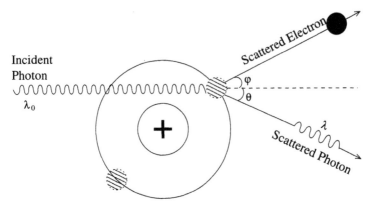

Figure 2.3.4: Compton scattering of a photon having energy $E_{\gamma 0} = hc/\lambda_0$ from a bound electron. Some of the energy of the incident photon goes into knocking the orbital electron out of its orbit.

Example:
Derive equation 2.3.10.

Solution:
For this derivation we will assume that the electron is not only at rest before the collision but is also not under the influence of any other potential. In other words it is free to move around. As we saw before, these two assumption are valid up to a good approximation even for electrons bound in an atom provided the incident photon has high enough energy.

The scattering process is depicted in Fig.2.3.5 where the momenta before and after collision have been broken down in their respective horizontal and vertical components. This will aid us in taking the *vector sum* of the momenta while applying the law of conservation of momentum.

Since electron is supposed to be at rest before the collision, therefore $p_{e0} = 0$ and the total momentum in horizontal direction before collision is simply photon's momentum $p_{\gamma 0}$. After the scattering there are two horizontal momenta corresponding to both the electron and the photon. Application of conservation of momentum in horizontal direction then gives

$$p_{\gamma 0} = p_\gamma \cos\theta + p_e \cos\phi.$$

Rearranging and squaring of this equation gives

$$p_\gamma^2 \cos^2\theta = (p_{\gamma 0} - p_e \cos\phi)^2 \tag{2.3.11}$$

To apply the conservation of momentum in the vertical direction, we note that before scattering there is no momentum in vertical direction. Hence we have

$$0 = p_\gamma \sin\theta - p_e \sin\phi,$$

where the negative sign simply shows that the two momenta are in opposite direction to each other. Rearranging and squaring of this equation gives

$$p_\gamma^2 \sin^2\theta = p_e^2 \sin^2\phi \tag{2.3.12}$$

By adding equations 2.3.11 and 2.3.11 together and using the trigonometric identity $\sin^2\theta + \cos^2\theta = 1$ we get

$$p_e^2 = p_{\gamma 0}^2 + p_\gamma^2 - 2p_{\gamma 0}p_\gamma \cos\theta \qquad (2.3.13)$$

To eliminate p_e from this equation we apply the conservation of total energy, that is we require that the total energy of the system before and after momentum must be equal. The electron, being initial at rest, does not have any kinetic energy and its total energy is only its rest mass energy m_0c^2. After scattering it attains a momentum, which according to special relativity is related to its energy by the relation

$$E_e^2 = p_e^2 c^2 + m_0^2 c^4.$$

the energy of a photon in terms of its momentum is given by

$$E_\gamma = p_\gamma c.$$

Application of the conservation of energy to the scattering process and using the above two energy relations gives

$$
\begin{aligned}
E_{\gamma 0} + E_{e0} &= E_\gamma + E_e \\
\Rightarrow p_{\gamma 0}c + m_0 c^2 &= p_\gamma c + \sqrt{p_e^2 c^2 + m_0^2 c^4} \\
\Rightarrow p_e^2 &= p_{\gamma 0}^2 + p_\gamma^2 - 2p_{\gamma 0}p_\gamma + 2m_0 c\left(p_{\gamma 0} - p_\gamma\right) \qquad (2.3.14)
\end{aligned}
$$

Equating 2.3.13 with 2.3.14 and some rearrangements yields

$$p_{\gamma 0} - p_\gamma = p_{\gamma 0}p_\gamma\left(1 - \cos\theta\right)$$

Using $p_\gamma = h/\lambda$ in the above equation we get the required relation

$$\lambda = \lambda_0 + \frac{h}{m_0 c}\left(1 - \cos\theta\right)$$

In terms incident and scattered photon energies, equation 2.3.10 can be written as

$$E_\gamma = E_{\gamma 0}\left[1 + \frac{E_{\gamma 0}}{m_0 c^2}\left(1 - \cos\theta\right)\right]^{-1}, \qquad (2.3.15)$$

where we have used the energy-wavelength relation $E_\gamma = hc/\lambda$.

This relation shows that energy of the scattered photon depends not only on the incident photon energy but also on the scattering angle. In other words the scattering process is in no way isotropic. We will see later on that this directional dependence is actually a good thing for spectroscopic purposes. Let us now have a look at the dependence of scattered photon energy on three extreme angles: 0^0, 90^0, and 180^0.

▶ **Case-1** ($\theta = 0$): In this case $\cos\theta = 1$ and therefore equation 2.3.15 gives

$$E_\gamma = E_\gamma^{max} = E_{\gamma 0}.$$

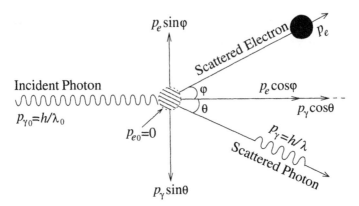

Figure 2.3.5: Initial and final momenta of the particles involved in Compton scattering. The momenta of the scattered electron and photon have been broken down into their horizontal and vertical components for ease in application of the law of conservation of momentum. The electron is assumed to be at rest before the collision.

That is, the scattered photon continues in the same direction as the incident photon and carries with it all of its energy. Of course this implies that the photon has not actually interacted with the electron and therefore this should not be regarded as a scattering process at all. However it should be noted that this case gives us the upper bound of the scattered photon energy. This may at first seem an intuitive conclusion but as it turns out there is a special process known as *reverse Compton scattering* in which the scattered photon energy is actually higher than the energy of the incident photon. In this process the electron is not at rest and therefore carries significant kinetic energy. During its interaction with the photon it transfers some of its energy to the photon, hence the term *reverse* Compton effect.

Certainly, this is a special case of Compton scattering, which is not generally encountered in laboratories. In the case of normal Compton scattering where the target electron is essentially at rest with respect to the incident photon, the scattered photon energy should never exceed the incident photon energy.

▶ **Case-2** ($\theta = 90^0$): In this situation the incident photon flies away at right angles to its original direction of motion after interacting with the electron. By substituting $\cos \theta = 0$ in equation 2.3.15 we can find the energy it carries away by the scattered photon.

$$E_\gamma = E_{\gamma 0} \left[1 + \frac{E_{\gamma 0}}{m_0 c^2} \right]^{-1}$$

The change in photon's wavelength in this as estimated from equation 2.3.10 is

$$\triangle \lambda = \frac{h}{m_0 c} = 2.432 \ fm$$

▶ **Case-3** ($\theta = 180^0$): It is obvious from equation 2.3.15 that a photon scattered at $\theta = 180^0$ will carry the minimum possible energy (since $1 - \cos \theta$ is maximum at $\theta = 180^0$). In this case, substituting $1 - \cos \theta = 2$ in equation 2.3.15 yields

$$E_\gamma^{min} = E_{\gamma 0} \left[1 + \frac{2E_{\gamma 0}}{m_0 c^2} \right]^{-1}. \qquad (2.3.16)$$

This equation can also be written as

$$E_\gamma^{min} = \frac{m_0 c^2}{2} \left[1 + \frac{m_0 c^2}{2E_{\gamma 0}} \right]^{-1}. \qquad (2.3.17)$$

To obtain a numerical result independent of the incident photon energy, let us assume that the incident photon energy is much higher than half the electron rest energy, such as $E_\gamma \gg m_0 c^2/2$. There is nothing magic about *half* the electron rest energy. It's been chosen because we want to eliminate the term containing $m_0 c^2/2E_\gamma 0$. In this situation the above equation reduces to

$$\begin{aligned} E_\gamma^{min} &\approx \frac{m_0 c^2}{2} \\ &= 255 \ keV. \end{aligned}$$

This is a very interesting result because it tells us that the electron will carry the maximum energy it could at any angle. This process resembles the simple head-on collision of two point masses in which the incident body completely reverses its motion and the target body starts moving forward. To determine the energy of the electron, we assume that *most* of the energy of the incident photon is distributed between the scattered photon and the electron. Hence the maximum energy of the scattered electron can be calculated from

$$E_e^{max} \approx E_\gamma - 255 \ keV.$$

This implies that in a γ-ray spectroscopy experiment one should see a peak at the energy $E_\gamma - 255 \ keV$. Such a peak is actually observed and is so prominent that it has gotten a name of its own: *the Compton edge* (see Fig.2.3.7). A consequence of this observation is that our assumption that even though the scattering is inelastic, however the energy imparted to the atom is not significantly high.

It is also instructive to plot the dependence of change in wavelength of the photon on the scattering angle. Fig.2.3.6 shows such a plot spanning the full 360^0 around the target electron. As expected, the largest change in wavelength occurs at $\theta = 180^0$, which corresponds to the case-3 we discussed above. The photon scattered at this angle carries the minimum possible energy E_γ^{min} as allowed by the Compton process.

A fair question to ask at this point is *how can the energy that the electron carries with it be calculated?* Though at first sight it may seem trivial to answer this question, the reality is that the *inelastic* behavior of this scattering process makes it a bit harder than merely subtracting the scattered photon energy from the incident photon energy. The reason is that if the electron is in an atomic orbit before

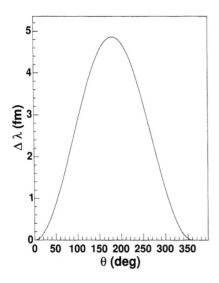

Figure 2.3.6: Angular distribution of change in wavelength of a Compton-scattered photon.

collision then some of the energy may also go into exciting the atom. A part of this energy, which is equal to the binding energy of the electron, goes into helping the electron break the potential barrier of the atom and get scattered. The remaining excess energy may not be large enough for the atom to be emitted as some sort of de-excitation process. To make life simple though, we can always assume that this energy is very small as compared to the energy carried away by the photon. In this case the energy carried away by the scattered electron can be estimated by subtracting the scattered photon energy E_γ and the atomic binding energy E_b of the electron from the incident photon energy.

$$E_e \approx E_{\gamma 0} - E_\gamma - E_b = \frac{E_{\gamma 0}^2}{mc^2} \left[\frac{1 - \cos\theta}{1 + \frac{E_{\gamma 0}}{mc^2}(1 - \cos\theta)} \right] - E_b \qquad (2.3.18)$$

Another simplification to this equation can be made by noting that the binding energy of electrons in low to moderate Z elements is several orders of magnitude smaller than the energy of the γ-ray photons emitted by most sources. In this case one can simply ignore the term E_b in the above equation and estimate the energy of the electron from

$$E_e \approx E_{\gamma 0} - E_\gamma = \frac{E_{\gamma 0}^2}{mc^2} \left[\frac{1 - \cos\theta}{1 + \frac{E_{\gamma 0}}{mc^2}(1 - \cos\theta)} \right]. \qquad (2.3.19)$$

Up until now we have not said anything about the dependence of the cross section on the scattering angle θ. Let us do that now. The differential cross section for Compton scattering can be fairly accurately calculated from the so called Klein-Nishina formula

$$\frac{d\sigma_c}{d\Omega} = \frac{r_0^2}{2} \left[\frac{1 + \cos^2\theta}{(1 + \alpha(1 - \cos\theta))^2} \right] \left[1 + \frac{4\alpha^2 \sin^4(\theta/2)}{(1 + \cos^2\theta)\{1 + \alpha(1 - \cos\theta)\}} \right]. \qquad (2.3.20)$$

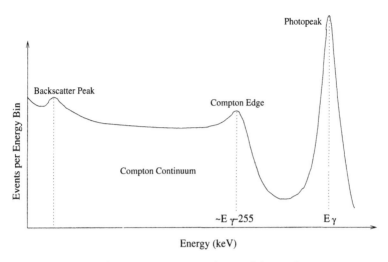

Figure 2.3.7:   Typical γ-ray spectrum obtained by a detector surrounded by heavy shielding (such as lead).

Here $r_0 = 2.818 \times 10^{-13}$ cm is the classical electron radius and $\alpha = h\nu/mc^2$ with ν and m being the wavelength of the incident photon and the electron rest mass respectively.

Example:
A photon incident on an atom scatters off at an angle of 55^0 with an energy of 150 keV. Determine the initial energy of the photon and the energy of the scattered electron.

Solution:
The wavelength of the scattered photon is

$$
\begin{aligned}
\lambda_s &= \frac{hc}{E_s} \\
&= \frac{\left(6.62 \times 10^{-34}\right)\left(2.99 \times 10^8\right)}{\left(150 \times 10^3\right)\left(1.6 \times 10^{-19}\right)} \\
&= 8.247 \times 10^{-12} \ m.
\end{aligned}
$$

The wavelength of the incident photon can be computed with the help of equation 2.3.10 as follows.

$$
\begin{aligned}
\lambda_i &= \lambda_s - \frac{h}{m_0 c}[1 - \cos\theta] \\
&= 8.247 \times 10^{-12} - \frac{6.62 \times 10^{-34}}{\left(9.11 \times 10^{-31}\right)\left(2.99 \times 10^8\right)}[1 - \cos 55^0] \\
&= 5.870 \times 10^{-12} \ m.
\end{aligned}
$$

This corresponds to an energy of

$$E_i = \frac{hc}{\lambda_i}$$
$$= \frac{\left(6.62 \times 10^{-34}\right)\left(2.99 \times 10^{8}\right)}{5.870 \times 10^{-12}}$$
$$= 3.372 \times 10^{-14} \ J$$
$$= 210.5 \ keV.$$

To compute the energy of the scattered electron we assume that the binding energy of the atom is negligibly small. Then the energy of the scattered electron would simply be equal to the difference in the energy of the incident photon and the scattered photon, that is

$$E_e \approx E_i - E_s$$
$$= 210.5 - 150 = 60.5 \ keV.$$

A.3 Thompson Scattering

Thompson scattering is an elastic scattering process between a free electron and a photon of low energy. By *low energy* we mean the energy at which the quantum effects are not significant. Therefore in order to derive kinematic quantities related to Thompson scattering, the concepts of classical electromagnetic theory suffice.

The differential and total cross sections for Thompson scattering are given by

$$\frac{d\sigma_{th}}{d\Omega} = r_e^2 \sin^2 \theta \tag{2.3.21}$$

$$\text{and} \quad \sigma_{th} = \frac{8\pi}{3} r_e^2 = 6.65 \times 10^{-29} \ m^2, \tag{2.3.22}$$

where θ is the photon scattering angle with respect to its original direction of motion and r_e is the classical electron radius.

A.4 Rayleigh Scattering

In this elastic scattering process there is very minimal coupling of photons to the internal structure of the target atom. The theory of Rayleigh scattering, first proposed by Lord Rayleigh in 1871, is applicable when the radius of the target is much smaller than the wavelength of the incident photon. A Rayleigh scattered photon has *almost* the same wavelength as the incident photon, which implies that the energy transfer during the process is extremely small. For most of the x-rays and low energy γ-rays, the Rayleigh process is the predominant mode of elastic scattering.

The cross section for this process is inversely proportional to the fourth power of the wavelength of the incident radiation and can be written as [13]

$$\sigma_{ry} = \frac{8\pi a^6}{3} \left[\frac{2\pi n_m}{\lambda_0}\right]^4 \left[\frac{m^2 - 1}{m^2 + 1}\right]^2. \tag{2.3.23}$$

Here λ_0 is the wavelength of the incident photon, a is the radius of the target particle, and $m = n_s/n_m$ is the ratio of the index of refraction of the target particle to that of the surrounding medium.

A.5 Pair Production

Pair production is the process that results in the conversion of a photon into an electron-positron pair. Since photon has no rest mass, while both the electron and the positron do, therefore we can say that this process converts energy into mass according to Einstein's mass energy relation $E = mc^2$. Earlier in the chapter we discussed the process of electron-positron annihilation, in which mass converts into energy. Hence we can think of pair production as the inverse process of the electron-positron annihilation. However there is one *operational* difference between the two processes: the pair production always takes part in a material while the electron positron annihilation does not have any such requirement. To be more specific, for the pair production to take place, there must be another particle in the vicinity of the photon to ensure momentum conservation. The process in the vicinity of a heavy nucleus can be represented as

$$\gamma + X \rightarrow e + e^+ + X^*, \tag{2.3.24}$$

where X and X^* represent the ground and excited states of a heavy nucleus.

Intuitively thinking, we can say that since energy is being converted into two particles that have discrete masses, therefore enough energy should be available for this process to take place. That is, the photon must have the energy equivalent of at least the rest masses of two electrons (electron and positron have equal masses).

$$
\begin{aligned}
E_{\gamma,thresh} &\geq 2m_e c^2 \\
\Rightarrow E_{\gamma,thresh} &\geq 1.022 \ MeV
\end{aligned}
\tag{2.3.25}
$$

Here m_e is the mass of an electron or a positron. Hence a photon carrying energy below $1.022 \ MeV$ can not convert into an electron-positron pair.

The actual threshold energy for the process in the vicinity of a heavy nucleus is given by

$$E_{\gamma,thresh} \geq 2m_e c^2 + \frac{2m_e^2 c^2}{m_{nuc}}, \tag{2.3.26}$$

Here m_{nuc} is the mass of the nucleus. We can also write the above equation in the form

$$E_{\gamma,thresh} \geq 2m_e c^2 \left[1 + \frac{m_e}{m_{nuc}}\right]. \tag{2.3.27}$$

Now since the mass of a nucleus is much greater than the mass of an electron ($m_{nuc} >> m_e$), we can neglect the second term in the parenthesis on the right hand side and get the threshold condition we derived earlier, such as

$$E_{\gamma,thresh} \geq 2m_e c^2.$$

The pair production can also occur in the vicinity of lighter particles, such as electrons. The process in the vicinity of an electron is generally referred to as *triplet production* and can be written as

$$\gamma + e \rightarrow e + e^+ + e. \tag{2.3.28}$$

The reader might be wondering why this process is termed as triplet production although only an electron and a positron are produced. It is true that only an electron-positron pair is generated, however since the original electron also scatters off therefore in the imaging detector three tracks become visible: two electrons and a positron. Hence the process is called triplet production not because three particles are produced but because of its unique signature of three particle tracks that it produces. To avoid confusion, some authors use the term *triplet pair production* for this process. The triplet production has a threshold of 2.044 MeV. Since there are only a few radioactive sources that emit γ-rays having energies higher than this threshold, therefore this process is not of much significance in usual radiation measurements.

The positrons produced during the process of pair production have very short mean lives. The reason is of course the occurrence of this reaction in materials, which always have electrons in abundance. Thus the positrons quickly combine with a nearby electrons to produce photons through the process of annihilation.

Example:
Determine the region of the electromagnetic spectrum for a photon to be able to generate an electron-positron pair through the process of pair production.

Solution:
To determine the threshold wavelength for the process, we use the pair production threshold relation 2.3.25 and $E_\gamma = hc/\lambda$.

$$E_{\gamma,thresh} \geq 2m_e c^2$$

$$\Rightarrow \frac{hc}{\lambda_{thresh}} \geq 2m_e c^2$$

$$\Rightarrow \lambda_{thresh} \leq \frac{h}{2m_e c}$$

Substituting $h = 6.625 \times 10^{-34}\ Js$, $m_e = 9.11 \times 10^{-31}\ kg$, and $c = 2.99 \times 10^8\ ms^{-1}$ in the above relation gives

$$\lambda_{thresh} \leq 1.2 \times 10^{-12}\ m = 1.2\ fm$$

Looking at the electromagnetic spectrum 1.6.1, we conclude that only γ-rays or high energy hard x-rays are capable of producing electron-positron pairs.

It is apparent that this process is not possible at all at low photon energies. If we compare this with the usual photon energy of a few tens of keV generally used in x-ray diagnostics, it is absolutely unnecessary to even consider this process as a possibility. However in some γ-decays, photons of energy greater than this threshold are emitted.

Now we turn our attention to the probability of occurrence of this process in any material. The pair production cross section for high energy photons $E \gg 20\ MeV$ has a roughly Z^2 dependence, Z being the atomic number of the material.

$$\sigma_{pair} \propto Z^2 \qquad (2.3.29)$$

This means that for heavy elements the pair production cross section is significantly higher than for lighter elements. The actual cross section can be written as

$$\sigma_{pair} = 4\alpha r_e^2 Z^2 \left[\frac{7}{9} \ln \left(\frac{183}{Z^{1/3}} \right) - \frac{1}{54} \right] \ cm^2/\text{atom}, \tag{2.3.30}$$

where α is the electron fine structure constant and r_e is the classical electron radius. Fig.2.3.8 shows behavior of σ_{pair} with respect to atomic number Z.

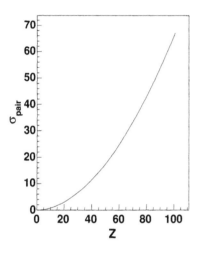

Figure 2.3.8: Dependence of pair production cross section of high energy photons ($E_\gamma \gg 20 \ MeV$) on atomic number Z as calculated from equation 2.3.30. σ_{pair} is in units of $barn/atom$.

It is also convenient to represent the pair production cross section in terms of radiation length X_0, which was described earlier in the chapter for passage of electrons through matter. For high energy photons, we mentioned that the process of pair production should be considered when defining the radiation length. Hence in similar terms we can define the radiation length for photons in a material as the depth at which a photon beam looses $1 - 1/e$ (or about 63%) of its energy through pair production. As before, e refers to the exponential factor. The formulas for radiation length for photons are the same as those for electrons since the processes of Bremsstrahlung and pair production are connected by underlying physics. In fact, the reader may readily recognize the similarity between the pair production cross section 2.3.30 and the radiation length formula 2.1.15. To write the relation 2.3.30 in terms of the radiation length, we can safely ignore the $1/54$ term. Hence σ_{pair} can also be written as

$$\sigma_{pair} = \frac{7}{9} \frac{A}{N_A} \frac{1}{X_0}, \tag{2.3.31}$$

where A is the atomic mass of the element and N_A is the Avogardo's number.

2.3.B Passage of Photons through Matter

Now that we have learned the basic processes that define the interactions a photon may undergo while passing through matter, we are ready to discuss the gross outcomes of such interaction. Luckily enough, although the underlying processes, as we saw earlier, are fairly complicated, but their overall effect on a beam of photons and the material through which it passes can be characterized by some simple relations. Before we go on to define these relations, let us have a qualitative look at the passage of a photon beam through matter.

A photon beam consists of a large number of photons moving in a straight line. The beam may or may not be monochromatic, such as, all the photons in the beam may or may not have the same energy. Of course the term *same* is somewhat loosely defined here since even a so called monochromatic photon beam has some variations in energy around its mean value. Depending on their energy, each photon in the beam may undergo one of the several interactions we discussed earlier. It is hard, even impossible, to say with absolute certainty that a photon with a certain energy will definitely interact with an atom and in a defined way. The good thing is that the gross interaction mechanisms of a large number of photons can be quite accurately predicted with the help of statistical quantities such as cross section.

In radiation measurements, a photon beam is relatively easier to handle as compared to a beam of massive particles. The reason is that the interaction of photons with matter is localized or discrete. That is, a photon that has not interacted with any other particle does not loose energy and remains a part of the beam. This means that the energy of all non-interacting photons in a beam remains constant as the beam passes through the material. However the intensity of the beam decreases as it traverses the material due to loss of interacting photons.

It has been found that at any point in a material, the decrease in intensity of a photon beam per unit length of the material depends on the intensity at that point, that is

$$\frac{dI}{dx} \propto -I$$

$$\Rightarrow \frac{dI}{dx} = -\mu_t I, \tag{2.3.32}$$

where dI is the change in intensity as the beam passes through the thickness dx (actually dI/dx represents the tangents of the I versus x curve at each point). μ_t is the total linear attenuation coefficient. It depends on the type of material and the photon energy. Integrating the above equation gives

$$I = I_0 e^{-\mu_t x}, \tag{2.3.33}$$

where I_0 is the intensity of the photon beam just before it enters the material and I is its intensity at a depth x.

It is interesting to note the similarity of the above equation with the radioactive decay law we studied in the previous chapter. In fact, in analogy with the half and mean lives of radioisotopes, mean free path λ_m and half-thickness $x_{1/2}$ have been defined for photon beam attenuation in materials.

$$\lambda_m = \frac{1}{\mu_t} \tag{2.3.34}$$

$$x_{1/2} = \frac{\ln(2)}{\mu_t} \tag{2.3.35}$$

The *term mean free path* was defined earlier in the chapter. Now, using equation 2.3.33 we can assign quantitative meaning to it. The reader can easily verify that the mean free path can be defined as the distance a photon beam will travel such that its intensity decreases by $1 - 1/e$, which corresponds to about 63% of the original intensity.

The total linear attenuation coefficient in the above relations determines how quickly or slowly a certain photon beam will attenuate while passing through a material. It is a function not only of the photon energy but also of the type and density of the material. Its dependence of these three parameters is actually a problem for tabulation and therefore instead of the total linear attenuation coefficient a related quantity, the mass attenuation coefficient, defined by

$$\mu_m = \frac{\mu_t}{\rho}, \tag{2.3.36}$$

is quoted. Here ρ is the density of the material. Since mass attenuation coefficient is independent of the physical state of the material, therefore one can easily deduce the linear attenuation coefficient by simply multiplying it by the density of the material. μ_t and μ_m are generally quoted in literature in units of cm^{-1} and cm^2/g respectively.

Just like the mean free path, we can also define the specific mean free path using the mass attenuation coefficient as

$$\lambda_p = \frac{1}{\mu_m}. \tag{2.3.37}$$

The reader is encouraged to compare this with the definition given in equation ??.

The total attenuation coefficient characterizes the probability of interaction of photons in a material. Since the cross section also characterizes the probability of interaction therefore these two quantities must be related. This is true since one can write the total linear attenuation coefficient μ_t in terms of total cross section σ_t as

$$\begin{aligned} \mu_t &= \sigma_t N \\ &= \sigma_t \frac{\rho N_A}{A}, \end{aligned} \tag{2.3.38}$$

where N represents the number of atoms per unit volume in the material having atomic number A and ρ is the weight density of the material. N_A is the familiar Avogadro's number. It is evident from the above relations that the units of μ_t will be cm^{-1} if σ_t is in $cm^2/atom$, N is in $atoms/cm^{-3}$, and ρ is in gcm^{-3}.

We can also write the total mass attenuation coefficient in terms of the total cross section as

$$\mu_m = \sigma_t \frac{N_A}{A}. \tag{2.3.39}$$

Example:
Determine the thickness of lead needed to block $17.4 \ keV$ x-rays by a factor of 10^4. The density and mass attenuation coefficient lead are $11.3 \ gcm^{-3}$ and $122.8 \ cm^2/g$.

Solution:
Blocking the beam by a factor of 10^4 implies that the intensity of the x-ray beam coming out of the lead should be 10^{-4} times the original beam intensity.

Therefore using $I = I_0 10^{-4}$ in equation 2.3.33 we can determine the required thickness d as follows.

$$
\begin{aligned}
I &= I_0 e^{-\mu_t d} \\
10^{-4} &= e^{-\mu_t d} \\
\Rightarrow \ln(10^{-4}) &= -\mu_t d \\
\Rightarrow \quad d &= -\frac{\ln(10^{-4})}{\mu_t}
\end{aligned}
$$

The total linear attenuation coefficient can be determined from the given mass attenuation coefficient using relation 2.3.36.

$$
\begin{aligned}
\mu_t &= \mu_m \rho \\
&= 122.8 \times 11.3 = 1387.64 \ cm^{-1}
\end{aligned}
$$

Hence the required thickness d of lead is

$$
\begin{aligned}
d &= -\frac{\ln(10^{-4})}{\mu_t} \\
&= -\frac{\ln(10^{-4})}{1387.64} \\
&= 0.0066 \ cm = 66 \ \mu m.
\end{aligned}
$$

This result clearly shows the effectiveness of lead in attenuating photons. Lead is commonly used for shielding purposes in radiation environments.

Let us now see how the attenuation coefficient for a particular material can be determined. Since there is direct relation between attenuation coefficient and the cross section therefore if we know the total cross section, we can determine the total attenuation coefficient. The total cross section σ_t can be determined by simply adding the cross sections for individual processes, that is

$$ \sigma_t = \sigma_{pe} + \sigma_c + \sigma_{ry} + \sigma_{pair} + \sigma_{trip} + ..., \tag{2.3.40} $$

where the subscripts pe, c, ry, $pair$, and $trip$ represent the photoelectric effect, Compton scattering, Rayleigh scattering, pair production, and triplet production respectively. The dots represent other processes such as Thompson scattering and photonuclear interactions. Then, according to equations 2.3.38 and 2.3.39, the total linear and the total mass attenuation coefficients can be determined from

$$ \mu_t = \frac{\rho N_A}{A} \left(\sigma_{pe} + \sigma_c + \sigma_{ry} + \sigma_{pair} + \sigma_{trip} + ... \right) \tag{2.3.41} $$

$$ \mu_m = \frac{N_A}{A} \left(\sigma_{pe} + \sigma_c + \sigma_{ry} + \sigma_{pair} + \sigma_{trip} + ... \right). \tag{2.3.42} $$

The tabulated values of photon attenuation coefficients found in literature are actually computed using this method. However, since such computations depend heavily on theoretically obtained formulas for cross sections, therefore sometimes it is desired that the coefficients are also experimentally measured. Discrepancies of around 5% between theoretical and experimental values are not uncommon.

Fig.2.3.9 shows the mass attenuation coefficients of potassium and lead. As one would expect, the attenuation coefficient decreases with increase in energy. However at certain energies there are abrupt jumps in attenuation coefficients. These jumps occur when the energy of the photon matches one of the binding energies of the atom. At that energy the absorption of the photons increases abruptly and the attenuation coefficient jumps to a higher value. These abrupt changes in attenuation coefficients has been exploited in radiation imaging techniques to enhance image contrast.

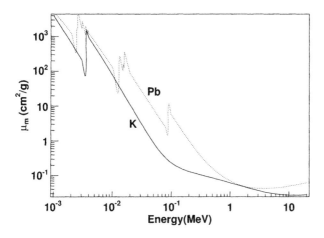

Figure 2.3.9: Dependence of mass attenuation coefficients of potassium and lead on energy.

B.1 Measuring Attenuation Coefficient

To see how one can determine the attenuation coefficient experimentally, we note that according to equation 2.3.33 the total linear attenuation coefficient can be written as

$$\mu_t = \frac{1}{d} \ln \left(\frac{I_0}{I} \right). \tag{2.3.43}$$

Hence, to determine the linear attenuation coefficient we just need to measure the incoming and the outgoing intensities of a photon beam that passes through a slab of thickness d of the material. Such an experimental setup is shown in Fig.2.3.10. The intensities I_0 and I going in and coming out of the material can be measured by two detectors installed before and after the material. Due to the energy dependence of the attenuation coefficient, either the source should be able to provide monochromatic beam or a monochromator setup should be used. The choice of the source depends on the application. For composite materials, such as body organs, radioactive sources are generally preferred due to the ease in their deployment and use. For more accurate measurements, high intensity x-ray beams, such as the ones provided by synchrotron sources are used. However, since the x-ray spectra of such sources is quite broad, therefore the setup must also include a monochromator to selectively use photon beam of the desired wavelength.

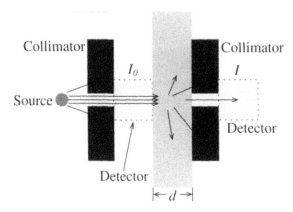

Figure 2.3.10: Sketch of a simple setup for determining the attenuation coefficient of a material. The incident and transmitted intensities (I_0 and I) are measured through two detectors before and after the slab of the material of which attenuation coefficient is to be determined.

B.2   Mixtures and Compounds

The attenuation coefficient of a compound or mixture at a certain energy can be obtained by simply taking the weighted mean of its individual components according to

$$\mu_t = \sum_i w_i \mu_t^i \quad \text{and} \tag{2.3.44}$$

$$\mu_m = \sum_i w_i \mu_m^i, \tag{2.3.45}$$

where w_i is the weight fraction of ith element in the mixture or compound having total linear and total mass attenuation coefficients of μ_t^i and μ_m^i respectively.

Fig.2.3.11 shows the dependence of mass attenuation coefficients of water and dry air on energy.

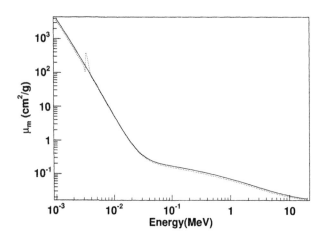

Figure 2.3.11: Dependence of mass attenuation coefficients of water (solid line) and dry air (dotted line) on energy.

Example:
Estimate the mean free path of 150 keV x-ray photons traveling in ordinary water kept under normal atmospheric conditions. The total mass absorption coefficients of hydrogen and oxygen are 0.2651 cm^2/g and 0.1361 cm^2/g respectively.

Solution:
The weight fractions of hydrogen and oxygen in water are

$$w_h \approx \frac{2}{2+16} = 0.1111$$

$$w_o \approx \frac{16}{2+16} = 0.8889$$

Hence according to equation 2.3.45, the mass attenuation coefficient of water is

$$\begin{aligned}
\mu_m^w &= w_h\mu_m^h + w_o\mu_m^o \\
&= 0.1111 \times 0.2651 + 0.8889 \times 0.1361 \\
&= 0.1504 \ cm^2 g^{-1}
\end{aligned}$$

Under normal conditions the density of water is 1 gcm^{-3}. Therefore the total linear absorption coefficient of water for 150 keV photons is

$$\mu_t^w = \rho_w\mu_m^w = 0.1504 \ cm^{-1}.$$

The mean free path is simply the inverse of total linear attenuation coefficient. Hence we get

$$\begin{aligned}
\lambda_m^w &= \frac{1}{\mu_t^w} \\
&= \frac{1}{0.1504} \approx 6.65 \ cm.
\end{aligned}$$

B.3 Stacked Materials

If the photon beam passes through a number of materials, the total attenuation of photons can still be determined by applying equation 2.3.33 for the consecutive materials. Let us suppose we have three materials stacked one after the other as shown in Fig.2.3.12. The intensity of photons at each junction of materials, as determined from 2.3.33 is given by

$$\begin{aligned}
I_1 &= I_0 e^{-\mu_t^1 d_1} \\
I_2 &= I_1 e^{-\mu_t^2 d_2} \\
&= I_0 e^{-\mu_t^1 d_1} e^{-\mu_t^2 d_2} = I_0 e^{-(\mu_t^1 d_1 + \mu_t^2 d_2)} \\
I_3 &= I_2 e^{-\mu_t^2 d_2} \\
&= I_0 e^{-(\mu_t^1 d_1 + \mu_t^2 d_2)} e^{-\mu_t^2 d_2} = I_0 e^{-(\mu_t^1 d_1 + \mu_t^2 d_2 + \mu_t^3 d_3)}.
\end{aligned}$$

Here the subscripts 1, 2, and 3 represent the three materials. Using this result we can write the equation for the intensity of a photon beam coming out of N materials in terms of total linear and total mass attenuation coefficients as

$$I_N = I_0 \exp\left(-\sum_{i=1}^{N} \mu_t^i d_i\right) \qquad (2.3.46)$$

$$= I_0 \exp\left(-\sum_{i=1}^{N} \mu_m^i \rho_i d_i\right), \qquad (2.3.47)$$

where ρ_i is the density of the ith material.

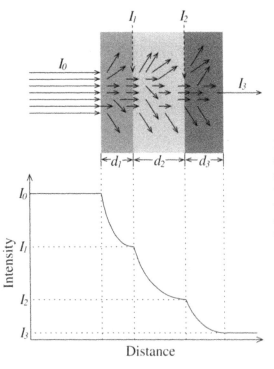

Figure 2.3.12: Depiction of passage of photons through three materials of different thicknessess and attenuation coefficients. The exponential variation of photon intensity in each material according to equation 2.3.33 is also shown.

Example:
Estimate the percentage of 100 keV photons absorbed in a cylindrical ionization detector. The photons enter the detector through a 100 μm thick aluminum window and are attenuated in the 6 cm thick bed of the filling gas (CO_2) kept at atmospheric temperature and pressure. The photons surviving the interactions leave the detector through another 100 μm thick aluminum window. The mass attenuation coefficients of 100 keV photons in aluminum, carbon, and oxygen are 0.1704 cm^2/g, 0.1514 cm^2/g, and 0.1551 cm^2/g respectively. The densities of aluminum and CO_2 are 2.699 g/cm^3 and 1.833×10^{-3} g/cm^3.

Solution:
According to equation 2.3.47, the intensity of the photons leaving the detector is given by

$$I_{out} = I_0 e^{-(\mu_t^a d_a + \mu_t^g d_g + \mu_t^a d_a)},$$

where the subscripts a and g stand for aluminum and fill gas respectively. The percentage of photons absorbed in the detector can then be written as

$$
\begin{aligned}
N_{abs} &= \frac{I_0 - I_{out}}{I_0} \times 100 \\
&= \left(1 - e^{-(\mu_t^a d_a + \mu_t^g d_g + \mu_t^a d_a)}\right) \times 100.
\end{aligned}
$$

The mass attenuation coefficient of CO_2 can be calculated by taking the weighted mean of the given mass attenuation coefficients of carbon and oxygen.

$$
\begin{aligned}
\mu_m^g &= w_c \mu_m^c + w_o \mu_m^o \\
&= \left(\frac{12}{44}\right) 0.1514 + \left(\frac{32}{44}\right) 0.1551 \\
&= 0.1541 \ cm^2 g^{-1}
\end{aligned}
$$

The required percentage of absorbed photons is then given by

$$
\begin{aligned}
N_{abs} &= \left(1 - e^{-(2 \times 0.1704 \times 2.699 \times 100 \times 10^{-4} + 0.1541 \times 1.833 \times 10^{-3} \times 6)}\right) \times 100. \\
&= 1.08\%.
\end{aligned}
$$

2.4 Interaction of Heavy Charged Particles with Matter

By heavy charged particles we mean particles with $A \geq 1$ such as protons and α-particles. The reason for differentiating between heavy and light charged particles is that the former, due to their heavier mass and higher charge, experience stronger Coulomb force of nuclei than the latter. Furthermore if they come close enough to nuclei, they can also be affected by the strong nuclear force, which on lighter particles, such as electrons, does not act. Consequently the heavy charged particles behave quite differently in matter than the light charged particles. At low to moderate energies the most important type of interaction for heavy charged particle s is the so called *Rutherford scattering*. We will look at it in some detail before we go on to the discussion of the overall effect of passage of such particles through matter.

2.4.A Rutherford Scattering

Rutherford scattering, first discovered by Lord Rutherford, refers to the elastic scattering of a heavy charged particle (such as an α-particle) from a nucleus. In his famous scattering experiment, Rutherford bombarded a thin gold foil by α-particles and studied how many of them deflected from their original direction of motion. He noticed that most of the α-particles passed through the foil undeflected while very few deflected at very large angles. This experiment proved that most of the space in the atom is empty and the positive charge is concentrated into a small space, which we now call nucleus. This pioneering work by Rutherford changed the way the atoms are visualized forever.

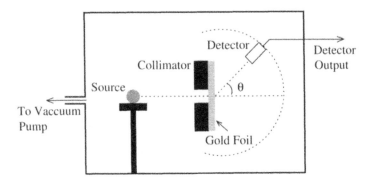

Figure 2.4.1: A simple setup to determine the angular distribution of Rutherford scattering.

Rutherford scattering is sometimes referred to as Coulomb scattering since the Coulomb force between the incident particle and the target nucleus is responsible for the interaction. A simplified diagram of a Rutherford scattering experiment is shown in Fig.2.4.1. A very thin gold foil is bombarded by α-particles from a radioactive source, such as Americium-241. The scattered particles are detected by a particle detector, which can be rotated at different angles. Such an experiment yields the following dependence of the number of scattered particles N_s to the scattering angle θ.

$$N_s \propto \frac{1}{\sin^4(\theta/2)} \tag{2.4.1}$$

This relationship is plotted in Fig.2.4.2 for $\theta = 0$ to $\theta = 180^0$. It is apparent that most of the incident particles pass through the target undeflected, hinting to the vast emptiness of atomic space.

The differential cross section of this process is given by the so called *Rutherford formula*

$$\frac{d\sigma}{d\Omega} = \frac{Z_i^2 Z_t^2 r_e^2}{4} \left[\frac{m_e c}{\beta p}\right]^2 \frac{1}{\sin^4(\theta/2)}. \tag{2.4.2}$$

Here Z_i and Z_t are the atomic numbers of the incident and target particles respectively and θ is the scattering angle. The impact parameter b and the scattering angle θ are as shown in Fig.2.4.3. These and other factors in the above definition of the differential cross section can be calculated from the following relations.

$$p = \frac{2Z_1 Z_2 e^2}{\beta cb\theta}$$

$$b = r\sin\phi$$

$$\beta = \left(1 - \frac{v^2}{c^2}\right)^{1/2}$$

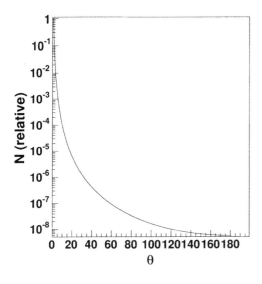

Figure 2.4.2: Angular dependence of the relative number of particles deflected through Rutherford scattering.

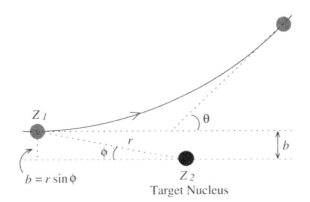

Figure 2.4.3: Rutherford scattering of a particle with charge eZ_1 by nucleus having charge eZ_2. The cross section of the process depends strongly the impact parameter b.

The above formula has been developed through quantum mechanical considerations. However if the incident particle can be considered non-relativistic (such as $v \ll c$) then classical mechanics can also be employed to derive a simpler from of the Rutherford formula given by

$$\frac{d\sigma}{d\Omega} = \left[\frac{Z_i Z_t e^2}{16\pi\epsilon_0 T}\right]^2 \frac{1}{\sin^4(\theta/2)}. \tag{2.4.3}$$

where T is the kinetic energy of the incident particle. The reader should readily realize that the angular dependence of differential cross section is the same in both cases and is also consistent with the experimental observation depicted in Fig. 2.4.2. The classical limit formula is the one that was originally derived by Rutherford.

The total cross section for Rutherford scattering can be determined by integrating the differential cross section over the whole solid angle Ω.

$$\sigma_{total} = \int_\Omega \frac{d\sigma}{d\Omega} d\Omega \qquad (2.4.4)$$

Generally one is interested in predicting the number of particles scattered in a certain solid angle (see example below). This can be done by simply multiplying the differential cross section by the element of the solid angle, such as

$$\triangle\sigma(\theta) = \frac{d\sigma}{d\Omega} \triangle\Omega. \qquad (2.4.5)$$

Example:
In his original scattering experiment, Rutherford observed the number of α-particles scattered off gold nuclei by counting the number of flashes on a scintillation screen in a dark room. Luckily, with the advent of electronic detectors, such painstaking work is now not needed. Let us replicate Rutherford's experiment but with an electronic detector having a surface area of 1 cm^2. Suppose the source emits 3 MeV α-particles with an intensity of 10^6 s^{-1} and the target is a 1 μm thick aluminum foil (the setup would be identical to the one shown in Fig.2.4.1). Assume that the detector can count individual α particles with an efficiency of 60%. Compute the number of counts recorded by the detector at 10^0 and 30^0 relative to the initial direction of motion of the α-particles. At both angles the distance of the detector from the interaction point remains 15 cm. Take the atomic density of gold to be 6×10^{28} m^{-3}.

Solution:
Let us begin by simplifying equation 2.4.3 for scattering of 3 MeV α-particles by gold nuclei.

$$\frac{d\sigma}{d\Omega} = \left[\frac{Z_i Z_t e^2}{16\pi\epsilon_0 T}\right]^2 \frac{1}{\sin^4(\theta/2)}$$

$$= \left[\frac{(2)(79)\left(1.602 \times 10^{-19}\right)^2}{16\pi(8.85 \times 10^{-12})(3 \times 10^6 \times 1.602 \times 10^{-19})}\right]^2 \frac{1}{\sin^4(\theta/2)}$$

$$= 3.597 \times 10^{-28} \frac{1}{\sin^4(\theta/2)} \ m^2/ster$$

Here *ster* stands for steradian, which is the unit of solid angle. Using this relation we can compute the differential scattering cross sections for the particles scattered at 10^0 and 30^0.

$$\frac{d\sigma}{d\Omega}\bigg|_{\theta=10^0} = 3.597 \times 10^{-28} \frac{1}{\sin^4(5^0)}$$

$$= 6.234 \times 10^{-24} \ m^2/ster$$

$$\frac{d\sigma}{d\Omega}\bigg|_{\theta=30^0} = 3.597 \times 10^{-28} \frac{1}{\sin^4(15^0)}$$

$$= 8.016 \times 10^{-26} \ m^2/ster$$

Now that we have the differential scattering cross sections at the two angles, we can multiply them by the solid angles subtended by the detector, which is given by

$$\Delta\Omega = \frac{A}{r^2}$$

$$= \frac{1}{15^2} = 4.44 \times 10^{-3} \ ster$$

to obtain the total cross sections for scattering at 10^0 and 30^0 as follows.

$$\Delta\sigma(\theta) = \frac{d\sigma}{d\Omega}\Delta\Omega$$

$$\Rightarrow \Delta\sigma(\theta = 10^0) = \left(6.234 \times 10^{-24}\right)\left(4.44 \times 10^{-3}\right) m^2$$

$$= 2.771 \times 10^{-26} \ m^2$$

$$\Delta\sigma(\theta = 30^0) = \left(8.016 \times 10^{-26}\right)\left(4.44 \times 10^{-3}\right) m^2$$

$$= 3.563 \times 10^{-28} \ m^2.$$

If we know the atomic density D and thickness t of the target material, we can estimate the number of particles scattered at an angle θ by

$$N_\theta = N_0 \, \Delta\sigma(\theta)Dt,$$

where N_0 is the number of incident particles. Therefore, for the two given angles we get

$$N_{10^0} = \left(10^6\right)\left(2.771 \times 10^{-26}\right)\left(6 \times 10^{28}\right)\left(10^{-6}\right)$$

$$\approx 1663 \ s^{-1}$$

$$N_{10^0} = \left(10^6\right)\left(3.563 \times 10^{-28}\right)\left(6 \times 10^{28}\right)\left(10^{-6}\right)$$

$$\approx 21 \ s^{-1}.$$

Since the detector is only 60% efficient, therefore the count rate recorded at the two angles will be

$$C_{10^0} \approx (1663)(0.6) = 998 \ s^{-1}$$

$$C_{30^0} \approx (21)(0.6) = 13 \ s^{-1}.$$

2.4.B   Passage of Charged Particles through Matter

There is a fundamental difference between the interaction of charged particle beams
and photon beams with matter. We saw earlier that whenever a photon interacts
with material, it is either absorbed or scattered and consequently removed from the
beam. On the other hand, when a charged particle interacts with matter it does
not get removed from the beam except for the rare cases where it gets scattered to
a very large angle.

There are a number of electronic and nuclear mechanisms through which charged
particles can interact with particles in the medium. However the net result of all
these interactions is a reduction in the energy of the particles as they pass through
the medium. Although the underlying interaction mechanisms are fairly complicated
but fortunately the rate of this energy loss can be fairly accurately predicted by a
number of semi-empirical relations developed so far.

The rate at which a charged particle loses energy as it passes through a material
depends on the nature of both the incident and the target particles. This quantity
is generally referred to in the literature as the *stopping power* of the material. It
should be noted that stopping power does not represent the energy loss per unit
time rather the energy that a charged particle loses per unit length of the material
it traverses. Generally speaking, any charged particle can have either electronic,
nuclear, or gravitational interaction with the particles of the material through which
it passes. However the gravitational interaction is too low to be of any significance
and is generally ignored. The total stopping power is then just the sum of the
stopping powers due to electronic and nuclear interactions.

$$S_{total} = -\frac{dE}{dx}$$

$$-\frac{dE}{dx} = S_{electronic} + S_{nuclear} \tag{2.4.6}$$

Here the negative sign signifies the fact that the particles lose energy as they pass
through the material. For most practical purposes the nuclear component of the
stopping power can also be ignored as it is generally only a fraction of the total
stopping power. For particles such as electrons, this statement is always valid since
they are not affected at all by the strong nuclear force. For heavy positive charges,
such as α-particles, this holds if the particle energy is not high enough for it to penetrate so deep into the atom that the short range nuclear forces of nuclear particles
become appreciable. Hence the stopping power can be written as a function of the
electronic component only.

$$-\frac{dE}{dx} \approx S_{electronic} \tag{2.4.7}$$

The first successful attempt to derive a relation for the energy loss experienced by
an ion moving in the electromagnetic field of an electron was made by Neil Bohr. He
argued that such an expression can be obtained by simply considering the impulses
delivered by the ion to the electron as it passes through its electromagnetic field.
This consideration led him to the following relation.

$$\left[-\frac{dE}{dx}\right]_{Bohr} = \frac{4\pi q^2 e^4 N_e}{m_e v^2} \ln\left[\frac{\gamma^2 m_e v^3 f(Z)}{q e^2}\right] \tag{2.4.8}$$

Here e is the unit electron charge,

  m_e is the mass of electron,

  N_e is the electron number density,

  q is the charge of the ion,

  v is the velocity of the ion,

  $f(Z)$ is a function of the atomic number Z of the material, and

  γ is the relativistic factor given by $\left(1 - v^2/c^2\right)^{-1/2}$.

Example:
Derive equation 2.4.8.

Solution:
Let us suppose that a heavy charged particle (such as an ion) is moving in x-direction in the electric field of an electron and define an impact parameter b as the perpendicular distance between the two particles (see figure 2.4.4). The rationale behind this definition is the fact that the impulse experience by the electron as the ion approaches it will be canceled by the impulse delivered by the receding ion and consequently the only contribution left will be perpendicular to the motion of the ion.

Supposing that the ion passes by the electron before it can move any significant distance, we can calculate the momentum transferred to the electron through the impulse delivered by the ion as

$$\Delta p = \int_{-\infty}^{\infty} \left(-e\bar{E}_\perp\right),$$

where $\bar{E}_\perp$ is the perpendicular component of the electric field intensity given relativistically by

$$\bar{E}_\perp = \frac{q e \gamma b}{(b^2 + \gamma^2 v^2 t^2)^{3/2}}.$$

Here $\gamma = 1/sqrt\left(1 - v^2/c^2\right)$ is the relativistic factor for the ion moving with velocity v.
Integration of the above equation yields

$$\Delta p = \frac{2 q e^2}{b v}.$$

In terms of energy transferred to the electron $\Delta E = \Delta p^2/2m_e$, this can be written as

$$\Delta E = \frac{2 q^2 e^4}{m_e v^2 b^2} \tag{2.4.9}$$

Let us now define a cylinder of radius b around the path x of motion of ion with a volume element of $2\pi b db dx$. If N_e is the electron number density then the total number of electrons in this volume element will be equal to $2\pi N_e b db dx$. Every electron in this volume element will experience the same impulse and therefore the total energy transferred to the electron when the ion moves a distance dx will be

$$-\frac{dE}{dx} = 2\pi N_e q \int \Delta E b db.$$

$$= \frac{4\pi N_e q^2 e^4}{m_e v^2} \ln\left(\frac{b_{max}}{b_{min}}\right).$$

Here we have evaluated the integral from minimum to maximum impact parameter (b_{min} to b_{max}) since integrating from 0 to ∞ would yield a divergent solution.

In order to determine a reasonable value of b_{min} we note that the minimum impact parameter will correspond to collisions in which the kinetic energy transferred to electron is maximum. The maximum energy that the ion can transfer to the electron turns out to be $E_{max} = 2m_e\gamma^2 v^2$. The reader is encouraged to perform this derivation (Hint: use law of conservation of linear momentum). Substitution of this in equation 2.4.9 gives

$$\Delta E|_{b_{min}} = \frac{2q^2 e^4}{m_e v^2 b_{min}^2} = 2\gamma.$$

Hence we find

$$b_{min} = \frac{qe^2}{\gamma m_e v^2}$$

Now we will use some intuitive thinking to come up with a good estimate of the maximum impact parameter. In order for the ion to be able to deliver a sharp impulse to the electron, it should move faster than the electron in the atomic orbit, such as

$$\frac{b_{max}}{\gamma v} \leq \frac{R_e}{v_e}.$$

Here R_e is the atomic radius and v_e is the velocity of the electron. Since this ratio is a function of the atomic number Z of the material, we can write

$$\frac{b_{max}}{\gamma v} \leq f(Z)$$

Substitution of b_{min} and b_{max} in 2.4.10 yields the required equation 2.4.8.

$$\left[-\frac{dE}{dx}\right]_{Bohr} = \frac{4\pi q^2 e^4 N_e}{m_e v^2} \ln\left[\frac{\gamma^2 m_e v^3 f(Z)}{qe^2}\right]$$

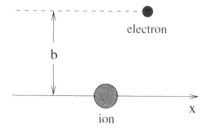

electron

b

X

ion

Figure 2.4.4: Definition of impact parameter for a charged particle (such as an ion) moving in the electric field of an electron.

Later on Bethe and Bloch derived another expression for the stopping power using quantum mechanics.

$$\left[-\frac{dE}{dx}\right]_{Bethe-Bloch} = \frac{4\pi N_A r_e^2 m_e c^2 \rho Z q^2}{A\beta^2}\left[\ln\left(\frac{W_{max}}{I}\right) - \beta^2\right] \qquad (2.4.10)$$

Here $N_A = 6.022 \times 10^{23}\ mole^{-1}$ is the Avogadro's number;

$r_e = 2.818 \times 10^{-15}\ m$ is the classical radius of the electron;

$m_e = 9.109 \times 10^{-31}\ kg$ is the rest mass of the electron;

q is the electrical charge of the ion in units of unit electrical charge;

ρ is the density of the medium;

A is the mass number of the medium;

I is the ionization potential of the medium;

β is a correction factor. It is generally calculated from the

relation: $\beta = \left[1 - \frac{E_0}{E_0+E/A_i}\right]^{1/2}$ where $E_0 = 931.5\ MeV$

is the rest mass energy per nucleon and E is the energy

of the incident particle having mass number A_i; and

W_{max} is the maximum energy transferred in the encounter.

It can be calculated from: $W_{max} = 2m_e c^2 \beta^2/(1-\beta^2)$.

The factor $4\pi N_A r_e^2 m_e c^2$ is constant and therefore can be permanently substituted in the above formula, which then becomes

$$\left[-\frac{dE}{dx}\right]_{Bethe-Bloch} = \frac{4.8938 \times 10^{-18}\rho Z q^2}{A\beta^2}\left[\ln\left(\frac{W_{max}}{I}\right) - \beta^2\right]\ Jm^{-1}. \qquad (2.4.11)$$

Although the units of Jm^{-1} are in standard MKS system, however in literature the stopping power is generally mentioned in units of $MeV cm^{-1}$. Therefore it is more convenient to write the above equation in the form

$$\left[-\frac{dE}{dx}\right]_{Bethe-Bloch} = \frac{0.30548\rho Zq^2}{A\beta^2}\left[\ln\left(\frac{W_{max}}{I}\right) - \beta^2\right] MeV\,cm^{-1}. \qquad (2.4.12)$$

One of the difficult parameters to evaluate in the above expression is the ionization potential I of the medium. For this a number of empirical formulas have been proposed, such as

$$\begin{aligned} I &= 12Z + 7, Z < 13 \\ I &= 9.76Z + 5.58Z^{-0.19}, Z \geq 13 \end{aligned} \qquad (2.4.13)$$

This equation has been corrected for two factors that become significant at very high and moderately low energies. One is the shielding of distant electrons because of the polarization of electrons by the electric field of the moving ion. This effect depends of the electron density and becomes more and more important as the energy of incident particle increases. The second correction term applies at lower energies and depends on the orbital velocities of the electrons. Both of these correction terms are subtractive and generally represented by the symbols δ and C respectively.

The modern form of the Bethe-Bloch formula for stopping power after applying the above corrections is given by

$$\left[-\frac{dE}{dx}\right]_{Bethe-Bloch} = \frac{4\pi N_A r_e^2 m_e c^2 \rho Zq^2}{A\beta^2}\left[\ln\left(\frac{W_{max}}{I}\right) - \beta^2 - \frac{\delta}{2} - \frac{C}{Z}\right] \qquad (2.4.14)$$

Bethe-Bloch formula can also be written in terms of mass stopping power, which is simply the stopping power as defined by 2.4.10 or 2.4.14 divided by the density of the medium.

$$\left[-\frac{1}{\rho}\frac{dE}{dx}\right]_{Bethe-Bloch} = \frac{4\pi N_A r_e^2 m_e c^2 Zq^2}{A\beta^2}\left[\ln\left(\frac{W_{max}}{I}\right) - \beta^2 - \frac{\delta}{2} - \frac{C}{Z}\right] \qquad (2.4.15)$$

It should be noted that the above expression for mass stopping power deals with a medium with unique atomic number and hence is valid for a pure element only. In case of a compound or a mixture of more than one element, we can use the so called *Bragg-Kleeman rule* to calculate the total mass stopping power.

$$\left[\frac{1}{\rho}\frac{dE}{dx}\right]_{total} = \sum_{i=1}^{n}\left[\frac{w_i}{\rho_i}\left(\frac{dE}{dx}\right)_i\right] \qquad (2.4.16)$$

Here w_i and ρ_i are the fraction by mass of element i in the mixture and its density respectively.

The Bragg-Kleeman rule can also be applied to compute stopping power of a compound material using

$$\left[\frac{dE}{dx}\right]_{total} = \sum_{i=1}^{n} w_i\left(\frac{dE}{dx}\right)_i. \qquad (2.4.17)$$

If we substitute the expressions for stopping power and mass stopping power in the above relations, we can find expressions for average ionization potential $< I >$,

$< Z/A >$, and $< \delta >$. However these expressions do not give reliable results due mainly to the increased bonding strength of electrons in compounds as compared to that in elements. The higher bonding energy of electrons in a compound means that a simple weighted mean of individual ionization potentials would be an underestimate. For more reliable estimates, one can use tables given in (52) and (51), which have been generated after including several correction.

Example:
Calculate the stopping power of 5 MeV α-particles in air.

Solution:
Let us assume that air is composed of 80% nitrogen and 20% oxygen. According to Bragg-Kleeman rule (equation 2.4.16), the total mass stopping power of air would be the weighted sum of the mass stopping powers of nitrogen and oxygen.
For simplicity, let us use the uncorrected Bethe-Bloch formula 2.4.10 (this assumption is valid since at this energy the correction factors δ and C are insignificantly small). For α-particles we have $Z = 2$ and $q = 2$. The other factors in equation 2.4.10 can be calculated as follows.

$$\beta = \left[1 - \frac{E_0}{E_0 + E/A_\alpha}\right]^{1/2}$$

$$= \left[1 - \frac{931.5}{931.5 + 5/2}\right]^{1/2}$$

$$= 0.05174$$

$$W_{max} = 2m_e c^2 \frac{\beta^2}{1 - \beta^2}$$

$$= 2(0.511)\frac{(0.05174)^2}{1 - (0.05174)^2}$$

$$= 2.743 \times 10^{-3} \ MeV$$

The mass attenuation coefficient, according to equation 2.4.12, can then be written as

$$\left[-\frac{1}{\rho}\frac{dE}{dx}\right]_{Bethe-Bloch} = \frac{0.30548 Z q^2}{A\beta^2}\left[\ln\left(\frac{W_{max}}{I}\right) - \beta^2\right].$$

$$= \frac{(0.30548)(2)(2)^2}{A(0.05174)^2}\left[\ln\left(\frac{2.743 \times 10^{-3}}{I}\right) - (0.05174)^2\right]$$

$$= \frac{912.893}{A}\left[\ln\left(\frac{2.743 \times 10^{-3}}{I}\right) - 2.677 \times 10^{-3}\right].$$

To calculate ionization potentials of nitrogen and oxygen we use equation 2.4.13.

$$I = 12Z + 7$$

$$\Rightarrow I_{nitrogen} = 12(7) + 7 = 91 \ eV$$

$$I_{oxygen} = 12(8) + 7 = 103 \ eV$$

The mass attenuation coefficients for nitrogen and oxygen are given by

$$\left[-\frac{1}{\rho}\frac{dE}{dx}\right]_{Bethe-Bloch}^{nitrogen} = \frac{912.893}{14}\left[\ln\left(\frac{2.743 \times 10^{-3}}{91 \times 10^{-6}}\right) - 2.677 \times 10^{-3}\right]$$

$$= 221.91 \; MeV\,cm^2 g^{-1}$$

$$\left[-\frac{1}{\rho}\frac{dE}{dx}\right]_{Bethe-Bloch}^{oxygen} = \frac{912.893}{16}\left[\ln\left(\frac{2.743 \times 10^{-3}}{103 \times 10^{-6}}\right) - 2.677 \times 10^{-3}\right]$$

$$= 187.11 \; MeV\,cm^2 g^{-1}$$

Now we can employ the Bragg-Kleeman rule 2.4.16 to compute the total stopping power of 5 MeV α-particles in air.

$$\left[-\frac{1}{\rho}\frac{dE}{dx}\right]_{Bethe-Bloch}^{air} = (0.8)\left[-\frac{1}{\rho}\frac{dE}{dx}\right]_{Bethe-Bloch}^{nitrogen} + (0.2)\left[-\frac{1}{\rho}\frac{dE}{dx}\right]_{Bethe-Bloch}^{oxygen}$$

$$= (0.8)(221.91) + (0.2)(187.11)$$

$$= 214.95 \; MeV\,cm^2 g^{-1}$$

2.4.C Bragg Curve

The Bethe-Bloch formulae for stopping power of charged particles we discussed in the preceding section have an implicit dependence on the energy of the particle through factors like β and W_{max}. As a heavy charged particle moves through matter it looses energy and consequently its sopping power changes. Since stopping power is a measures the effectiveness of a particle to cause ionization, therefore as the particle moves through matter its ionization capability changes. To understand this dependence, let us plot the Beth-Bloch formula 2.4.12 for α-particles with respect to their *residual* energy. By residual energy we mean the instantaneous energy of the particle retained by it as it travels through the material. For simplicity we will lump together all the terms that are constant for a particular material. Equation 2.4.12 can then be written as

$$\left[-\frac{dE}{dx}\right]_{Bethe-Bloch} = \frac{K}{\beta^2}\left[\ln\left(\frac{W_{max}}{10^{-4}}\right) - \beta^2\right] \; MeV\,cm^{-1}, \qquad (2.4.18)$$

where $K = 0.30548\rho Z q^2/A$ is a constant for a given material. Since we are only concerned with the shape of the curve and not its numerical value, therefore we have used $I = 10^{-4} \; MeV$ in the above expression. This value is typical of low Z materials. A plot of the above equation is shown in Fig.2.4.5. As one would expected, the stopping power increases with the residual energy of the particle. Hence as the particle looses energy it causes more and more ionization in its path until it reaches the highest point known as *Bragg peak*. After that point the particles have lost almost all of their energy and get quickly neutralized by attracting electrons from their surroundings.

The plot shown in Fig.2.4.5 is generally known as *Bragg curve*. A point to note here is that the range of a particle traveling through a material depends on its instantaneous energy. Hence one could in principle plot the stopping power with

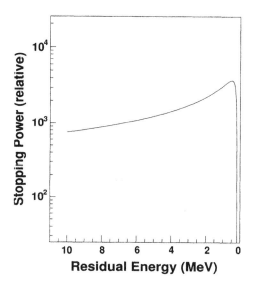

Figure 2.4.5:   Plot of equation 2.4.18 for α-particles having initial energy of 10 MeV passing through a material having an ionization potential of 100 eV. This variation of stopping power with respect to residual energy of the particles is generally known as Bragg curve.

respect to range as well and get the same Bragg curve. In fact, it is much easier to draw the Bragg curve in this way because most of the empirical relations between range and energy of particles can be used to derive very simple relations between stopping power and range. We will see how this is done shortly, but before that let us have a look at a couple of very important phenomena related to energy loss and range.

2.4.D   Energy Straggling

The stopping power equations presented above do not contain information about the statistical variations in the energy lost by the incident particles. In fact, due to this statistical effect, a mono-energetic beam of incident particles gets a finite width in its energy distribution as it travels through the medium. The effect is known as *energy straggling* and can be represented by a Gaussian distribution for thick absorbers.

$$N(E)dE = \frac{N}{\alpha\pi^{1/2}}e^{-(E-\bar{E})^2/\alpha^2} \qquad (2.4.19)$$

Here α is known as the *straggling parameter* It can be computed from

$$\alpha^2 = 4\pi q^2 e^4 N x_0 \left[1 + \frac{kI}{mv^2}\ln\left(\frac{2mv^2}{I}\right)\right],$$

where $k \approx 4/3$ is a constant, I is the ionization potential of the medium, q is the electrical charge of the ion (in units of unit electrical charge) having mass m and velocity v, e is the unit electronic charge, and x_0 is the thickness of the medium.

At lower absorber thicknesses, the energy straggling distribution becomes skewed and develops a tail at higher energies. For very thin absorbers it is best represented by a Landau distribution (31; 37). We will look at this distribution in the section on

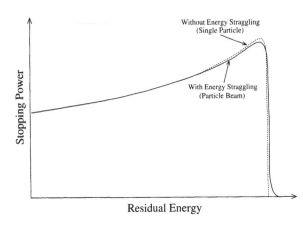

Figure 2.4.6: Bragg curves for a single particle (dotted line) and a particle beam (solid line). A beam of particles, which is originally mono-energetic, assumes a distribution as it travels through matter due to energy straggling.

electrons later in the chapter. The equations we will present there will be applicable to heavy charged particles passing through thin absorbers as well.

It is interesting to note that the Bethe-Block formulas 2.4.14 and 2.4.15 can not be used to describe the behavior of a single particle. Because of energy straggling, which is a stochastic process, these formulas actually represent the average stopping power.

Since this process is probabilistic in nature and the distribution is skewed specially for thin absorbers, therefore it is natural to look for the most probable energy loss as well. This parameter has higher relevance to detector calibrations than the mean energy loss. The reason is that the tail of the distribution generally gets buried in the background and it becomes difficult to define it.

A consequence of the energy straggling is shown in Fig.2.4.6. The shape of the Bragg curve slightly changes specially at the end of the particle track. Two differences are clearly visible: the Bragg peak is more profoundly rounded and the curve has a small tail at the end.

2.4.E   Range and Range Straggling

It is very tempting to try to compute the average distance a particle beam will travel (also called range) in a medium by integrating the stopping power over the full energy spectrum of the incident particles, such as,

$$R(T) = \int_0^T \left[-\frac{dE}{dx} \right]^{-1} dE. \tag{2.4.20}$$

However due to multiple Coulomb scattering, the trajectory of a charged particle in a medium is not a straight line. Rather the particle moves in small straight line segments. This implies that the range of a beam of particles would show statistical fluctuation around a mean value. In analogy with the energy straggling phenomenon, this fluctuation is termed as *range straggling*. It should be noted that there is a fundamental difference between loss of energy and range and their corresponding statistical fluctuations: the energy loss and energy straggling are differential quantities while range and range straggling are integral quantities. It is apparent that theoretical computation of range is a difficult task. A number of experimentalists therefore turned to empirical means of measuring this quantity and modeling the behavior on the basis of their results.

Bragg and Kleeman gave a formula to compute the range of a particle in a medium if its range is known in another medium.

$$\frac{R_1}{R_2} = \frac{\rho_2}{\rho_1}\left[\frac{A_1}{A_2}\right]^{1/2} \tag{2.4.21}$$

Here ρ and A represent density and atomic mass of the materials. If we have a compound material an effective atomic mass given by

$$\frac{1}{\sqrt{A_{eff}}} = \sum_i \frac{w_i}{\sqrt{A_i}}, \tag{2.4.22}$$

is used instead. Here w_i is the weight fraction of ith element having atomic mass A_i.

E.1 Range of α-Particles

Several empirical and semi-empirical formulae have been proposed to compute range of α-particles in air. For example, (56),

$$R_\alpha^{air}[mm] = \begin{cases} e^{1.61\sqrt{E_\alpha}} & \text{for } E_\alpha < 4\ MeV \\ \\ (0.05E_\alpha + 2.85)E_\alpha^{3/2} & \text{for } 4\ MeV \le E_\alpha \le 15\ MeV \end{cases} \tag{2.4.23}$$

and

$$R_\alpha^{air}[cm] = \begin{cases} 0.56E_\alpha & \text{for } E_\alpha < 4\ MeV \\ \\ 1.24E_\alpha - 2.62 & \text{for } 4\ MeV \le E_\alpha < 8\ MeV \end{cases} \tag{2.4.24}$$

Both of these equations yield almost same results as can be seen from Fig.2.4.7 which has been plotted for the α-particles having energy up to 8 MeV. Hence at least in this energy range one could use any one of these equations to compute the range in air. To compute the range in some other material, equation 2.4.21 can be used. For example, at normal pressure and temperature the range of α-particles in any material x can be determined from

$$R_\alpha^x = 3.37 \times 10^{-4} R_\alpha^{air} \frac{\sqrt{A_x}}{\rho_x}. \tag{2.4.25}$$

Here we have used the effective atomic number of air $A_{air} = 14.6$ and its density $\rho_{air} = 1.29 \times 10^{-3}\ g/cm^3$. Although it is very tempting to use Bragg-Kleeman rule, however it should be noted that the values thus obtained are only approximations and care must be taken while using them to draw conclusions.

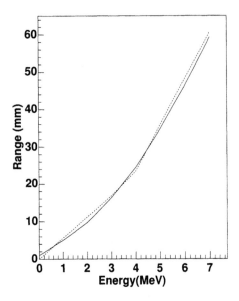

Figure 2.4.7: Range of α-particles in air as computed from equations 2.4.23 (solid line) and 2.4.24 (dashed line).

Example:
Compute the range of 4 MeV α-particles in air and tissue.

Solution:
Using equation 2.4.23 we get

$$\begin{aligned}
R_\alpha^{air} &= (0.05E_\alpha + 2.85)E_\alpha^{3/2} \\
&= 24.4 \; mm = 2.44 \; cm.
\end{aligned}$$

To calculate the range in tissue we can make use of the simplified Bragg-Kleeman identity 2.4.25

$$R_{tissue} = 3.37 \times 10^{-4} R_\alpha^{air} \frac{\sqrt{A_{tissue}}}{\rho_{tissue}}.$$

Substituting $\rho_{tissue} = 1 \; g/cm^3$ and $A_{tissue} \approx 9$ in the above equation, we get

$$\begin{aligned}
R_\alpha^{tissue} &\approx 3.37 \times 10^{-4} 2.44 \frac{\sqrt{9}}{1} \\
&= 10.1 \times 10^{-4} \; cm = 10.1 \; \mu m.
\end{aligned}$$

As an exercise, let us also compute the range in air using equation 2.4.24.

$$\begin{aligned}
R_\alpha^{air} &= 1.24E_\alpha - 2.62 \\
&= 2.34 \; cm
\end{aligned}$$

This shows that the relative difference between the ranges calculated from the relations 2.4.23 and 2.4.24 is only 4% for 4 MeV α-particles.

E.2 Range of Protons

There have been several theoretical and experimental studies of variation of range of protons with energy in several materials (see, for example (36; 9; 58; 45)). These studies have lead to the development of empirical relations specific for the material under investigation and within the energy range used in the experiment. Fortunately the number of such studies is so high that, together with some theoretical computations, several large databases of the proton range values are now available.

Let us have a look at the proton range relations for air and aluminum. In air the range of protons having energy E_p can be computed from (61)

$$R_p^{air}[m] = \left[\frac{E_p}{9.3}\right]^{1.8} \text{ for } E_p < 200 \ MeV, \tag{2.4.26}$$

while for aluminum, one can use the relation (8)

$$R_p^{Al}[mgcm^{-2}] = \begin{cases} 3.837E_p^{1.5874} & \text{for } 1.13 \ MeV < E_p \leq 2.677 \ MeV \\ \frac{2.837E_p^2}{0.68+\log E_p} & \text{for } 2.677 \ MeV \leq E_p \leq 18 \ MeV. \end{cases} \tag{2.4.27}$$

These two relations have been plotted in Fig.2.4.8. The reader should note that equation 2.4.27 gives range in mg/cm^2 and therefore the value must be divided by the density of aluminum to obtain the range in dimensions of length.

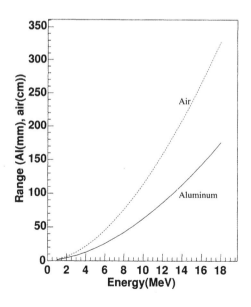

Figure 2.4.8: Range of protons in air (dashed line) and in aluminum (solid line) as computed from equations 2.4.26 and 2.4.27. For aluminum the equation has been divided by its density 2.7 gcm^{-3} to yield range in dimension of length.

2.5   Interaction of Electrons with Matter

The way an electron beam would behave when passing through matter depends, to
a large extent, on its energy. At low to moderate energies, the primary modes of
interaction are

▶ ionization,

▶ Moeller scattering,

▶ Bhabha scattering, and

▶ electron-positron annihilation.

At higher energies the emission of Bremsstrahlung dominates as shown in Fig.2.5.1.

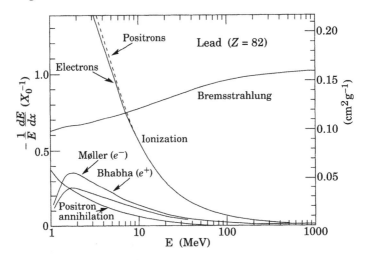

Figure 2.5.1: Fractional energy loss of electrons and positrons
per radiation length as a function of energy (19)

2.5.A   Interaction Modes

A.1   Ionization

If an incident electron departs enough energy to the atom, it may eject one of its
loosely bound electrons, resulting in the ionization of the atom. The energy of the
ejected electron depends on the incident electron energy as well as its binding energy.
If the energy carried away by the ejected electron is enough energy it can produce
secondary ionization in the same manner as the primary ionization. The process can
continue until the energy of the ejected electron is less than the ionization potential
of the atom. This process is graphically depicted in Fig.2.5.2. It should, however,
be noted that not all electrons that have energy higher than the ionization potential
of the atom produce subsequent ionization. The probability with which an electron
can cause ionization depends on its cross section, which to a large extent depends

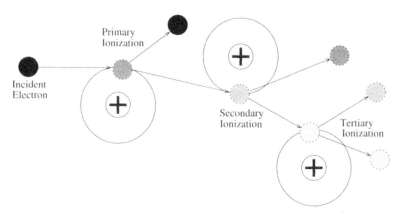

Figure 2.5.2: Depiction of electron-induced ionization processes. At each stage of ionization, if the ejected electron has an energy greater than the binding energy of the atom, it can cause another ionization.

on its energy and the type of target atom. It has been seen that, at each step of this ionization, only about one third of the electrons cause subsequent ionizations.

The ionization of atoms or molecules is a highly researched area due to its utility in material and physics research. Electrons have be ability to penetrate deep into the materials and can therefore be used to extract information about the structure of the material.

As can be seen from Fig.2.5.1, the ionization with electrons dominates at low to moderate energies. *Electron impact ionization* is a term that is extensively used in literature to characterize the process of ionization with electrons at relatively high energies. This useful process is routinely employed in spectroscopy of materials in gaseous state.

Symbolically, for an atom or a molecule X_q, with total positive charge q, the electron ionization process can be written as

$$e + X_q \rightarrow X_{q+1} + 2e. \tag{2.5.1}$$

A.2 Moeller Scattering

This refers to the elastic scattering of an electron from another electron (or a positron from another positron). The interaction can be symbolically described by

$$e + e \rightarrow e + e. \tag{2.5.2}$$

In quantum electrodynamics, Moeller scattering is said to occur due to exchange of virtual photons between the electrons. In classical electrodynamic terms one can simply call it a consequence of Coulomb repulsion between the two electrons.

A.3 Bhabha Scattering

It is the scattering of an electron from a positron. The reaction can be written as

$$e + e^+ \rightarrow e + e^+. \tag{2.5.3}$$

In terms of quantum electrodynamics, as with Moeller scattering, Bhabha scattering is also considered to be due to exchange of virtual photons between the electron and the positron. Classically, it can be thought to occur because of the Coulomb attraction between the two particles.

A.4   Electron-Positron Annihilation

The process of electron-positron annihilation has already been explained earlier in the chapter. In this process an electron and a positron annihilate each other and produce *at least* two photons. It is a perfect example of the notion that mass can be converted into energy.

We saw earlier that during this process, to conserve energy and momentum, at least two photons, each having $511\,keV$, are produced. However more than two electrons can be and in fact are produced. The cross section for this process at low electron energies in a material is not very high and decreases to almost zero at high energies. This low cross section is due to the very low abundance of positrons in materials. Fig.2.5.1 shows this behavior for electrons of energy from $1\,MeV$ to $100\,MeV$. Due to its lower cross section, this process does not contribute significantly to the total energy loss specially at moderate to high energies.

A point that is worth mentioning here is that the electron-positron annihilation process can also produce particles other than photons provided their center-of-mass energy before collision is high enough. For example, at very high energies (several GeV), the annihilation process produces quarks, which form mesons. Since discussion of such interactions is out of the scope of this book, the interested reader is referred to standard texts of particle physics and high energy physics.

A.5   Bremsstrahlung

We saw earlier in the chapter that the process of Bremsstrahlung refers to the emission of radiation when a charged particle accelerates in a material. For electrons we came up with a critical or cut-off wavelength below which no Bremsstrahlung photons can be emitted. This wavelength is given by

$$\lambda_{min} = \frac{hc}{eV},$$

where V is the potential experienced by the electron. For high Z materials, the process of Bremsstrahlung dominates other types of interactions above about 10 MeV.

As can be seen from Fig.2.5.1, the Bremsstrahlung process is the dominant mode through which the moderate to high energy electrons loose energy in high Z materials.

Example:
In an x-ray machine, electrons are accelerated through a potential of $40\,kV$. Compute the cut-off wavelength and energy of the emitted photons.

Solution

The cut-off wavelength is given by

$$\lambda_{min} = \frac{hc}{eV}$$

$$= \frac{(6.634 \times 10^{-34})(2.99 \times 10^8)}{(1.602 \times 10^{-19})(40 \times 10^3)}$$

$$= 3.095 \times 10^{-11} \ m = 30.95 \ fm.$$

The energy corresponding to this wavelength is given by

$$E_{max} = \frac{h\nu}{\lambda_{min}}$$

$$= \frac{(6.634 \times 10^{-34})(2.99 \times 10^8)}{3.095 \times 10^{-11}}$$

$$= 6.4 \times 10^{-15} \ J$$

$$= 40 \ keV. \tag{2.5.4}$$

A.6 Cherenkov Radiation

The concept of Cherenkov radiation has already been discussed earlier in the chapter. We saw that an electron, being a very light particle, can emit Cherenkov radiation when accelerated to high energies in a medium. Its energy should be so high that its velocity becomes higher than the velocity of light in that medium.

The velocity of light in a medium of refractive index n is given by c/n, where c is the velocity of light in vacuum. For Cherenkov radiation to be emitted, the velocity of the charged particle traversing the medium must be greater than this velocity, such as

$$v_{th} \geq \frac{c}{n}. \tag{2.5.5}$$

Similarly the threshold energy is given by

$$E_{th} = \gamma_{th} m_0 c^2, \tag{2.5.6}$$

where m_0 is the electron rest mass, c is the velocity of light in vacuum, and γ_{th} is the relativistic factor defined as

$$\gamma_{th} = \left[1 - \frac{v_{th}^2}{c^2}\right]^{-1/2}$$

$$= \left[1 - \beta_{th}^2\right]^{-1/2}. \tag{2.5.7}$$
$$\tag{2.5.8}$$

Here the term $\beta_{th} = v_{th}/c$ is an oft quoted condition for Cherenkov emission. From the above equations we can also deduce that

$$\gamma_{th} = \frac{n}{\sqrt{n^2 - 1}} \tag{2.5.9}$$

An important factor is the direction of emission of Cherenkov light. As the light is emitted in the form of a cone, we can define an angle of emission as the direction of the cone. This angle Θ_c of the Cherenkov for a particle moving in a medium of refractive index n is given by

$$\Theta_c = \arccos\left(\frac{1}{\beta n}\right). \tag{2.5.10}$$

This equation can be used to define the maximum angle Θ_c^{max} that one should expect to see in a medium. The maximum will occur when $\beta = 1$, that is, when the particle's velocity is approximately equal to the velocity of light *in vacuum*. Hence

$$\Theta_c^{max} = \arccos\left(\frac{1}{n}\right). \tag{2.5.11}$$

Example:
Compute the maximum angle of the Cherenkov cone one can observe in water having refractive index of 1.4.
Solution:
We use equation 2.5.11 to compute the desired angle.

$$\begin{aligned}
\Theta_c^{max} &= \arccos\left(\frac{1}{n}\right) \\
&= \arccos\left(\frac{1}{1.4}\right) \\
&= 0.775 \text{ radians} \\
&= \frac{0.775 \times 180}{\pi} = 44.4^0
\end{aligned}$$

The number of Cherenkov photons dN emitted by a particle having charge ze (e is the unit electrical charge) per unit length dx of the particle flight in a medium having refractive index n is given by

$$\frac{dN}{dx} = 2\pi\alpha z^2 \int \frac{1}{\lambda^2}\left[1 - \frac{1}{n^2\beta^2}\right] d\lambda. \tag{2.5.12}$$

Here α is the fine structure constant and λ is the wavelength of light emitted. It is interesting to note that the threshold of Cherenkov depends not only on the velocity of the particle but also on the refractive index of light in the medium. Since the refractive index is actually dependent on the wavelength, therefore by just looking at the refractive index we can deduce whether the photons of a particular wavelength can be emitted or not. Based on this reasoning it can be shown that most of the Cherenkov radiation emitted in water actually lies in the visible region of the spectrum. Therefore the above equation can be safely integrated in the visible region of the spectrum to get a good approximation of the number of photons emitted per unit path length. This yields

$$\frac{dN}{dx} = 490z^2 \sin^2\Theta \quad cm^{-1} \tag{2.5.13}$$

For an electron with $z = 1$, the above equation becomes

$$\frac{dN}{dx} = 490 \sin^2 \Theta \; cm^{-1}. \tag{2.5.14}$$

Example:
An electron moving in water emits Cherenkov radiation in a cone making an angle of 40^0 with electron's direction of motion. Compute the number of photons emitted per centimeter by the electron.

Solution:
Using equation 2.5.14 with $\theta = 40^0$, we get

$$\begin{aligned} \frac{dN}{dx} &= 490 \sin^2 \Theta \\ &= 490 \sin^2(40^0) \approx 272 \; cm^{-1}. \end{aligned}$$

Discriminating such a small number of photons from background is a very difficult task. Generally, the conic signature of the Cherenkov radiation is used to discriminate the Cherenkov photons from the background radiation.

2.5.B Passage of Electrons through Matter

As compared to heavy charged particles, electrons behave quite differently when passing through matter. The main reason for this difference is, of course, the very small mass of electrons as compared to heavy charged particles. Due to their low mass, electrons travel so fast that their velocity may become very close to the velocity of light. Since in such a situation the relativistic effects must be taken into account to deduce meaningful results, therefore the computations become more complicated than for the heavy charged particles.

We saw earlier that in certain situations an electron may even attain a velocity greater than the velocity of light in the same material. If this happens, a special kind of radiation, called Cherenkov radiation, with a specific cone signature is emitted. As electrons pass through matter they rapidly lose energy and hence decelerate. This deceleration gives rise to another type of radiation called Bremsstrahlung.

Whenever an electron beam passes through a material, the individual electrons in the beam can interact with the target atoms or molecules in a number of ways, most of which we have already discussed in the preceding sections. Fig.2.5.1 shows the contributions of various types of interactions on the stopping power of lead for electrons of various energies. It is interesting to note that except for the ionization process, the Bremsstrahlung remains the dominant mode of interaction from low to high energies. Therefore the radiative component of the stopping power can not be neglected in case of electrons.

At low to moderate energies the collisional energy loss of electrons is quite significant and up to a certain energy is higher than the radiative energy loss. Hence the stopping power of a material for electrons consists of two components: collisional and radiative.

$$S_{electron} = S_{collisional} + S_{radiative}$$

The collisional component not only includes the inelastic impact ionization process but also the other scattering mechanisms we discussed earlier, such as Moeller and Bhabha scattering. The analytic forms of the collisional and radiative components of the total stopping power for electrons are given by

$$\left[-\frac{dE}{dx}\right]_{collisional} = \frac{2\pi Z e^4 \rho}{m_e v^2}\left[\ln\left(\frac{m_e v^2 E}{2I^2(1-\beta^2)}\right) - \right.$$

$$\ln 2\left(2\sqrt{1-\beta^2} - 1 + \beta^2\right) + \qquad (2.5.15)$$

$$\left. \left(1-\beta^2\right) + \frac{1}{8}\left(1-\sqrt{1-\beta^2}\right)^2\right],$$

and

$$\left[-\frac{dE}{dx}\right]_{radiative} = \frac{Z(Z+1)e^4\rho E}{137 m_e^2 c^4}\left[4\ln\left(\frac{2E}{m_e c^2}\right) - \frac{4}{3}\right]. \qquad (2.5.16)$$

From these two equations we can deduce that the rate of energy loss of an electron through the collisional and radiative processes can be approximately expressed as

$$S_{collisional} \propto \ln(E) \text{ and}$$
$$S_{radiative} \propto E.$$

This implies that the losses due to radiative effects such as Bremsstrahlung increase more rapidly than the losses due to collisional effects such as ionization. This can also be seen from Fig.2.5.3, where the two effects have been plotted for electrons traveling through a slab of copper. The energy at which these two types of losses become equal is called the *critical energy*. A number of attempts have been made to develop a simple relation for this critical energy, the most notable of which is the one that uses the approximate ratio of equations 2.5.15 and 2.5.16 given by

$$\frac{S_{radiative}}{S_{collisional}} \approx \frac{(Z+1.2)E}{800}, \qquad (2.5.17)$$

where E is in MeV. From this equation we can find the critical energy E_c by equating the two types of loss rates. Hence

$$E_c \approx \frac{800}{Z+1.2} \ MeV. \qquad (2.5.18)$$

This definition was originally given by Berger and Seltzer (6). Although this is a widely used and quoted definition of the critical energy but there are other definitions as well that work equally well for most materials. For example, Rossi (48) has defined critical energy as the point where the ionization loss per unit radiation length becomes equal to the electron energy. This definition is also graphically depicted in Fig.2.5.3.

Equation 2.5.18 does not give accurate results for all types of matter because it does not take into account the state of the matter, that is, it gives same results whether the matter is in solid, liquid, or gaseous state. A much better approach is to parameterize the curve

$$E_c \approx \frac{a}{Z+b} \ MeV, \qquad (2.5.19)$$

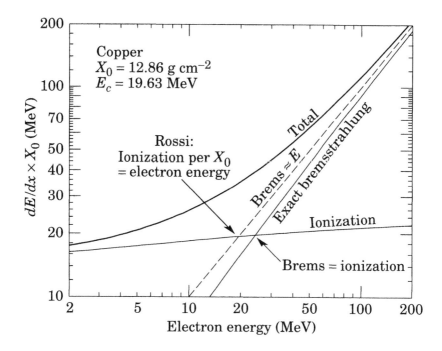

Figure 2.5.3: Energy loss per unit track length of electrons in copper as a function of energy. The plot also shows two definitions of the critical energy (19).

using the values obtained by the energy loss equations to find the constants a and b. Such computations have been shown to give the following values for solids and gases (19):

$$\text{Solids:} \quad a = 610, \quad b = 1.24$$
$$\text{Gases:} \quad a = 710, \quad b = 0.92$$

Equation 2.5.17 clearly shows that, for materials with low atomic numbers and low incident electron energies, the collisional component of the stopping power dominates. Hence most of the electrons in a beam of low energy electrons passing through a gas will loose their energy through collisions with the gas molecules. But for the same electrons passing through a high Z material (such as Lead), the radiative losses will be significant.

A very important point to note here is that these expressions are valid only for electrons. For positrons, the cross sections for the interactions are quite different, even though the underlying processes may be similar. The difference in cross sections for electrons and positrons are mainly due to the fact that the positrons passing through a material see an abundance of electrons with which they could combine and annihilate. On the other hand, the electrons only seldom encounter a positron along their paths.

Example:
Compute the critical energy for electrons in aluminum.

Solution:
To compute the critical energy, we use equation 2.5.19 with $Z = 13$, $a = 610$, and $b = 1.24$.

$$E_c \approx \frac{a}{Z+b}$$
$$= \frac{610}{13 + 1.24}$$
$$= 42.84 \; MeV$$

2.5.C   Energy Straggling

In case of heavy charged particles we saw that, for at least thick absorbers, the fluctuations in the energy loss can be described by a symmetric distribution. This is not the case for electrons, which, owing to their small mass, suffer more collisions as compared to heavy charged particles of same energy. As a consequence of such motion the energy distribution of these electrons gets skewed with a long tail towards high energy. Since for heavy charged particles passing through thin absorbers one can employ the Landau distribution to describe the energy straggling, therefore it is natural to use the same methodology for the electrons as well. This strategy works reasonably well for most materials and electron energies. The Landau distribution can be expressed as

$$f(x, \Delta) = \frac{\Phi(\lambda)}{\xi}, \tag{2.5.20}$$

where

$$\Phi(\lambda) = \frac{1}{\pi} \int_0^1 e^{-u \ln(u) - u\lambda} sin(\pi u) du \text{ and}$$

$$\lambda = \frac{1}{\xi} [\Delta - \xi (\ln(\xi) - ln(E) + 1 - \gamma)].$$

Here $\gamma = 0.577$ is known as Euler's constant. The term $\ln(E)$ in the above equation can be computed from

$$\ln(E) = \ln \left[\frac{I^2(1 - \beta^2)}{2mc^2\beta^2} \right] + \beta^2, \tag{2.5.21}$$

where I is the logarithmic mean excitation energy and $\beta = v/c$, v being particle's velocity. ξ is called the scale of the Landu distribution and is given by

$$\xi = \frac{2\pi N_A e^4 Z x}{mv^2 A} \text{ or} \tag{2.5.22}$$

$$\xi = \frac{0.1536 Z x}{\beta^2 A} \; keV, \tag{2.5.23}$$

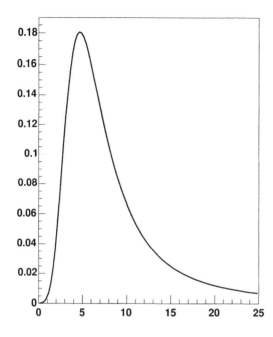

Figure 2.5.4: Typical shape of the Landau distribution. The most probable value for this distribution, corresponding to its peak, is 5. Axes' units are arbitrary.

where N_A is the Avogadro's constant, Z is the atomic number of the target atom, A is its atomic mass, and x is the thickness of the material given in the second expression in mg/cm^2.

Since Landau distribution is an asymmetric distribution therefore its most probable value is different from its average value. The most probably value is simply the value at which the distribution has a maximum. The average value, on the other hand, is much more complicated to determine. The reason is that in order to determine the average value one generally has to cut the tail of the distribution at some point, which may depend on some scheme or the personal bias. Since the average value is rarely used in practice and does not even have much physical significance, therefore we will concentrate here the most probable energy loss instead. The most probable energy loss, as obtained from the Landau distribution, can be computed from

$$\triangle_{mp} = \xi \left[\ln\left(\frac{\xi}{E} \right) + 0.423 \right], \tag{2.5.24}$$

which can also be written as

$$\triangle_{mp} = \xi \left[\ln\left(\frac{2mc^2\beta^2\xi}{I^2(1-\beta^2)} \right) - \beta^2 + 0.423 \right]. \tag{2.5.25}$$

Landau distribution is a skewed distribution with a long tail at the high energy side (see Fig.2.5.4). The degree of its skewness increases with the decrease in the thickness of the material. For very thin absorbers, the distribution no longer depicts reality even for electrons and should be replaced by some other distribution. As a matter of fact, in real practice, the use of Landau distribution is somewhat limited due to one or more of the following reasons.

▶ The Landau distribution is valid only if the maximum energy loss in a single collision is much larger than the typical energy loss. In the actual formalism of the distribution, it is assumed that the maximum energy transfer can be infinite.

▶ The Landau theory assumes that the typical energy loss is much larger than the binding energy of the innermost electron such that the electrons may essentially be considered free. This condition is not really fulfilled by most gaseous detectors, in which the average energy loss can be a few keV, which can be lower than the binding energy of the most tightly bound electrons of the gas atoms.

▶ It assumes that the velocity of the particle is constant, meaning that the decrease in particle's velocity is insignificantly small.

▶ If the Landau distribution is integrated, the result is an infinite value.

▶ Landau distribution is difficult to handle numerically since the energy loss computed from it depends on the step size used in the computations.

Another distribution describing the energy straggling phenomenon is the so called *Vavilov distribution*. With increase in the thickness of the material, the tail of the Landau distribution becomes smaller and the distribution approaches the Vavilov distribution. Because of this some authors prefer to call the Vavilov distribution a more general form of the Landau distribution. However, since one can not approximate the Landau distribution from the Vavilov distribution, we do not encourage the reader to make this assumption.

A major problem with Vavilov distribution is its difficult analytic form requiring huge numerical computation. Its use is therefore only warranted in situations where highly accurate results are needed and speed in computations is not an issue.

For general radiation measurements, we encourage the reader to concentrate on the Landau distribution as it gives acceptable results without computational difficulties.

2.5.D   Range of Electrons

As opposed to heavy charged particles, the range of electrons is very difficult to treat mathematically. The primary reason for the difficulty lies in the higher large-angle scattering probability of electrons due to their extremely low mass as compared to the heavy charged particles. However it has been found that the bulk properties of an electron beam can be characterized by relatively simple relations. The attenuation of an electron beam, for example has been seen to follow an approximately exponential curve given by

$$N = N_0 e^{-\mu x}. \tag{2.5.26}$$

Here N represents the number of electrons transmitted through a thickness x of the material. μ is the absorption coefficient of the material for the electrons and is also a function of the electron energy. For electrons having a continuous energy spectrum, it depends on their endpoint energy. Following the analogy of attenuation of photons in matter, here also we can define a path length or absorber thickness as

$$t = \frac{1}{\mu},\qquad(2.5.27)$$

the interpretation of which can be understood by substituting $x = t$ in the exponential relation above.

$$
\begin{aligned}
N &= N_0 e^{-1}\\
\Rightarrow \text{No. of electrons absorbed} &= N_0 - N = N_0(1 - e^{-1})\\
&= 0.63 N_0
\end{aligned}
$$

This implies that t is the thickness of the material which absorbs about 63% of the electrons of a certain energy. In Fig.2.5.5 we have plotted N versus x on both linear and semilogarithmic scales. Since the behavior of electrons is not perfectly logarithmic, therefore if one performs an experiment to measure the variation of electron intensity with respect to the thickness of the material, a perfect straight line on the semilogarithmic scale is not obtained. Such a curve is known as *absorption curve*. Absorption curves for specific materials are routinely obtained to determine the range of electrons in the material.

An experimental setup to obtain the absorption curve for a material simply consists of slabs of the material of varying thickness, a known source of electrons, and a radiation detector. The electrons from the source are allowed to pass through the material. The number of electrons transmitted through the material are then counted through the detector. This process is repeated for various thicknesses of the material. A plot of thickness versus log of number of counts gives the required absorption curve. The obtained curve looks similar to the second curve shown in Fig.2.5.5. However, since the simple exponential attenuation of electrons does not strictly hold for most of the materials, therefore the variation is not as linear as shown in the figure. The curve is actually seen to bend down at larger thicknesses. The determination of range from a perfectly linear variation is very simple as it can be done by extrapolating the line to the background level (see Fig.2.5.6). However if the curve shows a curvature with increasing thickness then the end point is generally taken as the range. Sometimes the experiment is not performed till the end, such as, till the detector stops seeing any electrons. In such a case the curve can be extrapolated as shown in Fig.2.5.6. It is interesting to note that the range as obtained from a *real curve* actually corresponds to the endpoint energy of electrons from a radioactive source.

Although the best way to determine the range of electrons in a material is to perform an experiment as described above, however it may not be always practical to do so. Fortunately enough, the range of electrons in any material can be fairly accurately determined from the following simple formulae given by Katz and Penfold (29).

$$
R_e^{sp}[kgm^{-2}] = \begin{cases} 4.12 E^{1.265-0.0954 \ln(E)} & \text{for } 10\ keV < E \leq 2.5\ MeV \\[2mm] 5.30 E - 1.06 & \text{for } E > 2.5\ MeV \end{cases}\qquad(2.5.28)
$$

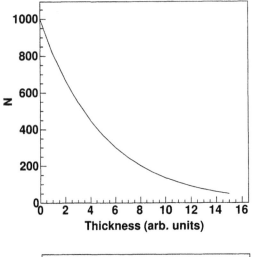

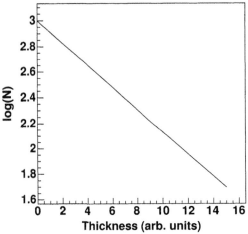

Figure 2.5.5:   Plot of number of electrons transmitted per unit thickness of an arbitrary material as computed from equation 2.5.26. For this plot we have arbitrarily taken the initial number of electrons to be 10^3 and $\mu = 1/5$. The plot is shown on both linear (upper) and semilogarithmic (lower) scales.

Here the energy of the electrons E should be taken in units of MeV. The superscript sp of R_e^{sp} specifies that the expression represents the *specific range* in units of kg/m^2. The range (or more specifically the specific range) as computed from these formulae are independent of the material. This, up to a good approximation, is a correct statement as can be seen from Fig.2.5.7. Here we have plotted the range using these functions as well as the ESTAR provided values for aluminum and silicon. ESTAR is an online database of parameters related to interaction of electrons in elements, compounds and mixtures (23). It is based on the values tabulated in the ICRU Report 37 (22). It is apparent from the figure that the variations are quite small and unless the application requires high accuracy, the above formulae can be safely used. The case of compounds and mixtures is a little bit complicated, though. For such materials, these expressions do not give reasonably accurate results specially at

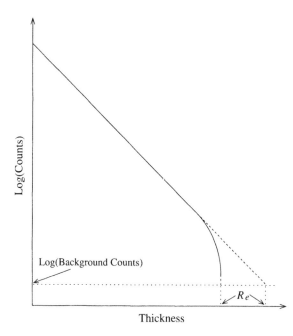

Figure 2.5.6: Typical absorption curve for electrons passing through a material. The curve deviates from the ideal straight line with increasing thickness. Since the extrapolation of the straight line overestimates the range, therefore normally the end of the curve is extrapolated to the background level to determine the true range. This corresponds to the electrons of endpoint energy.

higher energies. This can be seen from Fig.2.5.7, where we have plotted the ESTAR data for water and air along with the values computed from the above expressions.

It should be noted that the spectrum of electrons emitted from a radioactive source is continuous with an endpoint energy. In such a case generally one is interested in determining the range of the most energetic electrons. This can simply be done by substituting the endpoint energy of the spectrum into the above expression. Such computations are important, for example to determine the shielding necessary for a particular source in the laboratory.

The specific range as computed from the expressions 2.5.28 or obtained from some other data source (such as ESTAR) can be divided by the density of the material to determine the range in units of distance, such as

$$R_e = \frac{R_e^{sp}}{\rho}, \tag{2.5.29}$$

where ρ denotes the density of the material.

Example:
Compute the thickness of aluminum shielding required to completely stop electrons having an endpoint energy of 5 MeV.

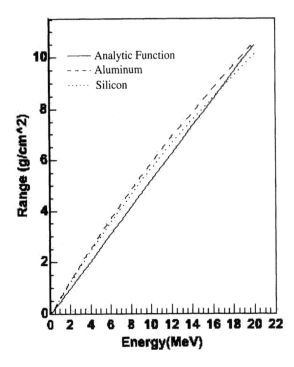

Figure 2.5.7: Variation of electron range in g/cm^2 with energy in MeV computed from the equations 2.5.28. Also shown are the ESTAR values for aluminum and silicon.

Solution:
We use equation 2.5.28 to compute the specific range of 5 MeV electrons.

$$
\begin{aligned}
R_e^{sp} &= 5.30E - 1.06 \\
&= 5.30 \times 5 - 1.06 \\
&= 25.44 \; kgm^{-2} \\
&= 2.544 \; gcm^{-2}
\end{aligned}
$$

The range of electrons in aluminum is then simply determined by dividing this number with the density of aluminum.

$$
\begin{aligned}
R_e &= \frac{R_e^{sp}}{\rho} \\
&= \frac{2.544}{2.699} = 0.94 \; cm
\end{aligned}
$$

Hence it will take less than 1 cm thick slab of aluminum to completely stop β-particles having an endpoint energy of 5 MeV.

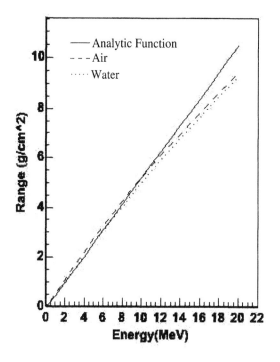

Figure 2.5.8: Variation of electron range in g/cm^2 with energy in MeV computed from the equations 2.5.28. Also shown are the ESTAR values for air and water.

2.6 Interaction of Neutral Particles with Matter

Neutrons and neutrinos are two chargeless particles that are found in abundance in nature. As neutrons have been found to be extremely useful, we will discuss here how they interact with matter in detail. The interaction mechanisms of neutrinos will be discussed briefly afterward.

2.6.A Neutrons

Since neutrons do not carry any electrical charge, they are not affected by the electric field of the atoms. This enables them to move swiftly through large open atomic spaces without interacting with atoms. However if they pass near the nuclei they encounter strong nuclear force. In such a situation, depending on their energy, they can interact with the nuclei in the following ways,

▶ Elastic scattering

▶ Inelastic scattering

▶ Spallation reaction

▶ Transmutation

▶ Radiative capture

A.1　Elastic Scattering

Elastic scattering is the principal mode of interaction of neutrons with atomic nuclei. In this process the target nucleus remains in the same state after interaction. The reaction is written as $A(n,n)A$ or

$$n + X_p^{n+p} \rightarrow n + X_p^{n+p}. \tag{2.6.1}$$

The elastic scattering process of neutrons with nuclei can occur in two different modes: potential elastic and resonance elastic. Potential elastic scattering refers to the process in which the neutron is acted on by the short range nuclear forces of the nucleus and as a result scatters off of it without touching the particles inside. In the resonance mode, a neutron with the right amount of energy is absorbed by the nucleus with the subsequent emission of another neutron such that the kinetic energy is conserved.

The cross section for elastic cross section for uranium-238 as computed from different models is shown in Fig.2.6.1 (21).

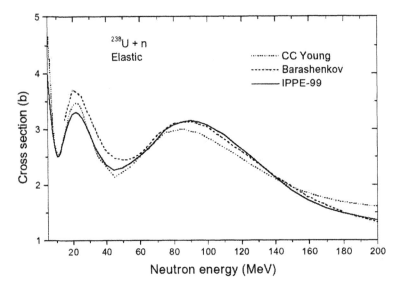

Figure 2.6.1:　Neutron elastic cross section for uranium-238 (21).

A.2　Inelastic Scattering

Unlike elastic scattering, the inelastic scattering leaves the target nucleus in an excited state. The reaction is written as $A(n,n)A^*$ or

$$n + X_p^{n+p} \rightarrow n + \left[X_p^{n+p}\right]^*. \tag{2.6.2}$$

In such a process the incoming neutron is absorbed by the nucleus forming a compound nucleus. The compound nucleus is unstable and quickly emits a neutron of lower kinetic energy. Since it still has some excess energy it goes through one or more γ-decays to return to the ground state.

A.3 Transmutation

It is a reaction in which an element changes into another one. Neutrons of all energies are capable of producing transmutations. For example, when a Boron-10 nucleus captures a slow neutron it transforms into Lithium-7 and emits an α-particle,

$$n + B_5^{10} \rightarrow Li_3^7 + \alpha.$$

A.4 Radiative Capture

Radiative capture is a very common reaction involving neutrons. In such a reaction, a nucleus absorbs the neutron and goes into an excited state. To return to stable state, the nucleus emits γ-rays. In this case no transmutation occurs, however the isotopic form of the element changes due to increase in the number of neutrons. The reaction is represented by $A(n, \gamma)A + 1$ or

$$n + X_p^{n+p} \rightarrow X_p^{n+p+1} + \gamma. \tag{2.6.3}$$

Radiative capture is generally used to produce radioisotopes, such as Cobalt-60

$$n + Co^{59} \rightarrow Co^{60} + \gamma.$$

A.5 Spallation

Spallation refers to the fragmentation of a nucleus into several parts when a high energy neutron collides with it. This process is important only with neutrons having energy greater than about 100 MeV.

A.6 Fission

This is perhaps one of the most important reactions a neutron can initiate. In this process a slow neutron is captured by a heavy nucleus, such as uranium-235, taking it into an excited state. The nucleus then splits up in fragments after a brief delay. Several neutrons and γ-ray photons are also emitted during this process. Fission of uranium-235 can be written as

$$n + U_{92}^{235} \rightarrow I_{39}^{139} + Y_{53}^{95} + 2n + \gamma. \tag{2.6.4}$$

It should be pointed out that although iodine and yttrium are the most probable elements produced during this fission process, however in a sample of large number of fission processes, fragments of varying atomic numbers can be found.

The fission process is the source of thermal energy produced in nuclear reactors. A nuclear reactor core is a controlled environment where neutrons are allowed to produce the so called *chain fission reaction*. In this process the neutrons emitted from the fissioning nucleus produce more fissions, which produce even more neutrons. As a result the fission initiated by a few neutrons spreads quickly to the whole fissioning material. The large number of fission fragments thus produced quickly loose their energy in the material due to their heavy masses. This energy is released in the form of heat, which is the main source of thermal energy in a nuclear reactor. The thermal energy is then converted into electrical energy through other processes.

A.7   Total Cross Section

The total neutron interaction cross-section is the sum of the cross-sections of all the processes described above.

$$\sigma_t = \sigma_{elastic} + \sigma_{inelastic} + \cdots\cdots \tag{2.6.5}$$

The total cross section for neutrons of energy up to 200 MeV is shown in Fig.2.6.2 (21). It is obvious that at very low neutron energies the variation in cross section with respect to energy is larger as compared to higher energies.

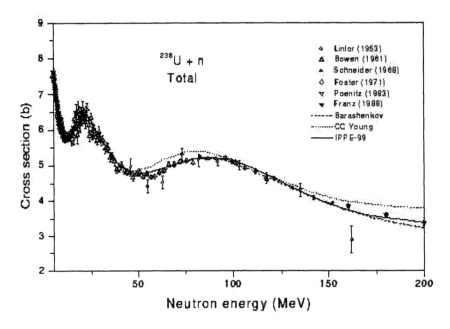

Figure 2.6.2: Total neutron cross section for uranium-238 (21).

A.8   Passage of Neutrons through Matter

A neutron interacts with the nuclear particles predominantly through the strong nuclear force. The strong force is extremely short range and therefore the particle must be very close to the nucleus to be affected by it. Neutrons, owing to their effective electrical neutrality, can get extremely close to the nucleus. In contrast, positively charged particles, such as protons and α-particles, experience Coulomb repulsion as they try to approach the nucleus. Unless their energy is fairly high, these charged particles can not penetrate deep enough to experience the strong nuclear force. Another major difference between neutrons (or for that matter, any other neutral particle) and charged particles is that the neutrons do not loose their energy through electromagnetic interactions with the material atoms. Hence they can penetrate deeper into the material as compared to the charged particles. This higher penetration capability is quite problematic in terms of developing effective radiation shields around neutron sources, such as nuclear reactors. Deeper penetration also carries advantages, though. For example, a neutron beam can be used for non-destructive testing of materials.

Just like photons, a beam of neutrons passing through a material also suffers exponential attenuation. The intensity of a neutron beam at a distance x from origin can then be evaluated from

$$I = I_0 e^{-\mu_n x}, \tag{2.6.6}$$

where μ_n is the attenuation coefficient of neutrons. It depends on the type of material as well as the neutron energy and is usually quoted in dimensions of inverse length. We can define the mean free path of neutrons by substituting

$$\lambda_n = \frac{1}{\mu_n}, \tag{2.6.7}$$

and $x = \lambda_n$ in the above exponential relation. This gives

$$\frac{I_0 - I}{I_0} \approx 63 \tag{2.6.8}$$

which implies that λ_n corresponds to the depth of the material that attenuates about 63% of the neutrons.

The attenuation coefficient can also be written in terms of the total nuclear cross-section σ_t, such as

$$\mu_n = N\sigma_t, \tag{2.6.9}$$

where N is the number density of nuclei in the material, which can be computed from $N = N_A \rho / A$, where N_A is the Avogadro's number, ρ is the weight density of the material, and A is its atomic weight. The attenuation coefficient can then be computed from

$$\mu_n = \frac{N_A \rho}{A} \sigma_t. \tag{2.6.10}$$

The above relation is valid only for a single element. In case of a compound with several elements or isotopes, generally an average attenuation coefficient is computed by taking weighted mean of total nuclear interaction cross-sections of all the isotopes present in the sample.

$$\mu_n^{av}(E) = \rho_n \left[\sum_{i=1}^{n} w_i \sigma_t^i(E) \right] \tag{2.6.11}$$

Here w_i is the fractional number of ith isotope in the sample of n isotopes and ρ_n is the *number* density of the sample.

Equation 2.6.6 can be used to experimentally determine the attenuation coefficient for an elemental isotope since we can write μ_n as

$$\mu_n = \frac{1}{x} \ln \left(\frac{I_0}{I} \right). \tag{2.6.12}$$

Such experiments are generally performed to determine the attenuation coefficients for materials at specific neutron energies.

Example:
Compute the thickness of a gold foil needed to remove 90% of the neutrons from a beam of thermal neutrons. The total cross section of thermal neutrons in gold can be taken to be 100 b. The density of gold is 1.9×10^4 kgm^{-3}.

Solution:
The attenuation coefficient of thermal neutrons in gold is

$$
\begin{aligned}
\mu_n &= \frac{N_A \rho}{A} \sigma_t \\
&= \frac{\left(6.022 \times 10^{23}\right)\left(1.9 \times 10^4\right)}{197 \times 10^{-3}} 100 \times 10^{-28} \\
&= 580.8 \ m^{-1}.
\end{aligned}
$$

Note that here we have used A in $kg/mole$. To compute the thickness of the foil we can use equation 2.6.12. For 90% removal of the neutrons we substitute $I/I_0 = 0.1$ and μ_n as calculated above in this equation to get

$$
\begin{aligned}
x &= \frac{1}{\mu_n} \ln\left(\frac{I_0}{I}\right) \\
&= \frac{1}{580.8} \ln(10) \\
&= 3.96 \times 10^{-3} \ m = 3.96 \ mm.
\end{aligned}
$$

2.7 Problems

1. The x-ray flux measured at a distance of $1.2 \ m$ from an x-ray machine is $3.6 \times 10^8 \ cm^{-2}s^{-1}$. Compute the flux at a distance of $3.5 \ m$.

2. Compare the mean free paths of thermal neutrons in silicon and aluminum. Assume the total cross sections for silicon and aluminum to be $2.35 \ b$ and $1.81 \ b$ respectively.

3. The mean free path of N_2 ions in nitrogen gas at $273 \ K$ and $1.2 \ atm$ is approximately $10^{-5} \ cm$. Compute the relative change in mean free path if the gas is brought to the standard conditions of 27^0C and $1 \ atm$.

4. A gas-filled detector is filled with 90% CO_2 and 10% CH_4 under atmospheric conditions of temperature and pressure. Compute the radiation length of high energy electrons passing through the detector.

5. Compute the maximum frequency of the photons emitted when an electron is accelerated through a potential of $30 \ kV$.

6. The cut-off Bremsstrahlung wavelength of an x-ray machine is found to be 35 fm. Estimate the electric potential applied across its electrodes.

7. Compute the threshold energy an α-particle must possess to emit Cherenkov radiation in light water with $n = 1.3$. At what angle would the light be emitted at this energy? Is it practical to build a Cherenkov detector for detecting α particles?

8. In a liquid filled Cherenkov detector, the maximum angle of Cherenkov cone ever observed was 48^0. Assuming that the particle emitting Chrenkov radiation traveled at a velocity of $0.9c$, estimate the refractive index of the liquid.

9. A fast moving electron in light water produces about 300 photons per cm of its travel path. Estimate the angle of the Cherenkov cone thus produced.

10. Compare the critical energies of electrons in lead and aluminum.

11. A photon beam of wavelength $80 \ nm$ ionizes hydrogen atoms. Compute the energy of the emitted electrons.

12. Compare the cutoff wavelengths of cesium in metallic and free states for photoelectric effect to occur. The ionization energy and work function of cesium are $3.9 \ eV$ and $1.9 \ eV$ respectively.

13. A metal is illuminated with light of $290 \ nm$, which results in the emission of electrons with a kinetic energy of $3.1 \ eV$. Compute the work function of the metal.

14. A photon having an energy of $200 \ keV$ scatters off of an atom at an angle of 35^0. Compute

   ▶ the wavelength of the scattered photon,

   ▶ the kinetic energy of the scattered electron, and

▶ the angle of the scattered electron.

15. An incident photon having an initial energy of 1.2 MeV undergoes Compton scattering. Estimate the energies of the scattered electrons if the photon scatters at 30^0, 60^0 and 90^0.

16. A 190 keV photon strikes an electron and scatters it off at an energy of 70 keV. Compute the energy and the angle of the scattered photon.

17. Determine the thickness of lead required to decrease the intensity of x-rays by a factor of 10^5.

18. Compute the mean free paths of 190 keV x-ray photons in lead.

19. In a Rutherford scattering experiment 1.2×10^4 α particles are detected per second at an angle of 35^0 with respect to their initial direction of motion. How many α-particles one should expect to observe at angles of 15^0 and 90^0?

20. Compute the stopping power of 6.5 MeV α-particles in air.

21. Compare the range of 2.5 MeV protons and α-particles in air.

22. Calculate the mean free path of thermal neutrons in cadmium having a density of 8.6×10^3 kgm^{-3}. The cross section of neutrons can be taken to be 2.4×10^4
    b. Estimate the thickness of cadmium needed to remove 99% of the neutrons from the beam.

23. A gas filled ionization chamber is constructed as shown in the figure below. The chamber is divided into two 1 cm wide regions filled with CO_2 under standard conditions. The central electrode and the windows are made up of 10 μm mylar foil. The windows are metalized only on their inner sides with 3 μm thick aluminum while the central electrode has 3 μm of aluminum deposited on both sides. Compute the percentage of the number of 100 keV photons that pass through the chamber without getting absorbed. Mylar can be assumed to be made up of carbon with a density of 1.4 g/cm^3.

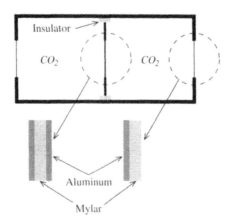

Bibliography

[1] Aridgides, A., Pinnock, R.N., Collins, D.F., **Observation of Rayleigh Scattering and Airglow**, Am. J. Phys., 44(3), 244-247, 1976.

[2] Aspect, A. et al., **Coherent and Collective Interactions of Particles and Radiation Beams: Varenna on Lake Como, Villa Monastero, International School**, Ios Pr. Inc., 1997.

[3] Attix, F.II., **Introduction to Radiological Physics and Radiation Dosimetry**, Wiley-Interscience, 1986.

[4] Balashov, V.V., Pontecorvo, G., **Interaction of Particles and Radiation with Matter**, Springer, 1997.

[5] Belokorov, V.V., Shirkov, D.V., Millard, P.A., **The Theory of Particle Interactions**, AIP Press, 1991.

[6] Berger, M.J., Seltzer, S.M., **Tables of Energy Losses and Ranges of Electrons and Positrons**, NASA-SP-3012, Washington, DC, 1964.

[7] Berne, B.J., Pecora, R., **Dynamic Light Scattering: with Applications to Chemistry, Biology, and Physics**, Dover Publications, 2000.

[8] Bichsel, H., **Experimental Range of Protons in Al**, Phys. Rev. 112(4), 1958.

[9] Blackett, P.M.S., Lees, D.S., Proc. Roy. Soc., 134, 658, 1931.

[10] Bohren, C.F., Huffman, D.R., **Absorption and Scattering of Light by Small Particles**, Wiley, New York, 1983.

[11] Born, M., Wolf, E., **Principles of Optics: Electromagnetic Theory of Propagation, Interference, and Diffraction of Light**, Cambridge Univ. Press, 1999.

[12] Chen, S.H., Kotlarchyk, M., **Interaction of Photons and Neutrons with Matter: An Introduction**, World Scientific Publishing Company, 1997.

[13] Cox, A.J., DeWeerd, A.J., Linden, J., **An Experiment to Measure Mie and Rayleigh Total Scattering Cross Sections**, Am. J. Phys. 70(6), June 2002.

[14] Drake, R.M., Gordon, J.E., **Mie Scattering**, Am. J. Phys., 53(10), 955-961, 1985.

[15] Eichholz, G.G., **Principles of Nuclear Radiation Detection**, Ann Arbor Science Publishers, 1979.

[16] Glaze, S., **The Interaction of Electromagnetic Radiation with Matter**, Health Sciences Consortium, 1979.

[17] Grard, R.J.L., **Photon and Particle Interactions with Surfaces in Space**, Springer, 1999.

[18] Gras-Marti, A. et al., **Interaction of Charged Particles with Solids and Surfaces**, Springer, 1991.

[19] Hagiwara, K. et al., Phys. Rev. 66, 010001, 2002.

[20] Ho-Kim, Q., Pham, X.Y., **Elementary Particles and Their Interactions: Concepts and Phenomena**, Springer, 2004.

[21] Ignatyuk, A.V. et al. **Neutron Cross Section Evaluations for U-238 up to 150** GeV, Nuclear Science and Technology, November 2000.

[22] **International Commission on Radiation Units and Measurements**, ICRU Report 37, Stopping Powers for Electrons and Positrons.

[23] Internet Source: http://physics.nist.gov/PhysRefData/Star/Text/method.html.

[24] Jackson, J.D., **Classical Electrodynamics**, Wiley, New York, 1975.

[25] Jansen, G.H., **Coulomb Interactions in Particle Beams**, Academic Press, 1990.

[26] Johnson, R.E., **Energetic Charged Particle Interactions with Atmospheres and Surfaces**, Springer, 1990.

[27] Johnson, C.S., Gabriel, D.A., **Laser Light Scattering**, Dover Publications, 1995.

[28] Kalashnikov, N., **Coherent Interactions of Charged Particles in Single Crystals**, Taylor & Francis, 1989.

[29] Katz, L., Penfold, A.S., Reviews of Modern Physics, 24, 28, 1952.

[30] Knoll, G.F., **Radiation Detection and Measurement**, Wiley, 2000.

[31] Leo, W.R., **Techniques of Nuclear and Particle Physics Experiments** Springer, 1994.

[32] Lehmann, C., **Interaction of Radiation with Solids and Elementary Defect Production (Defects in Crystalline Solids)**, Elsevier North-Holland, 1977.

[33] Leroy, C., Rancoita, P.G., **Principles of Radiation in Matter and Detection**, World Scientific Pub. Comp., 2004.

[34] Linderbaum, S.J., **Particle-Interaction Physics at High Energies**, Clarendon Press, 1973.

[35] Magill, J., Galy, J., **Radioactivity Radionuclides Radiation**, Springer, 2004.

[36] Mano, G., J. de Phys. et Rad., 5, 628, 1934.

[37] Matthews, J.L. et al., **The Distribution of Electron energy Losses in Thin Absorbers**, Nucl. Instrum. Meth., 180, 573, 1981.

[38] McNulty, P.J., **Interactions of Very Energetic Charged Particles with Matter: Final Report (AFCRL-TR-73-0143)**, Air Force Cambridge Research Labs, U.S. Air Force, 1973.

[39] Messel, H., Crawford, D.F., **Electron-Photon Shower Distribution Function**, Pergamon Press, Oxford, 1970.

[40] Mihailescu, L.C., Borcea, C., Plompen, A.J.M., **Measurement of Neutron Inelastic Scattering Cross Section for ^{52}Cr from Threshold up to 18 GeV**, Rad. Prot. Dosimetry, 115(1), 136-138, 2005.

[41] Mihailescu et al., **A New HPGe Setup at Gelina for Measurement of Gamma-Ray Production Cross Sections from Inelastic Neutron Scattering**, Nucl. Instr. Meth., A531, 375-391, 2004.

[42] Morton, E.J., **Radiation Sources and Radiation Interactions: 19 July 1999, Denver, Colorado (Proceedings of Spie–the International Society for Optical Engineering, V.3771)**, SPIE-International Society for Optical Engineering, 1999.

[43] Mozumder, A., **Charged Particle and Photon Interactions with Matter: Chemical, Physiochemical, and Biological Consequences with Applications**, CRC, 2003.

[44] Otsuki, Y., **Charged Beam Interaction with Solids**, Taylor & Francis, 1983.

[45] Parkinson, D.B. et al., **The Range of Protons in Aluminum and in Air**, Phys. Rev., 52, 75, 1937.

[46] Petty, G.W., **A First Course in Atmospheric Radiation**, Sundog Publishing, 2004.

[46] price1Price, W.J., **Nuclear Radiation Detection**, McGraw Hill Text, 1964.

[47] Rolnick, W.B., **The Fundamental Particles and Their Interactions**, Addison Wesley Publishing Comp., 1994.

[48] Rossi, B., **High Energy Particles**, Prentice-Hall, Inc., Englewood Cliffs, NJ, 1952.

[49] Rudie, N.J., **Interaction of Radiation with Matter (Principles and Techniques of Radiation Hardening)**, Western Periodicals Co., 1986.

[50] Rust, W.M., Donnelly, T.D., **Particle Size Determination: An Undergraduate Lab in Mie Scattering**, Am. J. Phys., 69(2), 129-136, 2001.

[51] Seltzer, S.M., Berger, M.J., Int. J. Appl. Rad. 35, 665, 1984.

[52] Sternheimer, R.M., Seltzer, S.M., Berger, M.J., **The Density Effect for the Ionization Loss of Charged Particles in Various Substances**, Atomic Data and Nuclear Data Tables, 30, 261, 1984.

[53] Tait, W.H., **Radiation Detection**, Butterworth-Heinemann, 1980.

[54] Tsai, Y.S., Rev. Mod. Physics, 46, 815, 1974.

[55] Tsang, L. et al., **Scattering of Electromagnetic Waves: Theories and Applications**, Wiley-Interscience, 2000.

[56] Tsoulfanidis, N., **Measurement and Detection of Radiation**, Hemisphere Publishing Corp., 1983.

[57] Turner, J.E., **Atoms, Radiation, and Radiation Protection**, Wiley-Interscience, 1995.

[58] Tuve, M.A., Heydenburg, N.P., Hafstad, L.R., Phys. Rev., 50, 806, 1936.

[59] Van Haeringen, H., **Charged-Particle Interactions: Theory and Formulas**, Coulomb Press Leyden, 1985.

[60] Van de Hulst, H.C., **Ligh Scattering by Small Particles**, Dover Publications, 1981.

[61] Wilson, R.R., **Range, Straggling, and Multiple Scattering of Fast Protons**, Phys. Rev. 71, 385-86, 1947.

[62] Young, A.T., **Rayleigh Scattering**, Phys. Today, 35(1), 208. 1982.

Chapter 3

Gas Filled Detectors

Radiation passing through a gas can ionize the gas molecules, provided the energy delivered by it is higher than the ionization potential of the gas. The charge pairs thus produced can be made to move in opposite directions by the application of an external electric field. The result is an electric pulse that can be measured by an associated measuring device. This process has been used to construct the so called *gas filled detectors*. A typical gas filled detector would consist of a gas enclosure and positive and negative electrodes. The electrodes are raised to a high potential difference that can range from less than 100 volts to a few thousand volts depending on the design and mode of operation of the detector. The creation and movement of charge pairs due to passage of radiation in the gas perturbs the externally applied electric field producing a pulse at the electrodes. The resulting charge, current, or voltage at one of the electrodes can then be measured, which together with proper calibration gives information about the energy of the particle beam and/or its intensity.

It is apparent that such a system would work efficiently if a large number of charge pairs are not only created but are also readily collected at the electrodes before they recombine to form neutral molecules. The choice of gas, the geometry of the detector, and the applied potential give us controlling power over the production of charge pairs and their kinematic behavior in the gas.

In this Chapter we will look at general design considerations of gas filled detectors and discuss their behavior in different conditions. Some of the extensively used special types of gaseous detectors will also be discussed in this Chapter.

3.1 Production of Electron-Ion Pairs

Whenever radiation interacts with particles in a gas, it may excite the molecules, ionize them, or do nothing at all. To complicate the matter further, there are different mechanisms through which these interaction could take place. In chapter 2 we visited some of these interaction mechanisms and found out that their gross outcomes can be fairly accurately predicted by using statistical quantities such as cross section and stopping power. Another quantity that is extremely important, at least for radiation detectors, is the average energy needed to create an electron-ion pair in a gas. This energy is referred to as the *W-value*. It would be natural to think that if the underlying radiation interaction processes are so complicated and dependent on energy and types of particles involved then the W-value would be different at each energy, for each radiation type, and for each type of gas. This is certainly true but only to a certain extent. In fact it has been found that the W-value depends only weakly on these parameters and lies within 25-45 eV per charge pair for most of the gases and types of radiation (see Table 3.1.1). An interesting point to note here is that the W-value is significantly higher than the first ionization potential for gases, implying that not all the energy goes into creating electron-ion pairs. Of course this is understandable since we know that radiation is not only capable of ionizing the atoms but can also just excite them.

The charges created by the incident radiation are called *primary* charges to distinguish them from the ones that are indirectly produced in the active volume. The production mechanism of these additional charge pairs are similar to those of primary charges except that they are produced by ionization caused by primary charge pairs and not the incident radiation. The W-value represents all such ionizations that occur in the active volume. For a particle that deposits energy ΔE inside a detector, the W-value can be used to determine the total number of electron-ion pairs produced by

$$N = \frac{\Delta E}{W}. \tag{3.1.1}$$

If the incident particle deposits all of its energy inside the detector gas, then of course ΔE would simply be the energy E of the particle. However in case of partial energy loss, we must use some other means to estimate ΔE. An obvious parameter that can be used is the stopping power dE/dx, which we discussed in chapter 2. In terms of stopping power, the above relation can be written as

$$N = \frac{1}{W}\frac{dE}{dx}\Delta x, \tag{3.1.2}$$

where Δx is the path covered by the particle. Sometimes it is more convenient, at least for comparison purposes, to calculate the number of electron-ion pairs produced per unit length of the particle track

$$n = \frac{1}{W}\frac{dE}{dx}. \tag{3.1.3}$$

As we saw in chapter 2, due to energy straggling, the stopping power fluctuates around its mean value. Similarly the W-values for different gases as measured by different experimenters suffer from significant uncertainties. Variations of as much as 30% in the reported values have been observed. These factors must be taken into consideration while estimating the total number of charges.

As mentioned above, the W-value represents all ionizations that occur inside the active volume of the detector. Sometimes it is desired to know the primary charge-pair yield as well. However, because of almost inevitable secondary ionizations that occur at nominal applied voltages, it is not always possible to determine this number experimentally. Nevertheless, a number of experiments have been performed and primary as well as total ionization yields have been reported by several authors (see Table 3.1.1).

Table 3.1.1: Ionization potentials I_e, W-values, stopping powers (dE/dx), primary ionization yield n_p, and total ionization yield n_t of different gases at standard atmospheric conditions for minimum ionizing particles (37) (ip stands for the number of electron-ion pairs).

Gas	Z	Density $(\times 10^{-4} g/cm^3)$	I_e (eV)	W eV/pair	dE/dx (keV/cm)	n_p (ip/cm)	n_t (ip/cm)
H_2	2	0.8	15.4	37	0.34	5.2	9.2
He	2	1.6	24.6	41	0.32	5.9	7.8
N_2	14	11.7	15.5	35	1.96	10	56
O_2	16	13.3	1.2	31	2.26	22	73
Ne	10	8.4	21.6	36	1.41	12	39
Ar	18	17.8	15.8	26	2.44	29	94
Kr	36	34.9	14.0	24	4.60	22	192
Xe	54	54.9	12.1	22	6.76	44	307
CO_2	22	18.6	13.7	33	3.01	34	91
CH_4	10	6.7	10.8	28	1.48	46	53

To determine the number of total and primary charge pairs in a gas mixture, a composition law of the form

$$n_t = \sum_i x_i \frac{(dE/dx)_i}{W_i} \qquad (3.1.4)$$

$$\text{and} \quad n_p = \sum_i x_i n_{p,i}, \qquad (3.1.5)$$

can be used. Here the subscript i refers to the ith gas in the mixture and x_i is the fraction by volume of gas i.

Example:
Compute the total and primary number of charge pairs produced in a mixture of 90% CO_2 and 10% CH_4.

Solution:
The total number of charge pairs, according to equation 3.1.4 is given by

$$
\begin{aligned}
n_t &= (0.9)\frac{(dE/dx)_{co2}}{W_{co2}} + (0.1)\frac{(dE/dx)_{ch4}}{W_{ch4}} \\
&= (0.9)\frac{3.01 \times 10^3}{33} + (0.1)\frac{1.48 \times 10^3}{28} \\
&\approx 87 \text{ charge-pairs}/cm.
\end{aligned}
$$

Similarly, the number of primary ion pairs can be computed from equation 3.1.5 as follows.

$$
\begin{aligned}
n_p &= (0.9)(n_{p,co2}) + (0.1)(n_{p,ch4}) \\
&= (0.9)(34) + (0.1)(46) \\
&\approx 35 \text{ charge-pairs}/cm.
\end{aligned}
$$

3.2 Diffusion and Drift of Charges in Gases

Both the electrons and ions produced as a result of passage of radiation quickly lose their energy by multiple collisions with gas molecules. The way these charges move in the gas depends largely on the type and strength of the force they experience. The behavior of the charges therefore with and without electric field differ significantly from each other.

3.2.A Diffusion in the Absence of Electric Field

In the absence of an externally applied electric field, the electrons and ions having energy E can be characterized by the Maxwellian energy distribution (39),

$$
F(E) = \frac{2}{\sqrt{\pi}}(kT)^{-3/2}\sqrt{E}e^{-E/kT}, \tag{3.2.1}
$$

where k is the Boltzmann's constant and T is the absolute temperature. The average energy of charges, as deduced from this distribution, turns out to be

$$
\bar{E} = \frac{3}{2}kT, \tag{3.2.2}
$$

which at room temperature is equivalent to about 0.04 eV. Since there is no externally applied electric field, there is no preferred direction of motion for the charges in a homogeneous gas mixture and therefore the diffusion is isotropic. In any direction x, the diffusion can be described by the Gaussian distribution

$$
dN = \frac{N}{\sqrt{4\pi Dt}}e^{-x^2/4Dt}dx, \tag{3.2.3}
$$

where N is the total number of charges and D is the diffusion coefficient. This relation simply represents the number of charges dN that can be found in an element

dx at a distance x from the center of the initial charge distribution after time t. D is generally reported in dimensions of cm^2/s and is an important quantity since it can be used to determine the standard deviation of the linear as well as the volume distribution of charges through the relations

$$\sigma_x = \sqrt{2Dt} \qquad (3.2.4)$$

$$\text{and} \quad \sigma_v = \sqrt{6Dt}. \qquad (3.2.5)$$

Electrons, owing to their very small mass, diffuse much faster. This can also be deduced by comparing the thermal velocities of electrons and ions, which usually differ by two to three orders of magnitude. The diffusion coefficient for electrons is, therefore, much different from that of the ions in the same gas. Since the diffusion coefficient has mass and charge dependence, therefore for different ions it assumes values that may differ significantly from each other. Further complications arise due to its dependence on the gas in which the ion is moving. Since in radiation detectors we are concerned with the movement of ions that are produced by the incident radiation therefore we restrict ourselves to the diffusion of ions in their own gas. Addition of admixture gases in the filling gas can also modify the diffusion properties, in which case the correct value of the diffusion coefficient, corresponding to the types and concentrations of the gases used, should be sought. The values of diffusion coefficients for different gases and gas mixtures have been experimentally determined and reported by several authors (see Table.3.2.1).

A.1   Diffusion in the Presence of Electric Field

In the presence of electric field the diffusion is no longer isotropic and therefore can not be described by a scalar diffusion coefficient. The diffusion coefficient in this case is a tensor with two non-zero components: a longitudinal component D_L and a transverse component D_T. For many gases, the longitudinal diffusion coefficient D_L is smaller than the transverse diffusion coefficient D_T (39).

3.2.B   Drift of Charges in Electric Field

In a gaseous detector, the Maxwellian shape of the energy distribution of charges can not be guaranteed. The reason is the applied bias voltage that creates electric field inside the active volume. The electrons, owing to their small mass, experience a strong electric force and consequently their energy distribution deviates from the pure Maxwellian shape. On the other hand, the distribution of ions is not significantly affected if the applied electric field is not high enough to cause discharge in the gas (39). The electrons behave quite differently than ions in the presence of electric field and therefore we must study the two types of charges separately.

B.1   Drift of Ions

In a gaseous detector the pulse shape and its amplitude depend not only on the motion of electrons but also of ions. The ions are positively charged and much heavier than electrons and therefore move around quite sluggishly. In most gaseous detectors, specially ionization chambers, the output signal can be measured from the positive or from the negative electrode. In both cases, however, what is measured is actually the change in the electric field inside the active volume. Hence the drift of electrons and ions both contribute to the overall output pulse. This implies that understanding the drift of positive charges is as important in a chamber as the negative electrons.

In the presence of externally applied electric field, ions move toward the negative electrode with a drift velocity that is much lower than that of electrons. The distribution of these ions can be fairly accurately characterized by a Gaussian distribution of the form

$$dN = \frac{N}{\sqrt{4\pi Dt}} e^{-(x-tv_d)^2/4Dt} dx, \qquad (3.2.6)$$

where v_d is the *drift* velocity of ions, which is actually the velocity of the cloud of ions moving along the electric field lines. This velocity is much lower than the instantaneous velocity of ions. t is the ion drift time. Drift velocity is an important parameter, since it tells us how quickly we should expect the ions to reach the cathode and get collected. It has been found that as long as no breakdown occurs in the gas, this velocity remains proportional to the ratio of electric field and gas pressure.

$$v_d = \mu_+ \frac{E}{P} \qquad (3.2.7)$$

Here E is the applied electric field, P is the pressure of the gas, and μ_+ is the mobility of ions in the gas. Mobility depends on the mean free path of the ion in the gas, the energy it looses per impact, and the energy distribution. In a given gas, it therefore remains constant for a particular ion. Table 3.2.1 gives the mobility, diffusion coefficient, and mean free paths of several ions in their own gases.

Table 3.2.1:   Mean free path λ, diffusion coefficient D, and mobility μ of ions in their own gas under standard conditions of temperature and pressure.

Gas	$\lambda(\times 10^{-5} cm)$	$D(cm^2/s)$	$\mu(cm^2 s^{-1} V^{-1})$
H_2	1.8	0.34	13.0
He	2.8	0.26	10.2
Ar	1.0	0.04	1.7
O_2	1.0	0.06	2.2
H_2O	1.0	0.02	0.7

A useful relationship between mobility and diffusion coefficient given by

$$\mu_+ = \frac{e}{kT} D_+, \qquad (3.2.8)$$

is known as Nernst-Einstein relation. Here k is the Boltzmann's constant and T is the absolute temperature.

For a gas mixture, the effective mobility can be computed from the so called *Blanc's law*

$$\frac{1}{\mu_+} = \sum_{j=1}^{n} \frac{c_j}{\mu_+^{ij}}, \tag{3.2.9}$$

where n is the number of gas types in the mixture, μ_+^{ij} is the mobility of ion i in gas j and c_j is the volume concentration of gas j in the mixture.

The drift velocity of ions is roughly two to three orders of magnitude lower than that of electrons. The slow movement of ions causes problems of space charge accumulation, which decreases the effective electric field experienced by the charges decreases. The resulting slower movement of ions has the potential of increasing the space charge and decreasing the pulse height at the readout electrode. This and other signal deterioration effects will be discussed later in the Chapter.

B.2   Drift of Electrons

If a constant electric field is applied between the electrodes, the electrons, owing to their small mass, are rapidly accelerated between collisions and thus gain energy. The energy that these electrons lose through collisions with gas molecules is not very large (again because of their small mass). Because of these collisions their mean energy increases and consequently the energy distribution can no longer be described by a Maxwellian distribution.

Along the electric field lines, the electrons drift with velocity v_d which is usually an order of magnitude smaller than the velocity of thermal motion v_e. However the magnitude of drift velocity depends on the applied electric field and finds its limits at the breakdown in the gas. The approximate dependence of drift velocity on the electric field E is given by (33)

$$v_d = \frac{2eEl_{mt}}{3m_e \bar{v}_e}, \tag{3.2.10}$$

where l_{mt} is the mean momentum transfer path of electrons. Using theory of electron transport in gases, more precise expressions for drift velocity and other related parameters have been obtained and reported by several authors (see (39) and references therein).

In early days of gaseous detector developments, a number of experimental studies were carried out to determine the drift velocities of electrons in the usual gases used in radiation detection. At that time the availability of computing power was a bottleneck in numerically solving complex transport equations needed to determine the drift velocities and therefore resort was made to experimental studies. Although we now have the capability to perform such computations, the published results of earlier experimental studies are still extensively used in modern detectors.

Fig.3.2.1 shows the variation of electron drift velocity in methane, ethane, and ethylene with respect to the applied electric field. It is apparent that only in the low field region, the drift velocity increases with the energy. Beyond a certain value of the electric field that depends on the type of gas, the velocity either decreases or stays constant. As evident from figures 3.2.2 and 3.2.3, this behavior is typical of gases that are commonly used in radiation detectors.

An important result that can be deduced from Fig.3.2.2 is the non-negligible dependence of electron drift velocity on the pressure of the gas. This, of course, can

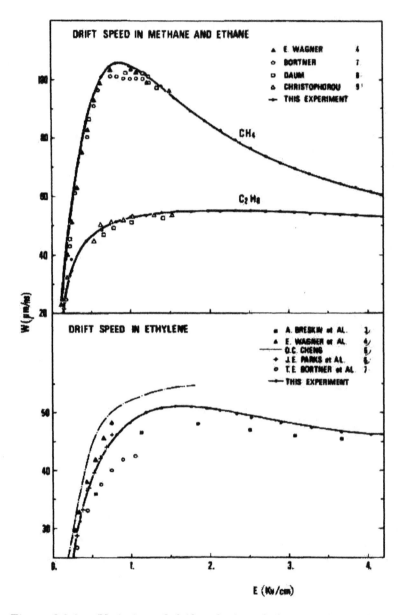

Figure 3.2.1: Variation of drift velocity of electrons in methane, ethane, and ethylene (14).

also be intuitively understood by noting that as the pressure of the gas increases the density of the target atoms also increases, thus forcing an electron to make more collisions along its track. Due to this dependence, many authors prefer to tabulate or plot the electron drift velocities with respect to the ratio of electric field intensity and pressure, that is, E/P. Such a curve is shown in Fig.3.2.4.

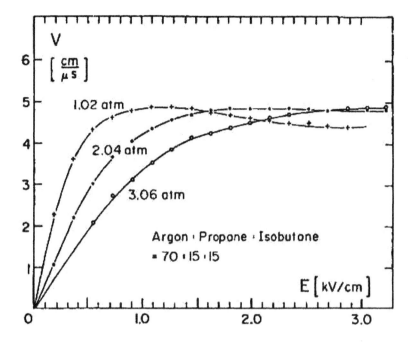

Figure 3.2.2: Variation of drift velocity of electrons in a mixture of argon, propane, and isobutane with respect to electric field strength. The curves have been drawn for different gas pressure values (23).

3.2.C Effects of Impurities on Charge Transport

In most applications, gaseous detectors are filled with a mixture of gases instead of a single gas. The ratio of the gases in the mixture depends on the type of detector and the application. In addition to these, the detector also has *pollutants* or *impurities*, which degrade its performance. Most of these pollutants are polyatomic gases, such as oxygen and air. Since they have several vibrational energy levels therefore they are able to absorb electrons in a wide energy range. Such agents are called *electronegative* and their electron attachment coefficients are generally high enough to be of concern. The main effect of these impurities is that they absorb electrons and result in degradation of the signal.

There are two methods by which electron capture occurs in gaseous detectors: resonance capture and dissociative capture. Resonance capture can be written as

$$e + X \rightarrow X^{-*}, \tag{3.2.11}$$

where X represents the electronegative molecule in the gas and $*$ denotes its excited state. To de-excite, the molecule can either transfer the energy to another molecule

$$X^{-*} + S \rightarrow X^- + S^*, \tag{3.2.12}$$

or emit an electron

$$X^{-*} \rightarrow X + e. \tag{3.2.13}$$

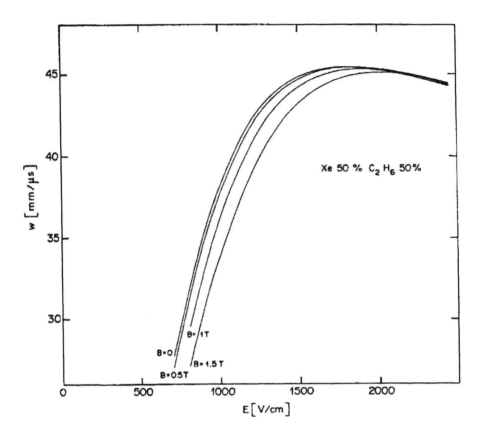

Figure 3.2.3: Variation of drift velocity of electrons in a mixture of xenon and C_2H_6 with respect to electric field strength. The curves have been drawn for different values of externally applied magnetic fields. The variation of drift velocity with magnetic field is very small and therefore except for very high field strengths it can be neglected for most practical purposes (31).

Here S can be any gas molecule in the gas but is generally an added impurity called *quench gas*. We will learn more about this later in the chapter. The process of electron emission is favorable for radiation detectors because the only effect it has is the introduction of a very small time delay between capture and re-emission of the electron. It does not have any deteriorating effect on the overall signal height.

If a constant electric field is applied between two electrodes, the number of electrons surviving the capture by electronegative impurities after traveling a distance x is given by

$$N = N_0 e^{-\mu_c x}, \tag{3.2.14}$$

where N_0 is the number of electrons at $x = 0$ and μ_c is the electron capture coefficient, which represents the probability of capture of an electron. It is related to electron's capture mean free path λ_c by

$$\mu_c = \frac{1}{\lambda_c}. \tag{3.2.15}$$

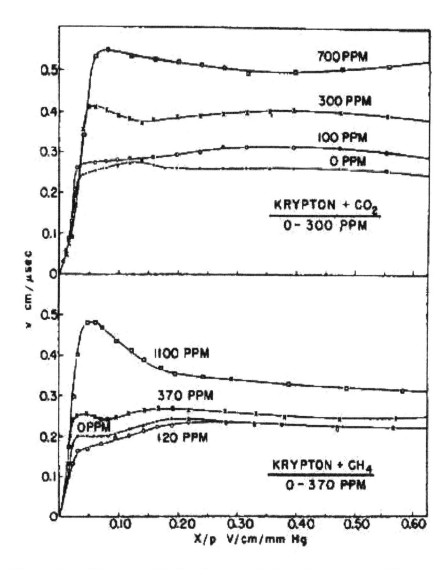

Figure 3.2.4: Variation of drift velocity of electrons in mixtures of (krypton + CO_2) and (krypton + CH_4) with respect to the ratio of the electric field strength and the gas pressure (10).

Using this, we can define λ_c as the distance traveled by electrons such that about 63% of them get captured. The capture mean free path depends on the electron *attachment coefficient* η, which characterizes the probability of electron capture in any one scattering event. If σ is the total electron scattering cross section of the electronegative gas, then $\eta\sigma$ will represent the attachment cross section. If N_m is the number density of the gas molecules (number of molecules per unit volume) and f is the fraction of its electronegative component, then the capture coefficient and the capture mean free path can be written as

$$\mu_c = f N_m \eta \sigma \tag{3.2.16}$$

$$\lambda_c = \frac{1}{f N_m \eta \sigma}. \tag{3.2.17}$$

Now since λ_c is the capture mean free path, we can simply divide it by the average electron velocity v to get its *capture mean lifetime* τ_c.

$$\tau_c = \frac{1}{f N_m \eta \sigma v} \tag{3.2.18}$$

The exponential relation 3.2.14 can then also be written in terms of time as

$$N = N_0 e^{-t/\tau_c}, \tag{3.2.19}$$

where N_0 is the initial electron intensity and N is the intensity at time t.

The factors h, σ, and v in the above relations depend on the electron energy while N_m has dependence on temperature and pressure. It is therefore not possible to find the values of these parameters in literature for all possible energy and working conditions. Further complications arise if there are more than one electronegative elements in the gas. This is because the charge exchange reactions between these elements can amount to significantly higher electron attachment coefficients as compared to the ones obtained by simple weighted mean. In such cases, one should resort to the experimentally determined values of λ_c or τ_c, which are available for some of the most commonly used gas mixtures.

Example:
Compute the percent loss of 6 eV electrons created at a distance of 5 mm from the collecting electrode in a gaseous detector filled with argon at standard temperature and pressure. Assume a 1% contamination of air with $\eta = 10^{-5}$. The total scattering cross section of air for 6 eV electrons is 5×10^{-14} cm^2.

Solution:
The given parameters are

$$f = 0.01,$$
$$\eta = 10^{-5},$$
$$\sigma = 5 \times 10^{-14} \ cm^2,$$
$$\text{and} \quad x = 0.5 \ cm.$$

To estimate N_m we note that the contamination level in argon is very low (1%) and therefore we can safely use the argon number density in place of the overall density of the gas. The number density of argon atoms can be calculated from

$$N_m = \frac{N_A \rho}{A},$$

where N_A is the Avogadro's number, ρ is the weight density, and A is the atomic mass. Hence for argon we have

$$N_m = \frac{(6.022 \times 10^{23})(1.78 \times 10^{-3})}{18}$$
$$= 5.97 \times 10^{19} \text{ atoms}/cm^3.$$

The percent loss of electrons n_{abs}, according to equation 3.2.14 is then given by

$$n_{abs} = \frac{N_0 - N}{N_0} \times 100 = \left[1 - e^{-fN_m\eta\sigma x}\right] \times 100.$$
$$= \left[1 - \exp\{-(0.01)(5.97 \times 10^{19})(10^{-5})(5 \times 10^{-14})(0.5)\}\right] \times 100$$
$$= 13.8\%$$

This example clearly shows how problematic small amounts of contaminants can be in a detector. For precision measurements, parasitic absorption of about 14% of the electrons could be enough to deteriorate the signal to unacceptable levels.

3.3   Regions of Operation of Gas Filled Detectors

Fig. 3.3.1 shows different regions of operation of a gas filled detector. Based on the applied bias voltage, a detector can be operated in a number of modes, which differ from one another by the amount of charges produced and their movement inside the detector volume. Choice of a particular mode depends on the application and generally detectors are optimized to work in the range of the applied voltage that is typical of that particular mode only. These operation regions are discussed here briefly.

3.3.A   Recombination Region

In the absence of an electric field the charges produced by the passage of radiation quickly recombine to form neutral molecules. At the application of the bias voltage some of the charges begin to drift towards the opposite electrodes. As this voltage is raised the recombination rate decreases and the current flowing through the detector increases. The recombination region depicted in Fig.3.3.1 refers to the range of applied voltage up to the value when the recombination is negligibly small. Because of appreciable recombination in this region, the current measured at the output of the detector does not honestly reflect the energy deposited by the incoming radiation. Consequently in terms of measuring the properties of radiation, it is useless to operate the detector in this region.

3.3.B   Ion Chamber Region

The collection efficiency of electron-ion pairs in the recombination region increases with applied voltage until all the charges that are being produced get collected. This

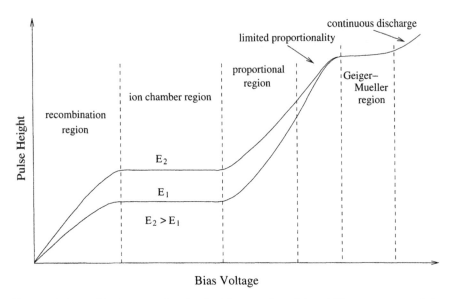

Figure 3.3.1: Variation of pulse height produced by different types of detectors with respect to applied voltage. The two curves correspond to two different energies of incident radiation.

is the onset of the so called ion chamber region. In this region further increasing the high voltage does not affect the measured current since all the charges being produced get collected efficiently by the electrodes. The current measured by the associated electronics in this region is called saturation current and is proportional to the energy deposited by the incident radiation. The detectors designed to work in this region are called ionization chambers.

It is almost impossible to completely eliminate the possibility of charge pair recombination in the ion-chamber region. However, with proper design, ionization chambers having plateaus of negligibly small slopes can be built.

3.3.C Proportional Region

In the previous chapter we studied the process of production of electron-ion pairs by the passage of radiation. This type of ionization is referred to as the primary ionization. If the charges produced during primary ionization have enough energy they themselves can produce additional electron-ion pairs, a process called secondary ionization. Further ionization from these charges is also possible provided they have enough energy. Obviously this process can occur only if a high enough electric potential exists between the electrodes so that the charges could attain very high velocities. Although the energy gained by the ions also increases as the bias voltage is increased, the electrons, owing to their very small mass, are the ones that cause most of the subsequent ionizations.

This *multiplication* of charges at high fields is exploited in the proportional detectors to increase the height of the output signal. In such a detector the multiplication of charges occurs in such a way that the output pulse remains proportional to the deposited energy. That is why these detectors are called proportional detectors. From figures such as Fig.3.3.1 it is sometimes concluded that in proportional counters the output pulse height is proportional to the applied bias. This is correct only up to an approximation, though. The correct reason for calling these devices proportional counters is that the total number of charges produced after multiplication is proportional to the initial number of charges. Let us now have a closer look at the process of charge multiplication.

C.1 Avalanche Multiplication

For a detector working in the proportional region, an electric field as high as several kV/cm is not uncommon. This high electric field not only decreases the charge collection time but also initiates a process called *avalanche multiplication*, which is a rapid multiplication of charges by primary charges produced from the incident radiation. This charge multiplication results in the increase in output pulse amplitude. Up to a certain bias voltage, the output pulse amplitude remains proportional to the bias voltage. A detector working in this region is therefore known as a *proportional counter.*

Due to the high electric field between the electrodes, the charges quickly gain energy between collisions. If the total energy of an electron or an ion becomes higher than the ionization potential of the gas atoms, it can ionize an atom, thus creating another charge pair.

If all of the conditions, such as electric field, temperature, and pressure remain constant and the electric field is uniform, then the change in the number of charge pairs per unit path length is simply proportional to the total number of charge pairs, that is

$$\frac{dN}{dx} = \alpha N. \tag{3.3.1}$$

Here N represents the total number of charge pairs and α is known as the *first Townsend coefficient*. The first Townsend coefficient represents the number of collisions leading to ionization per unit length of the particle track and is simply the reciprocal of the mean free path for ionization

$$\alpha = \frac{1}{\lambda}. \tag{3.3.2}$$

Here $\lambda = N_{mol}\sigma$ is the mean free path for ionization with N_{mol} being the number of gas molecules per unit volume and σ the total ionization cross section. α depends on the energy that an electron gains in a mean free path and the ionization potential of the gas. Solution of 3.3.1 as obtained by simple integration is

$$N = N_0 e^{\alpha x}. \tag{3.3.3}$$

If $\alpha > 0$ this equation guarantees exponential growth of number of charge pairs with distance. The multiplication of charges can be quantitatively described by *multiplication factor M* as follows.

$$\begin{aligned} M &= \frac{N}{N_0} \\ &= e^{\alpha x} \end{aligned} \tag{3.3.4}$$

The above equation is true only for a uniform electric field. In an non-uniform field, the Townsend coefficient becomes a function of x. In that case the multiplication factor for an electron that drifts from point r_1 to r_2 can be calculated from

$$M = \exp \left[\int_{r_1}^{r_2} \alpha(x)dx \right] . \tag{3.3.5}$$

Hence if we want to compute the multiplication factor, we must know the spatial profile of the first Townsend coefficient. Although it is quite challenging to determine this profile analytically, it has been shown that the reduced Townsend coefficient has a dependence on the reduced electric field intensity, that is

$$\frac{\alpha}{P} = f \left(\frac{E}{P} \right) , \tag{3.3.6}$$

where E is the electric field intensity and P is the gas pressure. Several authors have reported different forms of the first Townsend coefficient but a commonly used expression is the one originally proposed by Korff (19). It is given by

$$\frac{\alpha}{P} = A \exp \left(-\frac{BP}{E} \right) , \tag{3.3.7}$$

where the parameters A and B depend on the gas and the electric field intensity. These parameters have been experimentally determined for a number of gases (see Table.3.3.1).

Another simple expression for α that has been reported in literature is based on the intuition that since α is inversely related to the mean free path of electrons in a gas it should therefore be directly related to the molecular density N_m of the gas and the energy ξ of the electrons. This argument leads to the expression

$$\alpha = D_\alpha N_m \xi \tag{3.3.8}$$

where the proportionality constant D_α has been experimentally determined for several gases (see Table.3.3.1).

Table 3.3.1: Experimentally determined values of parameters appearing in equations 3.3.7 and 3.3.8 (37).

Gas	$A(cm^{-1}Torr^{-1})$	$B(Vcm^{-1}Torr)$	$D_\alpha(\times 10^{-17}cm^2V^{-1})$
He	3	34	0.11
Ne	4	100	0.14
Ar	14	180	1.81

An interesting aspect of avalanche is its geometric progression, which assumes the shape of a liquid drop because of the large difference between the drift velocities of electrons and ions (see, for example (37)). The electrons move much faster than ions and quickly reach the anode leaving behind a wide tail of positive ions drifting slowly towards the cathode (see Fig.3.3.2).

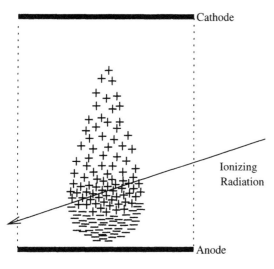

Figure 3.3.2: Typical droplet shape of avalanche in a gas filled detector. The incident radiation (shown by a solid line with an arrow) produces the charge pairs along its track. The charges start moving in opposite directions under the influence of the applied electric field. The electrons, being lighter than positively charged molecules, move faster and leave behind a long tail of positive charges drifting slowly towards cathode. Note that there is a time lag between the initial creation of charge pairs and the formation of droplet.

Example:
Calculate the first townsend coefficient for a helium filled chamber kept under 760 *torr* when an electric field of 10^4 $V\,cm^{-1}$ is established across the chamber electrodes. Also estimate the gas gain at a distance of 0.1 *cm* from the anode.

Solution:
The first townsend coefficient can be calculated from equation 3.3.7 using the values of A and B as given in table 3.3.1.

$$\alpha = AP \exp\left(-\frac{BP}{E}\right)$$

$$= (3)(760) \exp\left[-\frac{(34)(760)}{10^4}\right]$$

$$= 172.1 \; cm^{-1}$$

The multiplication factor at $x = 0.1$ *cm* is then given by

$$M = e^{\alpha x}$$

$$= e^{(172.1)(0.1)}$$

$$\approx 3 \times 10^7.$$

3.3.D   Region of Limited Proportionality

As we increase the bias voltage, more and more charges are produced inside the active volume of the detector. Now since heavy positive charges move much slower than the electrons, they tend to form a cloud of positive charges between the electrodes. This cloud acts as a shield to the electric field and reduces the effective field seen by the charges. As a consequence the proportionality of the total number of charges produces to the initial number of charges is not guaranteed. This region is therefore termed as the region of *limited proportionality* (see Fig.3.3.1). Since the loss of proportionality means loss of linearity, radiation detectors are not operated in this region.

3.3.E   Geiger-Mueller Region

Increasing the voltage further may increase the local electric field to such high values that an extremely severe avalanche occurs in the gas, producing very large number of charge pairs. Consequently a very large pulse of several volts is seen in the readout electronics. This is the onset of the so called *Geiger-Mueller region*. In this region, it is possible to count individual incident particles since each particle causes a breakdown and a large pulse. Since the output pulse is neither proportional to the deposited energy nor dependent on the type of radiation, the detectors operated in this region are not appropriate for spectroscopy.

There is also a significant *dead time* associated with such detectors. Dead time is the time during which the detector is essentially dead. This can happen due to the detector itself or the associated electronics. If there is a large accumulation of positive charges, it can reduce the internal electric field to a value that it can no longer favor avalanche multiplication. If radiation produces charge pairs during this time, the charges do not get multiplied and no pulse is generated. The detector starts working again as soon as most of the positive charges have been collected by the respective electrode. Dead time will be discussed in some detail later in the chapter.

The multiplication of charges in a GM detector is so intense that sometimes it is termed as *breakdown* of the gas. We will discuss this phenomenon in the following section.

E.1   Breakdown

The large number of ions created during the avalanche drift much slower than the electrons and therefore take longer to reach the cathode. When these heavy positive charges strike the cathode wall, they can release more ions from the cathode material into the gas. The efficiency γ of this process is generally less than 10%. γ is known as the *second Townsend coefficient*. At moderate voltages, γ is not high enough to make significant increase in charge population. However at higher voltages the secondary ion emission probability increases, deteriorating the linearity of the output pulse with applied voltage. Further voltage increase may start discharge in the gas. At this point the current goes to very high values and is limited only by the external circuitry, that is, the height of the pulse becomes independent of the initial number of electron ion pairs. Geiger tubes, that we will visit later in the chapter, are operated in this region.

To understand the breakdown quantitatively, let us write the equation for the multiplication factor M under steady state condition of discharge. It can be shown that when the discharge becomes independent of the ionization in the gas, the equation 3.3.4 should be replaced by (27)

$$M = \frac{e^{\alpha x}}{1 - \gamma \left(e^{\alpha x} - 1\right)},\tag{3.3.9}$$

where α and γ are the first and second Townsend coefficients respectively. The singularity in the above equation represents the breakdown (the value at which the current becomes infinite, at least theoretically), which occurs if the bias voltage is increased to very high values. The mathematical condition to start and sustain breakdown can therefore be written as

$$1 - \gamma \left(e^{\alpha x} - 1\right) = 0$$
$$\Rightarrow \gamma = \frac{1}{e^{\alpha x} - 1}.\tag{3.3.10}$$

Here γ is the *critical* value of the coefficient at which the breakdown starts. Furthermore as long as this condition remains fulfilled, the breakdown is sustained. Note that this value depends not only the first Townsend coefficient but also on the position x. As stated earlier, in most radiation detectors under normal operating conditions the second Townsend coefficient remains below 0.1, that is, the probability that a breakdown will occur is less than 10%.
Since the first Townsend coefficient in the above expression depends on the electric field intensity and gas pressure therefore we will see if we can derive an expression for the breakdown voltage. For this we first write the above relation as

$$\alpha x = \ln\left(1 + \frac{1}{\gamma}\right).$$

Substitution of equation 3.3.7 in this expression gives

$$(APx)\exp\left(-\frac{BP}{E}\right) = \ln\left(1 + \frac{1}{\gamma}\right)$$
$$\Rightarrow E = \frac{BP}{\ln\left[APx/\ln\left(1 + 1/\gamma\right)\right]}.\tag{3.3.11}$$

Let us now suppose that we have a parallel plate geometry in which the electrodes are separated by a distance $x = d$. To write the above equation in terms of voltage, we note that for such a geometry $E = V/d$. Hence the breakdown voltage V_{break} is given by

$$V_{break} = \frac{BPd}{\ln\left[APd/\ln\left(1 + 1/\gamma\right)\right]}.\tag{3.3.12}$$

This useful relation tells us that for a given gas the voltage at which the breakdown occurs depends on the product of pressure and electrode separation (that is, Pd). It is generally known as *Paschen's law* and a curve drawn from this equation between Pd and V_{break} is referred to as *Paschen curve*. Fig.3.3.3 shows such a curve for helium with arbitrarily chosen $\gamma = 0.1$. The point of minima in such a curve is called *Paschen minimum*, which is the voltage below which the breakdown is not possible.

An expression for the Paschen minimum can be derived by differentiating V_{break} with respect to Pd and equating the result to zero (see example below).

$$(Pd)_{min} = \frac{e}{A} \ln\left[1 + \frac{1}{\gamma}\right] \qquad (3.3.13)$$

Here $(Pd)_{min}$ is the value at which V_{break} is minimum and e is the natural logarithm number.

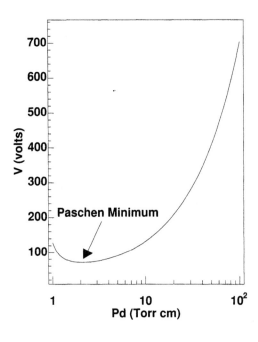

Figure 3.3.3: Paschen curve for helium enclosed between two parallel plate electrodes separated by distance d. The gas pressure is P and the second Townsend coefficient has been arbitrarily chosen to be 0.1.

Example:
Derive equation 3.3.13 for Paschen minimum.

Solution:
We will start with equation 3.3.12 for the breakdown voltage.

$$V_{break} = \frac{BPd}{\ln\left[APd/\ln\left(1 + 1/\gamma\right)\right]}$$

To simplify the mathematical manipulations with respect to Pd we first write this equation as

$$V_{break} = \frac{BPd}{\ln\left[A/\ln\left(1 + 1/\gamma\right)\right] + \ln(Pd)}.$$

$$\text{or } V_{break} = \frac{Bu}{\ln(v) + \ln(u)}.$$

where

$$u \equiv Pd, \qquad \text{and}$$
$$v \equiv \frac{A}{\ln(1 + 1/\gamma)}.$$

The minimum in the $V_{break} - u$ (that is, $V_{break} - Pd$) curve can be obtained by differentiating both sides of the above equation with respect to u and equating the result to zero.

$$\frac{dV_{break}}{du} = 0$$

$$\Rightarrow \frac{d}{du}\left[\frac{Bu}{\ln(v) + \ln(u)}\right] = 0$$

$$\Rightarrow -\frac{1}{[\ln(v) + \ln(u)]^2} + \frac{1}{\ln(v) + \ln(u)} = 0 \quad \text{since } B \neq 0$$

$$\Rightarrow \ln(v) + \ln(u) = 1$$

$$\Rightarrow \ln(uv) = 1$$

$$\Rightarrow uv = e$$

$$\Rightarrow u = \frac{e}{v}$$

This is the value of u at which V_{break} is minimum. Hence we can substitute $u = (Pd)_{min}$ and the actual expression for v in this expression to get the desired result

$$(Pd)_{min} = \frac{e}{A}\ln\left[1 + \frac{1}{\gamma}\right].$$

3.3.F Continuous Discharge

The breakdown process we studied in the previous section can further advance to the process of continuous discharge if the high voltage is raised to very high values. This continuous discharge starts as soon as a single ionization takes place and can not be controlled unless the voltage is lowered. In this region, electric arcs can produce between the electrodes, which may eventually damage the detector. It is apparent that radiation detectors can not be operated at such high voltages and therefore one must make sure that it remains below the threshold for this process.

 ## 3.4 Ionization Chambers

Ionization chambers are one of the earliest constructed radiation detectors. Because of their simplicity in design and well understood physical processes they are still one of the most widely used detectors.

3.4.A Current Voltage Characteristics

Fig.3.4.1 shows the current-voltage characteristics of an ionization chamber at different incident radiation intensities. Normally the chambers are operated in the middle of the plateau region to avoid any large variation of current with small variations in the power supply voltage. This ensures stability and softens the requirement of using very stable power supplies, which are normally very expensive.

As we stated earlier, the plateau of any chamber always has some slope but normally it is so small that it does not affect the signal to noise ratio in any significant way.

As shown in Fig.3.4.1, the form of the current-voltage characteristic curve of an ion chamber does not depend on the intensity of the incident radiation. Quantitatively, two differences arise: one is the onset of the plateau region and the second is the output current amplitude. The underlying reason for both of these differences is the availability of larger number of electron-ion pairs at higher intensities. If the rate of production of charge pair increases, then a higher electric field intensity will be needed to eliminate (or more realistically to *minimize*) their recombination. Hence the plateau in such a case will start at a higher voltage. Also, as we will see later, the output current is proportional to the number of charge pairs in the plateau region and therefore at higher intensities the plateau current is higher.

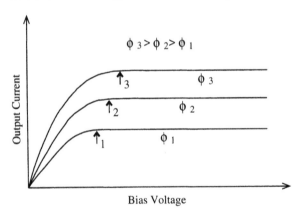

Figure 3.4.1: Current-voltage characteristic curves of an ionization chamber at different incident radiation intensities. The output signal as well as the onset of the plateau (indicated on the plot with 1, 2, and 3) increase with increasing intensity or flux of radiation.

3.4.B Mechanical Design

The mechanical design of an ion chamber consists of essentially three components: an anode, a cathode, and a gas enclosure. The particular geometry of these parts are application dependent. Fig.3.4.2 shows the two most common ion chamber geometries and the electric field inside their active volumes. Both of these designs have their own pros and cons, which we can only understand if we look at the production and behavior of electron-ion pairs.

B.1 Parallel Plate Geometry

This simple design consists of two parallel plates maintained at opposite electrical potentials (see Fig.3.4.3). Although, as we saw earlier, the curvature in the electric

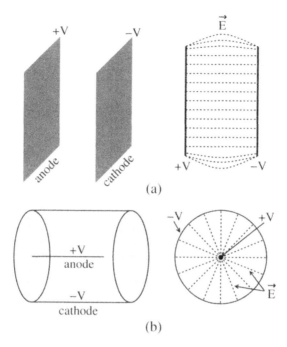

Figure 3.4.2: (a) Parallel plate ion chamber and a two dimensional view of electric field inside its active volume. The curved electric field at the sides may induce nonlinearity in the response. (b) Cylindrical ion chamber and a two dimensional view of radial electric field in its active volume. The increased flux of electric lines of force near the positively charged anode wire greatly enhances the electron collection efficiency.

lines of force at the edges of such a detector can potentially cause nonlinearity in the response, but with proper designing this problem can be overcome. In fact, very high precision parallel plate ionization chambers have been developed (see, for example (1)).

Let us see how the output voltage pulse from such a chamber looks. For this we note that the voltage pulse is actually the result of the perturbation in the electric potential caused by the movement of charge pairs towards opposite electrodes. This is because the electrons and ions generated inside the chamber decrease the effective electric field. The strength of this effective field varies as the charges move toward opposite electrodes, generating a voltage pulse at the output. The effective voltage at any time t inside the chamber can be written as

$$V_{eff}(t) = V_0 - V_{np}(t). \tag{3.4.1}$$

where V_0 is the static applied potential and $V_{np}(t)$ is the potential difference at time t caused by the electrons and ions inside the chamber.

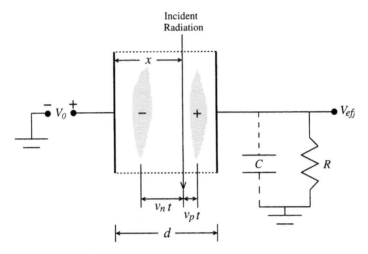

Figure 3.4.3: Simple parallel plate ionization chamber. In this geometry the output voltage V_{eff} depends on the point where charge particles are generated. Electrons owing to their small mass move faster than ions and therefore the distribution of electrons is shown to have moved a larger distance as compared to that of ions.

If we have N_0 ion pairs at any instant t, then the kinetic energy possessed by the electrons having average velocity v_n is given by

$$
\begin{aligned}
T_n(t) &= N_0 e E v_n t \\
&= \frac{V_0}{d} N_0 e v_n t,
\end{aligned}
$$

where we have assumed that the electric field intensity E, under whose influence the electrons move, is uniform throughout the active volume and can be written as $E = V_0/d$, d being the distance between the electrodes. Similarly the kinetic energy of ions having average velocity v_p can be written as

$$
\begin{aligned}
T_p(t) &= N_0 e E v_p t \\
&= \frac{V_0}{d} N_0 e v_p t.
\end{aligned}
$$

The potential energy contained in the chamber volume having capacitance C can be written as

$$
U_{ch} = \frac{1}{2} C V_{np}^2,
$$

while the total energy delivered by the applied potential V_0 is

$$
U_{total} = \frac{1}{2} C V_0^2.
$$

In the short circuit condition, this total energy should be equal to the total of kinetic and potential energy inside the chamber volume, that is

$$
\begin{aligned}
U_{total} &= U_{ch} + T_p + T_n \\
\Rightarrow \frac{1}{2}CV_0^2 &= \frac{1}{2}CV_{np}^2 + N_0 e E v_n t + N_0 e E v_p t.
\end{aligned} \tag{3.4.2}
$$

The above equation can be rearranged to give an expression for the effective potential $V_{eff} = V_0 - V_{np}$ as follows.

$$
\begin{aligned}
(V_0 - V_{np})(V_0 + V_{np}) &= \frac{2N_0 V_0 e}{Cd}(v_p + v_n)t \\
\Rightarrow V_{eff} &\simeq \frac{N_0 e}{Cd}(v_p + v_n)t
\end{aligned} \tag{3.4.3}
$$

Here we have used the approximation

$$
V_0 + V_{np} \approx 2V_0.
$$

Since the electrons move much faster than ions ($v_n \gg v_p$), therefore initial pulse shape is almost exclusively due to electron motion. If we assume that the charge pairs are produced at a distance x from the anode (see Fig.3.4.3), then the electrons will take $t_n = x/v_n$ to reach the anode. This will inhibit a sharp increase in pulse height with the maximum value attained when all the electrons have been collected by the anode. The ions, owing to their heavier mass, will keep on moving slowly toward cathode until time $t_p = (d - x)/v_p$, increasing the pulse height further, however at a much lower rate. The maximum voltage is reached when all the charges have been collected. Based on these arguments, we can rewrite the expression for the output pulse time profile for three distinct time periods as follows.

$$
V_{eff} \simeq
\begin{cases}
\frac{N_0 e}{Cd}(v_p + v_n)t & : \quad 0 \le t \le t_n \\[2mm]
\frac{N_0 e}{Cd}(v_p + x)t & : \quad t_n \le t \le t_p \\[2mm]
\frac{N_0 e}{C} & : \quad t \ge t_p
\end{cases} \tag{3.4.4}
$$

The above equation is graphically depicted in Fig.3.4.4. However an actual pulse measured through some associated electronic circuitry differs from the curve shown here due to the following reasons.

▶ An actual voltage readout circuit has a finite time constant.

▶ The charge pairs are not produced in highly localized areas.

In the above derivation leading to the pulse profile of Fig.3.4.4 we have not considered the effect of the inherent *time constant* of the detector and associated electronics on the pulse shape. Time constant is simply the product of resistance and capacitance of the circuit ($\tau = RC$). Every detector has some intrinsic capacitance as well as the cable capacitance. These capacitances together with the installed capacitor (if any) and load resistance of the output make up the effective time constant of the circuit. The difference between this time constant and the charge

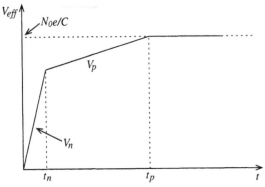

Figure 3.4.4: Pulse shape of an ideal parallel plate ion chamber. V_n and V_p are the voltage profiles due mainly to collection of electrons and ions respectively.

collection time characterizes the shape of the output pulse (see Fig.3.4.5). It is apparent that the quicker the pulse decays, the easier it will be to distinguish it from the subsequent pulse. On the other hand very small time constant may amount to loss of information and even non-linearity. Therefore considerable effort is warranted to tune the effective time constant according to the requirements. We will learn more about this in the chapter on signal processing.

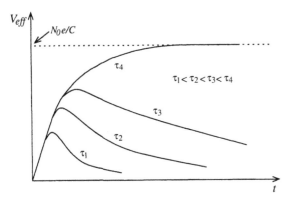

Figure 3.4.5: Realistic pulse shapes of an ion chamber with different time constants. The difference between the effective time constant of the detector and its charge collection time determines the shape of the pulse.

B.2   Cylindrical Geometry

Cylindrical ionization chambers offer a number of advantages over their parallel plate counterparts. Most notably their charge collection efficiency is much superior due to the presence of non-uniform electric field strength inside their active volumes. Such a chamber generally consists of a metallic cylinder and an anode wire stretched along the axis of the cylinder. The cylinder acts as the gas container as well as the cathode with the obvious advantage of large ion collection area and consequent high ion collection efficiency. As the anode is generally very thin, the electric lines of force around it are extremely dense and concentrated (see Fig.3.4.2). The electrons, therefore, travel toward the anode at much faster speeds than the ions moving toward the cathode. This increases the electron drift speed and improves the electron collection efficiency over the parallel plate geometry. Fig.3.4.6 shows a typical cylindrical ionization chamber.

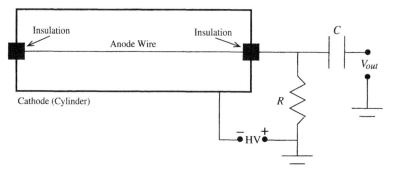

Figure 3.4.6: Sketch of a typical cylindrical ionization chamber.

Whenever an ionization interaction takes place inside the chamber, the charge pairs start moving to the respective electrodes under the influence of the applied voltage. This inhibits a change in the potential energy inside the chamber and causes a voltage pulse at the output electrode (generally anode), which can then be measured. We will now try to analytically study the time evolution of this signal. For this we first note that the change in the potential energy caused by the movement of a charge Q in the chamber by a small distance dr at r can be written as

$$dU = Q\frac{d\Phi(r)}{dr}dr. \tag{3.4.5}$$

For a cylindrical chamber the potential $\Phi(r)$ can be represented by

$$\Phi(r) = -\frac{C_l V_0}{2\pi\epsilon}\ln\left[\frac{r}{a}\right]. \tag{3.4.6}$$

Here ϵ is the permeability of the gas, C_l is the capacitance per unit length of the chamber, V_0 is the voltage applied across the electrodes, r is the radial distance from the center of the cylinder, and a is the radius of the cylinder.

The reader should note that, due to axial symmetry of the cylinder, the potential varies only in the radial direction. This is certainly not true for the potential at the two edges, since there the electric lines of force are no more uniform along the axial direction. However, the majority of practical ionization chambers have a long enough length to make this effect negligible.

Differentiating above equation with respect to r gives us the required potential gradient

$$\frac{d\Phi(r)}{dr} = -\frac{C_l V_0}{2\pi\epsilon}\frac{1}{r}. \tag{3.4.7}$$

Substituting this in equation 3.4.5 gives

$$dU = -Q\frac{C_l V_0}{2\pi\epsilon}\frac{1}{r}dr. \tag{3.4.8}$$

The potential energy U in the above equation is just the electrostatic energy contained in the electric field inside the chamber. For a cylindrical chamber of length l

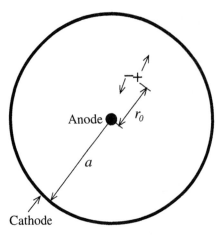

Figure 3.4.7: Production of an electron ion pair in a cylindrical ionization chamber at a distance of r_0 from the center of the anode wire.

this can also be written as

$$U = \frac{1}{2}lC_lV_0^2. \tag{3.4.9}$$

Differentiating both sides of this equation gives

$$dU = lC_lV_0dV. \tag{3.4.10}$$

Equations 3.4.8 and 3.4.10 can be equated to give

$$dV = -\frac{Q}{2\pi\epsilon l}\frac{1}{r}dr. \tag{3.4.11}$$

Let us now suppose that a charge pairs are produced at a radial distance r_0 (see Fig.3.4.7). The electrons and ions thus produced move in opposite directions under the influence of the electric potential. The change in potential due to the movement of electrons having total charge $-Q$ can be computed by integrating the above equation from $a + r_0$ to a. Hence we have

$$
\begin{aligned}
V^- &= \frac{Q}{2\pi\epsilon l}\int_{a+r_0}^{a}\frac{1}{r}dr \\
&= -\frac{Q}{2\pi\epsilon l}\ln\left[\frac{a + r_0}{a}\right].
\end{aligned}
\tag{3.4.12}
$$

Similarly, since the ions having the total charge Q move from $a + r_0$ to b, therefore the change in potential caused by them can be calculated as

$$
\begin{aligned}
V^+ &= \frac{Q}{2\pi\epsilon l}\int_{a+r_0}^{b}\frac{1}{r}dr \\
&= -\frac{Q}{2\pi\epsilon l}\ln\left[\frac{b}{a + r_0}\right].
\end{aligned}
\tag{3.4.13}
$$

To determine the total change in potential we must add V^- and V^+ together.

$$
\begin{aligned}
V &= V^- + V^+ \\
&= -\frac{Q}{2\pi\epsilon l}\ln\left[\frac{a+r_0}{a}\right] + -\frac{Q}{2\pi\epsilon l}\ln\left[\frac{b}{a+r_0}\right] \\
&= -\frac{Q}{2\pi\epsilon l}\ln\left[\frac{b}{a}\right]
\end{aligned}
\tag{3.4.14}
$$

Since the capacitance per unit length C_l for a cylindrical chamber is given by

$$
C_l = \frac{2\pi\epsilon}{\ln(b/a)},
\tag{3.4.15}
$$

therefore the total potential change V in equation 3.4.13 can also be written as

$$
V = -\frac{Q}{lC_l}.
\tag{3.4.16}
$$

This result clearly shows that for a cylindrical chamber the pulse height is independent of the point of charge generation, which proves its advantage over the parallel plate geometry. However, since the parallel plate chambers are easier to build as compared to the cylindrical chambers, therefore they are still widely used. The true advantage of the cylindrical geometry lies in its ability to produce radially non-uniform electric field and high field intensity near the anode wire, thus greatly enhancing the probability of gas multiplication. The process of gas multiplication is exploited in proportional and GM counters which we will visit later in the Chapter.

3.4.C Choice of Gas

Since the average energy needed to produce an ion pair in a gas (W-value) depends very weakly on the type of gas, therefore, in principle, any gas can be used in an ionization chamber. Ion chambers filled with air are also fairly common. However when it comes to precision detectors, the W-value is not the only factor that has to be considered since the precision of a detector depends heavily on the efficiency of charge collection. The charge collection efficiency depends not only on the detector geometry and the bias voltage but also on the drift and diffusion properties of the electrons and ions in the gas. Furthermore, small amounts of contaminants in the filling gas can severely deteriorate the performance of the chamber. The most troublesome of these contaminants are the so called *electronegative gases*, which parasitically absorb the electrons and produce non-linearity in detector's response. Since this effect is extremely important for the operation of a gas filled detector, we will revisit it in some detail later in the Chapter when we discuss the sources of errors in gaseous detectors.

The reader should again be pointed to the fact that the choice of gas is highly application dependent. For low resolution detectors where we are not concerned with fluctuations of a few percent in signal height, we can use any available gas. In fact, it is also possible to operate an ionization chamber filled with air. Such detectors are widely used in laboratories for educational purposes and generally consist of a cylindrical chamber with one end open.

3.4.D   Special Types of Ion Chambers

Ionization chambers are perhaps the most widely used radiation detectors. Because of their heavy usage and applicability in diverse applications, we will now visit some of the most commonly used variants of the standard ionization chamber geometry.

D.1   Parallel Plate Frisch Grid Chamber

The simple parallel plate geometry we discussed earlier has a major flaw: the pulse amplitude depends on the position of charge pair production (see equation 3.4.4). A cylindrical chamber, on the other hand, does not have such a dependence. This problem can be solved in parallel plate geometry as well by using the so called *Frisch grid* between the two electrodes. Fig.3.4.8 shows such a chamber. The grid (shown as dashed line) is simply another electrode kept at a potential that is intermediate to that of anode and cathode. Mechanically, the grid is made porous so that electrons could pass through it easily on their way to the anode. The chamber is designed in such a way that the incident radiation interacts only within the region enclosed by the grid and the cathode.

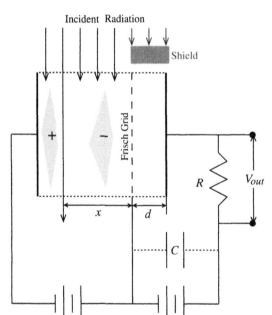

Figure 3.4.8:   Parallel plate ionization chamber with a Frisch grid. The incident radiation is directed with the help of shield to interact only within the volume bounded by the grid and the cathode.   In such a chamber the pulse shape does not depend on the position of charge pair production.

The output voltage in such a chamber is measured across a resistor between the anode and the cathode and is exclusively due to the motion of electrons. This implies that as long as there are no electrons moving between the grid and the anode, the output voltage remains constant. The point of interaction, therefore, does not have any effect on the shape of the signal. Using similar arguments as in the case of simple parallel plate ion chamber, we can deduce the expression for the output voltage. In this case, since the signal starts developing as soon as electrons pass through the grid, therefore it can be considered as a simple ionization chamber in which the charge pairs are always produced near the cathode (which in this case is actually the grid). Hence we can simply modify equation 3.4.4 to get

$$V_{out} \simeq \begin{cases} 0 & : \ 0 \leq t \leq t_{n1} \\ \frac{N_0 e}{Cd} v_n t & : \ t_{n1} \leq t \leq t_{n2} \\ \frac{N_0 e}{C} & : \ t \geq t_{n2}. \end{cases} \qquad (3.4.17)$$

where time t_{n1} and t_{n2} refer to the times the electrons take to reach the grid and the anode respectively. The shape of the pulse is shown in Fig.3.4.9.

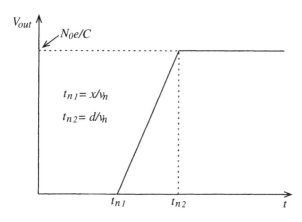

Figure 3.4.9: Output pulse shape of a Frisch grid ionization chamber. The signal is exclusively due to motion of electrons between the grid and the anode.

D.2   Boron-lined Ion Chamber

Standard ionization chambers are almost insensitive to neutron flux due to low neutron interaction cross sections in the usual filling gases. The neutrons, having no electrical charge, can hardly ionize atoms and almost exclusively interact with the nuclei. Therefore to detect neutrons one must use a material with which the neutrons could interact and produce another particle, which could then ionize the gas. Boron-10 is one such material, which absorbs a thermal neutron and emits an α-particle according to

$$n + B_5^{10} \rightarrow Li_3^7 + \alpha. \qquad (3.4.18)$$

The α-particle thus emitted can ionize the gas and thus a measurable signal can be produced. The boron-10 can be used either in gaseous form as BF_3 gas or in solid form. The BF_3 filled chambers are commonly used for neutron detection and have the same characteristics as the standard chambers. However the difficulty in purification of the BF_3 gas and its degradation with time is a major problem with these chambers. BF_3 counters are generally operated in the proportional region to obtain better signal to noise ratio. The other possibility is to use boron in solid form, which is relatively easier to purify and maintain. Such detectors, often called *boron-lined chambers*, are actually preferred over BF_3 chambers for this particular reason. In such a detector, boron is coated on the inside of the cylindrical chamber, which is filled with some conventional gas. In commercially available chambers the boron is enriched to about 20% by weight in boron-10, while in custom made chambers much higher concentrations of boron-10 are used. The thickness of the boron coating is kept smaller than the mean range of α-particles. This ensures that most of the α-particles enter the chamber volume and get detected. The detection mechanism of slow neutrons in a cylindrical boron-lined chamber is shown in Fig.3.4.10. The boron-lined chamber is also mostly used as a proportional counter but in high radiation environments it can be operated in the ionization chamber region.

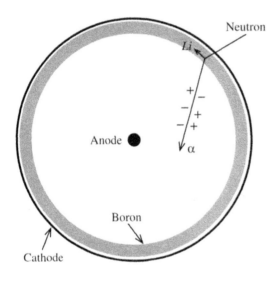

Figure 3.4.10: Principle of detection of a slow neutron from a boron-lined cylindrical chamber.

D.3 Compensated Ion Chamber

Determination of slow neutron dose in a nuclear reactor is difficult due to the presence of the accompanying high γ-ray flux. A simple boron-lined ion chamber would not work in such a situation since it can not differentiate between the two types of particles. The flux measured from such a detector is the sum of the neutron and the γ-ray responses and the elimination of the γ-ray background from the measurement is impossible. The trick that is often employed in such a situation is to simultaneously measure the total flux and just the γ-ray flux and then subtract the latter from the former. Such a system, consisting essentially of two separate or segmented ionization chambers, is referred to as a *compensated* ion chamber.

There are several possible designs of such a system including two separate detection systems for both fluxes. However the most commonly used design consists of a single but segmented ionization chamber for both measurements. One segment of this detector is boron lined sensitive to both neutrons and γ-rays while the other is an ordinary ion chamber capable of measuring only the γ-ray flux (see Fig.3.4.11). The current measured at the central electrode in this design is the current produced in the boron-lined segment minus the γ-ray induced current, which is proportional to the flux of neutrons.

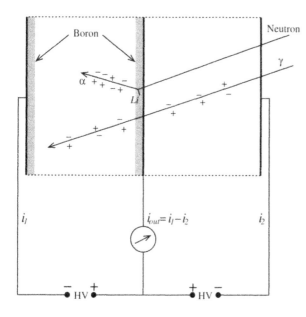

Figure 3.4.11: Working principle of the compensated ionization chamber. Here i_1 represents the current due to neu-
trons and γ-rays while i_2 is due
to γ-rays only.

3.4.E Applications of Ion Chambers

Ionization chambers are extensively used in a variety of applications due to their
simplicity in design and manufacturing, durability, radiation hardness, and low cost.
Some of their common applications include diagnostic x-ray measurements, portable
dose monitoring, radiation intensity monitoring, and use in smoke detectors.

3.4.F Advantages and Disadvantages of Ion Chambers

It should be noted that there is no such thing as a *universal detector* that could be
used in any application. Some detectors, such as ionization chambers, can be used
in a variety of applications, however most are designed and built according to the
particular applications and requirements. Therefore talking about advantages and
disadvantages of detectors is somewhat relative. Still, due to the versatility of ionization chambers we will have a general look at their advantages and disadvantages.
Let us first discuss some of their advantages.

▶ **Insensitivity to applied voltage:** Since the ionization current is essentially
independent of the applied voltage in the ion chamber region, therefore small
inevitable fluctuations and drifts in high voltage power supplies do not deteriorate the system resolution. This also implies that less expensive power supplies
can be safely used to bias the detector.

▶ **Proportionality:** The saturation current is directly proportional to the energy
deposited by the incident radiation.

▶ **Less Vulnerability to Gas Deterioration:** There is no gas multiplication
in ionization chambers and therefore small changes in the gas quality, such as
increase in the concentration of electronegative contaminants, does not severely

affect their performance. This is true for at least the low resolution systems working in moderate to high radiation fields.

Even though the ionization chambers are perhaps the most widely used detectors, still they have their own limitations, the most important of which are listed below.

- ▶ **Low current:** The current flowing through the ionization chamber is usually very small for typical radiation environments. For low radiation fields the current could not be measurable at all. This, of course, translates into low sensitivity of the system and makes it unsuitable for low radiation environments. The small ionization current also warrants the use of low noise electronics circuitry to obtain good signal to noise ratio.

- ▶ **Vulnerability to atmospheric conditions:** The response of ionization chambers may change with change in atmospheric conditions, such as temperature and pressure. However the effect is usually small and is only of concern for high resolution systems.

3.5 Proportional Counters

We saw earlier that the maximum pulse amplitude that can be achieved in a parallel plate ionization chamber is directly proportional to the number of charge pairs created by the incident radiation (see equation 3.4.4). This implies that for situations when the incident particle energy is not very large or the flux is small, the pulse amplitude may not be large enough to achieve acceptable signal to noise ratio.

Any increase in pulse amplitude is therefore tied to increase in the number of electron-ion pairs. The easiest way to achieve large number of charge pairs is to allow the primary charges produced by the incident radiation to create additional charges. We have seen this phenomenon in the section on Avalanche Multiplication. There we discussed that the primary charges are capable of producing secondary ionizations in the gas provided they achieve very high velocities between collisions. The process leads eventually to avalanche multiplication and consequently a large pulse at the output.

The basic requirement for the avalanche to occur is therefore application of very high electric potential between the two electrodes. Parallel plate geometry is very inefficient for this purpose because the electric lines of force near the anode and cathode have the same density. Even if we manage to operate a parallel plate chamber at the breakdown voltage, still it is not possible to attain acceptable proportionality between the applied voltage and the output pulse amplitude. The reason is of course the dependence of the pulse amplitude on the point of interaction of radiation. Cylindrical geometry solves both of these problems. Typically a cylindrical proportional counter is similar to a cylindrical ionization chamber, though with mechanics that can withstand higher electric potentials.

A typical proportional counter is shown in Fig.3.5.1(a). The anode in this chamber is in the form of a thin wire stretched across the center of the chamber while the wall of the cylinder acts as the cathode. This geometry ensures a higher electric field intensity near the anode wire as compared to the cathode. This non-uniformity in the electric field ensures, among other things, better electron collection efficiency

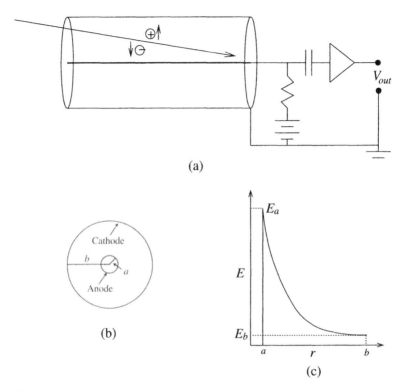

(a)

(b)

(c)

Figure 3.5.1: (a) Schematic of a cylindrical proportional counter. (b) Cross-sectional view of a cylindrical proportional counter. The thin central wire acts as the anode while the outer wall acts the cathode. (c) Radial electric field intensity profile inside a cylindrical chamber.

as compared to parallel plate geometry. The electric field intensity at any radial distance r in such a cylinder of radius b having center wire of radius a is given by

$$E(r) = \frac{1}{r} \frac{V_0}{\ln(b/a)}. \tag{3.5.1}$$

Here V_0 is the applied voltage. a and b are the radii of the anode wire and the cylinder respectively. This implies that the electric field intensity in radial direction has a $1/r$ behavior (see Figs.3.5.1(b and c)).

As stated earlier, the high electric field intensity in the vicinity of the anode ensures better electron collection efficiency. There is however another more profound effect of this, that is, the high field enables the electrons to initiate the process of avalanche multiplication, which we discussed earlier in the Chapter. For every counter geometry there is a unique range of applied voltage in which the number of charges produced in the avalanche is proportional to the number of primary charges produced by the incident radiation, that is

$$N = M N_0$$

where M is the multiplication factor, which for typical chambers lies between 10^3-10^4. Since the output signal is proportional to the total number of charges, therefore it is evident that such a chamber can amplify the signal considerably.

A detrimental effect of increasing the high voltage in proportional counters is the build up of space charge around the anode wire due to slower motion of heavy positive ions. The higher the voltage, the more ions are produced, which move much slower than electrons. The result is the screening of the electrodes and the consequent decrement in the effective electric field intensity inside the active volume. When this happens, the proportionality between the deposited energy and the pulse height can no longer be guaranteed. The proportional counters are therefore always operated below the onset of this region of *limited proportionality.*

3.5.A   Multiplication Factor

Determination of the multiplication factor is central to the operation of a proportional counter. We saw earlier in this chapter that for a uniform field the multiplication factor can be obtained from the relation 3.3.4

$$M = e^{\alpha x},$$

where α is the first Townsend coefficient. In a non-uniform field in which the Townsend coefficient has a spatial dependence, the relation 3.3.5, that is

$$M = \exp\left[\int \alpha(x)dx\right],$$

should be used instead. To evaluate this integral we need the spatial profile (more specifically radial profile for a cylindrical geometry) of α. We start with the simple relation 3.3.8 between α and the average energy gained by the electron between collisions ξ

$$\alpha = D_\alpha N_m \xi.$$

Since the electron is drifting in the influence of the electric field intensity E, therefore the energy it gains while traversing the mean free path $\lambda = 1/\alpha$ can be written as

$$\begin{aligned} \xi &= E\lambda \\ &= \frac{E}{\alpha}. \end{aligned} \tag{3.5.2}$$

Substituting this in the above expression for α gives

$$\alpha = (D_\alpha N_m E)^{1/2}. \tag{3.5.3}$$

Now we are ready to evaluate the multiplication factor using the relation 3.3.5. But before we do that we should first decide on the limits to the integral in that relation. We know that the initiation of the avalanche depends on the electric field strength. Therefore there must be a critical value of the field below which the avalanche will not occur. Let us represent this critical electric field intensity by E_c and the radial distance from the center of the cylinder at which the field has this strength by r_c. What we have done here is to essentially defined a volume around the anode wire inside which the avalanche will take place (see Fig.3.5.2). Outside of this volume

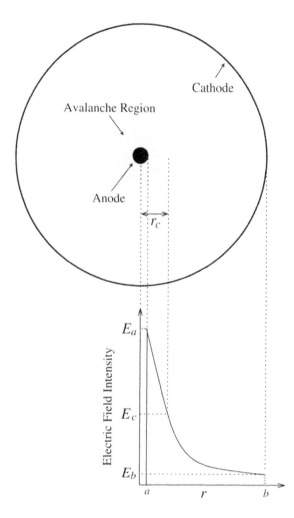

Figure 3.5.2: Depiction of avalanche region around the anode wire, the critical radial distance r_c, and critical electric field intensity E_c in a cylindrical proportional counter. In reality the avalanche region is very close to the anode wire.

there will not be any avalanche. The integral in relation 3.3.5 can then be evaluated from the surface of the anode wire a to r_c. Hence we have

$$
\begin{aligned}
M &= \exp \int_a^{r_c} \left[\frac{D_\alpha N_m V_0}{r \ln(b/a)} \right]^{1/2} dr \\
&= \exp \left[2 \left(\frac{D_\alpha N_m V_0 a}{\ln(b/a)} \right)^{1/2} \left(\sqrt{\frac{r_c}{a}} - 1 \right) \right].
\end{aligned}
\tag{3.5.4}
$$

The ratio r_c/a in the above expression can also be expressed in terms of applied voltage V_0 and the threshold voltage V_t. The threshold voltage is defined as the voltage applied at the anode below which there will not be any avalanche. Hence it can be evaluated by substituting $r = a$ and $E = E_c$ in equation 3.5.1.

$$
V_t = a E_c \ln \left(\frac{b}{a} \right)
\tag{3.5.5}
$$

Also, if we substitute $r = r_c$ in equation 3.5.1, we will get an expression for the critical field intensity E_c, that is

$$E_c = \frac{V_0}{r_c \ln(b/a)}. \tag{3.5.6}$$

By combining the above two relations we get

$$\frac{r_c}{a} = \frac{V_0}{V_t}. \tag{3.5.7}$$

The good thing about this expression is that the fraction on the right hand side can be easily determined since V_t is simply the voltage at which the avalanche multiplication begins. In other words V_t corresponds to the onset of proportional region. Substituting this ratio in equation 3.5.4 gives

$$M = \exp\left[2\left(\frac{D_\alpha N_m V_0 a}{\ln(b/a)}\right)^{1/2}\left(\sqrt{\frac{V_0}{V_t}} - 1\right)\right]. \tag{3.5.8}$$

This expression has been shown to be in good agreement with experimental results up to the moderate values of M (on the order of 10^4). At very high electric fields, the initial approximation for α we used in this derivation breaks down and therefore can not be used. However general proportional counters are operated such that the multiplication factor falls within the applicability range of this expression.

The above expression can also be written in terms of capacitance per unit length

$$C = \frac{2\pi\epsilon_0}{\ln(b/a)}.$$

Hence it can be shown that

$$M = \exp\left[2\left(\frac{D_\alpha N_m C V_0 a}{2\pi\epsilon_0}\right)^{1/2}\left(\sqrt{\frac{V_0}{V_t}} - 1\right)\right]. \tag{3.5.9}$$

Either equation 3.5.8 or 3.5.9 can be used to determine the multiplication factor for a cylindrical chamber at a certain voltage. As stated earlier, the values obtained from these expressions are good up to the usual range of applied voltages for proportional counters (37).

Example:
Plot the dependence of multiplication factor for a cylindrical proportional counter filled with argon under standard temperature and pressure on the applied voltage of up to 1000 V. The active volume of the counter has a diameter of 6 cm and the anode wire has a radius of 10 μm. Take the threshold voltage to be 500 V. The weight density of argon under standard conditions is 1.784 kg/m³.

Solution:
The given parameters are

$$a = 10^{-3} \ cm, \ b = 3 \ cm, \ \text{and} \ V_t = 500 \ V.$$

For argon $D_\alpha = 1.81 \times 10^{-17}$ cm^2/V (see Table 3.3.1) and its molecular number density can be calculated from

$$N_m = \frac{N_A \rho}{A},$$

where N_A is the Avogadro's number, ρ is the weight density, and A is the atomic mass. Hence for argon we have

$$\begin{aligned} N_m &= \frac{(6.022 \times 10^{23})(1.784 \times 10^{-3})}{18} \\ &= 5.97 \times 10^{19} \text{ atoms}/cm^3. \end{aligned}$$

Substituting all these values in equation 3.5.8 we get

$$\begin{aligned} M &= \exp\left[2\left(\frac{(1.81 \times 10^{-17})(5.97 \times 10^{19})(V_0)(10^{-3})}{\ln(3/10^{-3})}\right)^{1/2}\left(\sqrt{\frac{V_0}{500}} - 1\right)\right] \\ &= \exp\left[0.73\sqrt{V_0}\left(\sqrt{\frac{V_0}{500}} - 1\right)\right]. \end{aligned}$$

The required plot of this equation is shown in Fig.3.5.3.

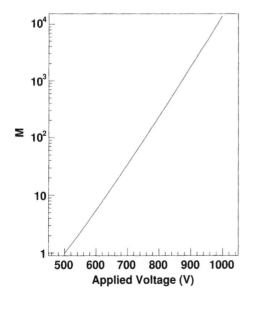

Figure 3.5.3: Variation of multiplication factor with applied voltage for an argon filled proportional counter.

3.5.B   Choice of Gas

As with ionization chambers, virtually any gas can be used in proportional counters as well. This follows from the fact that all gases and their mixtures allow the process of gas multiplication, which is the basic requirement for a detector to work as a proportional counter. However there are other factors that must be taken into account while deciding on a filling gas, some of which are discussed below.

B.1   Threshold for Avalanche Multiplication

During the discussion on the multiplication factor we realized that every gas at a certain pressure has a threshold electric potential below which the Avalanche does not take place. This threshold depends on the type of gas as well as its pressure. In Fig.3.5.4 we have plotted the first Townsend coefficient for three different types of gas mixtures. It is evident from this figure that not only the threshold for Townsend avalanche is different for each gas mixture but its voltage profile also differs somewhat. The reader should be warned that the published values of the first Townsend coefficients vary significantly from one source to another due to the difficulty in the associated measurements. Care should therefore be exercised while using the values available in literature specially for the design purposes.

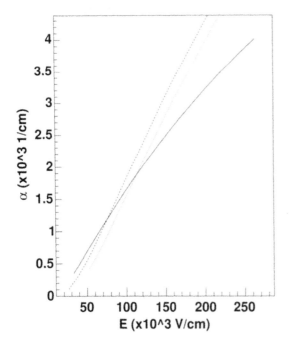

Figure 3.5.4: Dependence of the first Townsend coefficient on electric field intensity for three gas mixtures: $80\%Ne+20\%CO_2$ (solid line), $90\%Ar+10\%CH_4$ (dashed line), and $70\%Ar+30\%CO_2$ (dotted line).

Since the threshold for gas multiplication in noble gases is much lower than in polyatomic gases, therefore in general, the standard practice is to use a noble gas as the main component of the filling gas mixture. Perhaps the most commonly used noble gas in proportional counters is argon mainly because of its lower cost as compared to other inert gases.

B.2 Quenching

The avalanche multiplication in a proportional counter is a very localized process. However it has a byproduct that has the ability to cause additional localized avalanches. This byproduct is a photon with the wavelength in and around the ultraviolet region of the spectrum, which is produced during the avalanche process. The exact mechanism of the emission of these ultraviolet photons is the de-excitation of the gas molecules. For example, for argon we have

$$\gamma + Ar \quad \rightarrow \quad e + Ar^{+*} \qquad \text{- Ionization} \qquad (3.5.10)$$

$$Ar^{+*} \quad \rightarrow \quad Ar^{+} + \gamma_{uv} \qquad \text{- De-excitation.} \qquad (3.5.11)$$

Here γ in the first reaction above symbolically represents *any* ionizing radiation and γ_{uv} in the de-excitation process represents the ultraviolet photon emitted by argon. The minimum energy of these ultraviolet photons in argon is $11.6~eV$. This energy is unfortunately higher than the ionization potential of the metals commonly used in proportional counters. Hence when an ultraviolet photon strikes the cathode wall it may knock off an electron from the metal. If this electron enters the gas it gets accelerated between collisions with the gas molecules due to the high electric field inside the chamber. Eventually it may approach the anode and cause another avalanche. This process of secondary avalanche is graphically depicted in Fig.3.5.5. An obvious way to solve this problem is to add an agent in the gas that has high absorption coefficient for the photons in the ultraviolet region. Polyatomic gases (such as CH_4) fulfill this criteria since they have a number of closely spaced vibrational and rotational energy levels. Also after absorption of the photon a polyatomic molecule generally dissociates, which can be regarded as a radiationless process. The process of decreasing the probability of secondary discharges is called *quenching* and an agent used for this purpose is called *quencher*. The advantage of using a quencher in a proportional counter is evident from Fig.3.5.5.

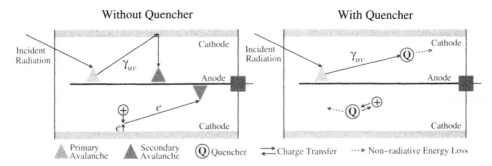

Figure 3.5.5: Positive effects of adding a quenching agent in a proportional counter.

Another effect that introduces nonlinearity in the response of a proportional counter is the emission of an electron during the process of ion-electron recombination near the cathode. We saw earlier that due to the slow movement of ions, a space charge of positive ions gets accumulated near the cathode. These ions attract free electrons from the surface of the metallic cathode and recombine with them to form neutral atoms. Although the atoms are neutralized but they are left in excited states due to the excess energy available to the ions and the electrons. The transition of such an atom to the ground state is generally accomplished through the emission of a photon, which may as well induce secondary electron emission from the cathode. Furthermore since there is a sheath of ions attracting electrons from the cathode, more electrons can be freed than required for the neutralization. All such electrons are potential avalanche initiators. This process, if not controlled, may lead to secondary avalanches in proportional counters. This problem is also solved by adding a polyatomic quenching gas in the main filling gas. The quencher molecules, having large number of rotational and vibrational energy levels, neutralize the ions through charge transfers. But since their own de-excitation processes are mostly non radiative therefore they considerably decrease the probability of secondary avalanches.

Although the polyatomic quenchers have obvious advantages we just discussed, however their use is not free from negative effects on the chamber. There are two main problems associated with the polyatomic quenchers: the buildup of polymers on anode and cathode surfaces and the decrease in the lifetime of the chamber. Both problems have the same cause, that is, the dissociation or polymerization of quenching molecules during the process of their de-excitation. The degradation of the electrodes with time is a serious problem because it may change the properties of the chamber with time. The decrease in lifetime of the chamber is a serious problem for sealed detectors. This can be circumvented by allowing the gas to continuously flow through the chamber.

B.3   Gas Gain

An important consideration while choosing filling gas for a proportional chamber is the maximum attainable gain or multiplication factor. We noted earlier that most proportional counters are operated with a multiplication factor on the order of 10^4. However sometimes it is desired to achieve higher gain before the Geiger breakdown, that is before the onset of multiple avalanches caused by a single primary avalanche. The quenching mechanism we just discussed serves this purpose to some extent. However, if the voltage is raised to high values to increase the gain, the free electrons can get enough energy to cause multiple avalanches. Therefore one must ensure that the active volume gets continuously depleted of these low energy free electrons. The best method to achieve this is by adding an electronegative impurity in the main filling gas. Freon is one such polyatomic gas. The good thing is that such gases act both as electronegative impurities as well as quenchers. The bad thing about them is their capability to parasitically capture the *good* electrons as well, thus suppressing even the primary avalanche processes. Certainly such an effect should be minimized as it can lead to appreciable decrease in detection efficiency.

3.5.C   Special Types of Proportional Counters

C.1 BF_3 Proportional Counter

BF_3 filled proportional counter is one of the most widely used neutron detectors. When a slow neutron interacts with boron-10, it produces an α-particle with two possible energies $(2.31MeV$ and $2.79MeV)$.

$$n + B_5^{10} \rightarrow Li_3^7 + \alpha$$

The α-particle thus produced has a very short range and therefore quickly interacts with gas molecules to produce electron-ion pairs. The electrons then under the influence of high electric field, initiate the avalanche. This usual gas multiplication process typical of proportional counters then ensures a large pulse at the readout electrode. BF_3 counters have a very good neutron discrimination capability due to good deposition of energy by the neutrons.

Fig.3.5.6 shows the energy spectra obtained from a very large and a typical BF_3 proportional counter. In a large counter, the electrons produced by the α-particles deposit their full energy in the active volume of the detector and thus only two well defined peaks corresponding to the two α energies are obtained. A typical counter is however not that large. In fact, the counter diameters are generally smaller than the range of energetic electrons.

A BF_3 counter can not be directly used to detect fast neutrons because of the low interaction cross section of boron-10 for fast neutrons. Therefore in order to detect fast neutrons some kind of moderator, such as paraffin, is used to first *thermalize* the neutrons. Thermalization or *moderation* is the process through which the neutrons quickly transfer their energy to the medium.

C.2 Helium Proportional Counters

A proportional counter filled with helium gas can be used to detect thermal neutrons. When an incoming neutron interacts with the helium nucleus, a proton is emitted. This proton creates secondary ionizations in the counter volume, which form the output pulse. The primary neutron interaction with a helium-3 nucleus can be written as

$$n + He_2^3 \rightarrow H_1^3 + p. \tag{3.5.12}$$

C.3 Multi-Wire Proportional Counters

A Multi-Wire Proportional Counter or simply MWPC consists of an array of closely spaced wires in a gas filled container. The wires are roughly 1 mm apart and act as anodes of individual proportional counters. Since each wire is read out through a separate electronic channel therefore the counter is used as a position sensitive detector. We will therefore defer a discussion on it to the chapter on radiation imaging.

3.6 Geiger-Mueller Counters

We mentioned earlier that, if the voltage is increased to very high values, the process of avalanche multiplication can spread throughout the detector to produce the so called breakdown in the gas. The spread is mainly caused by the ultraviolet photons emitted during the localized avalanches (see Fig.3.6.1). These photons have high enough energy to produce secondary electrons in the gas as well as in the electrodes and windows of the detector. The electrons thus produced drift towards the anode under the influence of the effective field inside the chamber. Since this field is very

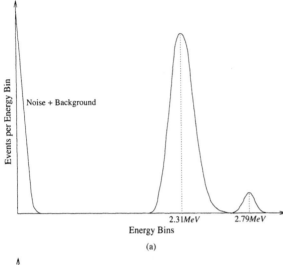

(a)

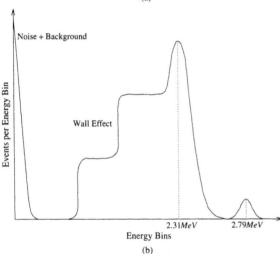

(b)

Figure 3.5.6: (a) Energy spectrum obtained from a very large volume BF_3 couner. (b) Response of a BF_3 counter having typical dimensions.

high, the electrons attain high enough energy between the collisions to produce secondary avalanches. The secondary avalanche may produce more ultraviolet photons, which may produce more avalanches and so on. This spread of avalanches throughout the detector volume is generally known as *Geiger breakdown* and the detector that behaves in this way is called a *Geiger-Mueller* or simply *GM* counter. In such a counter, whenever a single ionization takes place, it initiates an avalanche process which spreads so quickly that it causes the breakdown. The current flowing through the detector in this situation is extremely high and is limited only by the external circuitry. The voltage pulse is also quite high, generally of the order of several volts. This is a big advantage since it eliminates the need for amplification, something that is required in almost all the other types of detectors.

It is evident that the GM tubes can not provide any information about the particle energy since every particle causing an ionization in the gas produces the same pulse

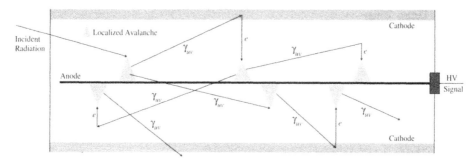

Figure 3.6.1: Spread of Geiger avalanche due to the ultraviolet photons in a GM counter.

amplitude irrespective of its energy. Hence these tubes are absolutely useless for spectroscopic purposes or for making any measurement to reveal properties of the incident radiation. This implies that the GM detectors can be used for particle counting purposes only.

3.6.A   Current-Voltage Characteristics

Let us go back to Fig.3.3.1 and have a closer look at the Geiger Mueller region. A closer view of this region is shown in Fig.3.6.2. It is evident that the pulse height at a certain bias voltage is independent of the energy delivered by the incident radiation. However the pulse height is not really independent of the applied voltage and a small positive slope is clearly visible. This occurs since at higher voltages it takes more time for the space charge to build up and decrease the effective electric field below the threshold for avalanche.

This slope, however, is of no significance as far as the operation of GM tubes is concerned. The reason is that these detectors are generally connected with a discriminator circuitry that increments an internal counter as soon as the pulse crosses its preset threshold that is chosen corresponding to the operational voltage.

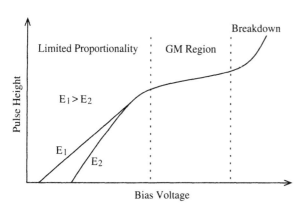

Figure 3.6.2: Pulse height variation with respect to applied voltage in and around the Geiger Mueller region of a gas filled detector.

3.6.B Dead Time

The physical process of a Geiger breakdown takes some time to subside in a GM counter. The output pulse therefore is not only large but also fairly long. The problem is that from the initiation of the breakdown until it has died and the pulse has been recorded, the counter remains dead for the subsequent ionization events. The time is called the dead time of the GM counter. Since it is not possible to eliminate the dead time therefore one must make corrections in the recorded count rate to account for the missed events. If N_c and N_t are the recorded and true count rates then, for a GM counter, they are related by

$$N_c = \eta N_t, \tag{3.6.1}$$

where η can be thought to represent the *effiiency* of the counter. The efficiency depends on many factors, for example the dead time due to the discharge process, the dead time due to electronics, and efficiency to discriminate the good events from noise etc. Let us now see if we can derive a simple relation for the efficiency using intuitive arguments. Suppose there is an average dead time τ during which the detector becomes unable to record any new ionizing event. We can assume this since under constant operating conditions the time it will take the avalanche to spread throughout the detector, cause breakdown, and then subside should not vary from pulse to pulse. Now, if C is the total number of counts recorded by the detector in a time t, then the recorded count rate will be

$$N_c = \frac{C}{t}. \tag{3.6.2}$$

Since τ is the dead time of the detector therefore the rate N_{lost} at which the true events are *not* recorded is given by

$$N_{lost} = \tau N_c N_t, \tag{3.6.3}$$

where N_t is the true count rate or the rate at which the detector will record pulses if it had no dead time. It is simply the sum of the recorded count rate and the lost count rate.

$$N_t = N_c + N_{lost} \tag{3.6.4}$$

Using the above two equations we can write

$$\begin{aligned} N_t &= N_c + \tau N_c N_t \\ \Rightarrow N_t &= \frac{N_c}{1 - \tau N_c}. \end{aligned} \tag{3.6.5}$$

Hence the efficiency of a GM tube can be written as

$$\eta = 1 - \tau N_c. \tag{3.6.6}$$

Typical GM tubes have a dead time of the order of 100 μs. It is obvious from the above relation that such a detector will have an efficiency of 50% if operated in a radiation field of 10 kHz. In other words, on the average, it will detect one particle out of two incident particles.

Example:
A GM detector having an efficiency of 67% is placed in a radiation field. On the average, it reads a count rate of 1.53×10^4 per second. Find the true rate of incident radiation and the dead time of the detector.

Solution:
The rate of incident radiation is the true count rate of equation 3.6.5. Hence we have

$$
\begin{aligned}
N_t &= \frac{N_c}{\eta} \\
&= \frac{1.53 \times 10^4}{0.67} = 2.83 \times 10^4 \ s^{-1}.
\end{aligned}
$$

For the dead time we use equation 3.6.6 as follows.

$$
\begin{aligned}
\eta &= 1 - \tau N_c \\
\Rightarrow \tau &= \frac{1 - \eta}{N_c} \\
&= \frac{1 - 0.67}{1.53 \times 10^4} \\
&= 2.15 \times 10^{-5} \ s = 21.5 \ \mu s
\end{aligned}
$$

Dead time for GM tubes is generally determined experimentally in laboratories by using the so called two-source method. This involves recording the count rates from two sources independently and then combined.

According to equation 3.6.5, the true count rates $N_{t,1}$, $N_{t,2}$, and $N_{t,12}$ of the two sources independently and combined are given by

$$
N_{t,1} = \frac{N_{c,1}}{1 - \tau N_{c,1}}
$$

$$
N_{t,2} = \frac{N_{c,2}}{1 - \tau N_{c,2}}
$$

$$
N_{t,12} = \frac{N_{c,12}}{1 - \tau N_{c,12}}
$$

Where $N_{c,x}$ with $x = 1, 2, 12$ represent the recorded count rates in the three respective configurations. Now, since the atoms in the two sources decay independent of each other therefore their true rates should add up, that is

$$
N_{t,12} = N_{t,1} + N_{t,2}. \tag{3.6.7}
$$

Substituting the true count rate expressions in this equations gives

$$
\frac{N_{c,12}}{1 - \tau N_{c,12}} = \frac{N_{c,1}}{1 - \tau N_{c,1}} + \frac{N_{c,2}}{1 - \tau N_{c,2}}
$$

$$
\Rightarrow \tau \approx \frac{N_{c,1} + N_{c,2} - N_{c,12}}{2 N_{c,1} N_{c,2}}. \tag{3.6.8}
$$

Note that, in order to determine the dead time using this method, one does not require the knowledge of the true count rates of either of the two sources. Hence any source with arbitrary strength can be chosen for the purpose. However one must make certain that the decay rate is neither very low nor very high to ensure that the efficiency of the detector does not fall too low.

Example:
In an attempt to find the dead time of a GM counter, three measurements are taken. The first two with separate radiation sources give average count rates of 124 s^{-1} and 78 s^{-1}. With both the sources in front of the detector, the count rate is found to be 197 s^{-1}. Estimate the dead time of the counter.

Solution:
We can use equation 3.6.8 to estimate the dead time as follows.

$$
\begin{aligned}
\tau &\approx \frac{N_{c,1} + N_{c,2} - N_{c,12}}{2N_{c,1}N_{c,2}} \\
&= \frac{124 + 78 - 197}{2(124)(78)} \\
&= 2.6 \times 10^{-4} \ s.
\end{aligned}
$$

3.6.C   Choice of Gas

As with proportional counters, the basic design criteria for the GM counters is that the filling gas should have low avalanche multiplication threshold. All inert gases fulfill this requirement and therefore can be used in GM counters. There is, however, an additional condition for the detector to operate in GM region, that is, it should allow the process of avalanche multiplication to cause breakdown in the gas. We discussed the process of breakdown earlier and noted that the condition for breakdown is governed by the so called Paschen's law, equation 3.3.12

$$
V_{break} = \frac{BPd}{\ln\left[APd/\ln\left(1 + 1/\gamma\right)\right]},
$$

where A and B are experimentally determined constants, P is the gas pressure and d is the separation of electrodes. Using this equation we arrived at the expression for the value of Pd at which the minimum V_{break} was possible.

$$
(Pd)_{min} = \frac{e}{A}\ln\left[1 + \frac{1}{\gamma}\right]
$$

This is a very useful relation as it can be used to determine the threshold value of the product of gas pressure and electrode gap for Geiger breakdown. Generally the physical dimensions of the detector are fixed by the engineering considerations. This means that the type of gas and its appropriate pressure can then be determined from the above relation.

　　Apart from the above condition, another thing to note is that sometimes due to safety considerations it is desired that the operating voltage is kept as low as possible.

However since at lower voltages the avalanche multiplication can not be achieved unless the pressure is lowered therefore in such situations the gas pressure is lowered to less than atmospheric pressure. The drawback to this approach as compared to the tubes that operate on atmospheric pressure is that they require specially designed containers and walls to be able to withstand the pressure gradients. In any case it must be ensured that the Paschen condition remains fulfilled.

We saw earlier that even an small amount of electronegative contaminant in the filling gas can drastically decrease the electron population due to parasitic absorption. In proportional counters this results in a decrease in the output signal strength. For GM counters the problem is not of signal strength, rather the spread and sustainment of the avalanche. With too much absorption of the electrons this process may die out soon enough and the resulting pulse may not be high enough to pass the discriminator threshold. Therefore a necessary condition for proper operation of a GM counter is that its filling gas should be as free from electronegative impurities as possible.

3.6.D   Quenching

The positive ions moving towards the cathode attract electrons from the cathode wall. The impact of these electrons on the ions may for them to emit more electrons. These new electrons can initiate another breakdown process. The result is a pulsating response of the detector after the main signal pulse has died. The two most commonly used quenching methods to reduce the dead time of a GM counter can be classified as *internal* and *external* quenching methods.

D.1   Internal Quenching

In this type of quenching a small amount of a polyatomic gas is added to the main fill gas in a concentration of around 5-10%. The main idea is to transfer the positive charge of the ion produced in the avalanche process to a molecule that, unlike the original ion, does not subsequently emit an electron thus suppressing the probability of another avalanche. This implies that there are two conditions that must be met by the added gas:

▶ It should have low enough ionization potential to ensure that the charge is efficiently transferred.

▶ It should not de-excite by emitting an electrons.

These two conditions are fulfilled by many polyatomic molecules. They have low ionization potentials and a number of closely spaced vibrational energy levels. Furthermore they generally de-excite by dissociating into simpler molecules.

D.2   External Quenching

The externally applied electric field helps the detector in multiple pulsing and is therefore the main cause of the dead time. An obvious method to solve this problem is then to decrease the high voltage rapidly so that the subsequent avalanches do not occur.

A good example of an external quenching circuit is the one that rapidly drops the anode voltage right after the beginning of a discharge. This is essentially equivalent to delivering a negative pulse of high amplitude to the detector, thus rapidly removing the space charge. The rise time of such a pulse is kept very small (on the order of several tens of a nanosecond) to ensure high efficiency in decreasing the dead time (8).

3.6.E   Advantages and Disadvantages of GM Counters

The advantages of GM counters are:

▶ **Simplicity in design:** GM counters are perhaps the easiest to build and operate in terms of readout electronics.

▶ **Invulnerability to environmental changes:** Since the magnitude of the output pulse in GM detectors is very high, therefore they work almost independent of changes in temperature and pressure.

Following are some of the disadvantages of GM counters.

▶ **Energy/Particle discrimination:** The pulse height of the GM counters is not proportional to the energy deposited by the radiation and therefore they can not be used to measure dose or discriminate between types of radiations or their energies.

▶ **Low dynamic range:** The dead time losses in GM counters increase with radiation strength. The effect can be reduced by decreasing the size of the chamber, though at the expense of reduced sensitivity. Due to this the dynamic range of GM counters is very limited.

3.7   Sources of Error in Gaseous Detectors

3.7.A   Recombination Losses

Ideally the measured ionization current in an ionization chamber should consist of all the electron-ion pairs generated in the active volume. However, due to different losses, the electrons and ions are not fully collected. For precision measurements, these losses must be taken into account. The recombination of electrons and ions is one of the major source of uncertainty in measurements especially at high fluxes of incident photons. Intuitively one can think that the recombination rate should depend directly on the concentration of charges. This suggests that the rate of change in the number of positive and negative charges should be proportional to the number of charges themselves, that is

$$\frac{dn^+}{dt} = S - \alpha n^+ n^- \tag{3.7.1}$$

$$\frac{dn^-}{dt} = S - \alpha n^+ n^-. \tag{3.7.2}$$

Here α is called the recombination coefficient and S represents the source of charges. The above two equations can be combined to give

$$\frac{d(n^- - n^+)}{dt} = 0$$

$$\Rightarrow n^- = n^+ + C_1, \tag{3.7.3}$$

where C_1 is the constant of integration and depends on the initial difference between the number of positive and negative charges. Substituting equation 3.7.3 into equation 3.7.2 gives

$$\frac{dn_-}{dt} = S - \alpha \left(n^-\right)^2 + \alpha C_1 n^-. \tag{3.7.4}$$

This is a first order linear differential equation with a solution of

$$n^- = \frac{r_1 - r_2 C_2 \exp\left(\sqrt{C_1^2 + rS/\alpha}\, t\right)}{1 - C_2 \exp\left(\sqrt{C_1^2 + rS/\alpha}\, t\right)}. \tag{3.7.5}$$

Here r_1 and r_2 are the roots of the quadratic equation on the right side of equation 3.7.4 given by

$$r_1, r_2 = \frac{1}{2}\left[C_1 \pm \sqrt{C_1^2 + 4S/\alpha}\right]. \tag{3.7.6}$$

Similarly the solution for positive charges can be obtained from equation 3.7.1. The roots in this case are given by

$$r_1, r_2 = \frac{1}{2}\left[-C_1 \pm \sqrt{C_1^2 + 4S/\alpha}\right] \tag{3.7.7}$$

The constants C_1 and C_2 can be determined by using the boundary conditions: $n = n_0$ at $t = 0$ and $n = n$ at $t = t$. Instead of solving this equation for a particular case, we note that these solutions represent complex transcendental behavior, which eventually reach the steady state value of r_2, that is

$$n_\infty \to r_2 \quad \text{as} \quad t \to \infty.$$

For a special case when the initial concentrations of positive and negative charges are equal, the constant C_1 assumes the value zero and consequently the steady state charge concentration becomes

$$n_\infty = \sqrt{\frac{S}{\alpha}}. \tag{3.7.8}$$

This shows that the equilibrium or steady state charge concentration is completely determined by the source producing electron-ion pairs and the recombination coefficient.

Application of electric field forces the charges to move toward respective electrodes thus reducing the recombination probability.

Example:
Estimate the steady state density of ions in a 0.5 atm helium filled ionization chamber if the ionization rate is 1.5×10^{11} $cm^{-3}s^{-1}$. The recombination

coefficient for helium at 0.5 atm is approximately $1.7 \times 10^{-7} \ cm^3 s^{-1}$.

Solution:
Assuming that the initial concentrations of electrons and ions are equal, we can estimate the required quantity using equation 3.7.8.

$$
\begin{aligned}
n_\infty &= \sqrt{\frac{S}{\alpha}} \\
&= \sqrt{\frac{1.5 \times 10^{11}}{1.7 \times 10^{-7}}} \\
&= 9.4 \times 10^8 \ cm^{-3}
\end{aligned}
$$

3.7.B   Effects of Contaminants

The gases used in radiation detectors are generally not free from contaminants. The most problematic of these contaminants are the electronegative molecules, which parasitically absorb electrons and form stable or metastable negative ions. Some of these impurity atoms are listed in Table 3.7.1. The most commonly found contaminants in gaseous detectors are oxygen and water vapors. It is almost impossible to purify a filling gas completely of oxygen. In fact, generally a few parts per million of oxygen is present at the filling of any gaseous detector. This concentration increases with time due to degassing of the chamber and, if the detector windows are very thin, by diffusion from outside.

The capture of electrons by these contaminants is not only a problem for proportional counters but also for high precision ionization chambers. On the other hand, for general purpose ionization chambers, capture of a few electrons by the contaminants is of not much concern as the nonlinearity caused by this generally falls within the tolerable uncertainty in detector response.

Table 3.7.1: Electro affinities of different molecules and ions (18).

Molecule	Electron Affinity (eV)	Negative Ion	Electron Affinity (eV)
O_2	0.44	O^-	1.47
C_2	3.54	C^-	1.27
Cl_2	2.38	Cl^-	3.61
OH	1.83	H^-	0.75

Often one is interested in comparing the probability of electron attachment for different gases. We can define this quantity with the help of two factors that have been measured for different gases: the mean lifetime τ_e for the electrons and their collision frequency ν_e. In terms of τ_e and ν_e the probability of capture in a single collision can be written as (37)

$$p = \frac{1}{\tau_e \nu_e}. \tag{3.7.9}$$

The values of these parameters for some common gases are shown in Table 3.7.2. It is apparent that the extremely small capture lifetimes of these gases can be a serious problem for at least high precision systems. Even small quantities of contaminants such as oxygen and water can produce undesirable non-linearity in detector response.

Table 3.7.2: Mean capture time τ_e, collision frequency ν_e, and probability of electron capture in a single collision p for some common contaminants and filling gases for radiation detectors (37). All the values correspond to electrons at thermal energies and gases under standard atmospheric conditions.

Gas	τ_e (s)	ν_e (s^{-1})	p
O_2	7.1×10^{-4}	2.2×10^{11}	6.4×10^{-9}
CO_2	1.9×10^{-7}	2.1×10^{11}	2.5×10^{-5}
H_2O	1.4×10^{-7}	2.8×10^{11}	2.5×10^{-5}
Cl_2	4.7×10^{-9}	4.5×10^{11}	4.7×10^{-4}

Let us now have a look at different mechanisms by which the contaminants capture electrons.

B.1 Radiative Capture

In this kind of capture, the capture of electron leaves the molecule in such an excited state that leads to the emission of a photon. Radiative capture can be symbolically represented as

$$\begin{aligned} e + X &\rightarrow X^{-*} \\ X^{-*} &\rightarrow X^- + \gamma, \end{aligned} \tag{3.7.10}$$

where (*) represents the excited state of molecule X. The radiative capture occurs in molecules that have positive electron affinity. Fortunately enough, for the contaminants generally found in filling gases of radiation detectors the cross section for this reaction is not significant (39).

B.2 Dissociative Capture

In this process the molecule that has captured an electron dissociates into simpler molecules. The dissociation can simply be the emission of an electron with an energy smaller than the energy of the original electron.

$$e + (XY..VZ) \rightarrow (XY..VZ)^{-*}$$
$$(XY..VZ)^{-*} \rightarrow (XY..VZ)^{*} + e' \text{ with } E_{e'} < E_e \qquad (3.7.11)$$

Here $(XY..VZ)$ represents a polyatomic molecule. In a proportional counter, the energy and the point of generation of the second electron may or may not be suitable to cause an avalanche. This uncertainty can therefore become a significant source of nonlinearity in the detector's response if the concentration of such molecular contaminants in the filling gas is not insignificant.

Not all polyatomic molecules emit secondary electrons during the process of de-excitation. Some molecules dissociate into smaller molecules such that each of the fragments is stable. Such a reaction can be written as

$$e + (XY..VZ) \rightarrow (XY..VZ)^{-*}$$
$$(XY..VZ)^{-*} \rightarrow (XY)^{*} + (VZ)^{-}\text{(stable)} \text{ or } \qquad (3.7.12)$$
$$(XY..VZ)^{-*} \rightarrow (XY..V)^{*} + Z^{-}\text{(stable)}. \qquad (3.7.13)$$

It is also possible for a fragmented part of the molecule to go in a metastable state and then decay to ground state by emitting an electron or by further dissociating according to

$$e + (XY..VZ) \rightarrow (XY..VZ)^{-*}$$
$$(XY..VZ)^{-*} \rightarrow (XY)^{*} + (VZ)^{-*}$$
$$(XY)^{*} + (VZ)^{-*} \rightarrow (XY)^{*} + (VZ)^{*} + e \text{ or } \qquad (3.7.14)$$
$$(XY)^{*} + (VZ)^{-*} \rightarrow (XY)^{*} + V^{*} + Z^{-}. \qquad (3.7.15)$$

B.3   Capture without Dissociation

In this process the polyatomic molecule captures an electron and instead of dissociating into simpler molecules, it transfers its excess energy to another molecule. This reaction can be written as

$$e + (XY) \rightarrow (XY)^{-*}$$
$$(XY)^{-*} + Z \rightarrow (XY)^{-} + Z^{*}$$
$$Z^{*} \rightarrow Z + \gamma. \qquad (3.7.16)$$

3.7.C   Effects of Space Charge Buildup

We noted earlier that the slow mobility of the positive charges towards cathode produces a sheath of positive charge cloud between the electrodes. This sheath moves slowly towards the cathode under the influence of the applied electric field. The most prominent effect of this charge cloud is that it decreases the effective electric field intensity and thereby affects the drift of electrons. To quantitatively understand this effect, let us suppose that we have a parallel plate chamber in which a cloud of positive ions is moving toward the cathode (see Fig.3.7.1).

Assuming that the positive charge density ρ_+ is constant throughout the sheath of charges having a width of δx, the equation of motion of this cloud of charges can

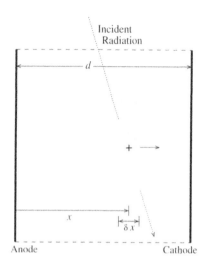

Figure 3.7.1: Schematic showing the drift of positive charge cloud towards cathode in a parallel plate chamber.

be written as

$$\bigtriangledown \cdot \vec{E} = \frac{\rho_+}{\epsilon} \tag{3.7.17}$$

$$\vec{J_+} = \rho_+ \vec{v_+} \tag{3.7.18}$$

$$\bigtriangledown \cdot \vec{J_+} = 0 \tag{3.7.19}$$

$$\vec{v_+} = \mu_+ \vec{E}. \tag{3.7.20}$$

Here $\vec{J_+}$ represents the current density of the charge flow, $\vec{v_+}$ is the drift velocity of the charges, and μ_+ is their mobility. These equations can be shown to yield

$$\bigtriangledown \cdot \left[\vec{E} \left(\bigtriangledown \cdot \vec{E} \right) \right] = 0. \tag{3.7.21}$$

For our case of parallel plate geometry this equation can be written in one dimension as

$$\frac{d}{dx} \left[E \left(\frac{dE}{dx} \right) \right] = 0. \tag{3.7.22}$$

This is justified if we assume that the distance between the electrodes is far less than the length and width of the electrodes. The correct way to say this is that we have assumed the electrodes to be infinitely long and wide, thus eliminating any need to consider the non-uniformity of electric field intensity at the ends of the chamber. In real detectors the electrodes are not that big but with proper designing (for example by using end rings) on can ensure that the edge effects are minimal. The solution to the above equation is

$$E = [2 (C_1 x + C_2)]^{1/2} , \tag{3.7.23}$$

where C_1 and C_2 are constants of integration. To determine these constants we use the following initial conditions.

$$E = -\frac{V_0}{d} - E_+ \quad \text{at } x = 0 \tag{3.7.24}$$

$$E = \frac{V_0}{d} - E_+ \quad \text{at } x = d \tag{3.7.25}$$

Here V_0 is the applied electric potential and E_+ is the electric field intensity due to the sheath of positive charges. Using these conditions in equation 3.7.23, the integration constants can be found to be

$$C_1 = -2\frac{V_0 E_+}{d^2} \quad \text{and} \tag{3.7.26}$$

$$C_2 = \frac{1}{2}\left[\frac{V_0}{d} + E_+\right]^2. \tag{3.7.27}$$

Hence equation 3.7.23 becomes

$$E = \left[-4\frac{V_0 E_+}{d^2}x + \left(\frac{V_0}{d} + E_+\right)^2\right]^{1/2}. \tag{3.7.28}$$

The dependence of the space charge induced electric field strength E_+ on the effective electric field intensity E for a parallel plate chamber of $d = 2$ cm at 2.5 cm, 3.5 cm, and 4.99 cm from the cathode has been plotted in Fig.3.7.2. The applied potential is 500 V. It is apparent that even an space charge induced electric field equal to 10% of the applied electric field can decrease the effective field intensity to unacceptable levels. Of course the value of E_+ depends on the number of positive charges in the space charge sheath and the effect will therefore depend on the number of charge pairs being created by the incident radiation. Now since we know that the number of charge pairs created depends on the energy deposited by the radiation therefore we can intuitively conclude that the effect of space charge will become more and more prominent as the energy and/or the intensity of the incident radiation increases. Since, due to very slow mobility of the positive ions, it is almost impossible to completely eliminate this effect, therefore the question that should be answered is that how much space charge can be tolerated. This depends on the application and the type of detector. In ionization chambers, where we have a plateau over a large range of applied potentials (and therefore effective electric field intensities), a small space charge effect is of not much significance. On the other hand, in a proportional chamber one must be careful in keeping the space charge to a minimum since it could lead to a decrease in the electric field intensity to lower than the avalanche threshold.

Since the effect of space charge induced electric field intensity is not negligible therefore we will now try to derive a relation for a simple case of monoenergetic photons incident uniformly over the whole detector. From the *uniformity of the flux* we mean that the incident flux can be factorized into two parts $\phi_0(y, z, t)$ and $\phi(x, t)$ such that only the second factor varies as the photons get absorbed in the detector. Let us now suppose that we want to find the number of photons absorbed in an element of thickness δx at a distance x (cf. Fig.3.7.1). If x' is a distance in this element, then according to the exponential law of photon absorption in matter, the number of photons being absorbed per unit time in x' from x is given by

$$\phi_{abs} = \phi_0(y, z, t)\phi(x, t)$$
$$= \phi_0(y, z, t)e^{-\mu\rho x}\left(1 - e^{-\mu\rho x'}\right). \tag{3.7.29}$$

Here μ and ρ are the mass absorption coefficient and density of the filling gas. The first exponential term in the above equation represents the number of photons

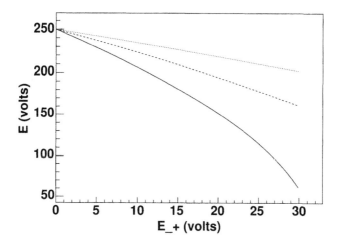

Figure 3.7.2: Dependence of space charge induced electric field E_+ inside a 2 cm long parallel plate chamber on the effective electric field intensity at three distances from the anode: 2.5 cm (dotted line), 3.5 cm (dashed line), 4.99 cm (solid line). Applied voltage is 500 V.

surviving the absorption after traveling a distance x. ϕ_0 is simply the incident photon flux. The term on brackets represents the number of photons being absorbed per unit time in a thickness x' from x.

Let us now suppose that the photon flux is constant in time. In this case the rate of creation of charge pairs will also be constant. The charge pairs thus created will move in opposite directions and constitute a current. The current can be integrated over time to determine the total charge. A good approximation of the total charge Q produced by the movement of charges by a distance $\triangle x$ is

$$Q = \frac{eN}{v} \triangle x \qquad (3.7.30)$$

where N represents the number of charges each having a unit charge e and moving with a drift velocity v. Using the relation we can estimate amount of space charge in the volume element δx to be

$$Q = \frac{eE_\gamma}{Wv_+} e^{-\mu\rho x} \int_{y=0}^{h} \int_{z=0}^{b} \phi_0(y, z, t) dy dz \int_{x'=x}^{x+\delta x} \left(1 - e^{-\mu\rho x'}\right) dx'. \qquad (3.7.31)$$

Here

e is the electronic charge,

E_γ is the photon energy,

W is the energy needed to create an electron-ion pair,

v_+ is the drift velocity of the ions,

h is the height of the space charge sheath, and

b is the breadth of the space charge sheath.

After integration the above equation yields

$$Q = \frac{eE_\gamma N_0}{W v_+} e^{-\mu\rho x} \left[\delta x - \frac{e^{-\mu\rho x}}{\mu\rho} \left(1 - e^{-\mu\rho\delta x} \right) \right],$$ (3.7.32)

where

$$N_0 = \int_{y=0}^{h} \int_{z=0}^{b} \phi_0(y,z,t) dy dz.$$

To simplify this relation we make a very valid assumption that the mean free path of the photons is much larger than the elemental length δx, that is

$$\lambda_\gamma = \frac{1}{\mu\rho} >> \delta x.$$

In this case the above relation for the space charge becomes

$$Q = \frac{eE_\gamma N_0}{W v_+} e^{-\mu\rho x} \delta x \left(1 - e^{-\mu\rho x} \right).$$ (3.7.33)

Now that we know the amount of charge in the sheath of space charge, we can compute the electric field intensity due to the whole space charge by using the Gauss's law. This law states that the net electric field intensity from a closed surface is drivable from the amount of charge Q enclosed by that surface, that is

$$\oint \vec{E} \cdot \vec{n} ds = \frac{Q}{\epsilon},$$ (3.7.34)

where ϵ is the permeability of the medium. From the beginning we have assumed that the variation in the number of charges is only in x direction. We can now extend this assumption to conclude that the components of the electric field intensity in y and z directions can be taken to be constant. Hence application of the Gauss's law on our case with Q given by equation 3.7.33 yields

$$E_{\delta x} \left[2 \int_{y=0}^{h} \int_{z=0}^{b} dy dz \right] = \frac{eE_\gamma N_0}{W \epsilon v_+} e^{-\mu\rho x} \delta x \left(1 - e^{-\mu\rho x} \right).$$ (3.7.35)

The factor 2 on the left hand side of this equation has been introduced to account for the fact that we have to integrate over both the left and the right surfaces out of which the electric lines of force are passing through. Hence the electric field intensity due to a sheath of charges is given by

$$E_{\delta x} = \frac{eE_\gamma N_0}{2W \epsilon v_+ hb} e^{-\mu\rho x} \delta x \left(1 - e^{-\mu\rho x}\right). \tag{3.7.36}$$

Now to obtain the space charge induced electric field at the edges of the volume enclosed by the electrodes we integrate the above relation over all whole x.

$$\begin{aligned} E_+ &= \frac{eE_\gamma N_0}{2W \epsilon v_+ hb} \int_{x=0}^{d} e^{-\mu\rho x} \left(1 - e^{-\mu\rho x}\right) dx \\ &= \frac{eE_\gamma N_0}{4W \epsilon v_+ \mu\rho hb} \left[1 + e^{-2\mu\rho d} - 2e^{-\mu\rho d}\right] \end{aligned} \tag{3.7.37}$$

Equations 3.7.37 and 3.7.28 can be used to determine the effective electric field inside a gaseous detector illuminated with a photon beam. The factors h and b in equation 3.7.37 can be determined from the cross sectional area of the photon beam, diffusion coefficient D of the filling gas for positive ions, and the charge integration time τ of the readout circuitry through the relations

$$h = h^* + 2\sqrt{6D\tau} \tag{3.7.38}$$
$$\text{and} \quad b = b^* + 2\sqrt{6D\tau}, \tag{3.7.39}$$

where h^* and b^* are the height and breadth of the incident photon beam.

3.8 Detector Efficiency

Now that we know about the different sources of error and their impact on detector performance, we can appreciate the fact that it is not practically possible to build a detector that is 100% efficient. If we could, such a detector would detect and measure the radiation *as it is* and not as it *sees it*. We can intuitively think that the detection efficiency of a gas filled detector would depend on many factors, such as detector geometry, type of filling gas, gas pressure and temperature, type of incident radiation, type of electronic circuitry etc. etc. The intuition is correct and leads to the problem that with so many parameters, it is fairly difficult, if not impossible, to analytically calculate the absolute efficiency of the system. However if we are really hard pressed to do that, the easier way to proceed would be to decompose the overall efficiency in components related to different parameters or sets of parameters. The individual efficiencies would then be easier to handle analytically. To make this strategy clearer, let us follow the path of a radiation beam through a parallel plate ionization chamber. As the beam passes through the entrance window of the chamber, a part of it gets parasitically absorbed in the window material. This implies that we can assign an efficiency to the system that tells us how effective the window is in *not* absorbing the radiation. Let us call it η_w. The beam of particles then passes through the filling gas of the chamber and deposits some of its energy. Now, a number of factors affect this absorption of energy by the gas molecules. Let us lump them together in an absorption efficiency and represent it by η_g. The absorbed energy can produce electron ion pairs through processes that we know are not 100% efficient (see chapter 2). Let us represent this efficiency by

η_p. The electrons and ions thus produced drift toward opposite electrodes under the influence of the applied electric potential. If the signal is measured at anode then we are mainly concerned with the drift of electrons (which, of course, also depends on the distortion of field due to the slow movement of heavy ions). We will call the efficiency of these electrons to reach the anode without being parasitically absorbed or getting escaped from the active volume by η_a. Lastly we will lump together the efficiencies of the electronic circuitry into a single parameter, η_e. The total efficiency will then be given by

$$\eta = \eta_w \eta_g \eta_p \eta_a \eta_e. \tag{3.8.1}$$

The efficiency associated with absorption in the window can be easily calculated for photons by recalling that a photon beam in matter follows exponential attenuation, that is

$$I = I_0 e^{-\mu_w x_w}, \tag{3.8.2}$$

where I_0 and I are the incident and transmitted photon intensities, μ_w is the attenuation coefficient of photons for the material of the window, and x_w is the thickness of the window. The efficiency is then given by

$$\eta_w = \frac{I}{I_0}$$
$$= e^{-\mu_w x_w}. \tag{3.8.3}$$

To calculate the absorption efficiency η_g, we again use the exponential attenuation term but since this time it is the absorption we are are interested in, the efficiency is given by

$$\eta_g = \frac{I_0 - I}{I_0}$$
$$= 1 - e^{-\mu_g x_g}, \tag{3.8.4}$$

where μ_g is the attenuation coefficient of the filling gas and x_g is the thickness of the active volume of the detector. The value is also referred to as the *quantum efficiency* of the detector. It depends on the attenuation coefficient, which is a function of the energy of the incident radiation and the type of the material. Therefore quantum efficiency has implicit energy and material dependence.

Let us now move on to the next efficiency factor, that is the efficiency of creating electron ion pairs by absorbed radiation. This process, though not 100% efficient, can still be considered so for the gases typically used in radiation detectors. The reason is that in these gases the possibility of loss of energy in the form of phonons (heat carrying particles) is fairly small. Also, the low levels of contaminants in these gases ensure that energy does not get parasitically absorbed. Hence for most practical purposes we can safely assume that

$$\eta_p \approx 1. \tag{3.8.5}$$

The efficiency of the electronic circuity highly depends on its type and therefore can vary significantly for values very close to unity for well designed systems to moderately low for systems with not so perfect designs. Therefore there is no general function or value that could be assigned to every type of electronic signal analysis and data acquisition chain. It should be mentioned that by electronic efficiency of pulse counting systems we essentially mean that the system does not miss any real pulse and does not count any false pulse. This is a very stringent requirement and is rarely met even by most sophisticated systems. One can however safely say that a well designed modern system has an efficiency that approaches unity. Let us assume that the efficiency of the electronic system is about The total efficiency of the system is then given by

$$
\begin{aligned}
\eta &= \eta_w \eta_g \eta_p \eta_a \eta_e \\
&= \left(e^{-\mu_w x_w}\right)\left(1 - e^{-\mu_g x_g}\right)(1)(1) \\
&= e^{-\mu_w x_w}\left(1 - e^{-\mu_g x_g}\right).
\end{aligned}
\tag{3.8.6}
$$

It should be stressed here that this is a somewhat simplified picture of the actual situation. We assigned perfect efficiencies to two factors. This might not be the case in a real system. But then the question is: *how accurately we actually want to know the overall efficiency?*. There is no general answer to this question since it is the application that dictates the answer.

The efficiency factor related to absorption in the detector window can also be, for most practical, purposes be assumed to be very close to unity. The reason is that the windows are generally made of very thin materials having low absorption cross sections in the energy range of interest. In such a case, the overall efficiency is simply given by the quantum efficiency of the system, that is

$$
\begin{aligned}
\eta \equiv QE &= \eta_g \\
&= 1 - e^{-\mu_g x_g}.
\end{aligned}
\tag{3.8.7}
$$

This is the reason why most experimenters concern themselves with the quantum efficiency of the system. Quantum efficiency actually sets a *physical limit* on the efficiency of the system. A system's efficiency can not be better than its quantum efficiency no matter how well the system has been designed.

Quantum efficiency, though very useful, does not tell us how efficiently the detector detects the incident particles. The reason is that it is only concerned with the efficiency of absorption of particles in the detection medium. A much more useful quantity is the so called *detective quantum efficiency* or DQE, which actually tells us how well the system works in terms of detecting and measuring radiation. Earlier in the chapter we looked at two types of gas filled detectors: integrating and quantum, though we didn't assign them these names. An ionization chamber is an integrating detector since it integrates the charges on an external capacitor for a predefined period of time and the resulting voltage is then measured by the readout circuitry. The voltage measured is proportional to the charge accumulated on the capacitor, which in turn is proportional to the energy deposited by the incident radiation. A quantum detector, such as a GM tube, on the other hand, counts individual pulses created by incident particles. Due to the difference in their modes of operation, these two types of detectors have different detective quantum efficiency profiles. For an integrating system, DQE is given by (43)

$$
DQE_{int} = \frac{QE}{1 + \frac{\sigma_m^2}{(QE)(N_{in})}},
\tag{3.8.8}
$$

where N_{in} is the number of incident photons and σ_m^2 represents the standard deviation of measurements. Note that this is not the standard deviation of the number of incident particles, which is given by $\sigma_{in} = \sqrt{N_{in}}$.

For a quantum detector, where the charges are not integrated and individual pulses are counted, this equation is not valid. For such detectors, the detective quantum efficiency is given by (43)

$$DQE_{quant} = QE\left[1 - \frac{(QE)(\tau)(N_{in})}{\Delta t}\right]^2 \exp\left(-\frac{(QE)(\tau)(N_{in})}{\Delta t}\right), \qquad (3.8.9)$$

where, as before, N_{in} is the number of incident photons, τ is detector's dead time, and Δt is the maesurement time.

Looking at equations 3.8.8 and 3.8.9 it becomes apparent that the two types of detectors have fairly different behaviors. The detective quantum efficiency of an integrating detector increases with incident photon intensity while the behavior of a quantum detector is quite the opposite (see Fig.3.8.1). This can also be understood by an intuitive argument: as the photon intensity increases, more and more photon pulses arrive within the dead time of a quantum detector, thus decreasing its detection efficiency. On the contrary, an integrating detector sees more pulses within the integration (measurement) time and hence its DQE increases.

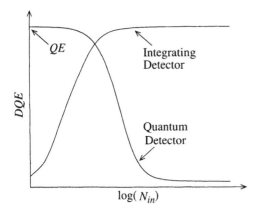

Figure 3.8.1: Variation of detective quantum efficiencies with respect to incident number of photons for integrating and quantum detectors.

The reader should again be reminded that quantum efficiency sets the physical limit of any detector. That is why the maximum DQE possible for integrating and quantum detectors, according to equations 3.8.8 and 3.8.9, is actually given by QE (cf. Fig.3.8.1).

Example:
An parallel plate ionization chamber is used to measure the intensity of a 5 keV photon beam. The detector, having an active length of 5 cm, is filled with dry air under standard conditions of temperature and pressure. For an input number of photons of 10^5, arriving at the detector within a specific integration time, the standard deviation of the measurements turns out to be 150 photons. Assuming that the absorption in the entrance window can be safely ignored, compute the quantum efficiency and detective quantum

efficiency of the detector.

Solution:
The quantum efficiency of the detector can be calculated from equation 3.8.7. However to use that equation, we need the value of the attenuation coefficient μ_g for dry air. For that we turn to the physical reference data made available by the National Institute of Standards and Technology (12). We find

$$\mu_{m,air} = 40.27 \ cm^2 g^{-1}$$
$$\text{and} \quad \rho_{air} = 1.205 \times 10^{-3} \ gcm^{-3},$$

where $\rho_{m,air}$ is the mass attenuation coefficient of dry air for 10 keV photons and ρ_{air} is the density of dry air under standard conditions. The attenuation coefficient is then given by

$$\mu_{air} = \mu_{m,air}\rho_{air}$$
$$= (40.27)\left(1.205 \times 10^{-3}\right)$$
$$= 4.85 \times 10^{-2} \ cm^{-1}.$$

We now substitute this and the length of the chamber into equation 3.8.7 to get the quantum efficiency.

$$QE = 1 - e^{-\mu_{air}x_{air}}.$$
$$= 1 - \exp\left[-\left(4.85 \times 10^{-2}\right)(5)\right].$$
$$= 0.21$$

To compute the detective quantum efficiency, we make use of equation 3.8.8.

$$DQE_{int} = \frac{QE}{1 + \frac{\sigma_m^2}{(QE)(N_{in})}}$$
$$= \frac{0.21}{1 + \frac{150^2}{(0.21)(10^5)}}$$
$$= 0.10$$

This shows that the detective quantum efficiency of the detector is only 10%, which may not be acceptable for most applications. However at higher incident photon fluxes the DQE will be higher and will eventually reach the quantum efficiency.

The detector in the previous example did not have a good quantum efficiency. But the good thing is that one has several options to increase the efficiency before building the detector. One can customize a detector according to the application to yield the maximum possible detective quantum efficiency. Perhaps the best way to increase the DQE is to increase the quantum efficiency. This can be done in may ways, such as by increasing the density of gas, using another gas, or by increasing the size of the detector.

Example:
A GM detector having a quantum efficiency of 0.34 and dead time of 50 μs is used in a low radiation environment. If within a measurement time of 1 s, 1000 photons enter the detector's active volume, compute its detective quantum efficiency. What will be the DQE if the number of photons increases to 10^5 within the same measurement time?

Solution:
Since a GM detector is a quantum detector, we can use equation 3.8.9 to compute the DQE.

$$
\begin{aligned}
DQE_{quant} &= QE \left[1 - \frac{(QE)(\tau)(N_{in})}{\Delta t} \right]^2 \exp \left(-\frac{(QE)(\tau)(N_{in})}{\Delta t} \right) \\
&= 0.34 \left[1 - \frac{(0.34)\left(50 \times 10^{-6}\right)(1000)}{1} \right]^2 \times \\
&\quad \exp \left(-\frac{(0.34)\left(50 \times 10^{-6}\right)(1000)}{1} \right) \\
&= 0.32
\end{aligned}
$$

Next we have to determine the detective quantum efficiency for the case when the same detector is used in a much more hostile radiation environment. Since all other parameters remain the same, we substitute $N_{in} = 10^5$ in the above equation to get

$$
DQE_{quant} = 0.2.
$$

3.8.A   Signal-to-Noise Ratio

Signal-to-Noise ratio, generally represented by SNR or S/R is one of the most widely used parameters to characterize detector response. As the name suggests, it is given by the ratio of the signal to noise, that is

$$
SNR = \frac{S}{N} \tag{3.8.10}
$$

where S and N represent signal and noise respectively. There are generally two types of $SNRs$ that are associated with detectors: input SNR and output SNR. The input SNR tells us what we should expect to see from the detector, while the output SNR represents the actual situation. Let us take the example of a photon detector. If we know the flux of photons incident on the detector window, we can calculate the number of photons expected inside the detector. This will be our *input* signal S_{in}. The noise of this signal is given by the statistical fluctuations in the number of photons, which can be calculated from

$$
N_{in} = \sqrt{S_{in}}. \tag{3.8.11}
$$

The signal-to-noise ratio will then be given by

$$
\begin{aligned}
SNR_{in} &= \frac{S_{in}}{\sqrt{S_{in}}} \\
&= \sqrt{S_{in}}.
\end{aligned}
\tag{3.8.12}
$$

Example:
A detector is exposed to a photon fluence of 10^5 cm^{-2}. If the detector's window has a radius of 3 cm, calculate the input signal-to-noise ratio.

Solution:
The number of photons seen by the detector are

$$
\begin{aligned}
S_{in} &= (\Phi)\left(\pi r^2\right) \\
&= \left(10^5\right)\left(\pi 3^2\right) \\
&= 2.83 \times 10^6 \text{ photons.}
\end{aligned}
$$

The expected noise from this signal is given by

$$
\begin{aligned}
N_{in} &= \sqrt{S_{in}} \\
&= \sqrt{2.83 \times 10^8} \\
&= 1.68 \times 10^3
\end{aligned}
$$

The input singal-to-noise ratio is

$$
\begin{aligned}
SNR_{in} &= \frac{S_{in}}{N_{in}} \\
&= \frac{2.83 \times 10^6}{1.68 \times 10^3} \\
&= 1683.
\end{aligned}
$$

The output signal-to-noise ratio is defined in a similar fashion but now the signal represents the actual particles seen by the detector and the noise includes all the sources of noise in the detector. If detetor's quantum efficiency is QE, the output signal is given by

$$
S_{out} = (QE)(S_{in}).
\tag{3.8.13}
$$

The output noise includes statistical noise in S_{out} as well as readout noise of the system, that is

$$
\begin{aligned}
N_{out} &= \left[\sigma_{stat}^2 + \sigma_{rout}^2\right]^{1/2} \tag{3.8.14} \\
&= \left[S_{out} + \sigma_{rout}^2\right]^{1/2} \\
&= \left[(QE)(S_{in}) + \sigma_{rout}^2\right]^{1/2}. \tag{3.8.15}
\end{aligned}
$$

Note that here readout noise includes all the noise sources in the detector and not just the electronics noise.

Example:
If the detector of the previous example has a readout noise of 200 photons and a quantum efficiency of 0.54, compute its output signal-to-noise ratio.

Solution:
The output signal is given by

$$
\begin{aligned}
S_{out} &= (QE)(S_{in}) \\
&= (0.54)\left(2.83 \times 10^6\right) \\
&= 1.53 \times 10^6 \text{ photons.}
\end{aligned}
$$

The output noise can be computed from equation 3.8.15.

$$
\begin{aligned}
N_{out} &= \left[S_{out} + \sigma_{rout}^2\right]^{1/2} \\
&= \left[\left(1.53 \times 10^6\right) + (200)^2\right]^{1/2} \\
&= 1.25 \times 10^3
\end{aligned}
$$

The output signal-to-noise ratio is then given by

$$
\begin{aligned}
SNR_{out} &= \frac{S_{out}}{N_{out}} \\
&= \frac{1.53 \times 10^6}{1.25 \times 10^3} \\
&= 1224. \tag{3.8.16}
\end{aligned}
$$

In the previous section we learned about detective quantum efficiency. The DQE can be calculated from the square of the ratio of the output to input signal-to-noise ratios, that is

$$
DQE = \left[\frac{SNR_{out}}{SNR_{in}}\right]^2. \tag{3.8.17}
$$

Example:
Compute the detective quantum efficiency of the detector described in the previous two examples.

Solution:
We use equation 3.8.17 to compute the detective quantum efficiency.

$$
\begin{aligned}
DQE &= \left[\frac{SNR_{out}}{SNR_{in}}\right]^2 \\
&= \left[\frac{1224}{1683}\right]^2 \\
&= 0.53. \tag{3.8.18}
\end{aligned}
$$

Problems

1. Compute the total and primary charge pairs created per *cm* in a mixture of 90% argon and 10% carbon dioxide.

2. If the concentrations of argon and carbon dioxide as given in the previous problem are reversed, how many charge pairs will be created per *cm*.

3. A gaseous detector is filled with 99% helium and 1% air (contaminant) at standard temperature and pressure. Assume that incident radiation produces 6 *eV* electrons at a distance of 10 *mm* from the collecting electrode. Compute the percent loss of these electrons at the electrode assuming that their total scattering cross section is 5×10^{-14} cm^2.

4. Determine the breakdown voltage for an argon filled chamber having electrode separation of 0.3 *cm* and first townsend coefficient of 100 cm^{-1} at standard pressure. What will be the breakdown voltage if the filling gas is replaced by helium?

5. The first Townsend coefficient for an argon filled chamber is desired to be around 300 cm^{-1} at a distance of 0.5 *mm* from the anode. If the gas must be kept at a pressure of 826 *torr*, estimate the electric field.

6. Determine the first Townsend coefficient for a gas at 0.01 *cm* from the electrode to achieve a multiplication factor in excess of 10^5. Also determine the second Townsend coefficient at the same location.

7. Derive the expression for Paschen minimum as a function of the Townsend coefficient.

8. Compare the Paschen minimum for helium filled chamber with that of an argon filled chamber at 0.02 *cm* from the positive electrode. Assume that the gases in both chambers are kept at 1.6 *atm* and the electric field intensity is 3×10^4 $V\,cm^{-1}$.

9. A collimated source of 6.5 *MeV* β particles is placed in a vacuumed enclosure in front of an argon filled ionization chamber having a depth of 1.7 *cm*. The window of the detector is made of 12 *μm* thick mylar sheet. If the filling gas pressure is 1.5 *atm*, calculate the energy deposited by β-particles in the detector.

10. A helium filled cylindrical chamber is operated under standard conditions of temperature and pressure. The radii of the chamber and its anode wire are 3.5 *cm* and 12 *μm* respectively. At what applied voltage you would expect this chamber to work as a proportional counter having a multiplication factor of at least 100?

11. A GM detector having a dead time of 85 *μs* measures count rates of 225 s^{-1} and 133 s^{-1} from two separate sources. If both the sources are placed in front of the detector at the same time, what count rate you would expect the GM counter to measure?

12. A GM counter records a rate of 2×10^3 counts per second when placed in a radiation field. Compute the efficiency of the detector if its dead time is 90 μs. Also compute the GM count rate you would expect if the detector was 100% efficient.

13. A gas filled ionization chamber is bombarded with a steady beam of photons having intensity of 10^{10} $cm^{-3}s^{-1}$. The photons produce electron ion pairs in the active volume of the detector. If the number density of electrons (or ions) is found to be in a steady state with a value of approximately 10^9 cm^{-3}, estimate the recombination coefficient of the filling gas.

14. Estimate the decrease in quantum efficiency of a photon detector if the entrance window absorbs 10% of incident photons instead of 1%. Assume that the absorption in the active volume remains 60% of the available photons in both cases.

15. An ionization chamber having an active depth of 5 cm is exposed to a photon flux of 1.5×10^5 $cm^2 s^{-1}$. The entrance window of the detector has an area of 6 cm^2 and the detector is filled with CO_2 under standard conditions of temperature and pressure. Compute the quantum efficiency of the detector. You can assume that the absorption in the window material is negligible.

16. For the detector in the previous problem, calculate the detective quantum efficiency if the standard deviation of the measurements turns out to be 400 photons within the integration time of 1 second. What should be the standard deviation of the measurement be if the detective quantum efficiency is to increase by 10%.

17. Derive expression for the detective quantum efficiency of an integrating detector (equation 3.8.8).

18. A quantum detector is exposed to a photon beam delivering 10^4 photons per second. Assume that the quantum efficiency of the detector is 0.6 and its dead time is 100 μs. Compute the detective quantum efficiency for a measurement time of 1 s. What will be the detective quantum efficiency if the measurement time is decreased by half?

19. A detector having a quantum efficiency of 0.4 sees a photon flux of 10^6 s^{-1}. The detector's window has an area of 5 cm^2. Assuming that the readout noise for the integration time of 100 ms is 350, compute the detective quantum efficiency of the system.

Bibliography

[1] Ahmed, S.N. et al., **High-Precision Ionization Chamber for Relative Intensity Monitoring of Synchrotron Radiation**, Nucl. Inst. Meth. in Physi. Res. A, 449, 2000.

[2] Almond, P.R. et al., **The Calibration and Use of Plane-Parallel Ionization Chambers for Dosimetry of Electron Beams: An Extension of the 1983 AAPM Protocol, Report of AAPM Radiation Therapy Committee Task Group 39**, Med. Phys. 21, 1251 - 1260, 1994.

[3] Assmann, W., **Ionization Chambers for Materials Analysis with Heavy Ion Beams**, Nucl. Inst. Meth. B 64(1-4), 267-271, 1992.

[4] Calhoun, J.M., **Radioactivity Calibrations with the 4π Gamma Ionization Chamber and Other Radioactivity Calibration Capabilities**, U.S. Dept. of Commerce, National Bureau of Standards, 1987.

[5] Constantinou, C., **Ionization Chambers Versus Silicon Diodes and the Effective Measuring Position in Cylindrical Chambers Used for Proton Dosimetry**, idwest Center for Radiological Physics, Medical Physics Division, University of Wisconsin, 1981.

[6] Cooper, P.N., **Introduction to Nuclear Radiation Detectors**, Cambridge Univ. Press, 1986.

[7] Delaney, C.F.G., Finch, E.C., **Radiation Detectors: Physical Principles and Applications**, Oxford Univ. Press, 1992.

[8] Eckey, H., A **Quenching Circuit of Very Short Rise Time for Geiger-Mueller Counters**, J. Sci. Inst. (Journal of Physics E), S.2, V.2, 1969.

[9] Eichholz, G.G., **Principles of Nuclear Radiation Detection**, Ann Arbor Science Publishers, 1979.

[10] English, W.N., Hanna, G.C., Can. J. Phys. 31, 768, 1953.

[11] Hubbell, J.H., **Photon Mass Attenuation and Energy-absorption Coefficients from 1 keV to 20 MeV**, Int. J. Appl. Radiat. Isot. 33, 1269 - 1290, 1982.

[12] Hubbell, J.H., Seltzer, S.M., **Tables of X-Ray Mass Attenuation Coefficients and Mass Energy-Absorption Coefficients 1 keV to 20 MeV for Elements Z = 1 to 92 and 48 Additional Substances of Dosimetric Interest**, Technical Report NISTIR 5632, NIST, Gaithersburg, MD 20899, 1995.

[13] IAEA, **The Use of Plane Parallel Ionization Chambers in High Energy Electron and Photon Beams: An International Code of Practice for Dosimetry (Technical Reports Series**, International Atomic Energy Agency, 1997.

[14] Jean-Marie, B., Lepeltier, V., L'Hote, D., Nucl. Instrum. Methods 159, 213, 1979.

[15] Kapoor, **Nuclear Radiation Detectors**, John Wiley and Sons Ltd., 1986.

[16] Kleinknecht, K., **Detectors for Particle Radiation**. Cambridge Univ. Press, 2001.

[17] Knoll, G.F., **Radiation Detection and Measurement**, John Wiley and Sons, 2000.

[18] Kondratev, V.N., **Energies of Chemical Bonds, Ionization Potentials, Electron Affinities**, Moscow, 1974.

[19] Korff, S.A., **Electrons and Nuclear Counters**, Van Nostrand, New York, 1946.

[20] Krithivas, G., Rao, S.N., **0.7 *cm* determination for a parallel-plate ion chamber**, Med. Phys. 13, 674 - 677, 1986.

[21] McEwan, A.C., Mathews, K.M., **The Effect of Scattered Radiation in ^{60}Co Beams on Wall Correction Factors for Ionization Chambers**, Medical Physics, 13(1): 17-118, 1986.

[22] Leroy, C., Rancoita, P.G., **Principles of Radiation in Matter and Detection**, World Scientific Pub. Comp., 2004.

[23] Ma, C.M. et al., MIT Technical Reports 129 and 130, 1982.

[24] Ma, C.-M., Nahum, A.E., **Effect of Size and Composition of Central Electrode on the Response of Cylindrical Ionization Chambers in High-Energy Photon and Electron Beams**, Phys. Med. Biol. 38, 267 - 290, 1993.

[25] Mattsson, L.O., Johansson, K.A., Svensson, H., **Calibration and Use of Plane-Parallel Ionization Chambers for the Determination of Absorbed Dose in Electron Beams**, Acta Radiol. Ther. Phys. Biol. 20, 385 - 399, 1981.

[26] Miller, D.G., **Radioactivity and Radiation Detection**, Gordon & Breach, 1972.

[27] Nasser, E., **Fundamentals of Gaseous Ionization and Plasma Electronics**, Wiley-Interscience, 1971.

[28] Ouseph, P., **Introduction to Nuclear Radiation Detectors (Laboratory Instrumentation and Techniques; V.2**, Springer, 1975.

[29] Price, W.J., **Nuclear Radiation Detection**, McGraw Hill Text, 1964.

[30] Pruitt, J.S., **Absorbed-Dose Calibration of Ionization Chambers in a ^{60}Co gamma-ray beam**, U.S. Dept. of Commerce, National Institute of Standards and Technology, 1990.

[31] Ramanantsizehena, P., Thesis, Universite de Strasbourg, CRN-HE 79-13, 1979.

[32] Reich, A.R. et al., **Flow Through Radioactivity Detection in HPLC**, VSP International Science Publishers, 1988.

[33] Rice-Evans, P., **Spark, Streamer, Proportional and Drift Chambers**, Richelieu, London, 1974.

[34] Rogers, D.W.O., **Calibration of Parallel-Plate Ion Chambers: Resolution of Several Problems by Using Monte Carlo Calculations**, Medical Physics 19, 889 - 899, 1992.

[35] Rogers, D.W.O., Bielajew, A.F., **Wall Attenuation and Scatter Corrections for Ion Chambers: Measurements Versus Calculations**, Phys. Med. Biol. 35, 1065 - 1078, 1990.

[36] Rubin, K.A., **Zapper: An ionization chamber for use on a time-of-flight system**, Thesis, Massachusetts Institute of Technology, Dept. of Physics, 1978.

[37] Sauli, F., **Principles of Operation of Multiwire Proportional and Drift Chambers**, CERN 77-09, May 1977.

[38] Sharpe, J., **Nuclear Radiation Detectors**, Wiley, 1964.

[39] Sitar, B. et al., **Ionization Measurements in High Energy Physics**, Springer-Verlag, 1993.

[40] Tait, W.H., **Radiation Detection**, Butterworth-Heinemann, 1980.

[41] Toivola, A., **A Comparative Study of the Saturation Characteristics of Air- and Hydrogen-Filled Ionization Chambers**, Annales Academiae Scientiarum Fennicae, Suomalainen Tiedeakatemia, 1966.

[42] Tsoulfanidis, N., **Measurement and Detection of Radiation**, Hemisphere Publishing Corp., 1983.

[43] Walenta, A.H., Besch, H.-J., **Gas-Filled Detectors in Medical and Industrial Imaging**, Proc. SPIE, X-ray and UV Detectors, Vol.2278, California, 1994.

[44] Wilkinson, D.H., **Ionization Chambers and Counters**, Cambridge University Press, 1950.

[45] Widdkaemper, F.W., Aalbers, A.H.L., Mijnheer, B.J., **Experimental Determination of Wall Correction Factors. Part II: NACP and Markus Plane-Parallel Ionization Chambers**, Phys. Med. Biol. 37, 995 - 1004, 1992.

Chapter 4

Liquid Filled Detectors

There is no reason why a liquid can not be used as an ionizing medium for detection of radiation. When radiation passes through a liquid, it produces charge pairs, which can be directed towards electrodes for generation of a pulse. If the liquid assures good proportionality between the energy deposited and the number of charge pairs generated, the height of the pulse would give a good measure of the energy deposited. As it turns out, there are a number of liquids that have fairly good proportionality and therefore can be used as detection media. Now, one would expect the charge recombination probability in a liquid to be much higher that in a typical gas. This is certainly true but we should also remember that the higher density ensures production of larger number of charge pairs as well. We will discuss these two competing factors later in the chapter, but the point to consider is that, in principle, liquids can be used as ionizing media to detect and measure radiation.

Apart from a section on *bubble chambers*, in this chapter we will concentrate on different types of electronic detectors that use liquids as detection media. The bubble chambers, as we will see later, do not work like conventional electronic detectors in which the voltage or current is measured at the readout electrode. Instead, the particles passing though them produce bubbles that are photographed and then visually inspected. There is also a class of detectors, called *liquid scintillation detectors*, in which the liquids produce light when their molecules are excited by incident radiation. Such devices will be discussed in the chapter on scintillation detectors.

The reason for devoting a whole chapter on this topic is that the use of liquids as active detection media is now gaining momentum in different fields, including medical imaging and high energy physics, which have traditionally relied on gas filled and solid state detectors.

4.1   Properties of Liquids

Before we go on to specific types of detectors, let us first discuss some properties of liquids that are important with respect to their use in detectors.

4.1.A   Charge Pair Generation and Recombination

The principle mechanism of a liquid filled detector is the same as a gas filled detector: a charge pair is created by the incident radiation and the resulting change in current or voltage across the electrodes is measured. Therefore the first thing to investigate is whether the generation of charge pairs in liquids and gases are analogous or not.

Unfortunately, unlike gases, in liquids the energy needed to create a charge pair does depend on the type of liquid. In the Chapter on gaseous detectors we introduced a term W-*value* to signify the energy needed to create an electron ion pair. The same terminology is used for liquids as well even though the values in liquid state are quite different from the ones in gaseous state. For liquid, another term that is extensively used is the G-*value*, which is defined as the yield of electrons for an energy of 100 eV. That is, the number of electrons (or the charge pairs) generated when the incident radiation deposits an energy of 100 eV. The reader should note that this is simply a conventional terminology and has nothing to do with the physics of pair generation in liquids. Certainly the W and the G values can be derived from one another through the relation

$$W = \frac{100}{G}, \tag{4.1.1}$$

where W will be in eV.

Table.4.1.1 gives the W and G values for various liquids that have been found to be suitable for use in radiation detectors. The reader would readily note that the energy needed to produce a charge pair in the liquefied noble gases are lower than the usual 30 eV for gases. This is encouraging for their use in radiation detectors since it would imply that more charge pairs are produced in liquids as compared to gases with the deposition of the same amount of energy. Another positive factor for liquids is their higher molecular density, because of which the total deposited energy per unit path length traversed by the radiation is also higher. The higher density in liquids implies spatial proximity of molecules, which increases the recombination probability of charges. This is a negative effect as far as radiation detectors are concerned since it introduces some uncertainty in the proportionality of measured pulse height with the deposited energy.

Liquefied argon is the most commonly used detection medium in large area detectors, such as liquid calorimeters for high energy physics experiments. Liquid xenon is generally used as a scintillation medium, that is, it produces light when its molecules are excited by the incident radiation. Liquid xenon filled detectors will be discussed in the Chapter on scintillators.

The basic principle of creation of a charge pair in a liquid is the same as in a gas. However since the energy states in liquid state are quite different than those in gaseous state therefore the process is a bit more complicated for the case of liquids. To understand this the reader is referred to Fig.4.1.1, which shows idealized energy level sketches of an element in gaseous and liquid states. The first point to note here is that the energy levels in a gas are discrete while in a liquid they are so closely spaced that they are said to form valence and conduction bands. In a liquid, the difference between the bottom of the conduction band to the top of the valence band is the band gap, which determines the energy required by an electron in the valence band to jump to the conduction band and become free to move around. In a gas this gap is much larger and therefore more energy is needed to force an electron in one of the valence energy levels to become free.

Up until now we have deliberately avoided to use the term electron ion pairs for liquids. The reason can be inferred from the energy level structure of liquids as shown in Fig.4.1.1. In a gaseous state, at least to a good approximation, each molecule can be regarded as an individual entity with its own discrete energy levels. In liquids the situation is not that simple since the spatial proximity of molecules makes them

Table 4.1.1: W and G values of several liquids used in radiation detectors (25; 9).

Liquid	W (eV)	G
Liquid Argon	23.7	4.2
Liquid Krypton	20.5	4.9
Liquid Xenon	16.4	6.1
Tetramethylsilane (TMS)	33.3	3.0
Tetramethylgermanium (TMG)	33.3	3.0
Tetramethyltin (TMT)	25.6	3.9
Hexamethyldisilane (HMDS)	50	2.0

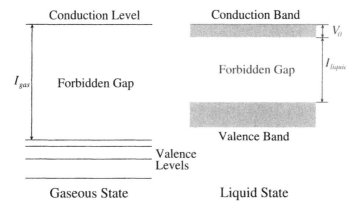

Figure 4.1.1: Comparison of energy levels in a gas and a liquid. Here I represents the ionization potential.

susceptible to each other's electromagnetic fields. The existence of energy bands is actually the result of this physical nearness. Hence, to understand the creation of a charge pair in a liquid, we must consider the whole liquid as an entity and not its individual molecules. Now let us suppose that we supply the liquid enough energy that it elevates one of its electrons from the valence band to the conduction band (see Fig.4.1.2). This process creates a vacancy in the valence band, which effectively produces a positive charge. This effective charge is generally referred to as a *hole* to signify the fact that it represents a vacancy in the valence band. The quantum mechanical treatment of this hole has shown that it can be regarded as a particle having an effective mass and unit positive charge. It is also able to move from one site to another by a reciprocal movement of an electron.

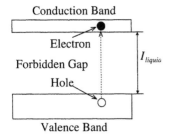

Figure 4.1.2: Creation of an electron-hole pair. Hole represents a vacancy created in the valence band created by migration of an electron to the conduction band. It has an effective positive charge.

The concept of hole is most exclusively used while treating movement of charges in solids, in which case the molecules share the electrons through bonding with neighboring molecules. The importance of such a treatment will become clear when we discuss the semiconductor detectors later on in this book. The case of liquids in a radiation detector is a little bit complicated as compared to gases and solids since we must consider holes as well as ions. The reason is that, apart from the main liquid molecules, there are other molecules in there too, which have their own energy level structures. These molecules, having lower density, can in fact be ionized. Let us suppose there is another molecule X in the filling liquid composed of molecules M. As the incident radiation passes through the liquid, it creates electron hole pairs as described earlier. These pairs can become involved in the following reactions.

▶ **Recombination:** The electron can fall back into the valence band and recombine with the hole either at the site of origination or at a spatially separated site. This would neutralize the molecule M but would leave it in an excited state. The excess energy contained by M^* will be equal to at least the difference of energy between the conduction band's lower edge and valence band's upper edge. M^* will eventually de-excite by emitting a photon, which can either escape from the detector or create another electron ion pair at another site. This process is graphically depicted in Fig.4.1.3.

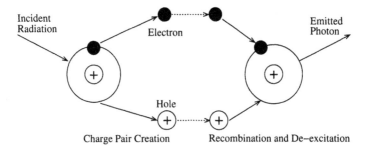

Figure 4.1.3: Creation of an electron hole pair and its subsequent recombination.

▶ **Charge Transfer:** As stated earlier, the hole produced by the incoming radiation can also transfer its charge to a *free* molecule (see Fig.4.1.4).

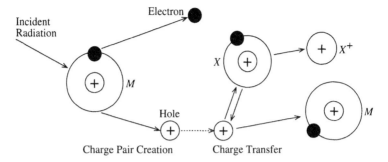

Figure 4.1.4: Creation of an electron hole pair in a molecule M with a subsequent transfer of positive charge to another molecule X.

▶ **Recombination of X^+:** The ion X^+ can also recombine with an electron to form a neutral molecule (see Fig.4.1.5). A photon may be emitted during the process as well.

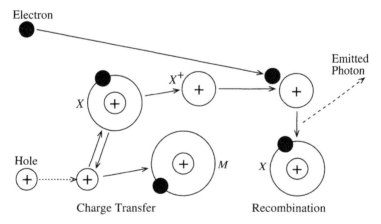

Figure 4.1.5: Recombination of X^+ with a free electron. During the process a photon may or may not be emitted depending on the type of molecule X. A multiatomic molecule having lots of vibrational and rotational energy states can also dissociate with a radiationless transition.

The recombination rate of electron hole pairs in a liquid is generally higher than electron ion recombination rate in gases due mainly to the spatial proximity of molecules. Also, as we saw earlier, the number of charge pairs created in liquids is lower than in gases for the same amount of deposited energy. Hence one should expect that the number of charges available in a liquid to create a measurable output pulse is lower than in gases. This makes the use of liquefied gases to detect and measure very low level fluxes somewhat impractical. Therefore apart from the liquid scintillators, the liquid filled detectors are generally used in moderate to high radiation fields.

Let us now turn our attention to the liquids that can be used as active detection media in detectors at room temperature. In Table 4.1.1 we mentioned four such compounds: tetramethylsilane, tetramethylgermanium, tetramethyltin, and hexamethyldisilane. One thing that can be readily observed is that the number of charge pairs created in these liquids is less than the liquefied noble gases. This implies that the signal obtained by such detectors will be weaker as compared to the conventional liquid filled detectors used in the same radiation field. Another problem that may be of concern is the creation of charge pairs through thermal agitation. However, since the difference between the valence and the conduction bands of liquids is higher than the thermal energy of molecules at room temperature, this effect can be safely ignored. In solid state detectors, such as silicon based detectors, the band gap is shorter and therefore creation of thermally agitated electron hole pairs can not be ignored. We will discuss this in detail in the Chapter on solid state detectors.

4.1.B   Drift of Charges

B.1   Drift of Electrons

While discussing the drift of charges in gases we noted that the drift velocity of electrons, to a very good approximation, is proportional to the applied electric field. In liquids the situation is not that simple, as can be deduced from Fig.4.1.6, which shows the variation of drift velocity with respect to the electric field strength for liquefied noble gases as well as their mixtures with nitrogen. It can be seen that at low electric field strengths the drift velocity can be fairly accurately described by the relation

$$v_d = \mu_e E, \qquad (4.1.2)$$

where μ_e is the mobility of electrons. However at higher field strengths, the electron drift velocity becomes less and less proportional to the field strength and at very high fields it becomes essentially independent of the field strength. The drift velocity of electrons at this stage is generally referred to as the *saturation velocity*. This nonlinear behavior of drift velocity is mainly due to the underlying nonlinearity in the energy gained by electrons through multiple collisions with increasing electric field strength.

An important thing that can be observed in Fig.4.1.6 is the independence of drift velocity on small amounts of nitrogen in the main liquid at low to moderate electric field strengths. The effect at high fields is an increase in the saturation velocity. However this behavior is not typical of all impurities or contaminants. In fact, some impurities have been seen to change the drift velocity even at very low electric field strengths (cf. Fig.4.1.7). In most cases the impurities change the drift properties of electrons significantly since their molecules act as scattering centers for the electrons. The electrons loose their energy through inelastic collisions and consequently their velocity distribution changes. In most cases, this effect is more pronounced at high field strengths when the energy gained by an electron becomes equal to or greater than the excitation energy of the impurity molecule.

Let us now turn our attention to the mobility of electrons in liquefied gases. Table.4.1.2 shows the mobilities and saturation velocities of commonly used liquefied noble gases at different temperatures. The values shown clearly demonstrate the sensitivity of electron transport on the temperature of the liquid. The drift velocity plots we saw earlier correspond to values obtained at specific temperatures. Small

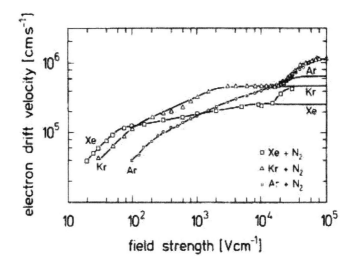

Figure 4.1.6: Variation of electron drift velocity with electric field strength for liquefied noble gases (in pure state and with addition of nitrogen). Liquid temperatures for argon, krypton, and xenon are 87 K, 12 K, and 165 K respectively (32).

fluctuations in temperature change the mobility, which changes the drift properties of electrons. This is a serious drawback of using liquefied noble gases in radiation detectors as it requires very careful monitoring and control of the temperature. Keeping the liquid at such low temperatures is generally accomplished by a liquid nitrogen flow system, which is costly as well as maintenance intensive. Even with these difficulties the liquefied noble gases are still the choice for most radiation detector developers. A number of room temperature liquids are now gaining exposure and although they have not yet been very successful but it looks like a matter of time for them to succeed their low temperature counterparts.

B.2 Drift of Ions

Unfortunately the transport of ions in liquids, that are commonly used in radiation detectors, has not been studied as rigorously as the transport of electrons. Since such studies are mostly need-related therefore we might be tempted to assume that for liquids the behavior of ions in liquids is not as important as in gases. This, however, is not a true statement because the output signal of the detector depends on movement of both negative and positive charges. The reader would recall our argument in the chapter on gaseous detectors that the output pulse is actually generated due to the change in the effective potential inside the chamber, which depends on how the positive and negative charges move under the influence of the applied electric field. The situation is not different for the liquid filled detectors either, even though here the output is almost always measured from the anode.

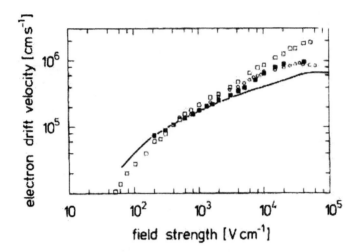

Figure 4.1.7: Variation of electron drift velocity with electric field strength for liquefied argon ($T = 87 \ K$) with additions ethane in concentrations of $5.5 \times 10^{19} \ cm^{-3}$ (circles), $8.7 \times 10^{19} \ cm^{-3}$ (filled squares), and $5 \times 10^{20} \ cm^{-3}$ (empty squares). Solid line represents pure argon (32).

Table 4.1.2: Electron mobilities μ_e and saturation velocities v_s of the three liquefied noble gases that are commonly used in radiation detectors at different absolute temperatures (32).

Liquid	$T \ (K)$	$\mu_e \ (cm^2 V^{-1} s^{-1})$	$v_s \ (cm \ s^{-1})$
Liquid Argon	87	400 ± 50	6.4×10^5
	85	475	7.5×10^5
Liquid Krypton	120	1200 ± 150	4.5×10^5
	117	1800	3.8×10^5
Liquid Xenon	165	2000 ± 200	2.6×10^5
	163	1900	2.9×10^5

As in case of gases, in liquids too the drift velocity is proportional to the applied electric field, that is

$$v_d = \mu_{ion} E, \tag{4.1.3}$$

where μ_{ion} represents the mobility of ions. Fig.4.1.8 shows an experimental curve of drift velocity of ions in liquid argon versus applied electric field intensity. It is apparent that the above equation holds up to a good approximation. The mobility in this case will be almost constant, which can be seen in Fig.4.1.9.

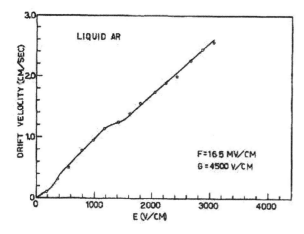

Figure 4.1.8: Variation of drift velocity of ions in liquid argon with respect to applied electric field. The parameters F and G are related to the experiment and are irrelevant for our discussion (15).

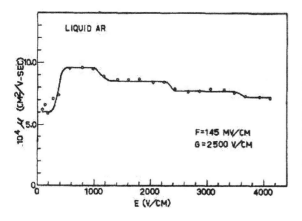

Figure 4.1.9: Variation of mobility of ions in liquid argon with respect to applied electric field. The parameters F and G are related to the experiment and are irrelevant for our discussion (15).

4.2   Liquid Ionization Chamber

Amongst the liquid filled detectors, the ionization chambers are the most widely used. Most of these detectors are filled with liquid noble gases such as liquid argon but other liquids are also now being investigated and used. Liquid filled ionization chambers have some very favorable characteristics that make it suitable for use in a variety of applications, some of which are listed below.

▶ **Stability:** Ionization chambers are stable over a longer period of time as compared to, for example liquid proportional counters. The main reason for this is their less vulnerability to small degradation of the liquid with time.

► **Proportionality:** In the Chapter on gas filled detectors we saw that one of the strongest points of ionization chambers is their excellent proportionality to the deposited energy. This is also true for liquid filled ion chambers.

► **Dynamic Range:** Liquid filled ionization chambers are not very well suited for low radiation fields due to the reasons that will become clear later in this section. However, from moderate to high fluxes, they do cover a wide dynamic range.

The basic principle of operation of a liquid filled ionization chamber is the same as a gas filled ionization chamber. In its most simple form, a sealed container having two electrodes is filled with a suitable liquid. The incident radiation produces charge pairs, which move in opposite directions under the influence of the applied electric field between the electrodes. The resulting current or the voltage pulse is measured at the anode.

4.2.A Applications of Liquid Filled Ion Chambers

Dosimetry is one of the applications where liquid filled ionization chambers are now beginning to be used. In particle accelerators used for radiation therapy, where the flux of particles is quite, such detectors are being successfully used. Liquid filled detectors have two characteristics that make them suitable for such applications: more radiation tolerant as compared to semiconductor detectors and smaller in size than gas filled detectors. For high radiation fields they offer good precision and can even be designed to provide a good degree of spatial resolution (29).

4.3 Liquid Proportional Counters

Let us now see if it is possible to develop a liquid filled proportional counter. We know from our discussions in the chapter on gas filled detectors that charge multiplication is a prerequisite for a detector to operate in the proportional region. This process can be achieved fairly easily in gases but in liquids it is neither easy to initiate nor to maintain for the reasons that will become clear shortly. Nevertheless several researchers have shown that such detectors can in fact be built.

4.3.A Charge Multiplication

As we saw earlier, in a liquid filled detector, the number of charge pairs produced is generally higher than in a typical gas filled detector. The reason is mainly the higher density of molecules and, in case of liquefied noble gases, the lower W-value. Although the larger number of charges produces, together with less diffusion of electrons, is a desirable factor specially for position sensitive detectors, but in certain situations it is also desired that the charge pair population is further increased. This can be accomplished by allowing the process of charge multiplication in much the same way as we saw in the gas filled proportional counters. That is, a very high potential can be applied in a chamber having suitable geometry such that the electrons could gain enough energy between collisions to create additional charge pairs. The most suitable geometry, as in case of gaseous proportional counters, is of course a cylinder with an anode wire stretched across its axis. Such a structure provides a very high electric field intensity near the anode wire for charge multiplication as well as better charge collection.

The basic underlying processes for charge multiplication in a liquid are the same as in the case of a gas. These processes were discussed at length in the chapter on gas filled detectors and therefore will not be repeated here. However, we will make references to the equations derived there and modify those accordingly.

The three most important parameters related to charge multiplication are its threshold voltage (or the electric field intensity), the first Townsend coefficient, and the gain. We referred to these quantities in the chapter on gas filled detectors through the terms V_t, α, and M respectively. In gases a gain of 10^4 can be quite easily achieved but in liquids it has been found that going beyond a few hundred is extremely difficult even with single wire chambers having very thin anodes. This, at first sight, may seem counterintuitive since one would expect the higher density of molecules in a liquid to favor the charge multiplication. However we should keep in mind that at each interaction not only electrons but also positive charges are produced. The ions thus produced move much slower than the electrons and produce a sheath of charges between the anode and the cathode. This cloud of charges decreases the effective electric field experienced by the electrons, thus suppressing the charge multiplication after some time.

The reader should note that even though the charge multiplication in liquids is not as large as in gases, since the initial number of charge pairs in liquids is much larger, the net effect is to increase the output signal height considerably.

The threshold for avalanche in liquids is higher than in gases, so cylindrical chambers are generally used to build liquid filled proportional counters. The first Townsend coefficient, in such a case, is a function of the position because the electric field intensity has a radial dependence inside the chamber. The growth of electron population is still exponential, that is

$$N = N_2 \exp\left[\int \alpha(r)dr\right], \tag{4.3.1}$$

where N_2 is the number of electrons initiating the avalanche and $\alpha(r)$ is the position dependent (or more specifically, electric field dependent) avalanche constant or, as it is generally known, the first Townsend coefficient. The reader can compare this equation with equation 3.3.3, which represents the avalanche multiplication in a uniform electric field environment. Here we have deliberately avoided the use of N_0 to represent the initial number of electrons, which we used in the case of gases. The reason lies in the fact that in liquids the recombination and parasitic absorption of electrons is non-negligible. We will therefore represent the initial number of electrons produced by the incident radiation by N_0 and the number of electrons that have survived the recombination by N_1. N_1 can be approximately computed from (see, for example (6))

$$N_1 = \frac{N_0}{1 + K/E(r)}, \tag{4.3.2}$$

where K is the recombination coefficient and $E(r)$ is the radial electric field intensity. We will discuss the process of recombination in the next section.

Now, the electrons that survive local recombination, encounter impurity molecules as they move toward the anode. This could result in the their parasitic capture by the impurities. If μ_c is the capture coefficient of the liquid, then after moving a distance r the number of electrons that survive the capture are given by

$$N_2 = N_1 e^{-\mu_c r}. \tag{4.3.3}$$

This, however, is an oversimplification of the actual situation since in reality the capture coefficient is itself a function of the electron energy, which in turn is a function of the electric field intensity. The above equation should then be replaced by

$$N_2 = N_1 \exp\left(-\int_a^r \mu_c(r)dr\right), \tag{4.3.4}$$

where a is the anode wire radius. Note that we have written the capture coefficient as a function of position since it depends on the electric field intensity, which is a function of position. It has been found that the capture coefficient varies approximately inversely with the electric field and can be written in a general form as (6)

$$\mu_c = A + \frac{B}{E(r)}, \tag{4.3.5}$$

where the constants A and B depend on the characteristics of the liquid and are determined experimentally.

The expression for the electric field intensity in the above equations depends on the geometry of the chamber. For parallel plate geometry the field is uniform throughout the active volume except at the edges. But such a geometry is not suitable for operation in proportional region (see also chapter on gas filled detectors). The reason is that the high field intensity needed to initiate the avalanche in a parallel plate chamber requires application of extremely high potentials at the electrodes. In liquid filled detectors the situation is even more demanding due to higher probabilities of electron recombination and capture as compared to gases. Therefore to ensure avalanche multiplication one should resort to cylindrical geometry. For a cylindrical chamber having radius b and anode wire radius a, the electric field intensity is given by

$$E(r) = \frac{V}{r \ln(b/a)}, \tag{4.3.6}$$

where V is the applied potential. Hence for a cylindrical proportional counter the capture coefficient can be written as

$$\mu_c = A + \frac{B}{V}r \ln\left(\frac{b}{a}\right). \tag{4.3.7}$$

Substitution of this expression into equation 4.3.4 yields

$$
\begin{aligned}
N_2 &= N_1 \exp\left[-\int_a^r \left\{A + \frac{B}{V}r \ln\left(\frac{b}{a}\right)\right\} dr\right] \\
&= N_1 \exp\left[-(r-a)\left\{A + \frac{(r+a)B}{2V} \ln\left(\frac{b}{a}\right)\right\}\right] \\
&\approx N_1 \exp\left[-r\left\{A + \frac{rB}{2V} \ln\left(\frac{b}{a}\right)\right\}\right], \tag{4.3.8}
\end{aligned}
$$

where we have used the approximation $r - a \approx r + a \approx r$ since $r >> a$. Substitution of N_1 from equation 4.3.2 into the above equation gives

$$N_2 = \frac{N_0}{1 + K/E(r)} \exp\left[-r\left\{A + \frac{rB}{2V}\ln\left(\frac{b}{a}\right)\right\}\right]. \qquad (4.3.9)$$

As a reminder, N_2 represents the number electrons that have survived the recombination and parasitic capture after traveling a distance r in the chamber. These electrons are able to cause avalanche multiplication provided they have gained enough energy between collisions. The number of electrons produced in the avalanche can be estimated by substituting N_2 in equation 4.3.1, that is

$$N = \frac{N_0}{1 + K/E(r)} \exp\left[-r\left\{A + \frac{rB}{2V}\ln\left(\frac{b}{a}\right)\right\} + \int_a^r \alpha(r)dr\right]. \qquad (4.3.10)$$

The multiplication factor M, which represents the ratio of the number of electrons produced in the avalanche to the initial number of electrons, is then given by

$$
\begin{aligned}
M &= \frac{N}{N_0} \\
&= \frac{1}{1 + K/E(r)} \exp\left[-r\left\{A + \frac{rB}{2V}\ln\left(\frac{b}{a}\right)\right\} + \int_a^r \alpha(r)dr\right]. \quad (4.3.11)
\end{aligned}
$$

The first and second terms in the brackets on right hand side represent the decay and growth of electron population respectively. The decay is due to the absorption of electrons by impurities and has a significant effect on the overall amplification. Therefore in order to predict the multiplication constant correctly one needs to evaluate this whole expression. However this requires the hard-to-find constants A, B, and K for the particular case under study. Although a number of experimental studies have been performed to determine these parameters but the fact that they depend heavily on small changes in impurity concentrations makes their use limited to the ones used in the experimental setup. For example the constant B, which is sometimes referred to as the *field dependent capture coefficient*, is so sensitive to the amount of impurity in the gas that it can range from almost zero for very low impurity levels up to several thousand V/cm^2 for impurity level of a few parts per million in liquefied noble gases. In gases the variation is not that dramatic and therefore small changes in impurity levels can be tolerated.

4.4   Commonly Used Liquid Detection Media

For liquid filled ionizing detectors, liquid argon is perhaps the most widely used detection medium. However, this does not mean that other liquefied gases do not possess favorable properties. In fact, there are a number of liquefied gases that have been successfully used to build ionizing and scintillating detectors. When it comes to scintillation detectors, liquid xenon is generally the choice due to its favorable scintillation light wavelength, refractive index, and Fano factor. In Table ?? we list some useful properties of the liquefied noble gases. It is interesting to note that all three gases listed in the table can act as scintillators, that is, they emit light after absorbing energy. Hence, in principle, all of them can be used to build

Table 4.4.1: Properties of liquefied noble gases.

Property	Argon	Krypton	Xenon
Z	18	36	58
A	40	84	131
Radiation Length (cm)	14.2	4.7	2.8
Critical Energy (MeV)	41.7	21.5	14.5
Fano Factor	0.107	0.057	0.041
Normal Boiling Point (K)	87.27	119.8	164.9
Liquid Density at Boiling Point (gcm^{-3})	1.4	2.4	3.0
Dielectric Constant	1.51	1.66	1.95
Scintillation Light Wavelength (nm)	130	150	175

scintillation detectors. However, as mentioned above, generally liquid xenon is used in scintillation detectors while liquid argon in ionizing detectors.

The reader might be thinking as to why one does not use regular liquids instead of the liquefied gases. There are several reasons why liquefied noble gases are preferred over regular liquids. For example the liquefied noble gases are dielectrics, which makes them suitable for free charge transport. Also, the large drift lengths of electrons make these liquids suitable for building large area detectors. As we will shortly see, the larger drift length is a consequence of lower recombination probability in liquefied noble gases as compared to molecular liquids.

Looking at Table 4.4.1, it becomes clear that xenon is the best choice for scintillating and ionizing detectors. Its stopping power is higher than argon and krypton due to its higher density. Higher stopping power allows quicker and higher deposition of energy, which means better timing and energy resolutions. Although liquid xenon is mostly used as a scintillating medium but it can, in principle, be used to build ionizing detectors as well. The biggest disadvantage of liquid xenon is its cost, which is much higher than liquid argon and krypton.

4.5 Sources of Error in Liquid Filled Ionizing Detectors

4.5.A Recombination

Recombination of electrons with the positive ions (or holes) degrades the initial electron population and therefore has the detrimental effect of signal weakening.

This can also introduce nonlinearity in the detector's response with respect to varying radiation levels. However with the introduction of high electric field across the electrodes, the recombination probability can be significantly decreased.

Most of the recombination occurs near the radiation track in the detector, where the charge pairs are produced. The electrons in a liquid quickly thermalize near their production point and in the absence of an electric field fall back to the valence band under the influence of the Coulomb field. This type of recombination is generally known as *geminate* recombination and is the major source of error in liquid filled detectors. The high rate of thermalization is due to the inelastic scattering of electrons near their point of generation. In the absence of any external electric field, the thermalization time could be as short as a fraction of a picosecond for molecular liquids at room temperature. This is a serious problem since it can lead to the loss of information and has the potential of introducing nonlinearity in detector's response. Liquids having such short recombination times are therefore not suitable for use as detection media. Fortunately, liquefied noble gases have recombination times on the order of a few hundred picoseconds, which make them suitable for use in radiation detectors. Table 4.5.1 lists the electron thermalization times in some liquefied gases commonly used in radiation detectors.

Table 4.5.1: Electron thermalization times in liquefied noble gases (26; 24; 3).

Liquid	Temperature (K)	Thermalization Time (ns)
Helium-4	4.2	0.4×10^{-3}
Argon	85	0.9 ± 0.2
Krypton	117	4.4 ± 0.2
Xenon	163	6.5 ± 0.5

The geminate recombination just described is not the only type of recombination that occurs in liquids. Apart from this localized recombination, there is also volume recombination effect that can happen anywhere in the liquid volume. Fortunately the probability of volume recombination is much lower than the geminate recombination simply because of the low concentration of ions away from the particle track. Also, under the influence of the applied electric field, the electrons move towards anode and the ions move towards cathode, which does not leave many ions or recombination centers for electrons to recombine with.

It was mentioned earlier that the number of electrons that survive the recombination can be calculated from (see equation 4.3.2)

$$N = \frac{N_0}{1 + K/E(x)},$$

where N_0 is the number of electrons produced by the incident radiation, $E(x)$ is the electric field intensity at position x in the detector, and K is a constant known as the *recombination coefficient*. Certainly for a parallel plate geometry the electric field intensity is constant at each point and therefore independent of x. For cylindrical geometry it generally suffices to use the radial component of the electric field intensity in the above relation.

The recombination coefficient in the above equation is determined experimentally. Unfortunately not much data is available for liquids that could potentially be used in radiation detectors. Also, one finds differences in reported values. Nevertheless, one can get an approximate idea of the recombination by using the reported values in the above equation (see example below).

Example:
A parallel plate liquid xenon filled ionization chamber is exposed to a flux of γ-rays, which is producing 5×10^3 charge pairs per second midway between the two electrodes. Compute the number of electrons that survive the recombination if the applied electric field is $1\ kV/cm$. Assume the recombination coefficient to be $100\ V/cm$.

Solution:
Since it is a parallel plate chamber, we can assume that the electric field intensity is constant throughout its active volume. We can then simply substitute the given values in equation 4.3.2 to get the desired result.

$$
\begin{aligned}
N &= \frac{N_0}{1 + K/E(x)} \\
&= \frac{5 \times 10^3}{1 + 100/1000} \\
&= 4.5 \times 10^3 \text{ charge pairs}
\end{aligned}
$$

4.5.B   Parasitic Electron Capture and Trapping

One of the major problems associated with using a liquid as the ionizing medium is the capture of electrons by the impurity molecules. The reader should note that the capture process is separate from the recombination effect we just studied. The recombination process involves the fall of an electron back into the conduction band of the liquid while in the capture process the electron gets captured by an impurity molecule. Another difference is that the recombination process generally occurs near the site of the charge pair production. On the other hand the capture process does not occur with this preference.

The impurity in a liquid filled detector can be of two types. One is the parasitic capture and trapping impurity and the other is the so called reversible attachment impurity. Most of the signal deteriorating effects in a detector are caused by the impurity of the first kind, in which once the electron gets captured, it does not get re-emitted and hence is said to have been trapped. The process of electron attachment to a trapping impurity molecule (XY) can be written as

$$
e + (XY) \rightarrow (XY)^{-*}. \tag{4.5.1}
$$

The ($*$) sign above represents the excited state of the molecule. The de-excitation could be by emission of a photon or, in case of multi-atomic complex molecules, by molecular dissociation into smaller fragments. The latter is a radiationless process. A good example of such an impurity is the oxygen molecule, which is very commonly found even in highly purified liquids.

There are also some molecular species that emit the electron after capturing it. Such a reaction is written as

$$e + (XY) \rightleftharpoons (XY)^-. \tag{4.5.2}$$

where the ($\rightleftharpoons$) sign represents the fact that the process of electron capture is reversible. A common example of such an impurity is the carbon dioxide molecule.

To understand the deteriorating effect of electron capture by an impurity molecule on the detector response, we first note that the capture introduces an effective negative charge on the molecule. The reader should recall that the output signal of an ionization chamber has two edges: a fast rising edge and a slow falling edge. The fast rising edge is almost exclusively described by the movement of negative charges towards the anode. The introduction of heavy and slow moving negative ions in the electron population produces larger slope in the rising edge and can even reduce the signal height (the signal height depends on the movement of both positive and negative charges). Hence for liquids, having higher molecular density, the electron capture process can be a source of nonlinearity in the detector response.

Let us now see how we can numerically describe this parasitic capture process. The reader may recall that earlier in the Chapter we had described the survival of electrons in a liquid by an exponential function of the form (cf. equation 4.3.4)

$$N = N_0 e^{-\mu_c x}, \tag{4.5.3}$$

where N_0 represents the initial number of electrons, N is the number of electrons that have survived after traversing a distance x. μ_c is the capture coefficient for electrons in the liquid. The capture coefficient depends not only on the type of the medium but also on the energy of electrons. In most situations the electrons are in thermal equilibrium with the liquid molecules and therefore we can use the capture coefficient at the mean thermal energy for the liquid under consideration. Using equation 4.5.3 we can define the probability of capture P_c and probability of survival P_s through the relations

$$
\begin{aligned}
P_c &= \frac{N_0 - N}{N_0} \\
&= 1 - e^{-\mu_c x} \tag{4.5.4} \\
P_s &= \frac{N}{N_0} \\
&= e^{-\mu_c x}. \tag{4.5.5}
\end{aligned}
$$

Since μ_c in the above relations has the units of inverse distance, we can define a term *capture mean free path* λ_c through the relation

$$\lambda_c = \frac{1}{\mu_c}. \tag{4.5.6}$$

Substituting $\lambda_c = x$ in equations 4.5.4 and 4.5.5 gives us the definition of the capture mean free path as

$$P_c = 1 - e^{-1} \approx 0.63$$
$$P_s = e^{-1} \approx 0.37,$$

that is, it represents the distance for which the capture and survival probabilities of an electron is about 63% and 37% respectively. In other words, if a population of electrons travels a distance λ_c then only about 37% of the electrons will survive the capture by molecules.

Up until now we have used the spatial variation of the electron flux due to the capture process. We can also describe the change in electron concentration with respect to time. For this, we argue that the rate of electron capture should be proportional to not only the impurity concentration but also the electron concentration in the liquid. In fact this has been observed by various experimenters. We can therefore write the rate equation as

$$\frac{dC_e}{dt} = -kC_eC_{imp}, \qquad (4.5.7)$$

where C stands for concentration (generally described in *number of particles per mole of the liquid*) with subscripts e and imp representing electrons and impurity molecules respectively. k is the constant of proportionality that depends on the type of impurity. This so called *reaction rate constant* can generally be found in literature in units of per mole per second (i.e. $M^{-1}s^{-1}$). The negative sign in the above expression describes the decrease in electron population with time.

If we now assume that the concentration of impurity in the liquid does not change with time then equation 4.5.7 can be solved to give

$$C_e = C_{e0}e^{-kC_{imp}t}, \qquad (4.5.8)$$

where we have used the initial condition $C_e = C_{e0}$ at $t = 0$. It should be noted that the constancy of C_{imp} with time is not a strictly valid assumption specially for trapping impurities since the loss of electrons also means loss of impurity molecules. The reason is that most molecular ions thus formed loose their ability to capture more electrons. However, since in liquid filled detectors that use high purity liquids the capture rate is generally quite low, we can safely ignore the time dependence of impurity concentration. For the reversible attachment impurity molecules this assumption holds up to a good approximation. Such impurities release the captured electron after a small time delay, which at most results in the longer transit time of electrons and increases the slope of the rising edge of the output pulse by a small amount. The signal height, however, is not affected.

Using equation 4.5.8 we can define the *mean electron lifetime* τ as

$$\tau = \frac{1}{kC_{imp}}. \qquad (4.5.9)$$

τ represents the time it takes an electron population to decrease by about 63% (see Example below).

Example:
Define the mean electron lifetime τ in terms of capture rate of electrons.

Solution:
We just saw that the electron concentration at any time t can be computed from equation 4.5.8, provided the impurity concentration does not change with time. Let us Substituting $t = \tau$ from equation 4.5.9 into this equation to determine the electron concentration after passage of one lifetime. This gives

$$C_e = C_{e0}e^{-1}$$
$$\Rightarrow \frac{C_e}{C_{e0}} \approx 0.37$$

This implies that after time τ the surviving electron population is about 37% of the initial population. In other words τ is the time taken by the impurity molecules to capture about 63% of electrons.

The reaction rate constant k is an important and widely used parameter to assess the effect of impurities in detectors. It tells us how quickly or *violently* an impurity captures the free electrons in the liquid, and therefore gives us a measure to determine which impurities can be tolerated in a certain environment. Another advantage of using k is that it can be used to determine the lifetime of electrons.

The reaction rate constant can be determined from the following relation

$$k \approx \int_0^\infty \sigma(E)f(E)dE, \tag{4.5.10}$$

where $\sigma(E)$ is the energy dependent attachment cross section for electrons in the liquid and $f(E)$ is the distribution of electrons in the liquid. For most practical purposes, we can safely assume that the electrons in a liquid can be described by Maxwellian distribution even in the presence of high electric field. This is in contrast to the case of gases where, as we saw in the previous chapter, the electrons do not retain their Maxwellian distribution in the presence of high electric field. As a reminder, the Maxwellian distribution is given by

$$f(E) = \frac{2}{\sqrt{\pi}k_B T}\left(\frac{E}{k_B T}\right)^{1/2} e^{-E/k_B T}, \tag{4.5.11}$$

where k_B is the familiar Boltzmann's constant and T is the absolute temperature. The distribution assumes that the particles are in thermal equilibrium with the surrounding. The average energy of the electrons in such a case is given by

$$<E> = \frac{3}{2}k_B T. \tag{4.5.12}$$

This energy can be increased by the application of electric field across the electrodes, something that is always done in liquid ionization and proportional chambers. As stated earlier, this increase in the average energy does not significantly affect the distribution of electrons and therefore the Maxwellian distribution can still be used to determine the reaction rate constant.

The reaction rate constants have also been determined for different impurities by several researchers. The data varies considerably from one specie to the next. for example for oxygen the value has been found to be $6 \times 10^{11} \ M^{-1}s^{-1}$ while for sulfur hexafluoride and carbon tetrachloride the constant assumes a value about three orders of magnitude higher than that for oxygen (see, for example (9)).

Because of the inverse relationship between the electron lifetime and the rate constant, the lower the rate constant the better for the detector since it would imply that more electrons survive to induce the potential change necessary to produce the output signal.

Example:
Determine the lifetime of electrons in a liquid argon filled detector having 2 *ppm* of oxygen as impurity.

Solution:
2 parts per million (ppm) of oxygen would be equivalent to a molar concentration of

$$C_{oxy} = 2 \times 10^{-6} \ M. \tag{4.5.13}$$

According to equation 4.5.9 the mean lifetime of electrons is given by

$$\tau = \frac{1}{kC_{oxy}}$$
$$= \frac{1}{(6 \times 10^{11}) (2 \times 10^{-6})}$$
$$\approx 0.8 \ \mu s.$$

Such a small lifetime can be detrimental to the performance of the detector specially in low radiation environments.

4.6 Cherenkov Detectors

In chapter 2 we discussed the process of production of Cherenkov radiation in transparent media. The detectors based on the exploitation of this phenomenon to detect particles are called Cherenkov detectors. Before we go on to the discussion of such detectors, let us first summarize the basic points related to the emission process of Cherenkov radiation.

▶ Whenever a charged particle in a medium moves faster than speed of light *in that medium*, it emits Cherenkov radiation.

▶ Cherenkov radiation is composed of photons with wavelengths mostly in the visible region of the electromagnetic spectrum.

▶ The radiation is emitted in the shape of a cone with the path of the particle as its axis. The angle θ_c of the cone is related to the velocity v of the particle and the refractive index n of the medium through the relation

$$
\begin{aligned}
\cos \theta_c &= \frac{c}{vn} \\
&= \frac{1}{\beta n},
\end{aligned}
\tag{4.6.1}
$$

where $\beta = v/c$ is the ratio of the velocity of the particle *in the medium* to the velocity of light *in vacuum*.

▶ The threshold velocity of the particle to produce Cherenkov light as obtained from the above relation is

$$
v_{th} = \frac{c}{n}.
\tag{4.6.2}
$$

▶ The angle of the cone is proportional to the energy of the particle producing the radiation.

The last three points above are very important in terms of building a detector based on Cherenkov emission. The unique conic geometry of the emitted radiation can be used not only to tag events (that is, differentiate them from the background) but also to estimate the energy of the particle producing the radiation.

A liquid filled Cherenkov detector uses a liquid to produce Cherenkov light either directly by the particle to be detected or by another particle interacting with it. The light thus produced is detected by detectors surrounding the liquid. Though any type of detector can be used to detect light photons but the general practice is to use photomultiplier tubes due to their high photon detection efficiency. We will learn more about these tubes in the chapter on photo detectors.

Example:
Calculate the critical angle of Cherenkov cone produced by the passage of electrons through light water. Assume the average velocity of the electrons to be $0.92c$, where c is the velocity of light in vacuum.

Solution:
Substituting $\beta = 0.92$ and $n = 1.33$ (for light water) in equation 4.6.1, we get

$$
\begin{aligned}
\cos \theta_c &= \frac{1}{\beta n} \\
&= \frac{1}{(0.92)(1.33)} = 0.817 \\
\Rightarrow \theta_c &= 35.2^0.
\end{aligned}
$$

Fig.4.6.1 shows a Cherenkov detector consisting of a spherical ball containing a liquid and photomultiplier tubes surrounding it. Such a detector has been built for detecting neutrinos at the Sudbury Neutrino Observatory (SNO) in Canada. The neutrinos, as we saw in chapter 1, have extremely low interaction cross section and therefore detecting them is quite a challenging task. The SNO detector uses heavy water to detect these elusive particles. There are three modes of interaction of neutrinos that are exploited at this facility. We will look at one of those to explain the concept of indirect detection through Cherenkov light. A neutrino can elastically scatter off an orbital electron in heavy water and set it free. Since the neutrinos coming from the sun have high enough energy, they provide the electron with so much energy that their velocity crosses the threshold of equation 4.6.2. As a result the electron produces Cherenkov light. The cone of light spreads and is ultimately detected by the photomultiplier tubes (see Fig.4.6.1).

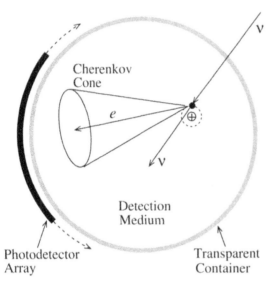

Figure 4.6.1: Sketch of the working principle of a neutrino detector. The neutrino is shown to scatter off an electron from the medium. If the scattered electron moves with a velocity higher than the velocity of light in that medium, it produces Cherenkov light in the form of a cone. The light photons thus produced are detected by an array of photomultiplier tubes installed around the spherical container of the detection medium. Such a detector has been built at the Sudbury Neutrino Observatory in Canada.

4.7 Bubble Chamber

Bubble chamber is one of the earliest and extremely successful imaging detectors. It was built for tracking particles in high energy particle collisions.

A conventional bubble chamber is made of a sealed container filled with a liquefied gas. The chamber is designed such that pressure inside can be quickly changed. The idea is to momentarily superheat the fluid when the particles are expected to pass through it. This is accomplished by suddenly lowering the pressure, which decreases the boiling point of the liquefied gas, thus converting it into a superheated liquid. When particles pass through this fluid they produce dense tracks of localized electron ion pairs. The energy delivered to the liquid during this process produces tiny bubbles along the particle's track. The whole chamber is then illuminated and photographed by a high definition camera. The photograph is then analyzed offline for particle identification and measurements. Bubble chambers were highly successful in the early days of high energy physics research, where application of external magnetic field allowed measurements of particle momenta and thus facilitated particle identification. Fig.4.7.1 shows a typical photograph obtained from a bubble chamber.

The obvious disadvantage of bubble chambers is that it is extremely difficult to use it for online analysis and triggering. The bubble chambers have now been replaced with other types of electronic trackers, most of which are based on sili-

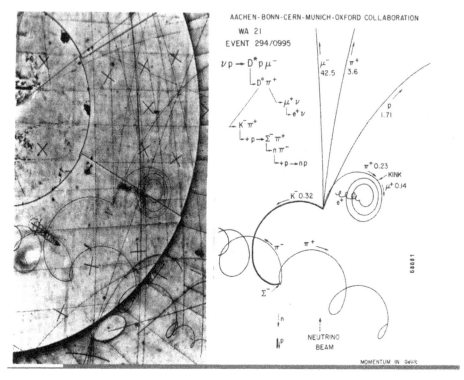

Figure 4.7.1: Typical particle tracks seen from a bubble chamber (left) and their interpretations (right). Courtesy of CERN.

con multistrip detectors. However some experimenters have recently proposed that specially designed bubble chambers can still be useful in detecting low energy and weakly interacting particles (2).

4.8 Liquid Scintillator Detectors

Liquid scintillators are another class of detectors that are now being extensively used in a variety of applications. They operate on the principle of light emission by scintillation media when exposed to radiation. Since in this book a whole chapter has been devoted to scintillators, we will not discuss such detectors here.

Problems

1. Loss of electrons through recombination is a serious problem in detectors that use liquids as ionizing media. One way to reduce recombination probability is to increase the electric field intensity. Derive the expression for the change in recombination rate with respect to the change in electric field intensity.

2. Determine the electric field intensity needed to keep the recombination rate of electrons below 1% in a parallel plate liquid xenon filled ionization chamber.

3. Compare the lifetimes of electrons in liquid argon and liquid xenon if these liquids contain 5 *ppm* of oxygen as impurity.

4. Calculate the threshold velocity of electrons to produce Cherenkov light in heavy water.

5. Compare the threshold energy of an electron to that of a proton for producing Cherenkov light in light water.

Bibliography

[1] Benetti, P. et al., A **Three-Ton Liquid Argon Time Projection Chamber**, Nucl. Instr. Meth. Phys. Res., A332, 1993.

[2] Bolte, W.J., **Development of Bubble Chambers With Enhanced Stability and Sensitivity to Low-Energy Nuclear Recoils**, arXiv:astro-ph/0503398 v1, March 2005.

[3] Broomall, J.R. et al., **Density Dependence of the Electron Surface Barrier for Fluid ^3He and ^4He**, Phys. Rev. B, Vol.14, No.7, 1976.

[4] Chang, J., Mageras, G.S., Ling, C.C., **Evaluation of Rapid Dose Map Acquisition of a Scanning Liquid-Filled Ionization Chamber Electronic Portal Imaging Device**, Int. J. Radiat. Oncol. Biol. Phys., 55(5), 2003.

[5] Collot, J., Jan, S., Tournefier, E., **A Liquid Xenon PET Camera for Neuroscience**, Proc. IX Int. Conference on Calorimetry in High Energy Physics CALOR2000, Annecy, France, October 2000.

[6] Derenzo, S.E., et al. **Electron Avalanche in Liquid Xenon**, Phys. Rev. A, Vol.9, No.6, 1974.

[7] Doke, R., **Fundamental Properties of Liquid Argon, Krypton and Xenon as Radiation Detector Media**, Portug. Phys., Vol.12, 1981.

[8] Eberle, K. et al., **First Test of a Liquid Ionization Chamber to Monitor Intensity Modulated Radiation Beams**, Phys. Med. Biol., 48, 2003.

[9] Engler, J., **Liquid Ionization Chambers at Room Temperatures**, J. Phys. G: Nucl. Part. Phys., 22, 1996.

[10] Engler, J. et al., **A Warm-Liquid Calorimeter for Cosmic-Ray Hadrons**, Nucl. Instr. Meth., A427, 1999.

[11] Essers, M. et al., **Transmission Dosimetry with a Liquid Filled Electronic Portal Imaging Device**, Inst. J. Radiat. Oncol. Biol. Phys., 34, 1996.

[12] Greening, J.R., **Saturation characteristics of Parallel-Plate Ionization Chambers**, Phys. Med. Biol., 9, 1964.

[13] Greening, J.R., **On Greening's Treatment of Saturation Characteristics of Parallel-Plate Ionization Chambers**, Phys. Med. Biol., 10, 1965.

[14] Holroyd, R.A. et al., **Free-Ion Yields for Several Silicon-, Germanium, and Tin-Containing Liquids and their Mixtures**, Phys. Rev. B, 43, 1991.

[15] Henson, B.L., **Mobility of Positive Ions in Liquefied Argon and Nitrogen**, Phys. Rev., Vol.135, No.4A, 1964.

[16] Johansson, B., Wickman, G., **General Collection Efficiency for Liquid Isooctane and Tetramethylsilane Used as Sensitive Media in Parallel-Plate Ionization Chamber**, Phys. Med. Biol., 42, 1997.

[17] Lavoie, L., **Liquid Xenon Scintillators for Imaging of Positron Emitters**, Med. Phys., 3, 1976.

[18] Louwe, R.J. et al. **The Stability of Liquid-Filled Matrix Ionization Chamber Electronic Portal Imaging Devices for Dosimetry Purposes**, Med. Phys., 31(4), 2004.

[19] Meertens, H., van Herk, M., Weeda, J., **A Liquid Ionisation Detector for Digital Radiography of Therapeutic Megavoltage Photon Beams**, Phys. med. Biol., Vol.30, No.4, 1985.

[20] Migliozzi, P., **Perspectives for Future Neutrino Oscillation Experiments with Accelerators: Beams, Detectors and Physics**, arXiv:hep-ph/0311269v1, Nov. 2003.

[21] Miyajima, M. et al., **Average Energy Expended per Ion Pair in Liquid Argon**, Phys. Rev. A, Vol.9, No.3, 1974.

[22] Muller, R.A. et al., **Liquid-Filled Proportional Counter**, Phys. Rev. Lett., Vol.27, 1971.

[23] Policarpo, A.J.P.L. et al., **Observation of Electron Multiplication in Liquid Xenon with a Microstrip Plate**, Nucl. Instr. Meth. Phys. Res., A365, 1995.

[24] Smejtek, P. et al., **Hot Electron Injection into Dense Argon, Nitrogen, and Hydrogen**, J. Chem. Phys., 59(3), 1973.

[25] Takahashi, T. et al., **The Average Energies, W, Required to Form an Ion Pair in Liquefied Rare Gases**, J. Phys. C: Solid State Phys., Vol.7, 1974.

[26] Ulrich, S. et al., **Hot-Electron Thermalization in Solid and Liquid Argon, Krypton, and Xenon**, Phys. Rev. B, Vol.25, No.5, 1982.

[27] van Herk, M., **Physical Aspects of a Liquid-Filled Ionization Chamber with Pulsed Polarizing Voltage**, Med. Phys., 18, 1991.

[28] Warman, J.W., **The Dynamics of Electrons and Ions in Non-Polar Liquids**, The Study of Fast Processes and Transient Species by Electron Pulse Radiolysis, Reidel, 1982.

[29] Wickman, G., **A Liquid Ionization Chamber with High Spatial Resolution**, Phys. Med. Biol., Vol.19, No.1, 66-72, 1974.

[30] Wickman, G., Nystrom, H., **The Use of Liquids in Ionization Chambers for High Precision Radiotherapy Dosimetry**, Phys. Med. Biol., 37, 1992.

[31] Wickman, G. et al., **Liquid Ionization Chambers for Absorbed Dose Measurements in Water at Low Dose Rates and Intermediate Photon Energies**, Med. Phys., 25, 1998.

[32] Yoshino, K. et al. **Effect of Molecular Solutes on the Electron Drift Velocity in Liquid Ar, Kr, and Xe**, Phys. Rev. A, Vol.14, No.1, 1975.

[33] Zaklad, H. et al., A **Liquid Xenon Radioisotope Camera**, IEEE Trans. on Nucl. Sci., NS-19. 1972.

[34] Zaklad, H. et al., **Liquid Xenon Multiwire Proportional Chamber for Nuclear Medicine Application**, Proc. of the First World Congress on Nuclear Medicine, Tokyo, 1974.

Solid State Detectors

We visited some of the most widely used gas filled detectors in the previous section. Although such systems have proved to be extremely useful in many applications, still their usage is somewhat limited due to a number of reasons. For example the small number of electron-ion pairs that can be generated in a gas is a serious problem for high resolution systems in low radiation environments. One of the reasons of this inefficiency lies in the number of target atoms per unit volume in the gas that the incident radiation sees. This implies that if we use liquids or solids instead, the probability of production of charge pairs would increase. However, as it turns out, the mechanism of charge pair production depends on many factors and not just the density. Nevertheless one type of solid has been found to have far superior charge pair production properties than gases. These so called *semiconductors* have electrical conduction properties between conductors and insulators. Diamond is another solid that has been found to have very good charge pair production capabilities. All of the detectors that use solids as active detection media are collectively called solid state detectors, a term that is sometimes exclusively used for semiconductor detectors. In this chapter we will discuss the mechanism of radiation detection and measurement using solids as active detection media. We will also survey some of the most widely used solid state detectors.

5.1 Semiconductor Detectors

Semiconductors are basically crystalline solids in which atoms are held together by covalent bonds. They are called semiconductors because their electrical conduction properties lie between those of insulators and conductors. Germanium (Ge) and Silicon (Si) are two of the most commonly used semiconductor materials. Up till now the majority of the semiconductor detectors have been made with silicon, a trend that may change in future as the search for more radiation tolerant semiconductors continues. $GaAs$ is one of the materials that have shown to be very promising as an alternative to silicon, although at a much higher cost. $GaAs$ has another property that makes it very desirable is its larger band gap. We will explore this further later in the chapter. In the following sections we will discuss the important characteristics of semiconductors and look at how they are employed as active media for detection of ionizing radiation.

5.1.A   Structure of Semiconductors

We know from Quantum Physics that electrons in an atom can occupy only discrete energy levels. In fact this discreteness or *quantization* is not in any way limited to isolated atoms. For example the covalent bonding between atoms in semiconductors creates allowed discrete energy levels. However these energy levels are lumped together in the so called two *bands*: valence band and conduction band. Valence band represents a large number of very closely spaced energy levels at lower energies as compared to the conduction band, which contains levels at higher energies. These two bands are separated by a *forbidden* gap: a region in the energy level diagram containing no energy levels. This essentially means that electrons can not assume any energy that lies in this band. We'll later see that this holds for only ideal semiconductors with no impurities and any practical semiconductor does have at least one energy level in the band gap.

The electrons in the valence band are tightly bound to the atoms and need at least an energy equal to the band gap to move to the conduction band. The conduction band electrons, on the other hand, are very loosely bound and are almost free to move around. These electrons take part in the electrical conduction process. In an ideally pure semiconductor in the ground state all the electrons would populate the valence band while conduction band would be empty.

Actually this band structure is not typical of just semiconductors. Insulators and conductors also have similar structures. The distinguishing feature between them is the band gap, since it represents the energy barrier that must be overcome by bound electrons to become free and take part in the electrical conduction process. Fig. 5.1.1 compares the three types of solids in terms of energy level diagram. The band gap in insulator and conductors are exactly opposite to each other, being very large for insulators and non-existent for conductors. Semiconductors, on the other hand, have a small band gap, so small that even a small thermal excitation can provide enough energy to electrons in the valence band to jump up to the conduction band.

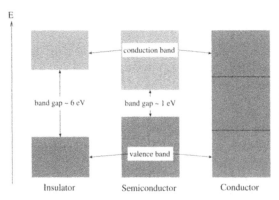

Figure 5.1.1: Simplified energy band structure diagrams for insulators, semiconductors, and conductors.

When an electron from the valence band jumps to the conduction band, it leaves a net positive charge behind. This effective positive charge, called a *hole*, behaves like a real particle and takes part in electrical conduction process. However it should be noted that by movement of a hole we mean the shift of a net positive charge from one site to another due to the movement of an electron. A hole should not be considered a localized positive charge having defined mass.

5.1.B Charge Carriers Distribution

The free charges in the bulk of a semiconductor crystal can occupy different energy levels with an occupancy that can be described by the so called *Boltzmann distribution*

$$f(E) = \frac{1}{1 + e^{(E-E_F)/k_B T}}, \tag{5.1.1}$$

where E is the energy of the electron, k_B is the familiar Boltzmann constant, T is the absolute temperature, and E_F is the Fermi level.

The Fermi function $f(E)$ actually gives the probability at which an available energy state E can be occupied by an electron. For intrinsic semiconductors, which have equal number of positive and negative charge carriers, the Fermi level lies exactly in the middle of the band gap. This is the level at which the probability of electron occupancy is exactly $1/2$, or in other words, half of the states are filled by electrons (see example below).

As can be inferred from the relation 5.1.1, the occupancy of charge carriers is a function of the absolute temperature. Of course the reason for this can be traced back to the few electron volt wide band gaps of semiconductors that are comparable to the energy of thermal agitations even at room temperatures. The temperature dependence is so strong that even small fluctuations in temperature can produce significant changes in the number of free charge carriers. We will see later that this effect is a serious problem in semiconductor detectors since it may cause nonlinear changes in the response of the detector.

Example:
Compute the probability for an electron to occupy the Fermi level in an intrinsic semiconductor.

Solution:
The required probability can be computed from the Fermi function 5.1.1

$$f(E) = \left[1 + e^{(E-E_F)/k_B T}\right]^{-1}.$$

Since we have to find the probability at the Fermi level, therefore we substitute $E = E_F$ in the above equation to get

$$\begin{aligned} f(E) &= \left[1 + e^{(E_F-E_F)/k_B T}\right]^{-1} \\ &= \frac{1}{2}. \end{aligned}$$

5.1.C Intrinsic, Compensated, and Extrinsic Semiconductors

The energy band structure shown in Fig.5.1.1 represents an ideal semiconductor. A crystal in which the impurities are either non-existent or they do not affect its conduction properties significantly is said to be *ideal* or *intrinsic*. The electrons and holes in such a material are in equilibrium with each other. This state of equilibrium is actually a consequence of similar temperature dependences of the density of states of conduction and valence bands. It has been found that the density of states of conduction band N_c and valence band N_v vary with absolute temperature according to

$$N_c \propto T^{3/2} \tag{5.1.2}$$

$$\text{and} \quad N_v \propto T^{3/2}. \tag{5.1.3}$$

The intrinsic charge concentration, however, has a much stronger temperature dependence, which certainly is a major operational problems of semiconductor detectors. The intrinsic charge concentration can be written as

$$n_i = [n_c n_v]^{1/2} \exp\left[-\frac{E_g}{2k_B T}\right], \tag{5.1.4}$$

where n_c and n_v represent the charge concentrations in conduction and valence bands respectively. This expression shows that the intrinsic charge concentration asymptotically reaches a saturation value, which is characterized by the density of states of conduction and valence bands only.

As states above, an intrinsic material is called an ideal semiconductor, which essentially means that it simply does not exist in nature. This is a true statement since, in reality, due to crystal defects and impurities, there are also other energy states within the forbidden gap that significantly change the electrical conduction properties of the material. These crystal imperfections lower the energy threshold needed for transitions and consequently the electron and hole densities change from an ideal semiconductor having equal number of free electron and hole pairs. Hence a naturally found or grown semiconductor does not possess intrinsic properties. However, through the process of impurity addition one can turn any semiconductor into an intrinsic type, which then is referred to as a *compensated* material.

Whenever impurity is added to a semiconductor, its electrical conduction properties change. The material is then referred to as an *extrinsic* semiconductor. The impurity addition or *doping* is an extremely useful process that dramatically improves the performance of semiconductors.

5.1.D Doping

The electrical conduction properties of semiconductors can be drastically changed by adding very small amounts of impurities, a process known as doping . In this process, another element with different number of electrons in its outer atomic shell than the semiconductor atom is added in a very small quantity to the bulk of the material. The net effect of this process is the creation of additional energy levels between valence and conduction bands of the crystal. The locations of these levels in the energy band diagram depend on the type of impurity added. The impurity that creates abundance of positive charges in the material is known as *acceptor* impurity and it creates energy levels near the valence band. The resulting material is known as p-type semiconductor. On the other hand a *donor* impurity makes the material abundant in negative charges and creates additional energy levels near the conduction band. Such a material is referred to as n-type semiconductor.

To use semiconductors as radiation detectors, generally very small quantities of impurities are added to the bulk of material. for example the typical ratio of impurity atomic density to the semiconductor atomic density for silicon is of the order of $10^{-10} cm^{-3}$. This means that for each impurity atom there are around 10^{10} semiconductor atoms. Although most of the semiconductor detectors are made with small impurity additions, there is also a special class of detectors that are made with *heavily doped* semiconductors. Typical impurity atomic concentration in such materials is $10^{20} cm^{-3}$ in the bulk semiconductor with a density of the order of $10^{22} cm^{-3}$.

An interesting aspect of doping agents is that their ionization energies depend on where their energy levels lie within the energy band structure of the bulk of the semiconductor material. This implies, for example that the ionization energy of boron impurity in silicon will be different than that in germanium. Typical ionization energies for the doping agents used in semiconductor detector materials range between $0.01\ eV$ and $0.1\ eV$. If we compare this with the typical ionization energies of several electron volts for the semiconductors, we can conclude that the doping agents should ionize very quickly after their introduction into the material. If the material did not have significant impurities beforehand then it can be said that the free charge carrier density in the bulk of the material is, to a large extent, characterized by the acceptor and donor impurity concentrations. If n_a and n_d are the acceptor and donor impurity concentrations, then the acceptor and donor charge concentrations can be written as

$$N_{a,-} \approx n_a \qquad (5.1.5)$$
$$N_{d,+} \approx n_d, \qquad (5.1.6)$$

where the $(-)$ and $(+)$ signs represent the ionization states of the impurity atoms. Note that N does not represent the free charge density, rather the number of ionized atoms. A donor gives off its electron and becomes positively ionized while an acceptor becomes negatively charged after accepting an electron.

Since there are always donor and acceptor impurities present in a material, therefore the characterization of a material as n or p type depends on the difference of charges, which as we just saw, depends on the difference of number density of ionized atoms. Hence we can say that a material is of type n if

$$n_n = N_{d,+} - N_{a,-} >> n_i, \qquad (5.1.7)$$

where n_n is the charge carrier density of the n-type material and n_i represents the charge carrier density of the pure material. This is the carrier density of the material before being doped. Similarly a p type material is defined as the one that satisfies the condition

$$n_p = N_{a,-} - N_{d,+} >> n_i. \qquad (5.1.8)$$

Here n_p is the charge carrier density of p type material.

The reader should note that, since the amount of impurity needed to modify the semiconductor into n- or p-type is very small, all the semiconductor crystals are naturally of either n- or p-type. An absolutely pure semiconductor material does not exist. However it is possible to dope the material such that its positive and negative charge carrier densities become nearly equal, that is

$$n_n \approx n_p. \qquad (5.1.9)$$

Such compensated materials show bulk properties similar to an ideally pure semiconductor. To develop semiconductor detectors, generally shallow doping is done.

That is, not the whole intrinsic material is doped, rather the doping is done only up to a certain depth. The rest of the material then remains intrinsic. We will learn more about this technique and its advantages later in the Chapter. Let us first have a closer look at the physical process involved in acceptor and donor doping.

D.1 Doping with Acceptor Impurity

If the element added has an electron less than the one in the semiconductor, it can either form one bond less than the semiconductor atoms or captures an electron from the semiconductor lattice to fit in the structure. In both the cases a net positive vacancy or *hole* is created. If a large number of such impurity atoms are added, the created holes outnumber the free electrons in the valence band. Interestingly enough, if an external electric field is now applied to the material, these holes start drifting and constitute an electric current. Such a semiconductor material with an acceptor impurity is called a p-type semiconductor .

The net effect of acceptor impurity addition is the shifting of Fermi level towards the valence band, which essentially means that the occupancy of free positive charges in the bulk of the material is larger than that of the negative charges.

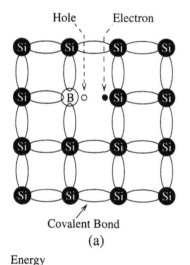

(a)

Figure 5.1.2: (a)Addition of boron in silicon lattice. Boron having three available electrons for bonding leaves a hole, which can be filled by a free electron in the lattice. (b)Addition of an acceptor impurity in a semiconductor shifts the Fermi level towards the valence band.

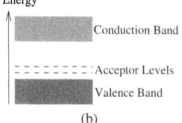

(b)

Boron is a common example of an acceptor impurity that can turn silicon into a p-type semiconductor. Fig.5.1.2 depicts the effect of this doping on a regular silicon lattice. A silicon atom in the semiconductor lattice has 4 electrons making covalent bonds with four other neighboring silicon atoms. If a boron atom having three outer shell electrons is added to this lattice, it tries to fit into the structure but since it has one electron less therefore it can form only three covalent bonds with silicon atoms. The fourth location can then be thought to have a positive hole since it has strong affinity to attract a free electron. In fact if this hole is filled by a free electron in the lattice it would essentially move the hole to the original site of the electron. Since in an intrinsic silicon lattice there are a very few free electrons therefore even a small number of boron atoms in the bulk of the lattice can make it abundant in p-type charges and turn the material into a p-type semiconductor.

The distribution function of acceptors in a semiconductor is given by

$$f_A(E_A) = \frac{1}{1 + 4e^{(E_A - E_F)/k_B T}},\tag{5.1.10}$$

where E_A represents the acceptor energy level. Note that this distribution is somewhat different from the Boltzmann distribution we saw earlier for for free charge carriers (cf. equation 5.1.1).

D.2 Doping with Donor Impurity

If the impurity has more electrons than the semiconductor atoms in the outermost shell, the excess electrons are not able to make covalent bonds with the lattice atoms and are thus free to move around. Such an impurity is called *donor* impurity since it donates free charge carriers to the bulk of the material. In the energy band structure the net effect is the lowering of the conduction band (see Fig. 5.1.3). Since the energy gap is reduced with more free electrons in the conduction band, the electrical conduction properties of the semiconductor are greatly enhanced. Such a material is known as n-type semiconductor due to the abundance of free negative charges.

As with acceptors, the distribution function of donors also differs from the Boltzmann distribution given in equation 5.1.1. In this case it is given by

$$f_A(E_D) = \frac{1}{1 + 0.5e^{(E_D - E_F)/k_B T}},\tag{5.1.11}$$

where E_D represents the donor energy level.

5.1.E Mechanism and Statistics of Electron-Hole Pair Production

Radiation passing through a semiconductor material is capable of causing the following three distinct phenomena to occur in the bulk of the material.

- ▶ **Lattice Excitation:** This occurs when the incident radiation deposits energy to the lattice increasing the lattice vibrations.

- ▶ **Ionization:** In semiconductors ionization means production of an electron hole pair.

- ▶ **Atomic Displacement:** This non-ionizing phenomenon is the major contributor to the bulk of the damage caused by radiation. We'll discuss this in detail later in the Chapter.

Electron

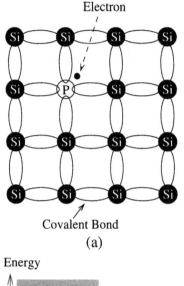

Covalent Bond

(a)

Figure 5.1.3: (a)Addition of phosphorus to the silicon lattice. Phosphorus having five available electrons for bonding leaves an extra electron, which can move freely in the lattice. (b)Addition of a donor impurity in a semiconductor creates new energy levels near to the conduction band, which effectively extends the valence band to lower energy. As a result the band gap is decreased with electrons as major charge carriers.

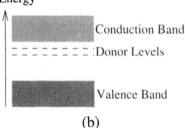

(b)

Although in semiconductor detectors we are mainly interested in the ionization process but the process of lattice excitations also has a significant impact on the statistics of electron-hole pair production. Before we look at the statistics of the charge pair production, let us first have a closer look at the ionization mechanism in semiconductors.

In a perfect semiconductor material at a temperature below the band gap energy, all the electrons are in valence band. The conduction band in such a situation is completely empty. The outer shell electrons, taking part in the covalent bonding between lattice atoms, are not free to move around in the material. However as the temperature is raised, some of the electrons may get enough thermal excitation to leave the valence band and jump to the conduction band. This creates an electron deficiency or a net positive charge in the valence band and is generally referred to as a *hole*. This process can also occur, albeit at a much higher rate, when radiation passes through the material. Any radiation capable of delivering energy above a material-specific threshold is capable of creating electron-hole pairs along its track in the material. This threshold is higher than the band gap energy of material as some of the energy also goes into crystal excitations. For silicon the threshold is very low (3.62 eV), which makes it highly desirable for use in radiation detectors (see Table 5.1.1).

The underlying physical processes involved in the creation of electron-hole pairs in semiconductors are the same as in other solids, which we have already discussed in Chapter 2 and therefore will not be repeated here. As in gases, in semiconductors too, the average energy needed to create an electron hole pair is independent of the type of radiation and depends on the semiconductor material and its temperature. The process is similar to the ionization process in gases except that the energy needed in semiconductors is approximately 4 to 8 times less than in gases. This implies that the number of charge carriers produced by radiation in a semiconductor is much higher than in gases. Although one would expect that the noise equivalent charge in semiconductors would also be higher by approximately the same amount, but as we will see later, this is not necessarily the case. Because of this property the semiconductor detectors are considered to be far superior than gaseous detectors in terms of resolution and sensitivity.

Table 5.1.1: Densities and average ionization energies of common semiconductor materials.

Material	Density (g/cm^3)	Ionization Energy eV
Silicon	2.328	3.62
Germanium	5.33	2.8
Silicon Dioxide	2.27	18
Gallium Arsenide	5.32	4.8

Fig.5.1.4 shows the mechanism of production of charge pairs by incident photons and by thermal agitation. Note that the energy levels in the forbidden gap produced by crystal imperfections and impurities enhance the production of charge pairs. This is certainly not a very desirable channel since it can produce non-linearities in the detector response through an effect called charge trapping. What happens is that an electron jumping to an impurity level may get trapped there for some time. This electron can then do two things: it can either jump up to the conduction band and complete the process of electron hole pair generation or it can fall back into the valence band and recombine with the hole. The former introduces a time delay in the charge pair production while in latter no charge pair is produced. The excess energy in this case is not enough to create an electron hole pair (since it will be equal to E_{imp} and not E_g) and is absorbed by the lattice.

To understand the statistics of the electron-hole pair production, let us assume that the energy deposited by the incident radiation goes into causing lattice excitations and ionization. If ϵ_i and ϵ_x represent the average energies needed to produce ionization and excitation respectively, then the total deposited energy can be written as

$$E_{dep} = \epsilon_i n_i + \epsilon_x n_x, \qquad (5.1.12)$$

where n_i and n_x represent the total number of ionization and excitations produced by the radiation. If we now assume that these processes follow Gaussian statistics,

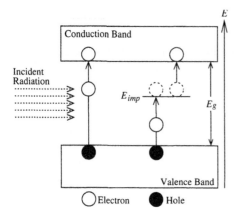

Figure 5.1.4: Production mechanisms of electron hole pairs by incident radiation in a semiconductor.

it would mean that the variance in the number of ionization and excitations can be written as

$$\sigma_i = \sqrt{n_i} \quad \text{and}$$
$$\sigma_x = \sqrt{n_x}.$$

These two variances are normally not equal because of difference in the thresholds for excitation and ionization processes. However if we weight them with their corresponding thresholds, they should be equal for a large number of collisions, i.e.,

$$\epsilon_i \sigma_i = \epsilon_x \sigma_x \quad \text{or}$$
$$\epsilon_i \sqrt{n_i} = \epsilon_x \sqrt{n_x}.$$

Combining this with equation 5.1.12 gives

$$\sigma_i = \frac{\epsilon_x}{\epsilon_i} \left[\frac{E_{dep}}{\epsilon_x} - \frac{\epsilon_i}{\epsilon_x} n_i \right]^{1/2} \tag{5.1.13}$$

Let us now denote the average energy needed to create an electron-hole pair by w_i. Note that this energy includes the contribution from all other non-ionizing processes as well. This means that it can be obtained simply by dividing the total deposited energy by the number of electron-hole pairs detected n_s. Hence we can write

$$w_i = \frac{E_{dep}}{n_s} \quad \text{or}$$
$$n_s = \frac{E_{dep}}{w_i}.$$

If we have a perfect detection system that is able to count all the charge pairs generated, then we can safely substitute n_s for n_i. In this case the above expression for σ_i yields

$$\sigma_i = \left[\frac{\epsilon_x}{\epsilon_i} \left(\frac{w_i}{\epsilon_i} - 1 \right) \left(\frac{E_{dep}}{w_i} \right) \right]^{1/2}. \tag{5.1.14}$$

Using $E_{dep}/w_i = n_s$, this can be written as

$$\sigma_i = \sqrt{F n_s}. \tag{5.1.15}$$

where

$$F = \frac{\epsilon_x}{\epsilon_i}\left(\frac{w_i}{\epsilon_i} - 1\right)$$

is called the *Fano factor*. It is interesting to note here that even though we assumed that the individual processes of ionization and excitations were Gaussian in nature, but the spread in the output signal can be described by the Poisson process only if we multiply it by another factor. The reason, of course, is that these processes are not uncorrelated as required by a strictly Gaussian process. The Fano factor was first introduced to explain the anomaly between the observed and expected variance in the signal (14). The simple calculations we performed above do not produce very accurate results and they were only meant to introduce the idea of the need for introducing the Fano factor. For detailed calculations the interested reader is referred to (2) or (43). The value of the Fano factor lies between 0 and 1: 0 for no fluctuations and 1 for perfect Poisson process. It has been found that for germanium and silicon $F = 0.1$ gives satisfactory results.

Example:
Determine the relative statistical fluctuations in the number of charge pairs produced in silicon if 2.5 MeV of energy is deposited by the incident radiation.

Solution:
For silicon we have $w_i = 3.62$ eV/charge-pair and $F = 0.1$. The absolute statistical fluctuations can be computed from substituting these values and the deposited energy in equation 5.1.15. Hence we get

$$\begin{aligned}
\sigma_i &= \sqrt{Fn_s} = \sqrt{F\frac{E_{dep}}{w_i}} \\
&= \left[(0.1)\frac{2.5 \times 10^6}{3.62}\right]^{1/2} \\
&= 262.8 \text{ charge-pairs}
\end{aligned}$$

If $N = E_{dep}/w_i$ represents the mean number of charge pairs produced, the corresponding relative fluctuations are

$$\begin{aligned}
\frac{\sigma_i}{N} &= \frac{\sigma w_i}{E_{dep}} \times 100 \\
&= \frac{(262.8)(3.62)}{2.5 \times 10^6} \times 100 \\
&\approx 0.04\%.
\end{aligned}$$

The amount of relative statistical fluctuations as computed above is a measure of the *physical* limit of the system resolution. Of course in a semiconductor detector there are a number of other sources of error that contribute to the measurement error too and therefore the actual uncertainty is much larger.

E.1   Intrinsic Energy Resolution

Equation 5.1.15 gives the observed spread in the number of electron hole pairs produced by the incident radiation. Since the number of charges produced is related to the energy delivered, therefore this equation can also be used to determine the *intrinsic* spread in the energy deposited by the incident radiation. The term intrinsic refers to the fact that here we are dealing with the uncertainty associated with the physical process of charge pair production. The energy resolution thus obtained characterizes the best possible resolution that the system can be expected to possess. In reality there are other factors, such as noise and the resolving power of the associated electronics, that may deteriorate the resolution significantly. The good thing about computing the intrinsic resolution is that it tells us the physical limits of the system.

Since $E_{dep} = w_i n_s$, therefore the intrinsic uncertainty in energy can be written as

$$\sigma_E = \sigma(w_i n_s) = w_i \sigma_i, \tag{5.1.16}$$

where we have made use of the constancy of w_i under non-varying working conditions. Hence according to equation 5.1.15, the spread in the energy is given by

$$
\begin{aligned}
\sigma_E &= \xi w_i \sigma_i \\
&= \sqrt{F E_{dep} w_i},
\end{aligned}
\tag{5.1.17}
$$

where we have used $n_s = E_{dep}/w_i$.

Now we are ready to compute the intrinsic energy resolution of the semiconductor material, which can be written as

$$
\begin{aligned}
R &= \xi \frac{\sigma_E}{E_{dep}} \\
&= \xi \sqrt{\frac{F w_i}{E_{dep}}}.
\end{aligned}
\tag{5.1.18}
$$

Here we have introduced a factor ξ, which can be thought of as a yardstick to decide whether two peaks could be resolved or not. Its value depends mostly on how the peak looks like, or in other words which distribution it seems to follow. for example for a perfectly Gaussian peak $\xi = 2\sqrt{2\ln(2)} = 2.355$. This value corresponds to the Full Width at Half Maximum (or FWHM) of a Gaussian peak. We will discuss this in more detail in the Chapter on data analysis.

Energy resolution is the most important factor for a radiation detector used for spectroscopic purposes since it characterizes the detector in terms of how well it can differentiate between closely spaced energy peaks in the spectrum. for example if a source emits two photons having an energy difference of 2 keV then the spread in the measured energy must be better than 2 keV for the two peaks to be detected separately. Otherwise the peaks will get superimposed on one another and become indistinguishable.

An important point to note is that the energy resolution varies inversely with $\sqrt{E_{dep}}$. Therefore a material that does not have good energy resolution at a certain energy might be more efficiently utilized at a higher energy.

Example:
Determine the energy resolution of a silicon detector for 520 keV photons using Fano factor and also by assuming a perfectly Poisson process.

Solution:
Assuming that the photons deposit their full energy in the active volume of the detector, we have $E_{dep} = 520\ keV$. For silicon we have $w_i = 3.62\ eV$ and $F = 0.1$. For ξ we will use the generally used value of 2.355. Substituting these values in equation 5.1.18 we get

$$R = \xi\sqrt{\frac{Fw_i}{E_{dep}}}$$

$$= 2.355\left[\frac{(0.1)\left(3.62 \times 10^{-3}\right)}{520}\right]^{1/2}$$

$$= 1.9 \times 10^{-3}.$$

If we assume the process to be perfectly Poisson, then the energy resolution will be given by

$$R = \xi\sqrt{\frac{w_i}{E_{dep}}}$$

$$= 2.355\left[\frac{3.62 \times 10^{-3}}{520}\right]^{1/2}$$

$$= 6.2 \times 10^{-3}.$$

E.2   Recombination

We saw earlier that electrons and holes can recombine through an intermediate energy state created by crystal imperfection or impurity. This is not the only recombination mechanism. In fact an electron in the conduction band can also directly recombine with a hole in the valence band. This process is similar to the process of electron-ion recombination we studied in the chapter on gas filled detectors except that in this case the positive charge is a hole and not an ion. The end result of the process is the removal of an electron from the conduction band and the hole from the valence band. A point to remember here is that the electron, having a defined mass, does not annihilate in this process, as is sometimes wrongly concluded. Recombination is *not* an annihilation process. The free electron simply gets trapped in the valence band but retains its identity and properties.

The process of recombination can occur in two distinct ways as described below.

▶ **Band-Recombination:** An electron in the conduction band can fall into the valence band to recombine with a hole. This is the simplest and most prevalent form of recombination that occurs in semiconductors. The difference in the energy of the electron in the two states is then emitted as shown in Fig.5.1.5. Now this energy can be absorbed by another electron in the conduction band, which then gets emitted and may even escape from the detector. This emitted electron is called Auger electron and the process is sometimes referred to as Auger recombination. The excess energy can also go into increasing the lattice vibrations of the crystal. These vibrations can travel through the crystal in the form of heat carrying particles generally known as *phonons.*

▶ **Trap-Recombination:** As we saw earlier, there are always crystal defects and impurities in semiconductors. These defects and impurities produce energy levels inside the forbidden gap. Such an energy level can act as a metastable electron trap such that when an electron falls into this level, it may remain there for some time before eventually falling into the valence band. The net effect is still recombination but the process is somewhat delayed as compared to the normal band-recombination. As shown in Fig.5.1.6, two photons can be produced during this process with the combined energy equal to the energy released during the band-recombination process. Also, as with band-recombination, the process could be radiationless such that the excess energy gets transferred to increase lattice vibrations. The trap-recombination process is generally known as Shockley-Hall-Read (SHR) recombination.

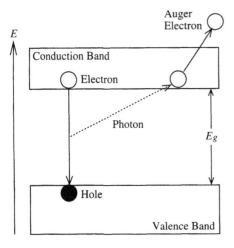

Figure 5.1.5:   Band-recombination with subsequent emission of an Auger electron.

The net effect of this recombination process is the removal of a charge pair from the free charge population. Crystal imperfections and impurities produce intermediate energy levels that greatly enhance the recombination process. Radiation damage to semiconductors, which we will discuss later in the chapter, also increase the recombination probability. To minimize the deteriorating effects of this process, it must therefore be ascertained that the material has very few imperfections and also has high radiation tolerance.

As described above, recombination of electrons and holes has different channels and is therefore a fairly complicated process. However its overall effect can be characterized by simple considerations of Poisson process. By *overall effect* we specifically mean the recombination rate, which has been seen to follow Poisson statistics. This implies that the rate of change of number of charge pairs is proportional to the

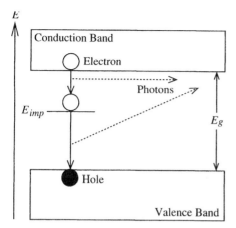

Figure 5.1.6: Trap-recombination with subsequent emission of two photons.

number of charge pairs present at that time. Mathematically, we can write

$$\frac{dN}{dt} \propto -N$$
$$= -k_r N, \tag{5.1.19}$$

where k_r is the proportionality constant, generally known as the *recombination rate constant*. The integration of this equation yields

$$N = N_0 e^{-k_r t}, \tag{5.1.20}$$

where we have used the initial condition $N = N_0$ and $t = 0$. The factor k_r in the above equation can be used to define carrier lifetime τ through the relation

$$\tau = \frac{1}{k_r}. \tag{5.1.21}$$

To understand the meaning of τ we substitute $t = 1/k_r$ is equation 5.1.20 and get

$$\frac{N}{N_0} = e^{-1} \approx 0.37. \tag{5.1.22}$$

This implies that the carrier lifetime τ represents the time it takes the carrier population to decrease by approximately 63%.

The value of k_r and τ for a particular charge carrier (electron or hole) depend on the type of material, the donor and acceptor impurities and the temperature. for example at $300\ K$ in a p-type silicon having acceptor density of $10^{19}\ cm^3$, the lifetime of an electron is approximately $0.1\ \mu s$ but if the acceptor density is reduced by two orders of magnitude the lifetime increases to about $10\ \mu s$. This trend is typical of all semiconductors.

Example:
Compute the percentage of holes lost within $4\ \mu s$ of their generation in an n-type silicon kept at $300\ K$. The number density of the dopant impurity

in the material is 10^{17} cm^3 and the lifetime of holes at this dopant level is 10 μs.

Solution:
The recombination rate constant of the holes having lifetime $\tau = 0.1$ μs is given by

$$k_r = \frac{1}{\tau}$$
$$= \frac{1}{10 \times 10^{-6}} = 10^5 \ s^{-1}.$$

The percentage of holes absorbed after 4 μs can be computed from equation 5.1.20 as follows.

$$\delta N = \frac{N_0 - N}{N} \times 100 = \left(1 - e^{-k_r t}\right) \times 100$$
$$= \left[1 - \exp\left\{-\left(10^5\right)\left(4 \times 10^{-6}\right)\right\}\right] \times 100$$
$$\approx 33\%$$

5.1.F Charge Conductivity

The free charges in a semiconductor can drift under the influence of an externally applied field. Just like metallic conductors, their movement can also be characterized by the parameter called *conductivity*, which quantifies the ability of the material to conduct electric current. However as opposed to metallic conductors, there are a number of mechanisms that can contribute to or suppress the conductivity of a semiconductor. Some of these factors are described below.

▶ **Electron-Hole Recombination:** An electron in the conduction band can fall into the intermediate energy state and then recombine with a hole in the valence band. The overall effect of this process is the removal of an electron-hole pair from the free charge density and decrease in the current.

▶ **Hole Emission:** An electron from the valence band can jump to the intermediate level. Since this removes the electron from the valence band and leaves behind a hole, it can also be viewed as the emission of a hole from the intermediate energy level to the valence band.

▶ **Electron Emission:** An electron in the intermediate energy level can proceed to the conduction band and become part of the free charge density thus increasing the current.

▶ **Electron Trapping:** An electron can fall into a slightly lower energy level from the conduction band and get trapped there for some finite amount of time. This trapping mechanism has the potential of introducing nonlinearity in the response due to the time lag involved in electron trapping and release.

F.1   Drift of Electrons and Holes

The semiconductor detectors are almost always operated in the so called *photocon-ductive mode*. Such an operation involves establishment of an electric field across the material. When the incident radiation produces electron hole pairs along its track in the detector, the charges start moving in opposite directions under the influence of the applied electric field. The velocity with which the charges move depends on the electric field. I has been found that for low fields the velocity increases almost linearly with the field strength, that is

$$v = \mu E, \qquad (5.1.23)$$

where μ is a proportionality constant called mobility. Its value depends on the type of the material, for example in silicon it is 1350 $cmV^{-1}s^{-1}$ for electrons and 480 $cmV^{-1}s^{-1}$ for holes.

As the field is further increased the velocity of the carriers starts showing deviation from the above relation and eventually saturates (see Fig.5.1.7). Most of the detectors are built such that the charges could very quickly (within a few pico seconds) attain the saturation velocity.

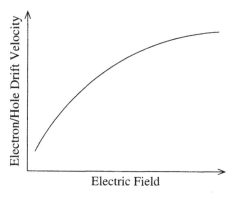

Figure 5.1.7: Typical variation of drift velocity with respect to the applied electric field intensity.

Based on the drift of electrons and holes, let us now discuss a simple scheme that could assure us the proportionality of a measurable quantity with the energy deposited by the incident radiation. As the charges move in opposite directions, they constitute an electric current with a current density J, which follows Ohm's law

$$J = \sigma E, \qquad (5.1.24)$$

where E is σ is a proportionality constant known as conductivity. Since current density can also be written as

$$J = \rho v, \qquad (5.1.25)$$

with ρ being the charge density, the above three equations can be combined to give

$$\sigma = \mu \rho. \qquad (5.1.26)$$

We saw earlier that the mobility of electrons μ_e differs significantly from that of holes μ_h. Hence the effective conductivity of the material has two separate components for each type of charge and hence we can write the above equation as

$$\sigma = e(\mu_e n_e + \mu_h n_h), \qquad (5.1.27)$$

where we have used the relations $\rho_e = en_e$ and $\rho_h = en_h$ with n_e and n_h being the number density of electrons and holes respectively .

If now this material is placed in an ionizing radiation field, electron hole pairs will be created and consequently the number of free charge pairs in the bulk of the material will increase. The result of this will be a change in the conductivity of the material. If n' represents the number of charge pairs created by the incident radiation, then the change in the conductivity will be given by

$$\delta\sigma = e(\mu_e + \mu_h)n'. \tag{5.1.28}$$

This change in conductivity is proportional to the energy delivered by the incident radiation provided all other conditions remain constant. Hence measuring the change in conductivity is equivalent to measuring the delivered energy if the detector has been properly calibrated. Such a measurement can be done by placing the detector in an external circuit, which could measure the change in current caused by change in conductivity.

It should be mentioned here that both electrons and holes take finite amount of time to recombine and hence the change in conductivity has a time profile that extends up to the lifetime of the slowest charge carrier. Therefore, even for a localized radiation interaction that could be represented by a delta function, the output signal actually has a shape with a finite rise and decay times.

5.1.G   Materials Suitable for Radiation Detection

Not all semiconductors can be used in radiation detectors. The choice depends on many factors such as resistivity, mobility of charges, drift velocity, purity, operating temperature, and cost. Silicon has traditionally been the most commonly used material in particle detectors, a trend that is now changing. Other commonly used materials are germanium (Ge), gallium arsenide ($GaAs$), and cadmium-zinc-tellurium ($CdZnTe$). The need of a new generation of radiation hard silicon detectors is now pushing the researchers to develop more complex semiconductor structures. In the following we will look at some of the commonly used semiconductor materials and study their properties relevant to their use as radiation detectors. The purpose of this activity is to supply the reader with enough information so that a good comparison of merits and demerits of these materials with respect to their use as detection media could be performed.

A quick comparison of basic properties of common semiconductor devices can be done from Table.5.1.2. However, the reader is encouraged to go through the details of each material as given in the following sections to develop a working knowledge of advantages and disadvantages associated with each device type. One should also note that there are other novel semiconductor devices besides the ones discussed here and more are being constantly designed and developed. It is not the intention here to give the reader a comprehensive list of materials available, rather to give a broader perspective of the basic properties that are essential for these devices to be used as efficient semiconductor detectors.

A very important property of semiconductor materials is their intrinsic carrier concentration since it can be used to estimate the signal to noise ratio at the room temperature. It is evident from Table.5.1.2 that $GaAs$ and $CdZnTe$ have intrinsic carrier concentrations that are several orders of magnitude lower than those of silicon and germanium. This property makes them suitable for operation at room temperature, which completely eliminates the need for colling system and is a big advantage in terms of operating cost. $GaAs$ and $CdZnTe$ based detectors have therefore gained a lot of popularity in recent years.

Table 5.1.2: Comparison of some basic properties at room temperature of semiconductor materials commonly used in radiation detectors (note that the actual values may differ slightly from these nominal values due to manufacturing and structural differences) (47).

Property	Si	Ge	$GaAs$	$CdZnTe$
Weight Density ($g\ cm^{-3}$)	2.329	5.323	5.32	5.78
Dielectric Constant	11.7	16	12.8	10.9
Energy Gap (eV)	1.12	0.661	1.424	1.56
Intrinsic Carrier Concentration (cm^{-3})	1×10^{10}	2×10^{13}	2.1×10^{6}	2.0×10^{5}
W-value (eV)	3.62	2.95	4.2	4.64
Intrinsic Resistivity ($\Omega\ cm$)	3.2×10^{5}	46	3.3×10^{8}	3.0×10^{10}

G.1   Silicon (Si)

For radiation detection, silicon is by far the most commonly used material. It is relatively cheaper than other semiconductor materials and is easily available in purified form. These factors and the fact that silicon has moderate intrinsic charge concentration and intrinsic resistivity makes it suitable for use as detection medium.

In Fig.5.1.1 we saw the simplified sketch of the band structure of a semiconductor material. In reality the energy levels are not so well behaved. Fig.5.1.8 shows the actual energy level diagram for silicon.

A good thing about silicon is that its forbidden energy gap is neither very low (as of germanium) nor very high (as of gallium arsenide). This makes it a good candidate for manipulation by adding impurities so that the desired properties, such as high resistivity, are achieved. As with all semiconductor materials, the energy gap for silicon has a moderate temperature dependence, which can be described by (47)

$$E_g = 1.17 - 4.73 \times 10^{-4} \frac{T^2}{T + 636}, \qquad (5.1.29)$$

where temperature T is in absolute units and E_g is in eV. This equation has been plotted in Fig.5.1.9. It is apparent that small changes in temperature can cause the band gap to shorten or widen. This is certainly not a desirable feature, since it could induce non-linearities in the detector response. Shortening of band gap means more electron hole pairs will be generated with the same deposited energy while a wider band gap would make it harder for the electrons in the valence band to jump to the conduction band.

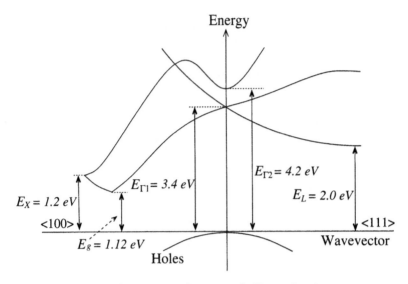

Figure 5.1.8: Band structure diagram of silicon showing energy versus wavenumber (reproduced from (47)). The subscripts of E represent different energy levels. The number in brackets (100 and 111) are the Miller indices. A Miller index represents the orientation of an atomic plane in a crystal lattice.

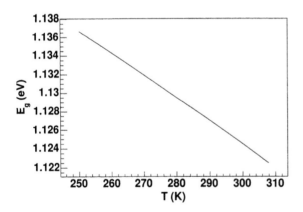

Figure 5.1.9: Variation of silicon band gap energy with absolute temperature.

Silicon detectors are generally operated at low temperatures of around -3^0 C to -10^0 C[1]. The reason is to suppress the thermal agitation, which can produce electron hole pairs even at room temperature. Lowering of the temperature has two effects: widening of the band gap and decrease in the thermal agitation. Both of these compliment one another to suppress the noise in the detector. It should, however be noted that, even though operating silicon detectors at low temperatures is a

---

[1]The choice of operating temperature is mainly based on noise considerations. Some silicon detectors are even operated at as low as -40^0 C.

general practice, still in principle one could operate a detector at room temperature at the expense of some added noise in the detector output.

Another parameter of interest for silicon is the intrinsic carrier concentration, given by

$$n_i = \sqrt{N_c N_v} e^{-E_g/2k_B T} \tag{5.1.30}$$

where N_c and N_v are the density of states in the conduction and valence bands respectively and k_B is the Boltzmann's constant. The density of states for silicon can be evaluated from (47)

$$N_c = 6.2 \times 10^{15} T^{3/2} \ cm^{-3} \tag{5.1.31}$$
$$N_v = 3.5 \times 10^{15} T^{3/2} \ cm^{-3}. \tag{5.1.32}$$

Except for compensated materials, the intrinsic carrier concentration is the major source of noise in the detector. Further complication arises from its strong temperature dependence, which stems from the low band gap energy in silicon. This is another reason to operate the silicon detectors at low temperatures. Substituting the expressions for N_c and N_v in equation 5.1.30 we get

$$n_i = 4.66 \times 10^{15} T^{3/2} e^{-E_g/2k_B T} \tag{5.1.33}$$

This equation, with E_g given by equation 5.1.29, has been plotted in Fig.5.1.10.

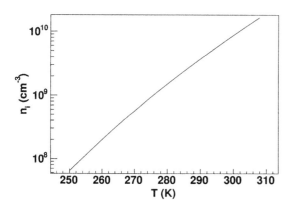

Figure 5.1.10: Dependence of intrinsic charge concentration in silicon on absolute temperature.

Up until now we have assumed that the band gap in silicon is completely empty. In reality the situation is not that simple because of the presence of impurities in the bulk of the material. These impurities could either act as donors or acceptors depending on where their energy levels lie in the forbidden gap. These energy levels can trap electrons and holes with the consequence of deterioration of the output signal. Table.5.1.3 lists some of the commonly encountered impurities in silicon with their position in the band gap. The reader should note that this list is in no way exhaustive and other impurities should also be expected in the material, though in lower concentrations. The fact that no two silicon wafers (wafers are thin slices of the bulk silicon that are used to produce detector modules) are exactly same, is well known by detector technologists and researchers. Even detector modules cut from the same wafer may show different behaviors. However this is not much of a problem since the general practice is to calibrate each detector module separately, which accounts for the small differences in impurity levels and other factors.

Table 5.1.3: Common impurities found in silicon with their positions in the band gap (47). Here E_v and E_c represent the highest valence band level and lowest conduction band level respectively.

Impurity	Symbol	Type	Position (eV)
Gold	Au	Donor	$E_v + 0.35$
		Acceptor	$E_c - 0.55$
Copper	Cu	Donor	$E_v + 0.24$
		Acceptor	$E_v + 0.37$
		Acceptor	$E_v + 0.52$
Iron	Fe	Donor	$E_v + 0.39$
Nickel	Ni	Acceptor	$E_c - 0.35$
		Acceptor	$E_v + 0.23$
Platinum	Pt	Donor	$E_v + 0.32$
		Acceptor	$E_v + 0.36$
		Acceptor	$E_c - 0.25$
Zinc	Zn	Acceptor	$E_v + 0.32$
		Acceptor	$E_c - 0.5$

Let us now turn our attention to the process of doping in silicon. Table.5.1.4 gives the commonly used dopers with their ionization energies. Other doping agents such as oxygen and copper are also sometimes used in detector technology, however boron and phosphorus are perhaps the most common choices. For detector fabrication the doping levels are kept very small so that the resistivity of the material remains high. High resistivity is important to suppress noise and can be afforded in silicon since its breakdown voltage is on the order of 10^5 V/cm.

Signal generation in a semiconductor depends on how the charges move in the bulk of the material. We saw earlier that diffusion coefficient and mobility are the two parameters than can be used to characterize the motion of electrons and holes in semiconductors. Extensive research has gone into understanding these parameters

Table 5.1.4: Common donor and acceptor elements used to dope silicon . Also given are their ionization energies (47).

Doping Agent	Symbol	Type	Ionization Energy (eV)
Arsenic	As	Donor	0.054
Phosphorus	P	Donor	0.045
Antimony	Sb	Donor	0.043
Aluminum	Al	Acceptor	0.072
Boron	B	Acceptor	0.045
Gallium	Ga	Acceptor	0.074
Indium	In	Acceptor	0.157

and producing the related quantities for use in computations required for detector development and operation (see Table 5.1.5).

Table 5.1.5: Mobilities (μ_e, μ_h,) velocities (v_e,v_h), and diffusion coefficients (D_e,D_h) of electrons and holes in silicon (47).

Property	Symbol	Value
Electron Mobility	μ_e	$\leq 1400\ cm^2V^{-1}s^{-1}$
Hole Mobility	μ_h	$\leq 450\ cm^2V^{-1}s^{-1}$
Electron Thermal Velocity	v_e	$2.3 \times 10^5\ m\ s^{-1}$
Hole Thermal Velocity	v_h	$1.65 \times 10^5\ m\ s^{-1}$
Electron Diffusion Coefficient	D_e	$\leq 36\ cm^2s^{-1}$
Hole Diffusion Coefficient	D_h	$\leq 12\ cm^2s^{-1}$

The reader might be wondering as to why in Table 5.1.5 only the upper bounds on the diffusion coefficient and mobility values have been given in Table 5.1.5. The reason is that these parameters depend on various factors, such as temperature, impurity type and concentration, and doping. This can be appreciated by looking at figures 5.1.11 and 5.1.12, which are the plots of electron and ion mobilities versus donor density. The plots have two interesting features. One is their non-linearity and the second is a range of donor density where the effects are most profound. It is apparent that increasing the donor density by three orders of magnitude decreases the electron and hole mobilities by approximately a factor of 10. Any change in donor density with time can therefore have serious effect on detector performance.

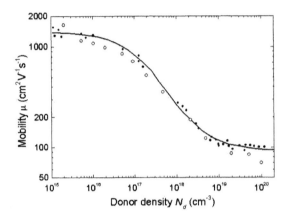

Figure 5.1.11: Dependence of electron mobility on the donor density in silicon at $300\ K$ (19).

We will see later in the chapter that the physical damage to silicon caused by radiation has the potential to change the intrinsic charge density. This is one of the reasons why prolonged deployment of silicon detectors in high radiation environments is associated with slow non-linearities in detector response. The practice, therefore, is to closely monitor the detector for changes and to use temperature lowering and bias increase to compensate for the deterioration.

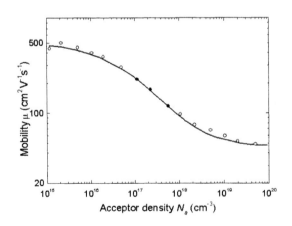

Figure 5.1.12: Dependence of hole mobility on the donor density in silicon at $300\ K$ (19).

Let us now examine how the mobilities of electrons and holes depend on temperature. It has been found that in silicon both types of charge carriers respond approximately in the same manner to temperature changes. The temperature dependence of electron and hole mobilities for silicon can be written as (44)

$$\mu_e \quad \propto \quad T^{-2.5} \tag{5.1.34}$$

$$\mu_h \quad \propto \quad T^{-2.7}. \tag{5.1.35}$$

Typical curves showing the dependence of temperature on electron and hole mobilities are shown in figures 5.1.13 and 5.1.14. We will see later that this behavior is not typical of only silicon and for almost all semiconductors the temperature dependence of mobility can be approximated by

$$\mu \propto T^{-n}, \tag{5.1.36}$$

where n is a real number that depends on the type of the semiconductor material and the particle (electron or hole). Of course the value of n can be significantly different from one material to the other. Even for the same material the value may differ for electrons and holes, as we will see later for germanium.

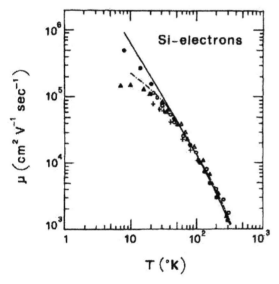

Figure 5.1.13: Dependence of hole mobility on the absolute temperature (9). Solid line represents the theoretical prediction while the points represent the experimental data.

Example:
Determine the percentage change in intrinsic charge concentration in silicon if the temperature is decreased from 27^0C to -10^0C.

Solution:
Let us first determine the band gap energies E_{g1} and E_{g2} at the two temperatures $T_1 = 300\ K$ and $T_2 = 263\ K$ using equation 5.1.29. The energy at 300

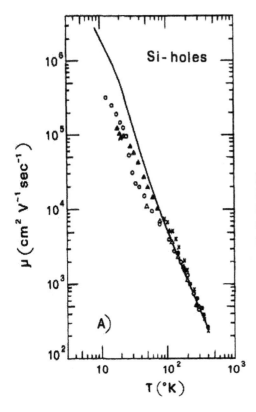

Figure 5.1.14: Dependence of hole mobility on the donor density in silicon at $300\ K$ (33). Solid line represents the theoretical prediction while the points represent the experimental data.

K is given by

$$E_g = 1.17 - 4.73 \times 10^{-4} \frac{T^2}{T + 636}$$

$$\Rightarrow E_{g1} = 1.17 - 4.73 \times 10^{-4} \frac{300^2}{300 + 636}$$

$$= 1.12\ eV$$

$$= 1.12 \times 1.602 \times 10^{-19} = 1.794 \times 10^{-19}\ J.$$

Similarly, the bandgap energy at $273\ K$ is

$$E_{g2} = 1.17 - 4.73 \times 10^{-4} \frac{263^2}{263 + 636}$$

$$= 1.13\ eV.$$

$$= 1.13 \times 1.602 \times 10^{-19} = 1.810 \times 10^{-19}\ J.$$

Now, according to equation 5.1.33, the percentage decrease in intrinsic charge concentration is given by

$$\triangle n = \frac{T_1^{3/2} e^{-E_{g1}/2k_B T_1} - T_2^{3/2} e^{-E_{g2}/2k_B T_2}}{T_1^{3/2} e^{-E_{g1}/2k_B T_1}} \times 100$$

$$= \frac{6.536 \times 10^{-7} - 6.40 \times 10^{-8}}{6.536 \times 10^{-7}} \times 100$$

$$= 90.2\%. \qquad (5.1.37)$$

This example clearly demonstrates the advantage of operating a silicon detector at low temperatures.

G.2   Germanium (Ge)

Use of germanium detectors in γ-ray spectroscopy is well established. Their high resolution and wide dynamic range make them highly suitable for spectroscopic purposes. However, in other applications, such as particle tracking, they are not preferred over silicon based detectors. In this section we will look at some of the important properties of germanium and compare them with those of silicon.

The crystal structure of germanium is the same as silicon but its atomic density is slightly lower. The most distinguishing feature of germanium is its low band gap energy (0.661 eV), which is almost half that of silicon. The energy band structure of germanium is shown in Fig.5.1.15.

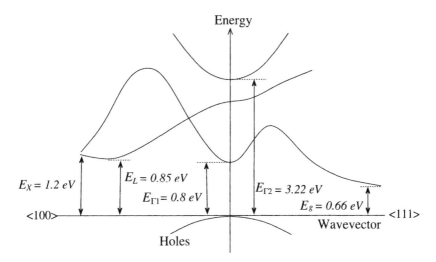

Figure 5.1.15: Band structure diagram of germanium showing energy versus wavenumber (reproduced from (47)). The subscripts of E represent different energy levels. The number in brackets (100 and 111) are the Miller indices. A Miller index represents the orientation of an atomic plane in a crystal lattice.

Because of the low band gap energy, intrinsic charge carrier concentration of germanium is about three order of magnitude higher than than of silicon. Certainly this is not a very desirable feature as far as radiation detection is concerned since it would imply larger intrinsic noise and the need to more aggressive cooling. The resistivity of the germanium is about four order of magnitude lower than that of silicon. The temperature dependence of germanium's energy gap is given by (47)

$$E_g = 0.742 - 4.8 \times 10^{-4} \frac{T^2}{T + 235}, \tag{5.1.38}$$

where T is the absolute temperature and E_g is in eV. This equation has been plotted in Fig.5.1.16. This figure when compared with that of silicon (Fig.5.1.9) does not reveal any dramatic difference between the temperature dependence of the band gap energies of the two materials. The only important thing here is that the band gap for germanium also increases with decrease in temperature.

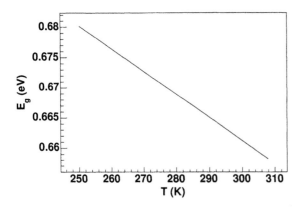

Figure 5.1.16: Variation of germanium band gap energy with absolute temperature.

Just like silicon, the intrinsic carrier concentration of germanium is also governed by equation 5.1.30. The temperature dependences of density of states in conduction and valence bands of germanium are given by (47)

$$N_c = 1.98 \times 10^{15} T^{3/2} \tag{5.1.39}$$
$$N_v = 9.6 \times 10^{14} T^{3/2}, \tag{5.1.40}$$

where T is the absolute temperature and the density of states are in cm^{-3}. Substituting these in equation 5.1.30 gives the expression for the intrinsic carrier concentration.

$$n_i = 1.38 \times 10^{15} T^{3/2} e^{-E_g/2k_B T} \tag{5.1.41}$$

Here, as before, n_i is in cm^{-3}, T is the absolute temperature, and k_B is the Boltzmann's constant. The plot of this equation (Fig.5.1.17) when compared with that of silicon (Fig.5.1.17) reveals that quantitatively there is a difference of several orders of magnitude between the intrinsic charge carrier densities of the two materials in the same temperature range. Of course this can be attributed to the lower band gap energy in germanium, which allows more electrons in the valence band to jump to the conduction band due to thermal agitation.

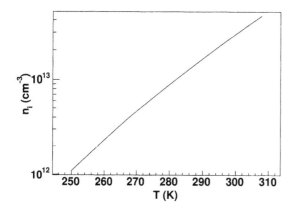

Figure 5.1.17: Dependence of intrinsic charge concentration in germanium on absolute temperature.

Fortunately, germanium can be obtained in extremely pure form[2]. The lower resistivity is therefore mainly due to low band gap energy. There are a number of elements that can be used to dope germanium to make it suitable for use as detection medium. The most common of such doping agents are listed in Table.5.1.6.

Table 5.1.6: Common donor and acceptor elements used to dope silicon . Also given are their ionization energies (47).

Doping Agent	Symbol	Type	Ionization Energy (eV)
Arsenic	As	Donor	0.014
Phosphorus	P	Donor	0.013
Antimony	Sb	Donor	0.010
Bismuth	Bi	Donor	0.013
Lithium	Li	Donor	0.093
Aluminum	Al	Acceptor	0.011
Boron	B	Acceptor	0.011
Gallium	Ga	Acceptor	0.011
Indium	In	Acceptor	0.012
Thallium	Tl	Acceptor	0.013

---

[2] *High Purity Germanium* or HPGe detectors are widely used for γ-ray spectroscopy.

Some of the important electrical properties of germanium are listed in Table 5.1.7. Comparison of these values with those of silicon (see Table 5.1.5) clearly shows that the overall charge collection efficiency in a detector based on germanium will be higher than in silicon. However charge collection efficiency is not the only criterion for selecting a material as detection medium. Other factors include intrinsic charge carrier density, crystal defects and imperfections, impurities, availability, and cost.

Table 5.1.7: Mobilities (μ_e, μ_h,) velocities (v_e,v_h), and diffusion coefficients (D_e,D_h) of electrons and holes in germanium (47).

Property	Symbol	Value
Electron Mobility	μ_e	$\leq 3900 \ cm^2V^{-1}s^{-1}$
Hole Mobility	μ_h	$\leq 1900 \ cm^2V^{-1}s^{-1}$
Electron Thermal Velocity	v_e	$3.1 \times 10^5 \ m \ s^{-1}$
Hole Thermal Velocity	v_h	$1.9 \times 10^5 \ m \ s^{-1}$
Electron Diffusion Coefficient	D_e	$\leq 100 \ cm^2s^{-1}$
Hole Diffusion Coefficient	D_h	$\leq 50 \ cm^2s^{-1}$

Let us now turn our attention to the temperature and electric field dependence of electrical conduction properties of germanium. Fig.5.1.18 shows the dependence of drift velocity on electric field intensity at two different temperature settings. The important thing to note here is the proportionality of the drift velocity to the electric field intensity at least up to moderate electric fields. Hence the relation

$$v_d = \mu_e E, \tag{5.1.42}$$

holds well up to about an electric field of 1000 V/cm. Here v_d is the drift velocity of electrons, E is the applied electric field, and μ_e is the mobility of electrons in germanium.

As in case of silicon, in germanium too the electron mobility has a temperature dependence, which can be approximately written as (44)

$$\mu_e \propto T^{-1.66}. \tag{5.1.43}$$

Fig.5.1.19 shows mobility of electrons in germanium as a function of absolute temperature. Though the graph does not cover a wide temperature range going up to the room temperature, but the trend of the graph is qualitatively similar to the one shown.

So far we are happy that the electrons in germanium behave in an orderly fashion with a drift velocity that is proportional to the applied electric field. But the output signal depends not only on electrons but also on how holes behave. Since movement of holes is coupled with the movement of electrons, we would expect that their drift

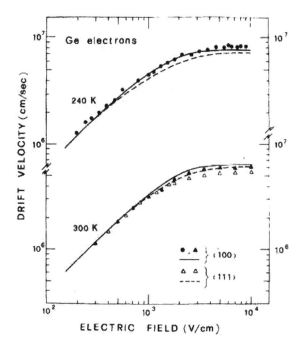

Figure 5.1.18: Dependence of electron drift velocity on electric field intensity in germanium at two different temperatures (20). Solid and dashed lines represent theoretical predictions while the points represent experimental data. 100 and 111 are the two crystallographic directions in which the electric field was applied.

velocity would also be proportional to the applied electric field. This is true to a certain extent, though. Actually the behavior of hole mobility differs from the behavior of electron mobility in germanium. Even though it decreases with increase in temperature but variation is not as linear on double logarithmic scale as it is for electron mobility. The hole mobility for germanium shows a temperature dependence given by (44)

$$\mu_h \propto T^{-2.33}. \tag{5.1.44}$$

G.3 Gallium Arsenide ($GaAs$)

Gallium arsenide is another semiconductor material that is extensively used as detection medium. The distinguishing feature of $GaAs$ is its higher photon absorption efficiency as compared to silicon, which has allowed the development of extremely thin (100-200 μm) x-ray detectors. Another advantage of $GaAs$ is that it can be operated at room temperature, which simplifies the detector design considerably and also cuts down the cost of development and operation.

The band structure diagram of gallium arsenide is shown in Fig.5.1.20 and its basic properties are listed in Table 5.1.2. It can be seen that, as far as atomic and weight densities are concerned, there is no significant difference between germanium and gallium arsenide. However since the band gap of $GaAs$ is more than twice that of germanium and significantly higher than silicon, therefore its intrinsic carrier concentration is several orders of magnitude lower than the two materials. The most dramatic difference is the intrinsic resistivity of gallium arsenide, which is about eight orders of magnitude higher than that of germanium and three orders of

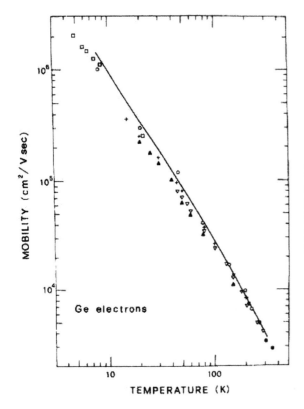

Figure 5.1.19: Dependence of electron mobility on absolute temperature in germanium (20). Solid line represents theoretical prediction while the points represent experimental data.

magnitude higher than that of silicon. This, of course, is a very desirable feature as far as its use as detection medium is concerned.

The temperature dependence of band gap of gallium arsenide is given by (47)

$$E_g = 1.519 - 5.405 \times 10^{-4} \frac{T^2}{T + 204}, \tag{5.1.45}$$

where T is the absolute temperature and E_g is in eV. This equation has been plotted in Fig.5.1.21. The reader would note that the variation in band gap energy with temperature for gallium arsenide is not much different than that for silicon or germanium. The distinguishing feature of $GaAs$ is the width of the band gap itself, which at each temperature is far higher than the other two materials.

The evaluation of intrinsic carrier density of gallium arsenide is not as simple as that of silicon or germanium. The reason can be understood by examining its band structure diagram (see Fig.5.1.20). The conduction band of $GaAs$ has two additional valleys (X and L) whose contribution to the overall density of states of the conduction band can not be ignored. The valence band density of states has a simple temperature dependence, though. The density of states of conduction and

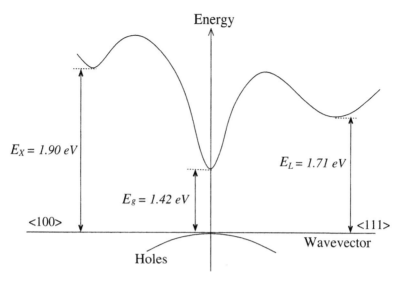

Figure 5.1.20: Band structure diagram of gallium arsenide showing energy versus wavenumber (reproduced from (47)). The subscripts of E represent different energy levels. The number in brackets (100 and 111) are the Miller indices. A Miller index represents the orientation of an atomic plane in a crystal lattice.

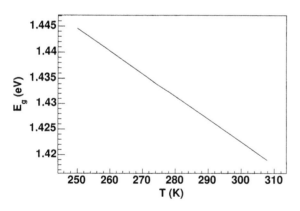

Figure 5.1.21: Variation of gallium arsenide band gap energy with absolute temperature.

valence bands in $GaAs$ can be written as (47)

$$N_c = 8.63 \times 10^{13} T^{3/2} \left[1 - 1.93 \times 10^{-4} T - 4.19 \times 10^{-8} T^2 \right.$$

$$\left. + 21 \exp\left(-\frac{E_L}{2k_B T}\right) + 44 \exp\left(-\frac{E_X}{2k_B T}\right) \right] \qquad (5.1.46)$$

and

$$N_v = 1.83 \times 10^{15} T^{3/2}, \qquad (5.1.47)$$

where both N_c and N_v are in units of cm^{-3}. The energy gaps E_L and E_X corresponding to L and X valleys respectively can be evaluated from (47)

$$E_L = 1.815 - 6.05 \times 10^{-4} \frac{T^2}{T + 204} \tag{5.1.48}$$

$$E_X = 1.981 - 4.60 \times 10^{-4} \frac{T^2}{T + 204}. \tag{5.1.49}$$

The intrinsic charge carrier density can now be calculated by substituting N_c and N_v from equations 5.1.46 and 5.1.47 into equation 5.1.30. Hence we get

$$n_i = 3.974 \times 10^{14} T^{3/2} \left[1 - 1.93 \times 10^{-4} T - 4.19 \times 10^{-8} T^2 \right.$$

$$\left. + 21 \exp\left(-\frac{E_L}{2k_B T}\right) + 44 \exp\left(-\frac{E_X}{2k_B T}\right) \right]^{1/2} e^{-E_g/2k_B T}, \tag{5.1.50}$$

where E_g, E_L, and E_X are given by equations 5.1.45, 5.1.48, and 5.1.49 respectively. The plot of the above equation (see Fig.5.1.22) can now be compared to the similar plots for silicon and germanium (see figures 5.1.10 and 5.1.17). It is clear that, in terms of intrinsic charge carriers, gallium arsenide is much superior than silicon and germanium. Such low intrinsic carrier concentration even at room temperature makes it possible to operate $GaAs$ based detectors without or with very minimal cooling.

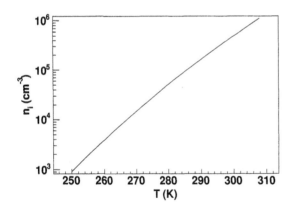

Figure 5.1.22: Dependence of intrinsic charge concentration in gallium arsenide on absolute temperature.

Let us now have a look at the electrical conduction properties of GaAs. It is apparent from Table 5.1.8 that the electron mobility in $GaAs$ is more than 20 times higher than the hole mobility. This behavior is in contrast with germanium and silicon where the mobilities differ by only about a factor of 2 to 3. However, interestingly enough, the temperature dependence of the hole mobility in $GaAs$ is about the same as in germanium, that is (44)

$$\mu_h \propto T^{-2.3}. \tag{5.1.51}$$

On the other hand the electron mobility in $GaAs$ follows (7)

$$\mu_e \propto T^{-0.66}. \tag{5.1.52}$$

Table 5.1.8: Mobilities (μ_e, μ_h,) velocities (v_e,v_h), and diffusion coefficients (D_e,D_h) of electrons and holes in gallium arsenide (47).

Property	Symbol	Value
Electron Mobility	μ_e	$\leq 8500 \ cm^2V^{-1}s^{-1}$
Hole Mobility	μ_h	$\leq 400 \ cm^2V^{-1}s^{-1}$
Electron Thermal Velocity	v_e	$4.4 \times 10^5 \ m \ s^{-1}$
Hole Thermal Velocity	v_h	$1.8 \times 10^5 \ m \ s^{-1}$
Electron Diffusion Coefficient	D_e	$\leq 200 \ cm^2s^{-1}$
Hole Diffusion Coefficient	D_h	$\leq 10 \ cm^2s^{-1}$

G.4   Cadmium-Zinc-Tellurium ($CdZnTe$)

Also referred to in short as CZT, this material has gained a lot of popularity in the recent years. It has several properties that make it highly desirable for demanding applications, such as spectroscopy. Its high detection efficiency, high resolution, low cost, and good signal to noise ratio at room temperature make it a product of choice for many applications. Its band gap is more than twice that of germanium at room temperature, which drastically decreases the number of intrinsic charge carriers (see Table.5.1.2). This makes it highly suitable for room temperature operation.

$CdZnTe$ is actually a ternary alloy of $CdTe$ and Zn. Its properties therefore depend on the concentration of zinc in the bulk as well as on the surface of the material. For radiation detection purposes the most important parameter is the band gap. Fortunately the band gap has been found to be very lightly dependent on the concentration of zinc. Fluctuations of a few percent in zinc concentration change the band gap by only a few meV (46), which is insignificant for estimation of most detector related parameters. In general the percentage of zinc in a typical CZT bulk is less than 10%.

Another important property of CZT material is that it can be formed into different shapes and sizes. Large area and complicated geometry CZT detectors can therefore be practically fabricated.

CZT detectors have fairly high stopping power and absorption efficiency due to high Z elements. This together with their high efficiency makes them well suited for use in imaging applications, such as medical x-ray imaging.

There are two main disadvantages associated with CZT materials: their low hole mobility (and hence low lifetime) and crystal defects. The later can be somewhat controlled by using techniques that yield less crystal defects. The low hole mobility, on the other hand, is caused by the hole trapping mechanisms and is more or less intrinsic to the material. The only way to increase the hole lifetime is by increasing the detector bias voltage. Note that crystal defects further deteriorate the hole mobility and should therefore be controlled as much as possible. As a reminder to the reader, the direct consequence of low charge carrier lifetime is the loss of signal. Since this loss is not linearly dependent on the amount of deposited energy, it results in nonlinear response of the detector. Another important point to note here is that the charge collection also depends on the depth of the material. for example if the charge is created near the collecting electrode of the detector, the loss of charge will be minimal. On the other hand, if the same charge is produced away from the collecting electrode, the signal loss will be higher.

Increasing the bias voltage might not always be practical or even desirable in certain applications. A novel method to *compensate* for the low hole mobility is to use the so called *ohmic contacts* at the electrodes. The advantage of this scheme is that the holes get recombined with the electrons released into the material by the ohmic contact. This hole recombination effectively stops the leakage current, while the signal current is carried predominantly by the electrons. Ohmic contacts therefore completely eliminate the need for operating the detector at high voltages or external circuitry to compensate for low hole mobility.

5.1.H   The pn-Junction

The n- and p-type semiconductors can be joined together to create the so called pn-junction (see Fig.5.1.23). These junctions have been found to be extremely useful not only for building semiconductor electronics but also for radiation detector technology. When a p- and a n-type semiconductors are brought together, a flow of charges automatically starts to compensate for the imbalance in charge concentrations across the junction. The electrons that are the majority charge carriers in the n-type semiconductors flow towards the p-type material. Similarly the holes move towards the n-type material. This process continues until the Fermi levels of the two materials coincide with each other, as shown in Fig.5.1.23(b). As the electrons and holes move in opposite directions and combine together to neutralize one another, a central region devoid of any electrical charges is created. This region, generally referred to as the depletion region, plays a central role in semiconductor radiation detectors since this is where the incident radiation creates electron hole pairs. These charges flow in opposite directions and constitute an electrical current that can be measured. However the junction in this configuration can not be very effectively used for radiation detection since firstly it is too thin and secondly the potential difference across it is very small. The trick then is to widen this gap somehow and establish a high enough electric field to allow the charges created by the radiation to flow and constitute a measurable current. This is done by applying a reverse bias across the junction. We will discuss the properties and characteristics of such a junction in the next section.

Bringing an n type material in contact with a p type material produces an effective electrostatic potential across the depletion region . The thickness of this depletion region can be calculated from

$$
\begin{aligned}
W &= x_{pd} + x_{nd} \\
&= \left[\frac{2\epsilon V_0}{q} \left(\frac{1}{N_A} + \frac{1}{N_D} \right) \right]^{1/2},
\end{aligned}
\tag{5.1.53}
$$

where x_{pd} and x_{nd} are the widths of the depletion regions on p- and n-sides respectively, N_A and N_D are the acceptor and donor doping densities, q is the unit charge of electron, ϵ is the permitivity of the medium and V_0 is the potential difference.

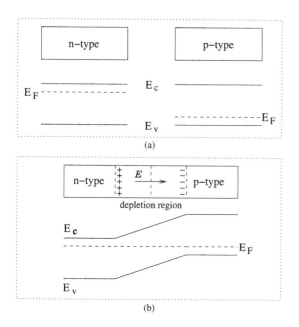

Figure 5.1.23: Semiconductors and their energy levels. (a) Separate n- and p-type crystals. (b) Formation of pn-junction. When n- and p-type crystals are brought in contact, flow of charges starts which continues till the Fermi levels E_F of the two materials do not coincide.

Before we go on any further an important point should be mentioned. The characterization of the depletion region as *devoid of any charges* is not entirely correct. No matter how good a semiconductor material is, there are always crystal imperfections and impurities, which introduce energy levels inside the band gap. Such energy levels can be exploited by electrons to jump out of the valence band and eventually go up into the conduction band. The result is the creation of an electron hole pair. Even if we assume that the material does not have any crystal imperfections and impurities, still some electrons can attain enough energy through thermal agitation to jump to the conduction band. In summary, the depletion region is not really completely devoid of free charges. However the number of such charges is very small and the corresponding current is extremely low. If a reverse bias is applied and there were no free charges in the depletion region, then there should not be any current flowing through the circuit. However since some free charge pairs are present, a very small current is always observed. This current is generally known as *dark* or *reverse* current (see Fig.5.1.24).

Let us now see what happens if we apply forward bias to the junction. In such a case, as our intuition suggests, a large current starts flowing, which increases rapidly with increasing voltage. This property of the pn junction or *semiconductor diode* is extensively used to design electronic devices such as switches and solar cells. For radiation detection purposes, the pn junction is always reversed biased.

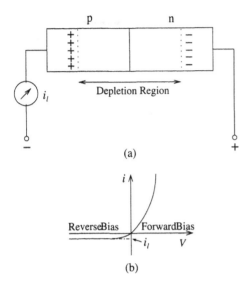

p   n

Depletion Region

i_l

(a)

i

ReverseBias   ForwardBias

i_l V

(b)

Figure 5.1.24: (a) Reverse biased pn-junction. (b) Current-voltage curve of a typical pn-junction. If the junction is reverse biased, a small leakage current i_l flows through it, which stays almost constant as the voltages is increased up to a point at which the potential is high enough to overcome the potential barrier (not shown here). At forward bias, however, the current increases with applied voltage.

H.1 Characteristics of a Reverse-Biased pn-Diode

In a semiconductor detector, the depletion region is used as the active medium for creating electron-hole pairs by incident radiation. This region is almost devoid of free charge carriers at operating temperatures and therefore very small leakage current flows through it in the absence of radiation. The charge pairs created by the radiation move in opposite directions under the influence of the effective junction electric field and constitute an electric current that can be measured. Under carefully maintained working conditions such as temperature, this current is proportional to the energy deposited by the radiation. In this respect a semiconductor detector has the same working principle as a gas filled chamber except that the number of charge pairs created in the former is far more than the latter and consequently the output signal is of higher strength.

The output signal of a semiconductor detector and its dynamic range depends on several factors, most notably the effective electric field strength, the capacitance, and the depth of the depletion region. For the typical planar geometry, these parameters can be fairly easily estimated using the Poisson's equation

$$\bigtriangledown^2 \Phi = -\frac{\rho}{\epsilon}, \qquad (5.1.54)$$

where Φ is the electric potential, ϵ is the permittivity of the semiconductor material, and ρ is the charge density profile in the depletion region. The permittivity in this equation can be written as a product of the dielectric constant of the material ϵ_r (also sometimes referred to as the relative permittivity) and the permittivity of free space ϵ_0, that is

$$\epsilon = \epsilon_r \epsilon_0. \qquad (5.1.55)$$

The dielectric constant or relative permittivity is a dimensionless constant and is extensively quoted in literature. Its value depends on the type of material and varies considerably from material to material. for example the dielectric constant for silicon is around 12 while that of germanium is about 16. For computations, this value must be multiplied by the permittivity of free space $\epsilon_0 = 8.854 \times 10^{-12}$ $C^2 N^{-1} m^{-2}$.

Moving back to our derivation, for the sake of simplicity, let us write the equation 5.1.54 in one dimension as

$$\frac{d^2\Phi}{dx^2} = -\frac{\rho(x)}{\epsilon}, \tag{5.1.56}$$

Although the charge density has a continuous profile inside the region, but to simplify the calculations we can approximate this with a step profile given by (see also Fig.5.1.25)

$$\rho(x) = \begin{array}{lll} eN_D & : & 0 \leq x < x_n \quad \text{n-side} \\[2mm] -eN_A & : & -x_p < x \leq 0 \quad \text{p-side}. \end{array} \tag{5.1.57}$$

Here N_D and N_A are the donor and acceptor impurity concentrations respectively, e is the usual unit electrical charge, and x_p and x_n respectively are the depths of junction on p and n sides. It should be noted that this charge density profile is not always a good approximation, specially in the so called *fully depleted* detectors or when the applied bias is very small; the two extremes corresponding to very large and very small depletion regions respectively. On the p-side $(-x_p < x \leq 0)$, it becomes

$$\frac{d^2\Phi}{dx^2} = \frac{eN_A}{\epsilon}. \tag{5.1.58}$$

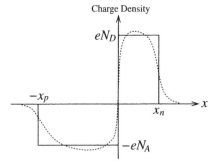

Figure 5.1.25: Realistic (dotted line) and idealized (solid line) charge density distributions in a pn junction. x_p and x_n are the depths of depletion regions on p and n sides respectively. In the majority of semiconductor detectors only one side is heavily doped making the depletion region very large on the opposite side.

Integrating this once gives us the electric field profile on the p-side

$$\begin{aligned} E(x) &= -\frac{d\Phi}{dx} = -\int \frac{d^2\Phi}{dx^2} dx \\[2mm] &= -\frac{eN_A}{\epsilon} \int dx \\[2mm] &= -\frac{eN_A}{\epsilon} x + A. \end{aligned} \tag{5.1.59}$$

To determine the integration constant A we note that the electric field E must vanish at the edge of the depletion region, i.e., $E(-x_p) = 0$. This gives

$$E(x) = -\frac{d\Phi}{dx} = -\frac{eN_A}{\epsilon}(x + x_p) \text{ for } -x_p < x \leq 0 \tag{5.1.60}$$

Similarly for the n-side we get

$$E(x) = -\frac{d\Phi}{dx} = \frac{eN_D}{\epsilon}(x - x_n) \text{ for } 0 \leq x < x_n \qquad (5.1.61)$$

Fig.5.1.26 shows these functions as well as the field profile in a realistic pn junction.

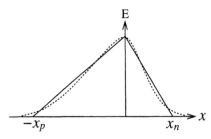

Figure 5.1.26: Electric field intensity profile of the idealized charge density shown in Fig.5.1.25 (solid line) together with a more realistic profile (dotted line).

To determine the profile of the electric potential and the depletion depth, we can integrate the above two equations again to get

$$\Phi(x) = \begin{cases} -\frac{eN_D}{\epsilon}\left[\frac{x^2}{2} - xx_n\right] + A_1 & : \quad 0 \leq x < x_n \quad \text{n-side} \\[2mm] \frac{eN_A}{\epsilon}\left[\frac{x^2}{2} + xx_p\right] + A_2 & : \quad -x_p < x \leq 0 \quad \text{p-side.} \end{cases} \qquad (5.1.62)$$

The integration constants A_1 and A_2 can be determined by noting that the applied reverse bias appears as a potential difference across the junction, which can be taken as 0 at $x = -x_p$ and V_0 at $x = x_n$. In such a case the potential profile inside the junction becomes

$$\Phi(x) = \begin{cases} -\frac{eN_D}{2\epsilon}(x - x_n)^2 + V_0 & : \quad 0 \leq x < x_n \quad \text{n-side} \\[2mm] \frac{eN_A}{2\epsilon}(x + x_p)^2 & : \quad -x_p < x \leq 0 \quad \text{p-side.} \end{cases} \qquad (5.1.63)$$

This potential has been plotted in Fig.5.1.27

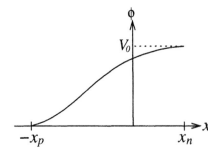

Figure 5.1.27: Variation of electric potential with respect to distance from the center of a pn junction.

An interesting result can be obtained if we use the condition that the potentials at $x = 0$ must be equal. This gives

$$V_0 = \frac{e}{2\epsilon}\left[N_A x_p^2 + N_D x_n^2\right] \qquad (5.1.64)$$

This expression can be used to determine the individual depletion depths provided we make use of another equation containing x_p and x_n. For this we can use the charge conservation relation

$$N_D x_n = N_A x_p, \tag{5.1.65}$$

which simply implies that the total charge remains constant no matter how much of it gets transferred between the two regions. The above two relations give the following depletion depths on p and n sides.

$$
\begin{aligned}
x_p &= \left[\frac{2\epsilon V_0}{e N_A (1 + N_A/N_D)} \right]^{1/2} \\
x_n &= \left[\frac{2\epsilon V_0}{e N_D (1 + N_D/N_A)} \right]^{1/2}
\end{aligned}
\tag{5.1.66}
$$

The total depletion depth d is then just the sum of these two depths, that is

$$d = x_p + x_n. \tag{5.1.67}$$

Usually in semiconductor detectors, the disparity in the dopant levels on p and n sides is so large that the depth on one side can be safely ignored. This greatly simplifies the expression for total depletion depth. To see this let us assume that the acceptor impurity level is much higher than the donor impurity level ($N_A \gg N_D$). In such a case the n side depletion depth will be much greater than the p side. This can also be seen from the charge conservation relation stated above, which for $N_A \gg N_D$ implies that $x_n \gg x_p$. In such a case the depletion depth on p side will be so small that it can simply be ignored. The total depletion depth as deduced from the relation 5.1.66 becomes

$$d \simeq \left[\frac{2\epsilon V_0}{e N_D} \right]^{1/2}. \tag{5.1.68}$$

Similarly for the case $N_D \gg N_A$ we get

$$d \simeq \left[\frac{2\epsilon V_0}{e N_A} \right]^{1/2}. \tag{5.1.69}$$

These relations show that once the junction has been physically established with fixed dopant levels, the applied voltage is the only parameter that can be varied to change the depletion depth. We will see later in the chapter that the depletion width in detectors exposed to high radiation environments decreases with time due to the damage caused by the radiation. We can see from the above relation that at fixed applied voltage, this would mean that the minority charge concentration has changed and the only way to compensate for this increase would be to increase the reverse bias. In fact this is what is routinely done. Since the depletion width is inversely proportional to the dopant level, therefore close monitoring of leakage current is done to see if the charge concentration has changed and based on the result the reverse bias is increased to increase the depletion width.

Sometimes it is more convenient to know the depletion depth in terms of resistivity and mobility since these parameters are generally known. The resistivity of a doped semiconductor is given by

$$\rho \simeq \frac{1}{e N \mu}, \tag{5.1.70}$$

where μ is the mobility of majority charge carrier and N is the dopant concentration. For the case $N_D \gg N_A$, the majority charge carriers are electrons and therefore the depletion depth in terms of resistivity becomes

$$d \simeq [2\epsilon\rho_n\mu_e V_0]^{1/2} . \qquad (5.1.71)$$

Similar expression can be derived for the case when acceptor impurity level is much higher than the donor level. Of course in such a situation the majority charge carriers will be holes.

In the absence of radiation, except for a minute leakage current, the depletion region of a pn junction essentially acts as an insulator sandwiched between positive and negative electrodes. The capacitance of this configuration can be easily estimated for the usual case of planar geometry if we assume the idealized charge density profile of Fig.5.1.25. This is the configuration of a simple parallel plate capacitor with a capacitance of

$$C = \epsilon\frac{A}{d}, \qquad (5.1.72)$$

where A is the surface area of the junction and d is the depletion width. A convenient parameter generally used for comparison is the capacitance per unit area $C_A = C/A$, which we can compute if we substitute the values of d from Equations 5.1.68 and 5.1.69 into the above expression. Hence we get

$$C_A = \begin{cases} \left(\frac{e\epsilon N_D}{2V_0}\right)^{1/2} & \text{for } N_A \gg N_D \\ \left(\frac{e\epsilon N_A}{2V_0}\right)^{1/2} & \text{for } N_A \ll N_D \end{cases} \qquad (5.1.73)$$

It should be noted that although the capacitance of a usual pn junction is very small, of the order of a few pico Farads (see example below), still together with the load resistor of the signal readout circuit, it can limit the frequency response of the detector. Therefore in practical detectors it is ensured that the capacitance is kept at a minimum. This can be done by simply increasing the reverse bias, as apparent from the above expressions for C_A.

Example:
Compute the capacitances per unit area of a silicon pn diode having donor and acceptor impurities of 10^{17} cm^{-3} and 10^{15} cm^{-3} respectively when a bias of 150 V exists across its junction. Also compute the absolute capacitance if the surface area of the diode is 0.01 cm^2.

Solution:
Since we have $N_D \gg N_A$, therefore according to equation 5.1.73 we can estimate the capacitance per unit area using only the acceptor impurity concentration. Since the material is silicon therefore we will assume that the

dielectric constant is 12. Hence we have

$$
\begin{aligned}
C_A &= \left[\frac{e\epsilon N_A}{2V_0}\right]^{1/2} \\
&= \left[\frac{\left(1.602 \times 10^{-19}\right)(12)\left(8.854 \times 10^{-12}\right)\left(10^{15} \times 10^{6}\right)}{(2)(150)}\right]^{1/2} \\
&= 7.5 \times 10^{-6} \; Fm^{-2}
\end{aligned}
$$

The absolute capacitance can be obtained by multiplying this value by the surface area of the diode, that is

$$
\begin{aligned}
C &= C_A A \\
&= \left(7.5 \times 10^{-6}\right)\left(0.01 \times 10^{-4}\right) \\
&= 7.5 \; pF.
\end{aligned}
$$

H.2 Signal Generation

We saw earlier that radiation passing through the depletion region produces free charge carriers that constitute a current under the influence of the externally applied electric field. This current can be estimated using Ramo's theorem, which for a planar geometry states that the instantaneous current can be obtained through the relation

$$
i = qv\frac{dV_w}{dx}, \tag{5.1.74}
$$

where q is the charge produced at position x and moving with a velocity v and V_w is the *weighting potential* that for a certain electrode is obtained simply by setting its potential to 1 and potentials on all other electrodes to 0. In terms of *weighting electric field* $E_w = dV_w/dx$, the above equation can be written as

$$
i = qvE_w. \tag{5.1.75}
$$

To use this theorem we need to know the velocity of the charge carriers. We saw earlier that the drift velocity of charges in a semiconductor is proportional to the electric field. Hence we can write

$$
\begin{aligned}
v &= \mu E \quad \text{or} \\
v &= \mu\frac{V_0}{d}, \tag{5.1.76}
\end{aligned}
$$

where μ is the mobility of the charge carrier, V_0 is the applied reverse bias, and d is the width of the depletion region.

If we apply unit potential to the electrodes where we are measuring the current, then the weighting field is given by

$$
E_w = \frac{1}{d} \tag{5.1.77}
$$

Hence according to Ramo's theorem the induced current will be

$$i = q\mu \frac{V_0}{d} \frac{1}{d}$$
$$= q\mu \frac{V_0}{d^2}. \qquad (5.1.78)$$

This expression can be used to deduce the corresponding total charge induced on the electrode provided we know the charge collection time. Because of linear dependence of the velocity of charge carriers on the electric field, we can find the transit time t_e of an electroncreated at a distance x from the collection electrode simply by using (see also Fig.5.1.28)

$$t_e = \frac{x}{v_e}$$
$$= \frac{x}{\mu_e E}$$
$$= \frac{xd}{\mu_e V_0}. \qquad (5.1.79)$$

The hole is also created at the same position but moves in opposite direction. Since it travels a distance $d - x$ before being collected by the opposite electrode, we can write its transit time t_h as

$$t_h = \frac{d - x}{v_h}$$
$$= \frac{d - x}{\mu_h E}$$
$$= \frac{(d - x)d}{\mu_h V_0}. \qquad (5.1.80)$$

The total charge induced by electrons Q_e can then be calculated from the relation $Q_e = i_e t_e$. Using the relations for the current and the transit time, we get

$$Q_e = i_e t_e$$
$$= q\mu_e \frac{V_0}{d^2} \frac{xd}{\mu_e V_0}$$
$$= q\frac{x}{d}. \qquad (5.1.81)$$

Similarly the charge induced by the hole is given by

$$Q_h = i_h t_h$$
$$= q\mu_h \frac{V_0}{d^2} \frac{(d - x)d}{\mu_h V_0}$$
$$= q\left(1 - \frac{x}{d}\right). \qquad (5.1.82)$$

It is apparent from above expressions that the amount of charges induced on the electrodes by electrons and holes depends on the position of the charge pair creation. Also since the mobility of electrons is approximately three times higher than that

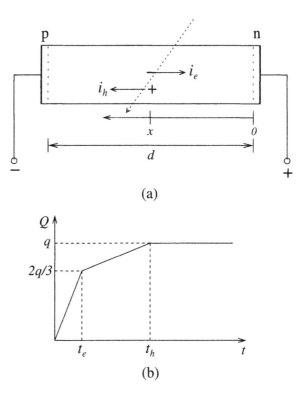

(a)

(b)

Figure 5.1.28: (a) An almost fully depleted pn junction. The incident radiation produces a charge pair in the middle of the junction. The electron and the hold move in opposite directions inducing charges on the electrodes. (b) Time profile of the charge induced by movement of an electron and a hole produced in the middle of a fully depleted pn junction. The electron mobility is almost three times larger than that of the hole and therefore only $1/3rd$ of the initial charge of $2q/3$ is due to the hole. The second part of the signal (t_e to t_h) is exclusively due to the hole.

of holes, the output signal at initial stages is almost fully due to electrons. To see this quantitatively let us assume that a charge pair is created in the center of the depletion region, i.e., $x = d/2$. The collection times of the electron and the hole in this case are

$$t_e = \frac{d^2}{2\mu_e V_0} \quad \text{and}$$

$$t_h = \frac{d^2}{2\mu_h V_0} \approx \frac{3d^2}{2\mu_e V_0} = 3t_e.$$

This shows that the hole takes approximately three times longer than the electron to reach the opposite electrode. The charge induced by the electron after time t_e is

$$Q_e = \frac{q}{2}. \tag{5.1.83}$$

The hole also keeps on moving during this time. The total charge induced by the hole during the electron drift is given by

$$
\begin{aligned}
Q_h &= i_h t_e \\
&= q\mu_h \frac{V_0}{d^2} \frac{d^2}{2\mu_e V_0} \\
&= q\frac{\mu_h}{2\mu_e} \\
&\approx \frac{q}{6},
\end{aligned} \tag{5.1.84}
$$

where we have used the relation $\mu_e \approx 3\mu_h$. The cumulative charge induced after time t_e is then

$$Q(t = t_e) = \frac{q}{2} + \frac{q}{6} = \frac{2q}{3}.$$

The hole keeps on moving even after the electron has been collected such that after time t_h the total induced charge is equal to q.

$$Q(t = t_h) = \frac{q}{2} + \frac{q}{6} + \frac{q}{3} = q$$

The time profile of the induced charge is shown in Fig.5.1.28(b).

H.3   Frequency Response

The way semiconductor detectors respond to different frequencies depends not only on the particular geometry and construction of the detectors but also their associated electronics. The discussion of the effect of electronic components on the frequency response will be deferred to the chapter on electronics. The detector related effects are mainly due to the transit time of charge pairs in the depletion region. This will be discussed here.

The response time of a detector depends on how quickly the charges are collected by the readout electrodes after their generation by the radiation. Most detectors are built and operated such that the charge carriers quickly attain saturation velocity after generation.

5.1.1   Modes of Operation of a pn-Diode

The pn-diodes can be operated in essentially two different modes: photovoltaic mode and photoconductive mode. Although the semiconductor detectors are almost always operated in the photoconductive mode but it is also theoretically possible and practically feasible to build a detector that operates in the photovoltaic mode. These two modes will be described in the next two sections. However, since the photovoltaic mode does not have much practical significance as far as radiation detectors are concerned, we will not spend much time discussing it.

I.1 Photovoltaic Mode

In this mode a very large load resistance is applied across the junction such that essentially no current flows through the circuit. The results in the creation of a potential difference across the diode. The incident radiation produces electron hole pairs inside the depletion region, which move in opposite directions under the influence of the junction potential. Let us call this current the radiation induced reverse current i_r. The consequence of this flow of charges is the reduction in the junction barrier, which starts another current but now in opposite direction to i_r. Let us call this current the forward current i_f. Since the diode is connected to essentially an open circuit (very large load resistance) therefore almost no current flows through the circuit. This implies that the forward current must be balanced by the reverse current, or

$$i_f = i_r. \tag{5.1.85}$$

The change in the energy barrier due to flow of the reverse current changes the potential difference across the diode. This potential difference can then be measured with a sensitive device. Note that this operation mode does not require the junction to be biased by an external source.

The good thing about this mode of operation is that essentially no leakage current exists across the junction and therefore the noise is extremely small. However noise is not the only important consideration for building a detection. There are other factors, such as small thickness of the depletion region and non-linearity in response, which make this mode undesirable for radiation detection purposes. The most problematic aspect of this mode is that the output voltage is not a linear function of the deposited energy and therefore the detector can not be easily used for spectroscopic purposes. Because of such reasons, as noted earlier, this mode of operation is not used for radiation detection.

Photovoltaic mode is most exclusively used to build solar cells.

I.2 Photoconductive Mode

This is the mode in which radiation detectors are generally operated. A high reverse bias is applied across the diode, creating a large depletion region in the middle of the junction. The incident radiation passing through the depletion region produces electron hole pairs along its track. These charges move in opposite directions and constitute an electrical current that can be measured. As the number of charge pairs created by the incident radiation depends on the deposited energy, the measured current is proportional to the energy carried by the radiation. For radiation detection purposes, there are three main advantages of reverse biasing the pn-junctions.

▶ Increase the size of the depletion region.

▶ Increase the signal to noise ratio.

▶ Decrease the capacitance.

Charge pairs are not exclusively produced in the depletion region by the incident radiation. In fact the electron hole pair production probability is the same whether the region is depleted or not. The advantage of depletion region is that it has very few free charge pairs and therefore the leakage current flowing through it in the absence of radiation is minimal. When radiation traverses through this region and produces a large number of charge pairs, the reverse current increases. The magnitude of this current is proportional to the energy deposited by the incident radiation. This proportionality is the most desirable factor for any detector since it enables one to derive meaningful quantities from the measurements about the incident radiation. The extremely low leakage current in a well designed semiconductor detector ensures that this proportionality remains a reality throughout the dynamic range of the detector. To get an idea, the typical leakage current is of the order of a few nano amperes, while the radiation induced current can be several orders of magnitude higher than this.

Fig.5.1.29 shows how the current increases with respect to increase in flux of incident radiation (the higher the flux the higher the energy deposited). Since the leakage current does not depend on the energy deposited by the radiation therefore its effect can be easily subtracted from the detector response, provided all other conditions including temperature remain constant.

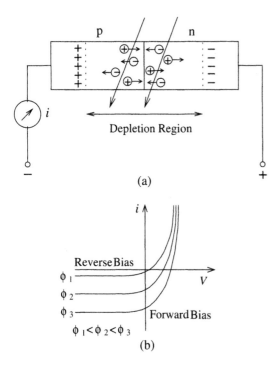

(a)

(b)

Figure 5.1.29: (a) Production of electron-hole pairs in the depletion region of a reverse biased pn-diode. (b) Typical current-voltage curve of a reverse biased pn-junction diode in the presence of radiation. As the flux of incident radiation increases, the number of electron-hole pairs in the depletion region increases, which in effect increases the measured current.

Since only the charge pairs produced in the depletion region contribute to the signal, therefore to make the active region large the depletion region is stretched as much as possible. This can be done in two ways: by increasing the reverse bias and by introducing an intrinsic semiconductor between the p and n materials. The former has already been discussed and the later will be explored later in this chapter.

Fig.5.1.30 shows a simple but realistic detector built by doping p- and n-type impurities on a bulk n-type semiconductor material such as silicon.

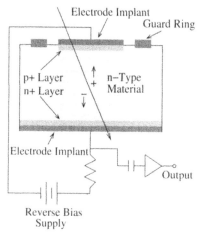

Figure 5.1.30: A simple but realistic semiconductor detector. Central light shaded region represents the sensitive volume that has been fully depleted by applying reverse bias. The bulk of the material is n-type, on which p+ and n+ regions have been created by the process of doping. A guard ring around the p+ layer guards against discharges due to high bias voltage.

5.1.J Desirable Properties

The desired properties of semiconductor detectors are highly application dependent as some of these have conflicting requirements. Semiconductor detectors are highly versatile devices that can be used in very high to very low radiation environments provided they are built for specifically for that radiation field. for example an avalanche photodiode can be used to detect single photons but it will be completely useless for detecting high intensity gamma rays.

Therefore we can safely divide semiconductor detectors in two categories depending on whether they are used in high or low radiation fields.

J.1 High Radiation Fields

There are three main characteristics of the detectors used in high radiation environments.

▶ **Charge collection efficiency:** At high incident radiation rate, very large number of charge carriers are produced, requiring efficient collection by electrodes. Any non-linearity in charge collection would reflect as non-linearity in the output signal. The deterioration in charge collection is caused by electron-hole recombination and charge trapping.

▶ **Fast response:** The associated electronics should be fast enough to avoid pulse pileup. Generally the electronics should be able to distinguish between pulses that are only a few tens of nanoseconds wide.

▶ **Radiation hardness:** The detector must be least susceptible to radiation damage. We will learn more about this later in this Chapter but at this time it suffices to say that radiation hardness is perhaps the most actively researched topics in semiconductor detector technology.

J.2 Low Radiation Fields

For low radiation environments the requirements are very different.

▶ **Charge yield:** The number of charge pairs created should be enough to yield a good signal to noise ratio. For very low fields, an internal amplification of charge pairs is sometimes done to enhance the signal.

▶ **Resolution:** Energy resolution is of prime importance in low radiation fields.

5.1.K   Specific Semiconductor Detectors

Semiconductor detectors can be build in different geometries with different structures and can be tuned according to applications. Their main disadvantage is that it is not possible to build large area detectors since the process of crystal growth has its size limitations. Nevertheless, large area position sensitive detectors have been built by integrating many smaller detectors together. Furthermore it has been shown that it is possible to use semiconductors to build detectors for not only very high radiation fields but also for extremely low radiation environments at the single photon level. Hence semiconductors have provided the long sought for versatility. Intense research has gone into and is still going on to build detectors with better operational characteristics. As a result state-of-the-art detectors have been built with qualities that far exceed their gaseous counterparts. In this section we will look at some specific types of semiconductor detectors that have been built and are in extensive use.

K.1   PIN Diode

Although a simple pn junction diode can be used as a radiation detector but it suffers from several disadvantages, some of which are described below.

▶ **Small Depletion Region:** Although the charge pairs are generated in the whole material but the ones generated inside the depletion region constitute the measurable current. Hence it is desired that the depletion region is as wide as possible to allow the radiation to deposit most of its energy. Increasing the depletion region in a simple pn diode requires increase in the bias voltage and therefore large depletion widths require very high external electric fields. Increasing the voltage to very high values may not always be possible due to the possibility of breakdown of the crystal structure. Hence for a simple pn-junction the depletion region can not be widened more than the value determined by the crystal structure's breakdown voltage. The work-around to this problem is to use such a thin material, which could be fully depleted at moderate voltages. This, in fact, is generally done in some detectors made with 100-300 μm thick wafers of semiconductor materials.

▶ **Large Capacitance:** Since the capacitance is inversely proportional to the width of the depletion region therefore simple pn diodes operating at moderate voltages have high capacitances. This is certainly not a desirable effect for using the junction as a detector.

▶ **Slow Response:** In general, the pn detectors are operated in fully depleted mode. But this is only possible for very thin detectors, since, as we saw earlier, increasing the depletion width requires introduction of high potential across the junction. For thicker detectors, which can not be fully depleted, the creation of electron hole pairs in the undepleted region is a major problem. It not only produces non-linearity in detector's response but can also decrease the response time due to diffusion of charges across the regions.

As we saw the increasing the bias voltage is not always possible to solve these problems. A convenient solution to increase the depletion width is to introduce an intrinsic layer of a semiconductor material between the p and n materials (see Fig.5.1.31). This so called PIN (P-type, Intrinsic type, N-type) structure is perhaps the most commonly used configuration of semiconductor detectors. The main advantage of introducing an intrinsic material in the middle is that it essentially fixes the depletion width to the width of the intrinsic region. In this way a large depletion width becomes available without the need for very high reverse bias. This solves most of the problems associated with simple pn junctions.

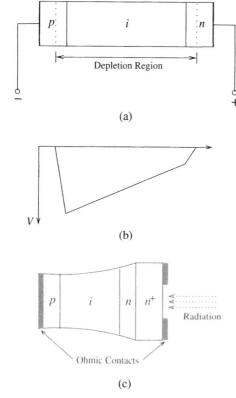

Figure 5.1.31: A simplified structure (a) of a reverse biased PIN diode and its idealized electric field profile (b). A realistic PIN structure is shown in (c).

K.2   Schottky Diode

Up until now we have only discussed the utilization of a junction, produced by p and n types of semiconductors, as a radiation detector. A similar junction can also be produced by bringing a semiconductor and a metal in contact with each other. Such a semiconductor-metal junction can also be exploited for radiation detection purposes.

Just like a semiconductor diode, a semiconductor-metal diode can be produced, which is generally referred to as Schottky diode. It is created by bringing a very thin metal layer in contact with a semiconductor material (see Fig.5.1.32). The process produces a junction near to the surface of the semiconductor.

The detectors made from Schottky diodes are generally used to detect photons. To ensure high transparency of the metal layer to the photons, the layer is kept as thin as possible. Since the active detection material essentially starts at the surface, therefore the signal loss is minimum and the charge collection efficiency is very high with reduced charge transit time. This type of detector works very well as a photodiode as it can operate at very high frequencies (in GHz) without causing high frequency roll off typical of pn junction diodes.

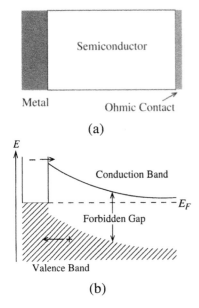

Figure 5.1.32: Sketch of a Schottky diode (a) with its energy band structure (b). The metal layer is made very thin to allow the photons to pass through it when used as a photodetector. Typical metallic surface is gold on p type silicon. The ohmic contact could be made by depositing aluminum on silicon surface.

K.3   Heterojunction Diode

It is sometimes desirable to build diodes that have multiple junctions of different types to optimize the generation of output signal with respect to the parasitic absorption of incident radiation. for example in photon detection where incident light level may be quite low, a considerable fraction of photons may get absorbed in the material before reaching the depletion region. The trick in this case would be to use a material for the surface layer that has wider band gap than the incident radiation. In this case most of the radiation would pass the surface layer without being absorbed and reach the main depletion region. Heterojunction diodes are mainly used for photon detection.

K.4 Avalanche Photodiode

Avalanche photodiodes utilize the concept of multiplying the number of charge carriers to increase the signal height. These detectors are highly efficient in detecting low light levels and in principle are capable of detecting single photons. We will have a look at APDs in the next chapter when we discuss the photodetectors.

K.5 Surface Barrier Detector

A surface barrier detector is usually made by evaporating a thin gold layer on an n-type highly pure silicon crystal. The contact on the other side is generally made by evaporating a thing layer of aluminum. The depletion region is created by applying reverse bias across the aluminum and the gold contacts, which produces a wide depletion region near the gold side and a thin dead region near the aluminum side. Since the gold thickness is very small, there is minimal parasitic absorption. The detector can therefore be used in low level radiation fields. The advantage of surface barrier detectors include high stability and superior energy resolution.

K.6 Position Sensitive Detectors

The onset of aggressive research and development in semiconductor detector technology can be traced back to the realization of the potential to develop position sensitive detectors of extremely high resolution and sensitivity. With today's technology, detectors having position sensitivity of better than 10 μm can be built. The main idea behind such detectors involves doping a wafer of semiconductor material by selectively masking arrays of desired geometry, thus developing an array of detectors on the wafer. The details of how this is done and how this array is read out depends on the particular design. Although we will discuss such detectors in the Chapter on position sensitive detectors but it is worthwhile to briefly describe some of the commonly used position sensitive detectors here as well.

Microstrip Detectors

Microstrip detectors are made by doping a large semiconductor wafer into very thin (typically 10 μm) parallel segments. The segments generally have a pitch (center to center distance between two consecutive segments) of around 50 μm. Each of these segments then behaves as a separate detector. The main advantage of such a geometry is high position sensitivity, which is not achievable from other conventional position sensitive detectors, such as wire proportional counters. With today's technology a position resolution of better than 10 μm is possible. The main complication with building such detectors is the incorporation of readout electronics. In a typical few centimeter wide detector module there are several hundred strips acting as individual detectors. Each of these detectors has to be read out through a unique analog electronics channel having its own preamplifier. Fortunately, there is technology available now to build electronic chips that can handle a large number of channels at the same time. These chips are installed on boards that are densely populated with interconnects and other electronic components, hence called High Density Interconnects or simply HDIs. The detectors are connected to the pads available on HDIs through wire bonds, a complicated process that is done through precision machinery.

Pixel Detectors

Pixel detectors are position sensitive devices having a two dimensional array of small rectangular detector elements or pixels. Each of these pixels has a linear dimension of less than a millimeter[3]. The technology is somewhat similar to the microstrip detectors we studied earlier. The pixel array is developed by selectively doping a semiconductor wafer.

Other Position Sensitive Detectors

A number of variants of the simple microstrip and pixel detectors we just discussed exist, some of which are listed below.

▶ Monolithic Pixel Detectors

▶ Hybrid Pixel Detectors

▶ Charged Couple Devices (CCDs)

▶ MOS CCDs

▶ Deep Depletion CCDs

▶ Drift Detectors

5.1.L   Radiation Damage in Semiconductors

A serious problem with semiconductors is their vulnerability to structural damage caused by radiation. Both instantaneous and integrated doses contribute to the overall damage. Although some of these effects can be compensated by increasing the reverse bias but eventually the bulk of the damage renders the detector material useless. Hence deployment of semiconductor detectors specially in hostile radiation environments, such as particle accelerators, needs careful consideration and regular monitoring.

L.1   Damage Mechanism and NIEL Scaling

The most damaging defects caused by radiation are the result of atomic displacements that destroy the lattice structure. These defects could be dispersed or clustered around the incident radiation track. The dispersed atomic displacements away from each other cause the so called *point defects*. A cluster of such displacements in close vicinity to each other is called *cluster defect*. The distinguishing feature of these clustered defects is that they render a large region of the bulk of the material unusable. The main damage mechanism is the displacement of an atom from its lattice site by an incident particle. This is called *Non Ionizing Energy Loss* or NIEL. As opposed to the ionizing energy loss, NIEL is not proportional to the total

---

[3]The choice of pixel dimension depends on the resolution requirements and engineering details. Practical systems with as small as 10 μm and as large as 200 μm pixels have been constructed.

energy absorbed in the semiconductor and is highly dependent on the type of radiation and its energy. It should however be mentioned that the atoms displaced by the incident radiation can also cause further damage. This damage is not a part of NIEL scaling. However since the energy spectrum of these recoil atoms determines the type of damage, this effect is not very pronounced for all types of incident radiation. for example charged incident particles generally produce recoil atoms having a distribution with a long tail at the low energy end, making the secondary point defect production more probable than the cluster formation. Hence in this case NIEL scaling will be a good approximation of the overall damage. In the case of high energy neutrons, however, the recoil spectrum is more skewed towards the high energy end and therefore NIEL scaling should be interpreted carefully.

Whatever the specific damage mechanism is, the end result is the performance deterioration of the detector. In the following sections we will discuss the most important of these deteriorating effects on the detector performance.

L.2 Leakage Current

A profound effect of radiation induced damage is the change in reverse bias current, the obvious cause of which is the increase or decrease in the number of free charge pairs in the depletion region. The decrease in leakage current, which is generally observed during initial irradiation, is primarily due to production of charge traps in the forbidden energy gap. The leakage current can show significant increase after prolonged irradiation as the probability of charge pair production increases by introduction of additional energy levels in the forbidden gap.

It has been found that the damage induced leakage current depends on the integrated radiation dose, the exposed volume of the detector, and its temperature. The integrated radiation dose, of course, depends on the particle fluence, which is simply the integrated radiation intensity. At a certain temperature, the dependence of change in leakage current $\triangle i_l$ on particle fluence Φ and volume V can be written as (25)

$$\triangle i_l = \alpha V \Phi, \tag{5.1.86}$$

where α is the so called *damage coefficient* and depends on the type of incident particle and its fluence. This equation can also be written in terms of leakage current before irradiation i_0 and after irradiation i_r

$$i_r = i_0 + \alpha V \Phi. \tag{5.1.87}$$

The temperature dependence of i_r, on the other hand, can be described by Boltzmann function

$$i_r(T) \propto T^2 e^{-E/2kT}, \tag{5.1.88}$$

where T is the absolute temperature and k is the Boltzmann's constant. E is the activation energy of the material, which is generally higher for irradiated material.

The change in leakage current has unwanted consequences on detector performance, most notably, increase in noise and consequent deterioration of signal to noise ratio. Fortunately the strong dependence on temperature can be easily exploited to compensate for the deterioration by decreasing the operating temperature. This is a common practice for detectors used in hostile radiation environments such as particle accelerators, where radiation induced damage is generally very high.

Example:
A $300\mu m$ thick silicon detector having a surface area of $8 \times 10^{-4}\ cm^2$ is placed in a 1 MeV neutrons field having intensity of $3 \times 10^6\ cm^{-2}s^{-1}$. Compute the change in leakage current relative to the initial value of 10 nA after one year of constant irradiation. Take the damage coefficient to be $2 \times 10^{-17}\ A/cm$ and assume that the detector was operated in fully depleted mode.

Solution:
In fully depleted mode we can assume that the active region is equal to the volume of the detector. Hence $V = (300 \times 10^{-4})(8 \times 10^{-4}) = 2.4 \times 10^{-5}\ cm^3$. The total fluence to which the detector remained exposes for one year is give by

$$\Phi = (10^6)(365 \times 24 \times 3600) = 3.15 \times 10^{13}\ cm^{-2}.$$

According to equation 5.1.87, the relative change in leakage current can be calculated as follows.

$$
\begin{aligned}
\delta i &= \frac{i_r - i_0}{i_0} \\
&= \alpha \frac{\Phi V}{i_0} \\
&= 2 \times 10^{-17} \frac{(3.15 \times 10^{13})(2.4 \times 10^{-5})}{100 \times 10^{-9}} \\
&= 0.15 \quad \text{or} \\
\delta i &= 15\%.
\end{aligned}
\tag{5.1.89}
$$

Hence after a year of irradiation the current will change by 15%. Whether this change is observed and attributed to the radiation damage or not depends on how well the detector is being monitored and calibrated.

L.3 Type Inversion

A special type of damage caused by high intensity radiation is the inversion of the material type, in which an n-type material may change into a p-type and vice versa after prolonged irradiation. The effect can be easily understood by noting that radiation damage has the overall effect of changing the effective dopant concentration in the bulk of the material by increasing the charge carriers of the opposite sign. With increase in integrated radiation dose, the original effective dopant concentration may be overcome by the charges of opposite sign. This effect has been observed specially in silicon detectors in hostile radiation environments. What happens is that the radiation creates acceptor sites, which capture electrons resulting in a decrease in effective doping level. This continues until the donor level becomes equal to the acceptor level, changing the material into intrinsic type. Further irradiation increases the acceptor level and consequently the types of material changes (see Fig.5.1.33).

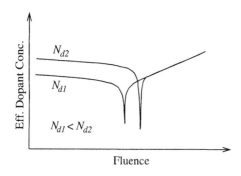

Figure 5.1.33: Dependence of effective dopant concentration on particle fluence in a silicon detector in high radiation environment for two different initial donor concentration levels.

L.4 Depletion Voltage

Generally, radiation detectors are operated in fully depleted mode. The bias voltage is kept at several tens of volts above the depletion voltage to ensure complete depletion. However with prolonged irradiation, the depletion voltage increases, requiring increase in bias voltage. The principle mechanism is the change in effective dopant concentration due to radiation damage. To overcome this effect, bias voltage is increased.

L.5 Charge Trapping and Carrier Lifetime

Charge collection efficiency of the electrodes is directly proportional to the number of free charges available. Perhaps the worst effect that incident radiation can cause in a semiconductor detector is the creation of energy levels that trap the charges for a long period of time. These could be either electron traps or hole traps according to whether they are near the valence band or conduction band respectively.

If a charge traps in the hole, it can no longer be a part of the current and would cause nonlinearity in the detector response. Note that the net result of traps is a decrement in the average carrier lifetime in the bulk of the material.

L.6 Annealing

Annealing is the *healing* process through which the radiation damage in a semiconductor material diminishes with time. The underlying process responsible for this reversal of damage is not fully understood. However it has been found that annealing has a strong temperature dependence. The defect concentration can be represented by an exponential of the form

$$N(t) = N_0 e^{-t/\tau}, \tag{5.1.90}$$

where N_0 is the initial defect concentration and τ is a function of the activation energy E_a and absolute temperature T. It is given by

$$\tau = A e^{E_a/kT}, \tag{5.1.91}$$

where k is the Boltzmann's constant. The constant A is determined experimentally.

5.2 Diamond Detectors

The most problematic thing with semiconductor detectors is their vulnerability to radiation damage. Radiation hard semiconductors, though available, are expensive to fabricate and have their own engineering difficulties. Another solid state material that has been shown to perform better than semiconductors in hostile radiation environments is diamond.

Table.5.2.1 lists some of the properties of diamond relevant for its use as an active medium for radiation detection. For comparison, the properties of silicon are also given. It can be seen that the energy needed to create an electron-hole pair in diamond is significantly larger than in silicon. This is due to its higher band gap energy, which is almost five times higher than silicon. This implies that the output signal from a diamond detector will be significantly less than from a silicon detector. However the higher band gap energy also has a positive side, that is, the leakage or dark current is also lower than in semiconductors. This implies less noise at the same bias voltage and therefore it can be expected that the signal to noise ratio will still remain acceptable. Another positive point is diamond's lower dielectric constant, which means lower capacitance as compared to silicon having similar dimensions.

Table 5.2.1: Comparison of some important properties of diamond with those of silicon (4).

Property	Diamond	Silicon
Mass Density $(gm - cm^{-3})$	3.52	2.33
Dielectric Constant	5.7	11.9
Displacement Energy $(eV/atom)$	43	13-20
W-value (eV)	13	3.6
Band Gap (eV)	5.5	1.12
Breakdown Field (V/cm)	10^7	3×10^5
Resistivity (Ωcm)	$> 10^{11}$	2.3×10^5
Electron Mobility $(cm^2 V^{-1} s^{-1})$	1800	1350
Hole Mobility $(cm^2 V^{-1} s^{-1})$	1200	480

The diamonds used as detector material are synthetically produced in laboratory through the process of Chemical Vapor Deposition (CVD). The diamond thus produced is referred to as a CVD diamond. Though going into the details of this process is out of the scope of this book but for the interest of reader we will give here a brief introduction.

CVD technique involves decomposition of hydrocarbon molecules in gaseous state and then their activation in a high temperature (or energy) environment. Specifically what is done is that a gas mixture consisting of about 98% hydrogen and 2% methane is led into the activation volume that is either kept at a very high temperature or has the ability to cause activation through the microwave-plasma process. The high energy departed to the molecules decomposes them and activates the chemical reactions necessary to fuse them into the diamond structure. The reactants are then transported to the deposition surface, where processes of nucleation and growth occur. This leads to development of carbon and diamond structures on the substrate. Since the diamond structures in the substrate are mixed with the graphite structures therefore they must be purified before use. This is accomplished by the process of gasification of the graphite by atomic hydrogen. Since diamond is more stable than graphite for this process therefore most of it survives.

Before we look at how a CVD diamond detector is built, let us study some of the important properties of this kind of diamond relevant to the detector technology.

5.2.A   Charge Pair Production

The process of charge pair production in diamond is essentially the same as in semiconductors. The only difference is the higher w-value of diamond owing to its wider band gap. In comparison to silicon, in a diamond detector the same amount of deposited energy would produce about three times less charge pairs. But if we compare this to gaseous detectors the number of charge pairs would still be about three times higher. The good thing about diamond is that the larger band gap also means lower leakage current because of lower probability of production of thermally agitated electron hole pairs.

5.2.B   Recombination

As with semiconductor materials, CVD diamonds also have crystal defects and impurities that produce additional energy levels within the forbidden gap. These levels facilitate the recombination of excess charge pairs. Furthermore the usual recombination characterized by fall of an electron from the conduction to the valence band also occurs in diamond. The lifetimes of charges recombining through these two processes depend mostly on the quality of the material.

The direct recombination of an electron in the conduction band with a hole in the valence band is generally known as intrinsic recombination. On the other hand the extrinsic recombination refers to the process that follows via an intermediate energy state. The lifetime for the intrinsic process can vary from a few μs to about 1 s while that for the extrinsic process can be as low as 0.1 to 10 ns. The average or effective lifetime, as in case of semiconductors, is given by

$$\frac{1}{\tau_{eff}} = \frac{1}{\tau_{int}} + \frac{1}{\tau_{ext}}, \tag{5.2.1}$$

where the subscripts eff, int, and ext refer respectively to effective, intrinsic, and extrinsic. It is apparent that for CVD diamond the effective lifetime is dominated by the extrinsic lifetime, which depends on the crystal defects and impurities. Hence we can write

$$\tau_{eff} \approx \tau_{ext}. \qquad (5.2.2)$$

5.2.C   Drift of Charge Pairs

The high mobility of free charges in diamond is a positive factor for radiation detection since it implies fast charge collection and consequently fast overall response of the detector. The electric field profile of drift velocity of charge pairs in diamond is qualitatively similar to semiconductors. The velocity increases with increase in electric field intensity and ultimately reaches a saturated value. Diamond detectors are operated at such high potentials that the charge carriers quickly attain their saturated velocities after their production by the incident radiation. Typical operating electric field is 1 kV/mm.

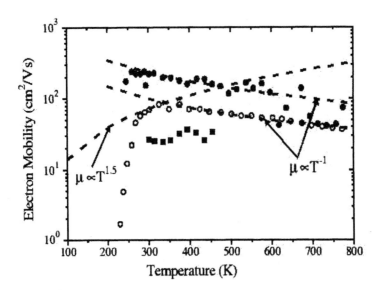

Figure 5.2.1: Variation of electron mobility in CVD diamond with respect to the absolute temperature at three impurity (phosphorus) levels: 100 ppm (■), 500 ppm (●), and 1000 ppm (○) (32).

Besides having a value higher than semiconductors, the mobilities of electrons and holes in diamond do not show strong temperature dependence at and around room temperature. Fig.5.2.1 shows the variation of electron mobility in diamond with temperature and concentration of impurity (phosphorus). The interesting thing to note here is that at temperatures lower than the room temperature, the mobility has a strong dependence on the impurity level. If the material is highly doped, the electron mobility decreases very rapidly, certainly not a very desirable effect as far as radiation detection is concerned. On the other hand, for temperatures higher than 300 K the electron mobility shows small variations even if the the dopant level changes by a factor of 5. The best fit to the data shows that the electron mobility in diamond has an approximate temperature dependence of the form (32)

$$\mu_e \propto \frac{1}{T}. \tag{5.2.3}$$

This profile is much softer than the ones we saw for semiconductor materials, which is a good thing in terms of detector design and operation specially because in this case the detector does not necessarily require temperature control and monitoring. However before we jump to any conclusion in this regard let us see how the holes behave at different temperatures in diamond. Fig.5.2.2 shows such a profile for natural diamond and boron doped CVD diamond.

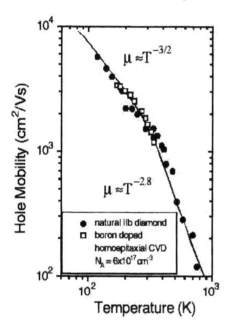

Figure 5.2.2: Variation of hole mobility in diamond with respect to the absolute temperature for natural diamond (•) and boron doped CVD diamond (□) (32).

According to Fig.5.2.2, the temperature dependence of hole mobility for CVD diamond is somewhere between

$$\mu_h \propto \begin{matrix} T^{-1.5} & \text{for} & T < 400\ K & \text{and} \\[2mm] T^{-2.8} & \text{for} & T \geq 400\ K. \end{matrix} \tag{5.2.4}$$

This certainly does not look as good as the mobility profile for electrons. However the point to note here is that the temperature dependence near and below the room temperature is still not so steep that one would argue for cooling the material.

A parameter that is routinely used to characterize the detectors made of CVD diamond is the drift length d_D defined as

$$d_D = (\mu_e \tau_e + \mu_h \tau_h)\, E, \tag{5.2.5}$$

where τ_e and τ_h are the *deep trapping lifetimes* of electrons and holes respectively and E is the electric field intensity. Of course here we are using our earlier conclusion that the effective lifetime of a carrier in diamond is dominated by its extrinsic lifetime (see equation 5.2.2). Therefore the lifetimes appearing in the above equation are actually the extrinsic lifetimes of electrons and holes. This equation is also sometimes written as

$$d_D = \mu \tau E, \tag{5.2.6}$$

where $\mu = \mu_e + \mu_h$ and

$$\frac{1}{\tau} = \frac{1}{\tau_e} + \frac{1}{\tau_h}.$$

The reader should note that defining μ as a sum of electron and hole mobilities hints towards the correct definition of d_D as the distance by which the electrons and hole move apart under the influence of applied electric field.

Now that we know the actual distance traveled by the charges, that is d_D, we can determine the charge induced at the readout electrodes through the relation

$$Q_{ind} = Q_0 \frac{d_D}{d}, \tag{5.2.7}$$

where d is the physical length of the material and Q_0 represents the total charge produced by the radiation. Since the charge collection efficiency of a detector is defined as the ratio of the total charge observed to the actual charge, therefore using the above equation we can write the efficiency as

$$\eta = \frac{Q_{ind}}{Q_0} = \frac{d_D}{d}. \tag{5.2.8}$$

The good thing about η is that it can be used to determine the optimum thickness of a material such that the efficiency is nearly perfect (100%). An important point to note here is that the CVD crystals grown in laboratories do not show spatially uniform properties. Small localized variations in trap densities have been observed in practical systems.

5.2.D   Leakage Current

At room temperature the resistivity of diamond is about 6 orders of magnitude higher than that of silicon (see Table 5.2.1). Due to such a high resistivity (and large band gap) the leakage current in a well grown CVD diamond crystal is extremely small. The variation of resistivity with temperature in diamond is shown in Fig.5.2.3. It can be seen that even at very high temperatures the resistivity of diamond still remains higher than most semiconductor materials kept at lower than room temperature.

5.2.E   Detector Design

As we noted earlier, because of the large band gap in diamond the leakage current is extremely small. The direct implication of this advantage with regard to radiation detection is that one does not need to establish a pn junction. The detector design is thus simpler as compared to semiconductor detectors where creating a junction and stretching it is a major task in itself.

A simple diamond detector can be built by simply establishing metal contacts on two sides of the material. These contacts act can then be used to apply bias

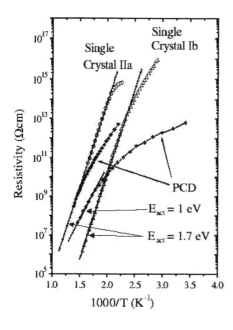

Figure 5.2.3: Variation of resistivity in different diamond crystals with respect to absolute temperature (32).

and to read out the detector output. Common materials used for this purpose are chromium-gold and titanium-gold alloys.

Fig.5.2.4 shows sketch of a simple diamond detector.

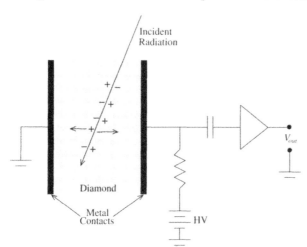

Figure 5.2.4: Sketch of a simple diamond based detector. Note that, in contrast to semiconductor detectors, there is no depletion region.

5.2.F   Radiation Hardness

The main advantage of preferring diamond detectors over the semiconductor detectors is their radiation hardness, that is, less vulnerability to radiation damage. It has been found that the radiation hardness of the detectors made of CVD diamond is about an order of magnitude higher than detectors made of silicon (1). For diamond the main damaging effect of radiation is the production of vacancies or energy levels inside the band gap. These vacancies can act as electron traps and thus introduce non-linearity and degradation in the detector response.

5.2.G Applications

As we have mentioned earlier, the biggest advantage of using diamond as a radiation detection medium stems from its radiation hardness. Hence the applications where semiconductor detectors have been seen to deteriorate with time due to accumulated dose are the best candidates of diamond based detectors. Accelerator based experiments of particle physics were the first where experimenters started exploiting this feature of diamond to think about replacing the conventional silicon detectors at places where radiation field was highest. The research is still going on in this direction and the results of prototype detectors look quite promising. However a full scale diamond detector completely replacing a silicon tracker is still not a reality.

The other field where radiation damage of detectors is a concern is the radiation therapy, which is done by high energy particles produced in small scale accelerators. The high frequency of treatment provided at these facilities requires radiation hard detectors for dosimetric purposes. Several authors have shown that diamond detectors can be used in such facilities more efficiently as compared to the conventional semiconductor detectors and gas filled ionization chambers (see, for example (10; 5; 6)).

Other applications where possible use of diamond detectors is being explored include

▶ synchrotron beam intensity monitoring,

▶ heavy-ion collision experiments, and

▶ neutron detectors for high radiation environments.

5.3 Thermoluminescent Detectors

The term thermoluminescence stands for thermally stimulated luminescence, which in simple words means emission of light be thermal stimulation. A thermoluminescent (TL) material is a kind of memory device that is capable of retaining radiation dose information for a long time or until its temperature is raised above a certain level. The usefulness of TL materials as radiation detectors was first realized in early 1950s. Since that time such detectors are being extensively used in personal dosimetry applications.

We will discuss different kinds of TL detectors in the chapter on radiation dosimetry but it is worthwhile here to discuss the principle of thermoluminescence specially because the principles involved are similar to those of semiconductors and diamond we studied in this chapter.

5.3.A   Principle of Thermoluminescence

During the discussion on semiconductors and diamond in the preceding sections we saw that the crystal defects and impurities in those materials can produce energy levels inside the forbidden gap. Since these levels can act as electron traps and distort the signal, therefore they are not desirable in radiation detectors. Though TL materials also have such levels but in this case they are advantageous since they can be used to store information.

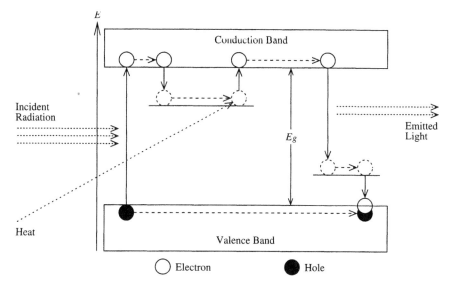

Figure 5.3.1:   A simple model of energy absorption in a TL material. The saved energy is released in the form of a light photon when the material is heated.

Fig.5.3.1 shows how the metastable electron traps in TL materials can be exploited to store and retrieve information. The TL material shown has two additional energy levels inside the forbidden gap. The incident radiation elevates an electron from the valence band to the conduction band and leaves behind a vacancy hole in the valence band. The high level of impurities in the TL materials makes it highly probable that this electron would quickly fall into the impurity level near the conduction band. Being a metastable site, this site retains the electron until energy is provided to the material externally, such as in the form of heat. The heat elevates the electron back to the conduction band. The electron can then fall into the impurity level near the valence band and emit a photon during this process. Since the energy difference between the lower end of the conduction band and the impurity level near the valence band corresponds to light photon energy therefore the emitted photon is in the visible region of the electromagnetic spectrum. The electron in this level eventually falls into the valence band and recombines with the hole.

Problems

1. Show that the probability for an electron to occupy the Fermi level in an intrinsic semiconductor is independent of the material's temperature.

2. Assuming equal charge concentrations in conduction and valence bands, compute the ratios of the intrinsic charge concentrations to the conduction band charge concentrations in germanium and silicon at $300\ K$.

3. Calculate the intrinsic carrier concentration of silicon and germanium at room temperature (23^0C) and at -10^0C.

4. Compare the statistical fluctuations and relative uncertainty in the number of charge pairs produced in silicon and gallium arsenide when γ-rays of $520\ keV$ are completely absorbed in the material.

5. Estimate the range of $6\ MeV$ protons in silicon.

6. A beam of $5\ MeV$ electrons passes through a $300\ \mu m$ thick silicon detector. Using Bethe-Bloch formula (see chapter 2), estimate the energy lost by the electrons when crossing $100\ \mu m$, $200\ \mu m$, and the full thickness of the detector.

7. Compare the change in intrinsic charge concentration in silicon with that in germanium if the temperature is increased from -5^0C to the room temperature of 23^0C.

8. An n-type germanium is doped with a donor concentration of $4.5 \times 10^{23}\ m^3$. Calculate the ratio of intrinsic to extrinsic conductivities at $300\ K$ and at $263\ K$.

9. Compute the energy resolution of a germanium detector for $680\ keV$ photons. What would be the resolution if the production of charge pairs were a perfectly Poisson process?

10. Repeat the previous exercise for silicon and compare the results for germanium.

11. Estimate the fraction of holes remaining in an n-type gallium arsenide having dopant level of $10^{14}\ cm^{-3}$. The lifetime of holes at this dopant level can be taken to be $3\ \mu s$.

12. We learned that the band gap in a semiconductor has temperature dependence. Plot the variation of band gap for silicon and germanium with respect to absolute temperature.

13. Calculate the capacitances per unit area of a germanium based pn diode having acceptor and donor impurities of $2 \times 10^{18}\ cm^{-3}$ and $5 \times 10^{16}\ cm^{-3}$ respectively when no bias is applied across its junction and when a potential of $200\ V$ is established.

14. How many electron hole pairs you would expect to be produced in a silicon detector when incident γ rays deposit $1\ MeV$ of energy.

15. Compare the drift velocity of electrons in silicon to that in germanium in the presence of an electric field intensity of $1.5\ kV cm^{-1}$.

16. A 300 μm thick silicon detector having an area of $0.05\ cm^2$ has an initial leakage current of 24 nA. It is placed in a radiation field of 3 MeV γ-rays for a period of two years where it sees a constant γ flux of $1.2 \times 10^7\ cm^{-2}s^{-1}$. Compute its leakage currents at each 6 months interval.

17. Calculate the ratio of the number of electron hole pairs generated in silicon by 5 MeV electrons to the ones created in an argon filled ionization chamber by electrons having the same energy.

18. Compare the number of charge pairs created in silicon to that in diamond if both are exposed to a beam of 6.5 MeV γ-rays.

19. Compute the ratio of the hole mobilities at 300 K and 450 K in CVD diamond.

Bibliography

[1] Adam, W. et al., **Pulse Height Distribution and Radiation Tolerance of CVD Diamond Detectors**, Nucl. Instrum. Methods A, 447, 2000.

[2] Alkhazov, G.D. et al., NIM 48(1), 1967.

[3] Bertolini, G., Coche, A., **Semiconductor Detectors**, Elsevier Science, 1968.

[4] Berdermann, E. et al. **Diamond Detectors for Heavy-Ion Measurements**, Proc. of the XXXVI Intern. Winter Meeting on Nucl. Physics, Bormio, 1998.

[5] Berdermann, E. et al., **First Applications of CVD-Diamond Detectors in Heavy-Ion Experiments**, Nucl. Phys. B, Proc. Sup. 78, 1999.

[6] Berdermann, E. et al., **The Use of CVD-Diamond for Heavy-Ion Detection**, Diam. Rel. Materials, 10, 2001.

[7] Blakemore, J.S., **Semiconducting and Other Major Properties of Gallium Arsenide**, J. Appl. Phys., Vol.53, No.10, 1982.

[8] Bock, E., **Bibliography on Semiconductor Detectors: A Compilation of Selected Literature Abstracts as Guide to Recently Available R. & D. Publications in Semiconductor Techniques covering the period 1963-1967**, Directorate Dissemination of Information, Center for Information and Documentation, 1967.

[9] Canali, C. et al., **Electron Drift Velocity in Silicon**, Phys. Rev. B, Vol.12, No.4, 1975.

[10] Cirrone, G.A.P. et al., **Natural and CVD Type Diamond Detectors as Dosimeters in Hadrontherapy Applications**, Nucl. Phys. B, Proc. Sup. 125, 2003.

[11] Debertin, K., Helmer, R.G., **Gamma- and X-Ray Spectroscopy with Semiconductor Detectors**, North Holland, 1988.

[12] Deme, S., **Semiconductor Detectors for Nuclear Radiation Measurement**, Wiley-Interscience, 1971.

[13] Dulinski, W. et al., **Diamond Detectors for Future Particle Physics Experiments**, CERN PPE/94-222, 1994.

[14] Fano, U., Phys. Rev., 72(26), 1947.

[15] Fink, R.W., **Semiconductor Electron and X-Ray Detector System for Fluorescence and Auger Investigations**, School of Chemistry, Georgia Institute of Technology, 1973.

[16] Fortunato, E. et al., **Large Area Position Sensitive Detector Based on Amorphous Silicon Technology**, Mat. Res. Soc. Symp. Proceedings, 297, 1993.

[17] Hartjes, F. et al., **Parameterization of Radiation Effects on CVD Diamond for Proton Irradiation**, Nucl. Phys. B (Proceedings Supplement), 78, 1999.

[18] Hayase, M., Arita, H., **Full-Contact Type Linear Image Sensor by Amorphous Silicon**, Mat. Res. Soc. Symp. Proceedings, 192, 1990.

[19] Jacoboni, C. et al., A **Review of Some Charge Transport Properties of Silicon**, Solid-State Electronics, Vol.21, Issue 2, 1977.

[20] Jacoboni, C. et al., **Electron Drift Velocity and Diffusivity in Germanium**, Phys. Rev. B, Vol.24, No.2, 1981.

[22] James, R.B., **Semiconductors for Room-Temperature Radiation Detector Applications II: Symposium Held December 1-5, 1997, Boston, Massachusetts, U.S.A**, Materials Research Society, 1998.

[22] James, R.B. et al., **Hard X-Ray, Gamma-Ray, and Neutron Detector Physics: Proceedings of Spie 19-23 July 1999 Denver, Colorado**, Society of Photo Optical, 1999.

[23] Knoll, G.F., **Radiation Detection and Measurement**, Wiley, 2000.

[24] Lachish, U., **CdTe and CdZnTe Semiconductor Gamma Detectors Equipped with Ohmic Contacts**, Nucl. Inst. Meth. A, 436, 1999.

[25] Lindstroem, G. et al., **Toward Super Radiation Tolerant Semiconductor Detectors for Future Elementary Particle Research**, J. Optoelectronics Adv. Materials, 6(1), March 2004.

[26] Lindstroem, G., **Radiation Damage in Silicon Detectors**, Nucl. Instr. Meth. A, 512, 30, 2003.

[27] Lutz, G., **Semiconductor Radiation Detectors**, Springer, 1999.

[28] Manasreh, M.O., **Semiconductor Quantum Wells and Superlattices for Long-Wavelength Infrared Detectors**, Artech House Publishers, 1992.

[29] McCrary, V.R., **Proceedings of the First International Symposium on Long Wavelength Infrared Detectors and Arrays: Physics and Applications and the Nineteenth State**, Electrochemical Society, 1995.

[30] Meier, D. et al., **Proton Irradiation of CVD Diamond Detectors for High Luminosity Experiments at the LHC**, Nucl. Instr. and Meth. A, 426, 1999.

[31] Moritz, P. et al., **Diamond Detectors for Beam Diagnostics in Heavy Ion Accelerators**, DIPACIII, Frascati, 1997.

[32] Nevel, C.E., **Electronic Properties of CVD Diamond**, Semicond. Sci. Technol. 18, 2003.

[33] Ottaviani, G. et al., **Hole Drift Velocity in Silicon**, Phys. Rev. B, Vol.12, No.8, 1975.

[34] Pidgeon, C.R., **Suppression of Non-Radiative Processes in Semiconductor Mid-Infrared Emitters and Detectors**, Elsevier Science, 1998.

[35] Reggiani, L. et al., **Hold Drift Velocity in Germanium**, Phys. Rev. B, Vol.16, No.6, 1977.

[36] Rogalski, A., **Selected Papers on Semiconductor Infrared Detectors**, Society of Photo Optical, 1993.

[37] Schnetzer, S. et al., **Tracking with CVD Diamond Radiation Sensors at High Luminosity Colliders**, IEEE Trans. on Nucl. Science, Vol.46, 1999.

[38] Spieler, H., **Semiconductor Detector Systems**, Oxford University Press, USA, 2005.

[39] Sze, S.M., **Physics of Semiconductor Devices**, John Wiley & Sons Inc., 1981.

[40] Sze, S.M., **Semiconductor Sensors**, Wiley-Interscience, 1994.

[41] Taylor, J.M., **Semiconductor Particle Detectors**, Butterworths, 1963.

[42] ten Kate, W.R.T., **The Silicon Microstrip Detector**, Technische Universiteit Delft, 1987.

[43] van Roosbroeck, W., Phys. Rev., 139(A1702), 1963.

[44] Wang, S., **Fundamentals of Semiconductor Theory and Device Physics**, Prentice Hall, Englewood Cliffs, 1989.

[45] Willardson, R.K., Weber, E.R., **Semiconductors for Room Temperature Nuclear Detector Applications**, Academic Press, 1995.

[46] Yang, J., Zidon, Y., and Shapira, Y., **Alloy Composition and Electronic Structure of** $Cd_{1-x}Zn_xTe$ **by Surface Photovoltage Spectroscopy** J. Appl. Phys., Vol.91. No.2, 2002.

[47] Zegrya, G. et al., **Electronic Archive of New Semiconductor Materials: http://www.ioffe.ru/SVA/NSM/**, Ioffe Physico-Technical Institute, Russia, 2005.

Chapter 6

Scintillation Detectors and Photodetectors

In Chapter 3 we saw that the passage of radiation through some materials causes them to emit photons. The basic process is very simple: when atoms of incident radiation interacts with atoms of these so called scintillation materials, they transfer some of their energy to the atoms. As a result these excited atoms go into short lived excited states. When they return to ground state they emit photons, mostly in visible and ultraviolet regions of the spectrum. This provides an alternate to the ionization mechanism to detect and measure radiation.

The basic steps involved in scintillation detection of radiation are:

▶ Interaction of radiation with scintillation material.

▶ Transfer of energy to the bound states of the material.

▶ Relaxation of the excited states to the ground state resulting in the emission of light photons.

▶ Collection of photons by the photodetector.

▶ Detection of the photodetector signal by associated electronics.

The efficiency of a typical scintillation material to emit light after absorbing radiation ranges from 10% to 15%, implying that scintillation is not a very efficient process. However, as we will see later, even with such an *inefficient* material, one can develop a highly efficient radiation detector. The trick is to use a photon detector that has high photon collection and counting efficiency. In this chapter we will also visit two distinct types of detectors, namely photomultiplier tubes and photodiodes, that are used to detect scintillation photons.

The basic working principle of a photomultiplier tube or PMT involves conversion of the scintillation photon into an electron and then multiplication of this electron into a very large number of electrons. The photon-electron conversion, which is basically the photoelectric effect we studied in Chapter 2, takes place in a thin material called *photocathode*. The electrons produced in the photocathode are accelerated towards a metallic structure called *dynode* which releases a large number of electrons due to the impact. These secondary electrons are then accelerated towards another dynode, which also multiplies their number. The process is repeated several times by letting the electrons pass through a series of dynodes. The end result of this process is an output pulse with an amplitude large enough to be easily measured by the associated electronics. The gain of a PMT can be higher than 10^5, which makes it very attractive for sensitive measurements. The downside is the low quantum efficiency of the photon-electron conversion process, which for a typical photoelectrode is around 20%. This implies that 80% of the scintillation photons are wasted away. Although more efficient photocathode materials have been developed but this is still one of the most problematic things with regard to operation of PMTs in low field environment.

Another type of detector, called photodiode detector, can be used to detect photons produced by scintillation materials. A photodiode, as we saw in Chapter on semiconductor detectors, is made of a semiconductor material that has been appropriately doped. In such a detector, the scintillation photon produces electron-hole pairs instead of just electrons as in a PMT photocathode. The quantum efficiency of a photodiode can be as high as 80%, which certainly makes it very attractive for use in low level radiation environments. For highly sensitive measurements in extremely low radiation fields, another special kind of photodiode detector, called the *avalanche photodiode*, can be used. In such a device the electron-hole pairs produced by the incident radiation are multiplied through an avalanche process. The process is analogous to the electron multiplication in a PMT except that in this case there are no dynode-like mechanical structures involved, making the system less susceptible to damage caused by mechanical shocks. The end result of the electron hole multiplication is the transformation of a very low level signal into a pulse large enough to be measured by the electronics.

Photodetectors are an integral part of scintillation detectors and therefore deserve careful and detailed attention. Therefore later in this Chapter we will spend some time discussing the working principles, structures, advantages, and disadvantages of some commonly used photodetectors.

6.1 Scintillation Mechanism and Scintillator Properties

Although scintillation materials come in all forms and types but, broadly speaking, they can be divided into two categories: organic and inorganic. Later in the Chapter we will look at the properties of the materials belonging to these two categories. We will see that how specific applications dictate the preference of a certain scintillator over the others. But before we do that, let us have a general look at the important properties of scintillation materials that are relevant to their use in radiation detectors.

6.1.A Basic Scintillation Mechanism

Scintillators are insulators having wide gap between their valence and conduction bands. Within this gap they also have the so called luminescence centers, which play the central role in producing scintillation light. As shown in Fig6.1.1, a luminescence center is generally composed of two energy levels with a difference that is equal to the energy of photons in and around the visible region of electromagnetic spectrum. If an electron jumps from the higher energy level of this center to the lower level, a scintillation photon can be emitted. *Can be* because there is also a possibility of non-radiation transfer in which the energy is dissipated by the heat carrying particles called *phonons*. If this happens, the process is said to have been *quenched* and the information is lost.

When radiation traverses a scintillation material it departs some of its energy along its track to the particles of the medium. If the energy is greater than the band gap of the material, electrons in the valence band jump up to the conduction band. The vacancy left behind in the valence band by the electron produces an effective positive charge called *hole*. Both the electron in the conduction band and the hole in the valence band are then free to move around in the material. Eventually the electron in the conduction band falls to an energy level below the lower level of the conduction band. If this energy level is the luminescence center of the material, the electron jumps further down to the lower luminescence center and either emits a scintillation photon or dissipates its energy though a non-radiative means. From there the electron jumps to the valence band and combines with the hole.

The electron in the conduction band can also jump to a so called *electron trap*. These traps are metastable energy states that are formed by impurities and defects (in case of crystalline scintillators). An electron trapped there can remain there for an extended period of time that can be as long as more than an hour and as short as a few nanoseconds. Eventually the electron jumps back to the conduction band after receiving enough energy by thermal agitation or some other means. From there it can jump to the luminescence center and cause scintillation light to be emitted. These delayed photons constitute the so called *delayed light* or *phosphorescence*.

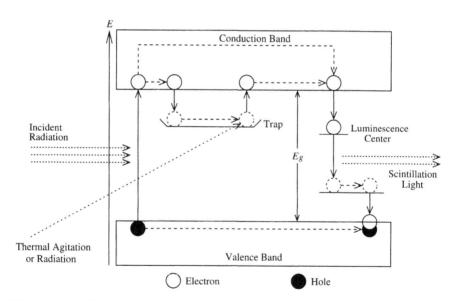

Figure 6.1.1: Principle of production of prompt and delayed scintillation light by incident radiation.

Now that we know how scintillation light is produced, we will proceed to the discussion of the important parameters related to use of scintillation materials as radiation detection media.

6.1.B   Light Yield

This is perhaps the most important parameter for any scintillation material. The reason is that if the light output is very low, the overall signal-to-noise ratio of the subsequent photodetector may not be acceptable. This is specially true for detectors used for spectroscopic purposes where good photon statistics is necessary to yield results well above systematic uncertainties of the system.

The light yield is usually measured in number of photons per MeV of absorbed radiation. For commonly used scintillators a light yield of 20,000-30,000 photons per MeV is not uncommon. In high energy physics it is also customary to characterize light yield in units of number of photoelectrons per minimum ionizing particle. The photoelectrons are produced in the photomultiplier tubes that are used to detect photons produced by the scintillators. The photons from the scintillator are guided to the photomultiplier tube where they are converted into electrons through the process of photoelectric emission in the photocathode material. The combined efficiency of the scintillator and the photocathode is what is generally used in high energy physics to define the light yield. Based on the physics considerations, system electronics, and calibration requirements, a threshold light yield is set to decide on the type of scintillator and the photocathode. 5-10 photoelectrons per minimum ionizing particles is the usual threshold set by experimenters. We will learn more about photomultiplier tubes and other types of photodetectors later in the Chapter.

The three main factors on which the light yield of a scintillator depends are

▶ the scintillation material,

▶ the type of incident particles,

▶ the energy of particles, and

▶ temperature.

The dependence of light yield on the material type can be appreciated from Fig.6.1.2, which shows the relative light output for various commonly used scintillators with respect to electron energy. Note the considerable difference between the qualitative and quantitative energy dependence of the materials shown. The dependence of light output on energy can also change considerably if the impurity concentration in the material is changed.

A troubling aspect of energy dependence of light yield as seen in Fig.6.1.2 is that it degrades the energy resolution of the system. To understand this, let us assume that an incident particle produces electron hole pairs along its track. The electrons move around in the bulk of the material and in the process produce more electrons through excitations. This results in broadening of the electron energy spectrum. Since light yield depends on the particle energy therefore the measured energy will also have a broad spectrum and consequently the energy resolution of the system will degrade. An important point to note here is that even though we assumed the energy of the original electrons to be single valued, the resulting energy spectrum broadened due to the energy dependence of light yield. In practice the electrons created by the incident radiation have their own energy spectrum and therefore this effect is even more pronounced. The degradation in energy resolution is intrinsic to the scintillation materials and has actually been observed. Fig.?? shows the energy

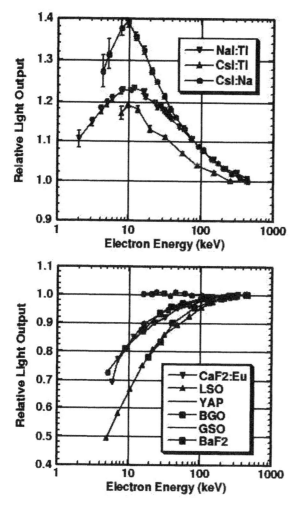

Figure 6.1.2: Relative light yield of various scintillators at different electron energies (34).

resolution of different commonly used scintillation materials as a function of their light output.

A point that should be kept in mind is that the production of scintillation light is a very inefficient process since most of the energy delivered by the incident radiation goes into radiationless transitions. for example the light yield of anthracene is only about 20,000 photons per MeV of energy transferred by high energy particles. This corresponds to an efficiency of approximately 5% for blue light (see example below). However not all scintillators have such a low efficiency. Plastic scintillators are obvious exceptions with an efficiency that is about 10 times that of anthracene. The light yield of a scintillator is generally given in literature with reference to either anthracene or NaI. The light yield of NaI is about twice that of anthracene.

It should be pointed out that even though light yield is an extremely important parameter but it is not the only criterion for selecting a scintillator for a specific

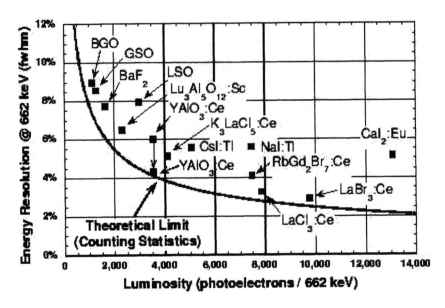

Figure 6.1.3: Energy resolution of some commonly used scintillators for 662 keV γ-rays as a function of their luminosity (34).

application. for example the range of the wavelengths of the emitted photons must also be matched with the sensitivity of the photodetector.

Example:
Convert the light yield of anthracene into efficiency for producing blue scintillation light.

Solution:
The light yield of anthracene is 20,000 photons per MeV. The efficiency can be defined as

$$\eta = \frac{\text{Total Energy of Photons Produced}}{\text{Total Energy Delivered}} \times 100$$

The wavelength of blue photons is about $\lambda = 0.47 \times 10^{-6}~\mu m$. This corresponds to an energy of

$$
\begin{aligned}
E_\gamma &= \frac{hc}{\lambda} \\
&= \frac{\left(6.63 \times 10^{-34}\right)\left(2.99 \times 10^{8}\right)}{0.47 \times 10^{-6}} \\
&= 4.22 \times 10^{-19}~J \\
&= \frac{4.22 \times 10^{-19}}{1.602 \times 10^{-19}} \\
&= 2.63~eV.
\end{aligned}
$$

Hence the efficiency of anthracene is given by

$$\eta = \frac{(20,000)(2.63)}{1 \times 10^6} \times 100$$
$$= 5\%. \tag{6.1.1}$$

6.1.C   Rise and Decay Times

The time profile of a typical scintillation light pulse is shown in Fig.6.1.4. As shown, the pulse rises very quickly with a typical rise time of less than 1 nanosecond. The decay of the pulse is rather slow. In fact, the slow pulse decay in scintillators poses a major problem for their use in detectors since it can decrease the overall efficiency in high rate situations. The decay time of typical scintillation pulse can be as low as a few nanoseconds and as high as several milliseconds.

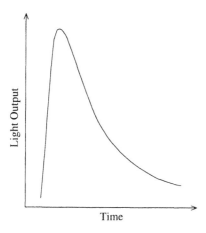

Figure 6.1.4: Typical time profile of scintillator output. Most scintillators produce pulses with very fast rise times (less than 1 nanosecond).

To model this pulse we first split the pulse into rising and decaying edges, both of which can be characterized by exponential functions. The time dependence of the rise of the pulse rise can be written as

$$L \propto 1 - e^{-t/\tau_r}, \tag{6.1.2}$$

where L is the intensity of light in any convenient units and τ_r is a constant we will describe later. Similarly the decay of the pulse can be written as

$$L \propto e^{-t/\tau_d}, \tag{6.1.3}$$

with τ_d is another constant to be described shortly. Combining the two profiles we get

$$L = L_0 \left(e^{-t/\tau_d} - e^{-t/\tau_r} \right), \tag{6.1.4}$$

where L_0 is the maximum height of the pulse.

Let us now see how the constants τ_r and τ_d can be defined. We start by writing only the rising part of the above equation and substituting $t = \tau_r$ in it. This gives

$$
\begin{aligned}
L &= L_0 \left(1 - e^{-t/\tau_r} \right) \\
&= L_0 \left(1 - e^{-1} \right) \\
&= 0.63 L_0.
\end{aligned}
$$

This implies that τ_r is the time taken by the pulse to reach 63% of the maximum height of the pulse. We will call this constant the *rise time constant*.

In a similar fashion, for the other constant τ_d we use only the decaying part of the pulse profile 6.1.4 with the substitution of $t = \tau_d$.

$$
\begin{aligned}
L &= L_0 e^{-t/\tau_d} \\
&= L_0 e^{-1} \\
&= 0.37 L_0.
\end{aligned}
$$

This implies that τ_d represents the time for the light pulse to decay to 37% of its maximum height. τ_d is generally known as the *decay constant* and is one of the most sought after parameters for any scintillator. The rise time constant, on the other hand, does not get much attention merely because the typical rise time of scintillation pulses is extremely small.

Each material has its own characteristic decay time, which is determined experimentally. The difference between the decay profile of materials can be quite significant and therefore the profile of one material can not be used to deduce any meaningful conclusion about the behavior of another material. The dependence of pulse decay time on material is demonstrated in Fig.6.1.5, which has been plotted for three different kinds of scintillators.

The decay time of a pulse produced by a scintillator depends not only on the scintillation material but also on the type and energy of the incident particle. The dependence on the type of incident particle arises due to the difference in the stopping powers of different types of radiation. for example the particles having high dE/dx, such as α-particles, fill more long lived states than particles having lower stopping power. The consequence of trapping more electrons in the metastable states is significant emission of delayed light. Thus the decay component of the pulse is slower and is characterized by a longer decay constant.

In the preceding paragraph we argued that a scintillation material may have more than one decay constant for different types of incident radiation. The main reason behind this difference, as we discussed, is the availability of more than one metastable energy states. Now since all scintillation materials have such energy states due to intrinsic and added impurities therefore we can conclude that, in principle, every scintillation material should have more than one decay constant. This, in fact, has been found to be the case. However since the difference between these decay constants is generally not significantly large therefore an average decay constant or an average decay time can be used to characterize the material. The decay constants found in literature are actually averages.

We will see shortly that all the decay constants of a material have temperature dependence. Generally the decay constant decreases with increase in temperature. Since lower decay constant would mean slower decay of the pulse therefore operating a scintillator at low temperatures is more suitable.

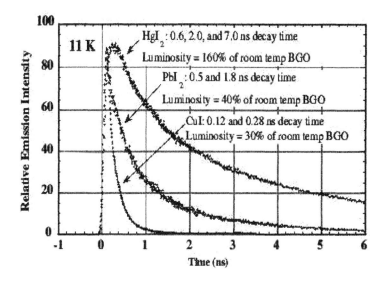

Figure 6.1.5: Time profile of scintillation pulse for three different scintillation materials (34).

6.1.D Quenching

Quenching describes the effect of decrease in light output of a scintillator due to change in some parameter such as temperature, energy of the incident radiation, impurity, or concentration of its constituents. Although the effects of these parameters will be discussed shortly but here we will very briefly describe the quenching mechanisms.

D.1 Self Quenching

The scintillation materials used in detectors are actually mixtures of two or more materials. The light yield of the scintillators depends to a large extent on the concentration of their constituents. In most scintillators the light output increases with increase in the concentration of the primary fluor but then after reaching a certain value it gets saturated.

D.2 Impurity Quenching

Some types of scintillation materials are highly vulnerable to amount of impurities they contain. The impurities can not only decrease their light output but can also affect their optical properties. This type of quenching is so problematic in some scintillators that all of their constituents must be used in highly purified form.

D.3 Thermal Quenching

When we discuss the dependence of temperature on scintillators, we will see that some scintillation materials are highly vulnerable to large temperature changes. The decrease in light output due to temperature increase is generally referred to as thermal quenching. However only a few scintillators show a strong enough dependence on temperature to be worried about.

D.4 Energy Quenching

The light output of a scintillator depends on the energy delivered by the incident radiation. However this process is not linear and eventually saturates at a certain stopping power that depends on the material. Delivering more energy to the scintillator beyond this point does not affect the light output. In such a state the material is said to have suffered energy quenching. Since this effect is most pronounced in organic scintillators, therefore it will be discussed in more detail later in the Chapter.

6.1.E Density and Atomic Weight

Scintillation detectors are used to detect all types of radiation including γ-rays, electrons, neutrons, α-particles, and neutrinos. Efficient detection of these particles depends on how well and how quickly they loose their energy in the scintillation medium. These factors are characterized by the stopping power of the material, which depends not only on the atomic weights of the constituents of the material but also on their densities. for example detection of light particles such as photons requires the material to have high effective atomic weight and high density.

Density and atomic weight are therefore important factors that are used to assess the effectiveness of scintillation material for a particular kind of radiation and the energy range of interest. There is no universal material that is good for all types of radiation.

Although density and atomic weight are important parameters, but the reader should note that they are not the only parameters that need to be considered. Other factors such as opaqueness of the material to the scintillation photons may make an otherwise *ideal* scintillation material worthless for use in a scintillation detector.

6.1.F Mechanical Properties and Stability

Vulnerability of scintillators to atmospheric conditions is well known. for example due to moisture and temperature fluctuations the optical properties of some scintillator deteriorate. Similarly changes in pressure may cause cracks. Hence mechanical stability of solid scintillators is a major concern for detection systems.

Another important point is that the solid scintillators have to be cut, machined, and polished after being produced. If they are not mechanically stable, they can not be properly prepared for use in detectors. As we will see later, the plastic scintillators have the kind of mechanical stability needed for mechanical manipulations.

6.1.G Optical Properties

Most scintillators are developed such that they produce light in the visible region of the electromagnetic spectrum. Since the photons thus produced have to be guided to a photodetector to be detected therefore they must be able to travel through the material without significant attenuation. Attenuation of photons causes loss of information and therefore can induce non-linearities in the detector response. Hence the optical properties of the material play a very important role in the overall efficiency of the system. The parameter of interest here is the refractive index of the material for the wavelength of the light emitted by the scintillation material.

Unfortunately most of the material that have very good scintillation efficiencies produce ultraviolet photons. These photons have very short attenuation lengths and therefore quickly get absorbed in the material, resulting in the loss of information. The trick that is employed to solve this problem is to mix homogeneously another scintillator in the bulk of the material that absorbs these photons and emits visible light photons. Since the visible photons have longer attenuation lengths, their lifetime in the material increases and consequently the overall efficiency of the detector is considerably increased. The second scintillators are generally known as *wavelength shifters* or *secondary fluors*.

6.1.H   Phosphorescence or Afterglow

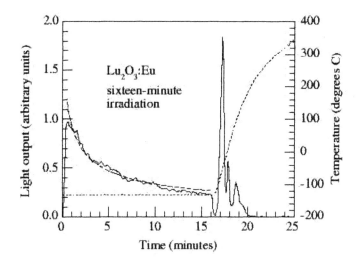

Figure 6.1.6:   Fluorescence and phosphorescence from $Lu_2O_3 : Eu$ after 16 minute irradiation (solid line). The curve also shows the variation of sample temperature (28).

Scintillation materials produce not only prompt scintillation light but also delayed light. The delayed light, as we discussed earlier, is emitted by transitions made by electrons trapped in the long lived energy states. This is obviously not a very desirable trait of a scintillation material since it can not only stretch the decaying part of the scintillation pulse but could also produce a significant afterglow if the temperature is raised. Fig.6.1.6 shows the time profile of light emission after 16 minute irradiation of a ceramic scintillator $Lu_2O_3 : Eu$. The second peak shown in the figure occurred after the sample was heated. Providing thermal energy to the electrons in the long lived states elevate them to the conduction band. From there they can follow the scintillation levels and eventually end up in the valence band.

In the process, they can emit light such as shown in Fig.6.1.6. Since production of delayed light with a prominent peak involves providing thermal energy to the material, the phenomenon is called *thermoluminescence*. There are many materials that are exceptionally good in producing this kind of light with minimal prompt scintillation. The phenomenon of thermoluminescence is exploited in the detectors called Thermo-Luminescent Detectors or TLDs, which are memory devices used to monitor absorbed dose over a period of time. We learned a little bit about such detectors in the Chapter on solid state detectors and will discuss them in detail in the Chapter on dosimetry.

In common scintillation detectors thermoluminescence is not of much concern provided the temperature of the sample is not allowed to increase with time. However saturation of metastable traps can sometimes cause undesirable results including broadening of the prompt pulse.

Let us now turn our attention to the long decay time of scintillation that is observed in most scintillators. The longer the pulse takes to decay the more afterglow the material is said to have. Intuitively we can think that the cause of this afterglow is the slow transitions of electrons trapped in impurity levels. However since we know that elevating electrons from the traps is more favorable at higher temperatures therefore we would expect the afterglow to increase at with temperature. In the next section we will see that this is actually what is observed.

6.1.1   Temperature Dependence

Almost all of the parameters we studied in the preceding sections depend on the temperature of the material. That is why the values given in literature always correspond to operation at a certain temperature or in a temperature range.

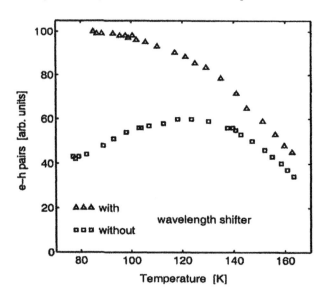

Figure 6.1.7: Relative yield of electron hole pairs in *CsI* with and without addition of a secondary fluor (2).

Let us first see how the production of electron hole pairs by the incident radiation is affected by a change in temperature. Based on our discussion on semiconductors in the previous Chapter we would expect to see some variation. Fig.6.1.7 shows the behavior of CsI crystal at different temperatures. The two curves represent the scintillator with and without addition of a wavelength shifter. The difference can be attributed to the change in the energy band structure after addition of the second scintillator.

Since the charge pair yield changes with temperature we would expect the light yield to also show temperature dependence. In fact, the temperature dependence of light yield has been found to be very profound in most of the scintillation materials. As shown in Fig.6.1.8, generally the light output is seen to decrease with increase in temperatures.

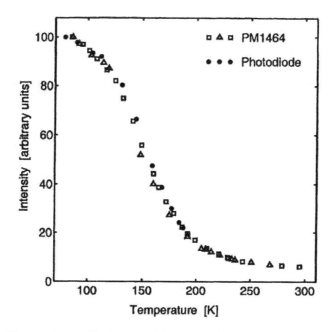

Figure 6.1.8: Variation of light yield of pure CsI with respect to temperature. The data was obtained with two different photodetectors (2).

Another important parameter that has strong temperature dependence is the decay constant. In most scintillation materials a time plot of the pulse decay at different temperatures produces significantly different decay rates, as can be seen in Fig.6.1.9. Since the decay of scintillation pulse depends on the thermal energy therefore one would expect the decay constant to depend on the thermal energy of the particles in the material through the relation

$$\tau_d = \tau_{d0} e^{E/k_B T}, \tag{6.1.5}$$

where τ_{d0} is a material dependent constant, E is the mean activation energy, k_B is the Boltzmann's constant, and T is the absolute temperature.

As we discussed earlier, every scintillator has more than one decay constant and therefore generally an average of these values is given in literature. Although this

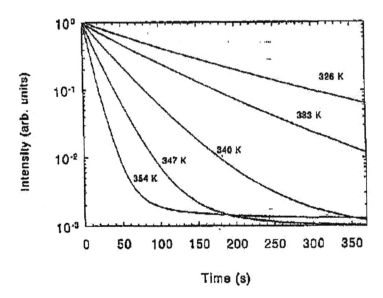

Figure 6.1.9: Pulse decay of Lu_2SiO_5 : Ce crystal at different temperaturs (17).

technique gives acceptable results for most applications but in high resolution systems extra care should be exercised while extrapolating the given average decay constant to another temperature may not be very accurate. The reason is that the temperature profiles of the individual decay constants can be quite different from one another. Fig.6.1.10 shows the temperature dependence of two decay constants of pure CsI. It can be seen that the two curves have different slopes and although they can be individually extrapolated using a straight line fit but the extrapolation of the average would require a non-linear fit.

6.1.J Radiation Damage

The scintillators used in hostile radiation environments are prone to damage caused by high instantaneous and integrated radiation doses. Monitoring the gradual degradation of the detector response with time is therefore a standard practice in such laboratories. Not all materials behave in the same way with respect to radiation damage, though. In fact, a number of newly developed scintillators have shown good resistance to radiation damage and therefore such *radiation hard* materials are preferred over the conventional ones in applications involving high radiation fields.

Although all types of scintillators show radiation induced damage but generally liquid scintillators have been found to be least susceptible to radiation. Plastic scintillators, on the other hand, are highly prone to radiation damage. We will discuss the damage mechanisms of different types of scintillators in their respective sections.

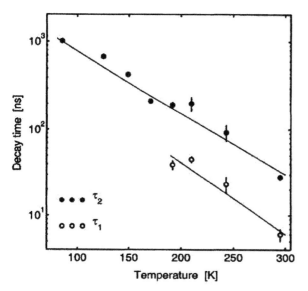

Figure 6.1.10: Temperature dependence of two decay constants of pure CsI (2).

6.1.K   Scintillation Efficiency

Scintillation efficiency is a term that is used to characterize the efficiency by which the energy lost by the incident particle is converted into scintillation photons. Generally it is described as a ratio of the total energy of scintillation photons and the total energy deposited by the incident radiation, that is

$$\eta = \frac{\text{Total Energy of Scintillation Photons}}{\text{Energy Deposited by Incident Radiation}}$$
$$= \frac{E_s}{E_i}, \tag{6.1.6}$$

where the subscripts s and i stand for scintillation and incident respectively.

In order to determine the energy of the scintillation photons in terms of the energy deposited by the incident radiation we first realize that scintillation is a complex and multi-step process. The reason is that the energy transfer to the luminous centers (that is the energy traps that are responsible for emission of scintillation light) follows the charge pair production by the incident radiation. The complexity of the scintillation process can be appreciated by looking at the following simplified version of it.

$$\gamma_i + X \rightarrow X^+ + e$$
$$X^+ + X \rightarrow X_2^{+*} + \text{heat}$$
$$X_2^{+*} + e \rightarrow 2X + \gamma_s \tag{6.1.7}$$

Here X and X^* symbolically represent the scintillator in ground and excited states respectively. γ_i and γ_s are the incident and scintillation photons respectively, though the incident particle does not necessarily have to be a photon. The first step of the scintillation process depicted here involves production of charge pairs. We already know that the number of charge pairs produced can be characterized by the so called W-value, that is, the energy needed to create a charge pair. Hence the number of charge pairs created by the radiation can be written as

$$N_{ep} = \frac{E_i}{W}. \tag{6.1.8}$$

The energy lost by the radiation is gained by the ions. Now, the ions thus produced attract other ions to produce excited molecular ions, which in turn attract electrons to again dissociate into their constituents. This process can be visualized as the transfer of energy from the excited ions to the luminous centers. Unfortunately this is not a very efficient process since the ions can also loose energy through non-radiative processes, such as by collision with impurity molecules. Let us represent the efficiency of this process by ξ. The number of luminous centers activated by the ions can then be estimated from

$$
\begin{aligned}
N_{lum} &= \xi N_{ep} \\
&= \xi \frac{E_i}{W}. \tag{6.1.9}
\end{aligned}
$$

The luminous centers are now ready to emit scintillation light. However there are also other non-radiative processes possible through which they can loose energy. The efficiency of scintillation emission by the luminous centers is generally represented by q and most authors prefer to call it the *quantum efficiency* of luminous emission. Mathematically it is equal to the ratio of number of scintillation photons N_s to the number of luminous centers N_{lum}, that is

$$q = \frac{N_s}{N_{lum}}. \tag{6.1.10}$$

The number of scintillation photons emitted can be determined by substituting N_{lum} from equation 6.1.9 into the above equation.

$$
\begin{aligned}
N_s &= q N_{lum} \\
&= q\xi \frac{E_i}{W} \tag{6.1.11}
\end{aligned}
$$

The scintillation photons thus emitted are not monochromatic but their spectrum has a well defined peak. The energy corresponding to this peak is the *most probable* energy. We will represent it by E_{max}. To simplify the computations we can assume that each of the scintillation photon carries this much energy. Hence the total energy of the scintillation light can be written as

$$
\begin{aligned}
E_s &= N_s E_{max} \\
&= N_s \frac{hc}{\lambda_{max}} \tag{6.1.12}
\end{aligned}
$$

where λ_{max} is the wavelength of the photons corresponding to the peak of the scintillation spectrum. Substituting N_s from equation 6.1.11 into the above equation gives

$$
\begin{aligned}
E_s &= q\xi E_{max} \frac{E_i}{W} \\
&= q\xi \frac{E_i}{W} \frac{hc}{\lambda_{max}}. \tag{6.1.13}
\end{aligned}
$$

Using this and the equation 6.1.6 we can write the expression for the scintillation efficiency as

$$\eta = q\xi\frac{E_{max}}{W} \tag{6.1.14}$$

$$= q\xi\frac{hc}{W\lambda_{max}}. \tag{6.1.15}$$

Although this expression does not take into account the energy dependence of different parameters involved, still it gives a good estimation of the efficiency for most scintillators. When the efficiencies q and ξ are not known, one can estimate the maximum scintillation efficiency by substituting $q = 1$ and $\xi = 1$ in the above expression.

$$\eta_{max} = \frac{hc}{W\lambda_{max}} \tag{6.1.16}$$

This expression is most suitable for estimating the physical limit on the scintillator efficiency. For crystalline scintillators it can be transformed into a more convenient form by noting that the W values for such a material is about 3 times its band gap energy. In general, one can write

$$W = \gamma E_g, \tag{6.1.17}$$

where E_g is the band gap energy and γ is a factor that depends on the material. For most crystals its value is around 3 but can be as high as 8 for certain materials. The maximum efficiency of crystalline scintillators can therefore also be written as

$$\eta_{max} = \frac{hc}{\gamma E_g \lambda_{max}}. \tag{6.1.18}$$

Since the light output of a scintillator is directly related to its scintillation efficiency therefore one strives to maximize the efficiency by adding impurities to the material or by operating it at a different temperature. Unfortunately the scintillation efficiency for most scintillators is quite poor and can range from a few percents to about 30%. The low efficiency points to the fact that most of the energy delivered by the incident radiation goes into increasing the thermal agitations and non radiative transitions. However even with such low efficiencies, scintillators can be very effectively used for precision measurements. The reason is that the light output is generally sufficient to obtain good signal-to-noise ratio. for example the light output of thallium activated NaI crystal is about 40,000 photons per MeV of energy delivered. If a photomultiplier tube is used to detect these photons, even with a modest 20% efficiency of the photocathode, the tube will see enough photons to produce a pulse that is large enough to be measured.

Example:
Determine the maximum efficiency of thallium activated NaI and CsI scintillators. The wavelength of maximum emission for NaI:Tl is 410 nm and that of CsI:Tl is 550 nm while their band gaps are approximately 5.9 eV and 6.2 eV respectively.

Solution:
The maximum efficiencies of the two crystals can be obtained from substituting
the given parameters into equation 6.1.18. Hence for $NaI{:}Tl$ crystal we get

$$\eta_{max}[NaI{:}Tl] = \frac{hc}{\gamma E_g \lambda_{max}}$$
$$= \frac{\left(6.63 \times 10^{-34}\right)\left(2.99 \times 10^8\right)}{(3)\left(5.9 \times 1.6 \times 10^{-19}\right)\left(410 \times 10^{-9}\right)}$$
$$= 0.17$$
$$= 17\%.$$

Similarly for $CsI{:}Tl$ we get

$$\eta_{max}[CsI{:}Tl] = \frac{hc}{\gamma E_g \lambda_{max}}$$
$$= \frac{\left(6.63 \times 10^{-34}\right)\left(2.99 \times 10^8\right)}{(3)\left(6.2 \times 1.6 \times 10^{-19}\right)\left(550 \times 10^{-9}\right)}$$
$$= 0.12$$
$$= 12\%.$$

It should be noted that these are not the efficiencies one should expect the
scintillators to exhibit in a practical system since the computations were based
on the assumption that the materials had 100% luminous center emission
and quantum efficiencies. Of course in a practical system this should not be
expected due to the possibility of energy loss through other non-scintillation
transitions and transfers. Another point is that the scintillation efficiency
also has temperature and energy dependence, which we have neglected. An
actual crystal can have a scintillation efficiency that is up to 50% of the value
obtained from equation 6.1.18.

6.2   Organic Scintillators

Organic scintillators are extensively used in radiation detectors. They are found in
solid, liquid, and gaseous states. One of the biggest advantage of organic scintillators is that they can be produced in virtually any geometry and therefore can be
customized to specific applications.

6.2.A   Scintillation Mechanism

The basic scintillation mechanism of organic scintillators is the *fluorescence* or prompt
emission of light following the S-state electronic transitions. To better understand
this process, let us have a look at typical energy level diagram of organic scintillators
as shown in Fig.6.2.1. The incident radiation transfers energy to the atoms causing
electronic transitions from S_0 ground levels to different vibrational levels in the S_1
band (higher band transition are of course also possible depending on the energy of
the incident radiation). The induced instability in the system forces the electrons

in different S_1 vibrational levels to go to the S_1 ground level through radiationless transitions. These transition are very fast and are completed within a few tenths of a nanosecond. From there the electrons fall into the S_0 ground and vibrational levels with decay times on the order of a few nanoseconds. The excess energy of the electrons is emitted in the form of fluorescence photons, which for most materials lie in the ultraviolet or visible region of the electromagnetic spectrum. The decay of electrons from the S_1 vibrational states to S_1 base level is a favorable process specially for scintillation detectors. The reason is that during such decays the electrons loose some of their energy. Consequently the energy emitted during their transitions from the S_1 ground level to the S_0 levels is less than the energy absorbed by them during the radiation induced transitions from S_0 ground level to S_1 levels. This essentially implies that the absorption and emission spectra of such materials do not match, hence disfavoring the re-absorption of the scintillation light in the material. This phenomenon was first observed by Sir George G. Stokes, who later on formulated a *Stoke's law* that says that the wavelength of fluorescence light is always greater than the wavelength of absorption. Of course, this law does not really hold for all wavelengths and all materials since there is always some overlap of emission and absorption spectra. However the so call *Stoke's shift* is a reality, which simply characterizes the fact that the peak of emission spectrum is shifted from the peak of absorption spectrum. The amount of Stoke's shift can be defined as

$$\triangle\lambda_{Stokes} = \lambda_{max,s} - \lambda_{max,a}. \tag{6.2.1}$$

Here λ_{max} represents the peak of the spectrum with subscripts a and s referring to the absorption and scintillation (fluoroscence) respectively.

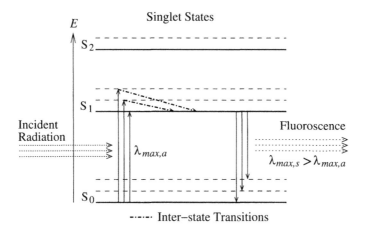

Figure 6.2.1: Typical singlet energy level diagram of organic scintillators. The incident radiation transfers the electrons from S_0 ground level to S_1 levels. These electrons first decay into S_1 ground level and then fall into the S_0 levels and as a result emit scintillation light. The most probable wavelength of emission $\lambda_{max,s}$ is greater than the most probable wavelength of absorption $\lambda_{max,a}$.

A typical emission and absorption spectra of an organic scintillator is sketched in Fig.6.2.2. In general, the higher the Stoke's shift the smaller the probability of re-absorption of scintillation light since the overlap area will be smaller. Hence Stoke's shift can be used as a measure of appropriateness of a scintillator for certain application.

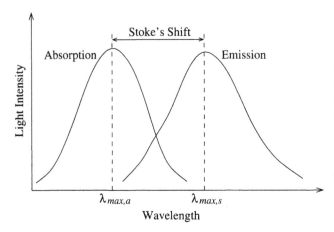

Figure 6.2.2: Typical absorption and emission spectra of an organic scintillator. The wavelength difference of the peaks of the two spectra is generally used to quantify the amount of Stoke's shift.

The above mentioned S band transitions are not the only transitions possible in organic scintillators. Another mode of electron relaxation is through the vibrational levels in the triplet T_1 band. The process is graphically depicted in Fig.??. As before, the incident radiation transfers the electrons from the S_0 ground level to the S_1 vibrational levels. These electrons first decay into the S_1 ground level through radiationless transitions. Now, instead of falling into the S_0 level directly, the electrons can also first go to the available T_1 levels. These triplet levels are much more stable than the singlet levels and consequently the electrons can be thought of being trapped there for an extended period of time. From the T_1 ground level, into which all electrons eventually decay, they fall into the S_0 levels. This also results in the emission of light but in this case it is called *phosphorescence* or *delayed fluorescence*, since it is emitted after a substantial delay, which is more than 100 *ms* for typical organic scintillators.

Now that we understand the basic scintillation mechanism in organic scintillators, let us discuss the different forms in which organic scintillating materials are available. In radiation detectors the following three forms of organic scintillators are widely used.

▶ Plastic scintillators

▶ Liquid scintillators

▶ Crystalline scintillators

In the following sections we will discuss each of these in some detail.

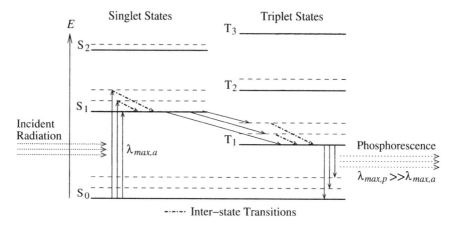

Figure 6.2.3: Typical energy level diagram of organic scintillators. The incident radiation transfers the electrons from S_0 ground level to S_1 levels. These electrons first decay into S_1 ground level and then either fall into the S_0 levels directly (see Fig.6.2.1) or through the triplet T_1 state. The later transition results in the emission of phosphorescence photons after a delay that depends on the lifetime of the T_1 state. The wavelength of phosphorescence photons $\lambda_{max,p}$ is much larger than the absorption wavelength $\lambda_{max,a}$.

6.2.B Plastic Scintillators

Plastic scintillators are extensively used in a variety of applications. They are synthetically produced through a process that is time consuming and highly labor intensive. The basic idea, however, is quite simple: add the scintillator in a convenient base and then polymerize the mixture. The practical situation is not that simple, though. First of all, all the constituents must be in highly purified form and secondly there are waiting periods involved between the different steps to let the material settle down to the required configuration. The base material that is generally used for the mixture is acrylic, polystyrene, or polyvinyltoluene monomer. A convenient scintillator is then added to this base and mixed homogeneously. The concentration of this main scintillator is about 1% by weight of the mixture, which is sufficient to produce a high light yield. Most of the time another scintillator is also added to the mixture. The individual purposes of all these materials will be discussed shortly. The whole mixture is then polymerized to form the plastic.

A plastic scintillator produced by this method has several attractive qualities, such as

▶ it is chemically stable,

▶ it has high degree of optical homogeneity, and

▶ it can be cut and machined into virtually any shape.

We have deliberately not said anything about their light yield, decay constants, and other important parameters. The reason is that the commercially available plastic scintillators have such varied properties that it may be misleading to quote one of them as a representative of all plastic scintillators.

Although we can not use absolute numbers of scintillation and optical properties of plastic scintillators to represent most of the materials but there are certain
trends that these materials follow. Unfortunately these trends point towards their
disadvantages than advantages as is obvious from the following list.

▶ **Nonlinearity of Light Output:** The major problem associated with plastic
  scintillators is their nonlinear behavior. That is, their light output per unit
  length has a nonlinear dependence on the energy loss per unit length of the
  particle's track. This behavior is characterized by the so called *Birk's formula*
  (6)

  $$\frac{dL}{dx} = \frac{AdE/dx}{1 + kdE/dx},$$ (6.2.2)

  where dL/dx represents the light output per unit length, dE/dx is the energy
  lost by the particle per unit length of its path, A is the absolute scintillation
  efficiency, and k is a parameter that relates the density of ionization centers
  to the energy loss. dE/dx in the above equation is usually determined by
  the Bethe-Bloch formula, that we studied in Chapter 2. We will discuss this
  behavior a little bit more shortly.

▶ **Delayed Fluoroscence:** Some of the fluorescence modes of plastic scintillators
  have long lived components that do not decay exponentially. This may result
  in emission of delayed light or afterglow. Such an afterglow is undesirable in
  radiation detectors since it not only introduces uncertainty in the measurements
  but also makes the material unsuitable for use in high radiation fields.

▶ **Dependence of Light Yield on Gas Pressure:** The light yield of plastic
  scintillators has a significant dependence on pressure. It has been observed
  that the light yield of typical plastic scintillators can decrease by up to 10%
  with increase in partial pressure of oxygen and other gases in the environment.

▶ **Dependence of Light Yield on Magnetic Field:** The light yield of plastic
  scintillators has a nonlinear dependence on magnetic field. However the effect
  is not significant in moderate magnetic fields and one could essentially ignore
  it in typical applications.

▶ **Decrease in Light Yield with Time:** The degradation of scintillation properties of plastics is well known and has been thoroughly studied. It has been
  found that the light yield of typical plastic scintillators decreases with the passage of time.

▶ **Mechanical Instability:** Although plastic scintillators are chemically stable
  but their mechanical structure is somewhat fragile. If they are not handled
  carefully, they can develop small cracks. The most dramatic effect of these
  cracks is the degradation in optical properties of the material.

▶ **Vulnerability to Foreign Elements:** Any foreign material, such as oil and
  dust, may degrade its optical and chemical properties.

▶ **Vulnerability to Radiation Damage:** Plastic scintillators are highly vulnerable to radiation induced damage. We will discuss this in detail later in the
  Chapter.

The reader should note that not all the disadvantages listed here are typical of only plastic scintillators. As we will see later, other types of scintillators also suffer from one or more of such performance related issues. The decision to use a particular type of scintillator is highly application specific. That is why, even with these disadvantages, plastic scintillators are still extensively used in a variety of applications ranging from medical dosimetry to high energy physics.

Before we start our discussion on the details of plastic scintillators, let us first spend some time understanding the Birk's formula as presented above and given by

$$\frac{dL}{dx} = \frac{AdE/dx}{1 + kdE/dx}.$$

This formula is extensively used to characterize response of plastic scintillators and therefore deserves some attention. The left hand side of the formula represents the light yield per unit length of the particle track. On the right hand side there are three parameters: A, k, and dE/dx. The constant A is called the *absolute scintillation efficiency* of the material. It is a function of the scintillator material and is generally available in literature for most of the scintillators. The constant k on the other hand, which is sometimes referred to as the *saturation parameter*, is unfortunately available only for a few materials. The practice is, therefore, to use the available value for a material that closely resembles the material under consideration. In general the value of k is of the order of 0.01 $gMeV^{-1}cm^{-2}$. The third parameter is the *stopping power* of the material for the type of incident particle. It depends not only on the type of the material but also on the type of incident radiation. In Chapter 2 we discussed the stopping power at length and saw that it can be fairly accurately determined from the Beth-Bloch formula.

Some authors prefer to write the Birk's formula for the total light yield $\triangle L$ in terms of the total energy lost by the particle $\triangle E$, that is

$$\triangle L = \frac{A \triangle E}{1 + kdE/dx}. \tag{6.2.3}$$

The reader can readily verify that this representation is equivalent to the formula 6.2.2.

Let us now try to understand what the Birk's formula physically represents. When particles pass through a scintillator they loose energy with a rate that can be determined by the Bethe-Bloch formula. The energy lost by the particles goes not only into exciting molecules but a part of it also increases the lattice vibrations. The de-excitation of the molecules leads to the emission of scintillation photons with a high probability. Now, if we increase the stopping power (for example by increasing the energy of the incident particles), more molecules will get excited per unit track length, thus producing more scintillation light. But this behavior can not persist indefinitely since the sample has a finite number of molecules that can be excited. This implies that the number of molecules available for scintillation actually decreases with increase in the total energy delivered by the radiation. Taking this argument a step further we can conclude that there must be a value of stopping power at which all the molecules have been excited. The sample in this state will be said to have reached a state of saturation. After reaching this state, delivering more energy to the material would not increase the light output. This statement is actually what is referred to as the *Birk's law* and is mathematically represented by the Birk's formula. The behavior of the light output with respect to the stopping power as represented by the Birk's formula is graphically depicted in Fig.6.2.4.

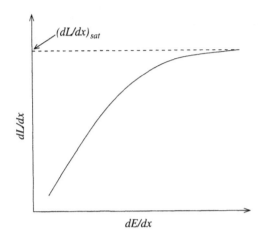

Figure 6.2.4: Plot of Birk's formula in arbitrary units. The saturation effect of light output is called Birk's law.

As stated earlier, the plastic scintillators are manufactured by dissolving one or more fluors to a base plastic material. Fig.6.2.5 shows the specific purpose of each of these materials. Also shown are the typical weight concentrations of primary and secondary fluors in the plastic base. The plastic base absorbs the energy of the incident radiation and emits ultraviolet photons. If there were no other materials the photons thus emitted would get quickly absorbed due to their very short attenuation lengths of the order of a few mm. Since such prompt absorption of photons is not desirable for practical photodetectors therefore another material is added to the base plastic to enhance its scintillation capabilities. This material is generally called the primary fluor and has the ability to emit ultraviolet photons of larger attenuation lengths. Since, even with higher attenuation length, the plastic is still not fully transparent to the photons, another material is added to shift their wavelength. This secondary fluor is also a scintillator and is generally known as the wavelength shifter. Its function is to absorb ultraviolet photons and emit light photons. Most wavelength shifters emit blue light.

As we just learned, the plastic base, though a scintillator itself, can not be effectively used without addition of other scintillation materials. Addition of the primary fluor in a concentration of about 1% increases not only the photon attenuation length but also the total light yield. In this configuration the plastic base does not really act as a scintillator but as an energy transfer medium. It absorbs incident radiation and then transfers the energy to the primary fluor through resonant dipole-dipole interactions. This mechanism is generally known as Forster energy transfer.

The light yield of plastic scintillators mainly depends on the following parameters.

▶ Type of material

▶ Type of radiation

▶ Energy of radiation

▶ Temperature

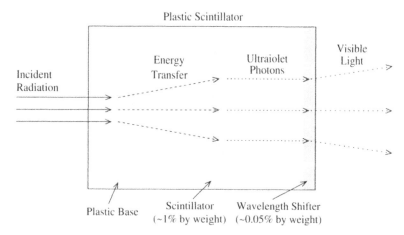

Figure 6.2.5: Working principle of a plastic scintillator based on two scintillation materials, which are homogeneously mixed in a base plastic material.

▶ Pressure

The cause of particle dependence of light yield has already been discussed in the beginning of the chapter and therefore will not be repeated here. But as a reminder, this dependence is actually the manifestation of the difference in stopping powers of the material for different particles. Now, since the stopping power depends also on the type of material as well as the energy of radiation therefore for most materials studying the dependence of light output on the stopping power suffices to characterize it in terms of its suitability as a detection medium. Even though the light yield for different materials and incident particles can vary considerably from one another, still for plastic scintillators there is a general trend that can be identified. for example we see that the light yield decreases with the particle mass and increases with particle energy. Fig.6.2.6 shows these general trends for plastic scintillators.

Earlier in the chapter we discussed the temperature dependence of scintillators and saw two graphs that showed the dependence of the charge pair production and the light yield on temperature for an inorganic scintillator (CsI). Plastic scintillators are no exception and show similar behavior.

Now we turn our attention to radiation damage in plastic scintillators. As stated earlier, most of the plastic scintillators are highly vulnerable to radiation damage. One of the factors most affected by long term accumulated dose is the decrease in attenuation length of scintillation light in the bulk of the material. An example of such deterioration is shown in Fig.6.2.7, which was observed in a plastic scintillator used in a high energy physics experiment called ZEUS. Here light transmission was measured using a source at different positions in the detector at two points in time. The effect of radiation damage is clearly visible with transmission getting poorer after ten years of irradiation. Since light transmission is directly related to the attenuation length therefore one can conclude that the attenuation length decreases with radiation exposure.

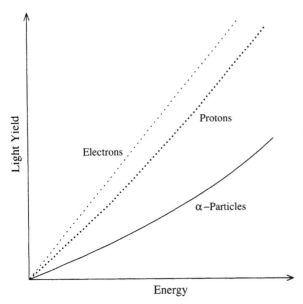

Figure 6.2.6: Variation of light output of plastic scintillators with respect to energy for electrons, protons, and α-particles.

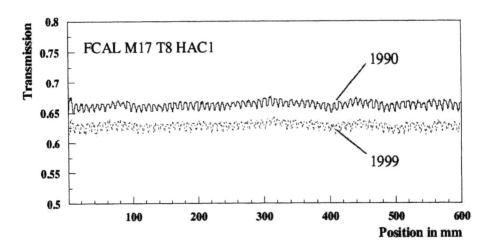

Figure 6.2.7: Decrease in transmission observed in a plastic scintillator (SCSN-38) after ten years of irradiation (7)

The scintillation and optical properties of plastic scintillators can be adjusted by changing the concentrations of primary and secondary (if any) fluors in the base. Therefore most manufacturers offer different scintillators based on varying concentrations of solvents so that the user could make the selection according to the system requirements. Table.6.2.1 lists the light emission properties of some plastic scintillators but it should be noted that these values are not typical of most plastic scintillators.

Table 6.2.1: Wavelength of maximum emission λ_{max} and decay constant τ of some organic plastic scintillators (39).

Liquid	λ_{max} (nm)	τ (ns)
Polystyrene with 36 gm/l of p-terphenyl	355	≤ 3.0
Polystyrene with 16 gm/l of 1,1,4,4-terphenyl-1,3-butadiene	450	4.6
Polystyrene with 36 gm/l of p-terphenyl and 0.2 gm/l of 1,1,4,4-terphenyl-1,3-butadiene	445	4.0
Polyvinyl-toluene with 16 gm/l of 1,1,4,4-terphenyl-1,3-butadiene	450	4.6
Polyvinyl-toluene with 36 gm/l of p-terphenyl and 0.2 gm/l of 1,1,4,4-terphenyl-1,3-butadiene	445	4.0

6.2.C Liquid Scintillators

There are a number of liquids that produce scintillation light and can therefore be used in detectors. In general liquid scintillators used in detection systems are composed of the following three organic components.

▶ **Primary Fluor:** The primary fluor or primary scintillator is the main scintillator in the mixture. It has high scintillation efficiency but produces light in the ultraviolet region of the electromagnetic spectrum.

▶ **Wavelength Shifter:** The UV photons produced by the primary fluor have short attenuation lengths in the liquid. Since the liquid is not completely transparent to them therefore most of them can not escape from the material. The purpose of the wavelength shifter is to shift the wavelength of these photons to the visible region. The wavelength shifters are also sometimes known as secondary fluor since they are also scintillators that absorb UV photons and give off visible light photons. Most of the practical wavelength shifters produce blue light.

▶ **Base Liquid:** The above two mentioned scintillators are mixed homogeneously in a liquid that has good light transmission properties.

Since the basic building blocks of liquid scintillators are essentially similar to the plastic scintillators, the requirements of purity for their constituents are also similarly quite stringent. The base, primary fluor and the wavelength shifter, all must be used in highly purified forms. Even addition of small amounts of water can contaminate the scintillator and significantly deteriorate its light transmission properties.

A large number of pure liquids and liquid mixtures have been identified as good scintillators having properties relevant to their use for radiation detection. These liquids have not been assigned any generic names but most of them are available commercially from different manufacturers under their own names.

Most liquid scintillators emit light with a peak somewhere between the wavelengths of 400 nm and 500 nm. The spectrum is, however, somewhat skewed at higher wavelengths as shown in Fig.6.2.8.

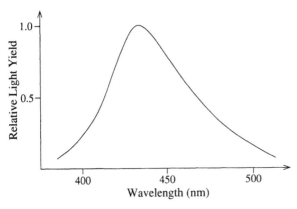

Figure 6.2.8: Light yield of a typical liquid scintillator as a function of wavelength.

Emission spectra of the kind shown in Fig.6.2.8 are very favorable to develop liquid scintillator detectors because most base liquids used in such liquids have optimal transmission only above a wavelength of 400 nm. This is illustrated in Fig.6.2.9, which is a plot of some suitable transmission property with respect to wavelength of photons (see also (20)). The reader should bear in mind that light transmission is not the only criterion for the usability of a scintillator in a specific application. The scintillation photons must also be efficiently detected by another detector such as a photomultiplier tube. Since photodetectors have their own limitations with respect to their sensitivity to different wavelengths, therefore the liquid scintillator should be chosen such that it not only has high transmission capability for the photons it emits but the photodetector also has high efficiency to detect them. Most photomultiplier tubes work best for wavelengths in the green region of the electromagnetic spectrum and therefore scintillators that have peak in this region can be used very efficiently. As a reminder the middle of the green region lies at approximately 510 nm. Some scintillators emit light in the blue region and therefore require a wavelength shifter to produce green light. The wavelength shifter can be a liquid dissolved in the base along with the scintillator or an external solid. The external wavelength shifters will be discussed later in the Chapter.

Most liquid scintillators show remarkable stability with respect to small temperature variations. Less than 1% variation per decade in temperature in Celcius is not uncommon (20). The light yield of liquid scintillators covers a broad range but mot manufacturers produce products that have light output from 50% to 80% of anthracene. Such a varied light yield is due to the strong dependence of the concen-

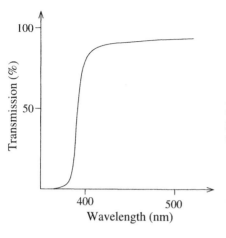

Figure 6.2.9: Typical light transmission capability of base liquids commonly used for liquid scintillation detectors.

tration of dye in the liquid. It has therefore become a general practice to determine the optimum value of the dye concentration based on the system requirements.

Organic liquid scintillators have been quite successfully used in large volume detectors. for example in experiments involving neutrino detection the weight of the liquid can be several hundred tons.

We mentioned earlier that one of the major problems with liquid scintillators is their contamination from foreign elements. Some of these contaminants can deteriorate the scintillation properties of the liquid while others can affect the transmission properties of the scintillation light. The most commonly encountered contaminant of the second kind is water. Small amounts of dissolved water in a liquid scintillator can significantly deteriorate its light emission properties. Fortunately water does not dissolve easily in most liquid scintillators, specially the ones that are based on mineral oil. However, since the effect of water contamination is significant, therefore every effort should be made to avoid any contact of the liquid with water. A common problem encountered during handling of a liquid scintillator is the moisture present in the inside of the container. If the container is not properly cleaned and dried, water and other contaminants can slowly dissolve in the liquid and decrease its efficiency.

Oxygen in air can also contaminate liquid scintillators and decrease its light output. Even small amounts of air contamination has been seen to produce a decrease in light output of 20-30%. Since it is extremely difficult, if not impossible, to avoid air or oxygen exposure to scintillators, therefore a normal practice is to de-oxygenate the liquid at least once before its use in a detector. The process is quite simple and involves slowly passing dry nitrogen through the liquid, which bubbles out any oxygen in the liquid. After this process is complete, the liquid should be kept in an atmosphere of highly purified nitrogen, argon, or any other inert gas to avoid any subsequent contact with atmospheric oxygen.

Some of the liquid scintillators can not only be highly flammable but can also damage skin. Therefore extra care must be exercised while handling such liquids. These problems are however do not make liquid scintillators any less desirable for detectors than their solid counterparts. In fact a number of extremely sensitive large volume liquid scintillator detectors are already in operation and others are being planned (see, for example (9; 1)).

One of the features of liquid scintillators is their high dependence on the relative densities of different solvents in the base liquid. Generally the amounts of primary and secondary fluors are selected according to the application requirements. Since the properties of organic liquid scintillators vary considerably from one type to another and from one fluor concentration to another therefore it is not possible to assign any typical values to them. Table.6.2.2 lists properties of some on of the organic liquid scintillators that have been investigated. The reader should, however, be warned that these values should not be taken to represent most organic liquid scintillators.

Table 6.2.2: Wavelength of maximum emission λ_{max} and decay constant τ of some organic liquid scintillators (39).

Liquid	λ_{max} (nm)	τ (ns)
Toluene with 5 gm/l of p-terphenyl	355	2.2
Toluene with 5 gm/l of p-terphenyl and		
0.02 gm/l of 2-(1-naphthyl)-5-phenyloxazole	415	≤ 3.2
Toluene with 3 gm/l of 2,5-diphenyloxazole	382	≤ 3.0
Phenylcyclobexane with 3 gm/l of p-terphenyl	355	< 2.9
Phenylcyclobexane with 3 gm/l of p-terphenyl and		
0.01 gm/l of 1,6-diphenyl-1,3,5-hexatriene	450	< 8.0

6.2.D   Crystalline Scintillators

Organic scintillating crystals have been found to be advantageous over their inorganic counterparts due to the following two reasons.

▶ Non-hygroscopicity

▶ Small back-scattering

We will see later that absorption of moisture from atmosphere is one of the major problems of inorganic crystalline scintillators. Organic crystals do not have this vulnerability to moisture and therefore do not required to be stored and used in sealed containers. The other good thing is that organic crystals are mostly made of hydrogen, which makes the probability of backscattering extremely small. Also presence of dense hydrogen makes them suitable for use for neutron spectroscopy.

In the following we will look at some commonly used crystalline organic scintillators and discuss their properties and applications.

D.1 Anthracene ($C_{14}H_{10}$)

Anthracene is perhaps the most widely used organic scintillator crystal. Its wide popularity has actually made it a standard for comparing the properties of other scintillators. The light yield of scintillators is sometimes given relative to the light yield of anthracene. The absolute efficiency of anthracen, as we saw earlier during the discussion on light yield of scintillators, is about 5% for blue light. Since the light output of NaI is superior than this, therefore some researchers prefer to compare all scintillators with respect to NaI rather than $C_{14}H_{10}$. However the standard practice is to compare organic scintillators with respect to anthracene and inorganic scintillators with respect to NaI.

The peak of the pure anthracene light spectrum fall at the blue light, which has long attenuation length in the crystal. Therefore with carefully produced crystals one can have very good quantum efficiency. In fact, it has been observed that for UV photons, its quantum efficiency approaches unity. The light spectrum can be shifted by addition of a suitable impurity or more accurately a wavelength shifter in the bulk of the material. For example addition of small amounts of naphthacene shifts the spectrum such that the peak occurs for green photons (see Fig.6.2.10). To remind the reader, the wavelength shifter is also a scintillator that absorbs the photons of the primary scintillator and emits photons having a different wavelength.

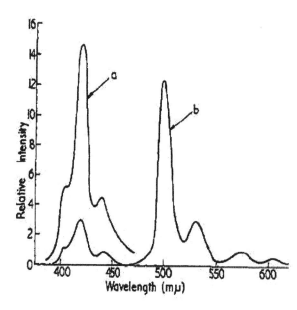

Figure 6.2.10: Output light spectra of pure (a) and doped anthracen (b). The doping was done with 10^{-4} mole fraction of naphthacene (11).

Anthracene, being one of the earliest discovered detector-ready crystals has undergone extensive studies related to effects of radiation. The fact that it has seen a number of applications in hostile radiation environments has more with hygroscopicity of inorganic scintillators to do than its radiation resistance. As can be seen in Fig.6.2.11 the light yield of anthracene whether doped or not decreases significantly with integrated dose.

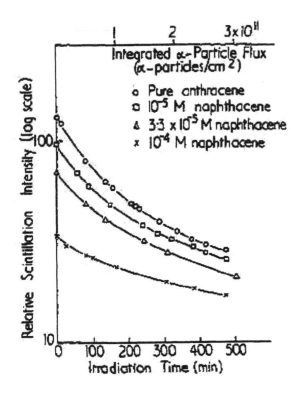

Figure 6.2.11: Relative light yield of pure and doped anthracene with respect to irradiation time and integrated flux of α-particles (11).

D.2   P-Terphenyl ($C_{18}C_{14}$)

The most desirable characteristic of p-terphenyl crystal is its very short decay time, making it suitable for spectroscopy of high intensity radiation. However the light yield of this crystal is quite bad. This shortcoming is overcome by doping the material with some suitable impurity, which has been shown to increase the light output to up to 5 times of pure $C_{18}H_{10}$.

D.3   Stilbene ($C_{14}H_{12}$)

Like anthracene, stilbene has also been thoroughly investigated for its possible use for radiation detection. Although its light output is not as good as anthracene, still it has found some applications in detection of heavy charged particles. The light yield of stilbene has been compared with that of anthracene in Fig.6.2.12. Note that the abscissas in the figure have been reduced by MZ^2, with M and Z being the mass and charge of the incident particles. This has eliminated the dependence of light yield on the type of charged particle on the scale of the graph.

6.3   Inorganic Scintillators

Most of the inorganic scintillators have crystalline structures. These materials are generally more dense and have higher atomic number than organic scintillators. This makes them attractive in applications where high stopping power for the incident

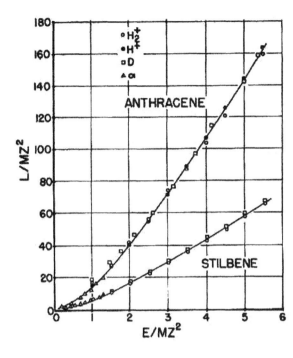

Figure 6.2.12: Light yield of anthracene and stilbene to different charged particles (40). The absicssas have been reduced by MZ^2, with M and Z being the mass and charge of the incident particles.

radiation is desired. Another advantage is their higher light output than organic scintillators.

6.3.A Scintillation Mechanism

A.1 Exciton Luminescence

When ionizing radiation passes through an inorganic scintillator crystal it produces electron hole pairs. Depending on the energy departed to the molecule the pair can become free to move around or remain *partially* bound by the Coulomb attraction. If the energy is high enough for the electron to jump to the conduction band then it becomes essentially free. In such a case the hole, having an effective positive charge, becomes free to move around as well. However if the energy departed is not that high then the electron becomes partially bound to the hole. This can be viewed as jumping of the electron to a band just below the conduction band (see Fig:6.3.1). The bondage between the electron and hole is not very strong, though, since a small amount of additional energy transferred to the electron can elevate it to the conduction band. In this bound state, the electron and hole are said to form a system called *exciton*. Exciton has an interesting property that it can move around in the material as an entity and can get trapped by an impurity or a defect site. If this site constitutes a luminous center, it can lead to the emission of scintillation photons when the electron falls into the lower level. This process, generally known as self trapping of excitons, is graphically depicted in Fig.6.3.1. The trapping of excitons can also occur through another process called *charge transfer*, in which case the luminescence is called charge transfer luminescence.

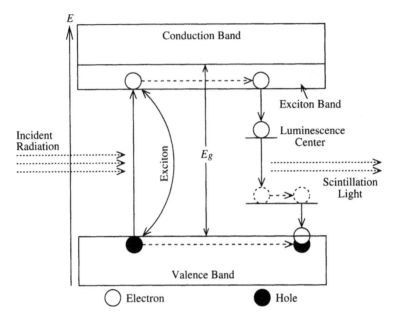

Figure 6.3.1:   Principle of exciton luminescence from a partially bound exciton in an inorganic scintillator. If the electron moves up to the conduction band, the exciton thus created will be free to move around. In either case the electron can eventually get trapped by a luminescence center through self trapping or charge transfer process.

A.2   Dopant Luminescence

Some scintillators are loaded with an impurity or *dopant* to enhance its scintillation properties. If an electron gets trapped in a dopant level and from there falls into the lower luminescence level, scintillation light is emitted. The process is shown in Fig.6.3.2.

A.3   Core Valence Band Luminescence

If the incident radiation deposits enough energy along its track into the lattice, it can elevate electrons from the deep core valence band to the conduction band (see Fig.6.3.3). An electron leaving the core valence band leaves behind a vacancy or hole. To stabilize the system a valence band electron quickly fills this vacancy. This results in the emission of scintillation light. The process is generally known as core valence band luminescence.

6.3.B   Radiation Damage

Let us now discuss the effects of radiation on inorganic scintillators. Although in general the properties of inorganic scintillators deteriorate with absorbed dose but special materials have been developed that show significant radiation hardness.

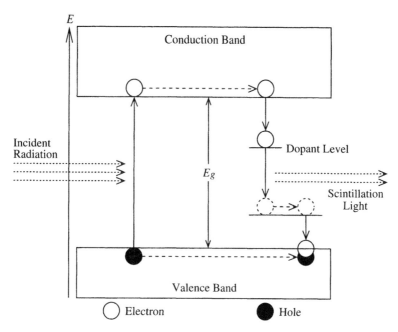

Figure 6.3.2: Principle of dopant luminescence in an inorganic scintillator.

A common example of radiation vulnerable detector is cesium iodide. CsI has been extensively used in high energy physics experiments and its radiation damage properties have been well observed. For example Fig.6.3.4 shows the decrease in gain of CsI based detector after irradiation. Such a significant decrease in gain is not specific to this material only as other materials such as $PbWO_4$ have also shown similar behavior.

Lead tungstate $PhWO_4$, as we will also see later, is one of the inorganic scintillators that have been found to possess significant radiation resistance. However even with such a relatively high degree of radiation hardness, $PhWO_4$ is not completely free from harmful effects of large integrated doses. For plastic scintillators we saw that the dose integrated over a period of several months changes the transmission properties of the detectors. This has also been observed with CWO under high irradiation (see Fig.6.3.5).

6.3.C Some Common Inorganic Scintillators

Some inorganic scintillators and their properties are listed in Table.6.3.1. The two properties that attract most attention of detector developers are the wavelength of maximum emission λ_{max} and the light yield dN/dE. The choice of λ_{max} is mainly driven by the efficiency of the photon counting detector, which is wavelength dependent. for example in photomultiplier tubes the efficiency of the photocathode depends on the wavelength of incident photons. This implies that a good match between the wavelength of scintillation photons and efficiency of the photocathode is necessary to ensure high overall efficiency and good signal-to-noise ratio. In practice,

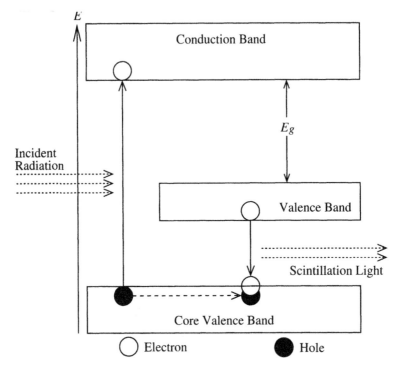

Figure 6.3.3: Principle of core valence band luminescence in an inorganic scintillator.

however, the choice is generally a compromise since a scintillator that is suitable due to some other properties may not have λ_{max} for which the photocathode has highest efficiency. Light yield is certainly another factor specially if the detector is to be used in a low radiation environment.

In the following subsections we will list the advantages and disadvantages of some inorganic scintillators that are commonly used in radiation detectors.

C.1   Thallium Doped Sodium Iodide (NaI:Tl)

Sodium iodide doped with thallium is the most widely used scintillator. Its key features include

▶ high light yield (41,000 photons per MeV),

▶ blue emission that coincides with the requirement of most PMTs,

▶ very low self absorption of scintillation light,

▶ good spectroscopic performance,

▶ easy availability and low production cost, and

▶ possibility to produce large area crystals.

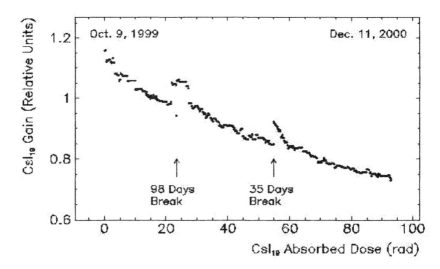

Figure 6.3.4: Decrease in gain of a detector based on undoped CsI (19).

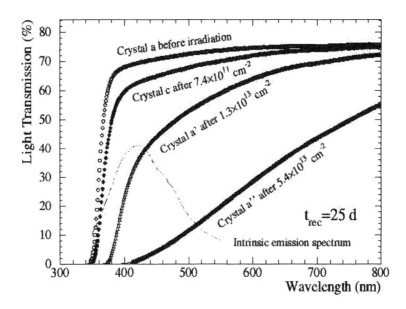

Figure 6.3.5: Effect of proton irradiation on light transmission in $PbWO_4$ (26).

The biggest problem with $NaI{:}Tl$ is its vulnerability to moisture. Because of this hygroscopicity it must be used in a sealed assembly. This, however, is not a bottleneck for using it in real systems since the size of the detector is usually small and the light can be guided through optical fibers.

Table 6.3.1: Density ρ, wavelength of maximum emission λ_{max}, decay time τ, and light yield dN/dE of some commonly used inorganic scintillators ((10) and other references therein).

Crystal	ρ $(g\ cm^{-3})$	λ_{max} (nm)	τ (μs)	$(\times 10^3)dN/dE$ MeV^{-1}
$NaI{:}Tl$	3.67	410	0.23	41
$CsI{:}Tl$	4.51	550	0.8-6	66
$CsI{:}Na$	4.51	420	0.63	40
$LaCl_3{:}Ce$	3.86	330	0.025(65%)	46
$LaBr_3{:}Ce$	5.3	358	0.035(90%)	61
$Bi_4Ge_3O_{12}$(BGO)	7.1	480	0.30	9
$CaHfO_3{:}Ce$	7.5	390	0.04	10
$CdWO_4$(CWO)	7.9	495	5	20
$YAlO_3{:}Ce$(YAP)	5.5	350	0.03	21
$LuAlO_3{:}Ce$(LuAP)	8.3	365	0.018	12
$Lu_2Si_2O_7{:}Ce$(LPS)	6.2	380	0.03	30
$Lu_2SiO_5{:}Ce$(LSO)	7.4	420	0.04	26
$Gd_2SiO_5{:}Ce$(GSO)	6.7	440	0.06	8
$Gd_2O_2S{:}Pr$(UFC)	7.3	510	3	50
$Gd_2O_2S{:}Tb$	7.3	545	1000	60

$NaI{:}Tl$ has been used in a variety of applications including medicine, physics, and environmental science. In fact, it has been termed as the standard scintillation material for a long time. Since, as we saw earlier, the properties of scintillators can be changed by varying doping levels therefore $NaI{:}Tl$ is usually tuned according to the application.

C.2 Sodium Doped Cesium Iodide ($CsI:Na$)

$CsI : Na$ is also one of the most commonly used scintillators. Its distinguishing features include

▶ high light yield (40,000 photons per MeV),

▶ blue emission that coincides with the requirement of most PMTs, and

▶ less hygroscopicity as compared to $NaI:Tl$.

C.3 Thallium Doped Cesium Iodide ($CsI:Tl$)

$CsI:Tl$ is not as widely used as the other two we studied earlier but its following features make is desirable in certain applications.

▶ High absorption efficiency.

▶ High light yield (66,000 photons per MeV).

▶ 550 nm emission that coincides with the requirement of most photodiodes, thus eliminating the need to use bulky and mechanically unstable photomultiplier tubes.

▶ Non-hygroscopic.

▶ Mechanically stable and shock resistant.

▶ Can be cut and shaped as required.

▶ Resistant to radiation induced damage.

Photodiodes are semiconductor detectors that are now beginning to replace the PMTs. We will learn more about them later in the Chapter. Their main advantage is that they do not require very high potentials as PMTs and can also be used in high magnetic fields. $CsI : Tl$ is then a good alternative over more commonly used scintillators in applications where using PMTs is difficult or the radiation field is very high.

C.4 Bismuth Germanate (BGO)

The chemical composition of BGO is $Bi_4Ge_3O_{12}$. Its advantages include

▶ high absorption efficiency,

▶ high energy resolution,

▶ short decay time

▶ high radiation resistance,

▶ large crystals can be produced, and

▶ mechanically stable and strong.

Because of its high γ-ray absorption efficiency, BGO is commonly employed in applications involving γ-ray spectroscopy.

C.5   Cadmium Tungstate (CWO)

$CdWO_4$ or simply CWO is not as widely used as GBO even though it has the following desirable properties.

▶ Significantly higher light yield than BGO.

▶ Low intrinsic background.

▶ Low afterglow.

The main problem with CWO is its large decay time, which makes it unsuitable for most applications. However it has been successfully used in low radiation spectroscopic applications such as spectrometry of very low activity radioactive substances.

C.6   Lead Tungstate (PWO)

$PbWO_4$ or simply PWO, with the following properties, is generally used in high radiation fields.

▶ Fast response.

▶ High resistance to radiation induced damage.

▶ Fast decay time, and

▶ Very low radiation length.

Lead tungstate is highly suitable for high radiation fields due to its extremely fast response and radiation hardness. It is mechanically stable and can be cut and shaped according to requirements. Its research and development has been mostly geared towards applications in high energy physics.

C.7   Cerium Doped Gadolinium Silicate (GSO)

$Gd_2SiO_5{:}Ce$ or simply GSO has the following characteristics.

▶ Fast response.

▶ Good temperature stability.

The main drawback of GSO is its relatively low light yield (8,000 photons per MeV) while its biggest advantage is its high temperature stability. Because of this it can be used in environments where small fluctuations in temperature can not be avoided.

C.8   Cerium Doped Lutetium Aluminum Garnet ($LuAG{:}Ce$)

The chemical composition of $LuAG : Ce$ is $Lu_3Al_5O_7{:}Ce$. Its advantages include

▶ fast pulse decay.

▶ emission wavelength suitable for most photodiodes,

▶ high density,

► chemically stability,

► high mechanical stability, and

► temperature stability.

Because of its mechanical stability, $LuAG{:}Ce$ can be used in imaging applications and has shown good spatial resolution. It has also been successfully used in PET scanners.

C.9 Cerium Doped Yttrium Aluminum Perovskite ($YAP{:}Ce$)

$YAlO_3{:}Ce$ or simply $YAP{:}Ce$ is one of the few scintillators that have properties desired for imaging applications, such as

► fast response,

► high mechanical stability,

► chemical stability, and

► low secondary emission.

The high mechanical stability of this material has been exploited by many researchers to design and develop high precision imaging detectors.

C.10 Liquid Xenon

Xenon is an inert gas having good scintillation properties. It is one of the few liquid scintillation materials that have been successfully used in detectors. Some of its properties relevant to its use in radiation detectors are listed below.

► **High density and high atomic weight:** Liquid xenon has much higher density ($2.98\ g\ cm^{-3}$ than the organic liquid scintillators. Its atomic weight is also much higher (54) than the effective atomic weight of the organic liquid scintillators. These two factors make it more aggressive in terms of stopping the incident radiation quickly, thus generating more scintillation light.

► **Light Yield:** The light yield of liquid xenon approaches that of NaI scintillator.

► **Response:** The response of xenon to radiation is quite fast which makes it suitable for use in timing applications.

► **Transparency:** Most of the scintillation photons emitted by liquid xenon are in the ultraviolet region. For these photons the liquid is highly transparent because the energy of such a photon lies well below the excitation energy of xenon.

The main disadvantage associated with using liquid xenon as scintillation medium is the dramatic degradation of its optical properties due to dissolved contaminants. During the discussion on organic liquid scintillators we saw that small amounts of water and oxygen can deteriorate their optical properties considerably. The same is true for liquid xenon. It has been observed that water or oxygen contamination at a few parts per million level can substantially reduce its transparency to the ultraviolet scintillation photons (4). This implies that a liquid xenon based detector can only be reliably operated if the purity of liquid xenon at all times is maintained. This points to the operational difficulty of such detectors since it involves not only regular purification but also continuous monitoring. Certainly for small volume detectors in which the highly purified liquid xenon can be kept in an air tight container, the contamination from the environment is not an issue.

The scintillation mechanism in xenon is a multi-step process as described below.

1. **Ionization:** This is the first step in which the incident radiation ionizes the xenon atoms through

$$\gamma + Xe \rightarrow Xe^+ + e, \tag{6.3.1}$$

   where a γ-ray photon is shown to knock an electron off of the xenon atom.

2. **Molecule Formation:** The xenon ion created during the ionization process attracts a neutral xenon atom to form a xenon molecular ion.

$$Xe^+ + Xe \rightarrow Xe_2^+ \tag{6.3.2}$$

3. **Electron Capture:** The xenon molecular ion captures a free electron, which not only dissociates the xenon atoms but also leaves on of the atoms in a highly excited state.

$$Xe_2^+ + e \rightarrow Xe + Xe^{**} \tag{6.3.3}$$

4. **De-excitation:** The highly excited xenon atom releases some of its energy to the liquid in the form of heat, which does not completely de-excites it.

$$Xe^{**} \rightarrow Xe^* + \text{heat} \tag{6.3.4}$$

   Here the $(*)$ represents the excited state.

5. **Excited Molecule Formation:** The excited xenon atom attracts a neutral xenon atom to form a xenon molecule in an excited state.

$$Xe^* + Xe \rightarrow Xe_2^* \tag{6.3.5}$$

6. **Radiative Decay:** This is the last step of the scintillation process. Here, the excited xenon molecule produced in the earlier step dissociates into two neutral atoms and as a result emits a scintillation photon in the ultraviolet region of the electromagnetic spectrum.

$$Xe_2^* \rightarrow 2Xe + \gamma_{uv} \tag{6.3.6}$$

6.4 Transfer of Scintillation Photons

The scintillators we studied in the previous sections produce photons of certain wavelengths that normally fall within the visible region of the electromagnetic spectrum. To detect these photons, photodetectors are used. We will study different types of photodetectors later on in the Chapter. First we will see how the scintillation photons are transferred from the scintillator to the photodetector.

The simplest way to do this is to directly attach a scintillator to a photodetector having an area greater than the scintillator to avoid loss of photons. However, the geometries of scintillators and photodetectors normally do not fulfill this requirement and the situation is the other way round. That is the scintillators have areas larger than the photodetectors. Hence building a detector in such a configuration is generally not possible. A more practical approach is to use a *light guide* to connect the photon emitting surface to the photodetector such that the scintillation photons reach the photodetector with minimal loss.

A light guide can be constructed in different configurations but all of them are based on the principle of reflection. Their main purpose is to guide the photons from the scintillator to the photodetector, hence the term *light guide*. Light guides can be constructed to work on simple reflection as well as total internal reflection. Since simple reflection is associated with loss of light intensity due to refraction and absorption therefore the detectors based on such a device lack good precision and sensitivity. On the other hand the devices based on total internal reflection offer minimal photon loss and consequently better sensitivity and precision. However it should be noted that it is impossible to build a light guide that guarantees total internal reflection at all angles. Therefore sometimes another material is used in conjunction with such a light guide to guide the refracted photons towards the photodetector.

A number of materials have been identified with good photon transmission properties, some of which are listed below.

► Glass and plexiglass

► Fiberglass

► Clear plastic

► Liquid

All these materials have been used to construct light guides but optical fiber is the most popular and, in most instances, the best choice. The good thing about optical fiber is that it offers a fair bit of flexibility in arranging the photodetector with respect to the scintillator.

6.4.A Types of Light Guides

The basic function of a light guide is to transmit scintillation photons to the photodetector with minimum loss. A device that performs this function, no matter on what principle it is based, can be called a light guide. The task of designing and constructing such a device can be fairly involved. This holds specially if the geometries of the scintillator and the photodetector differ considerably from each other.

In the following sections we will discuss different types of light guides that are commonly used in radiation detectors.

A.1 Simple Reflection Type

A simple reflection type of light guide can be constructed such that both scintillator and photodetector are enclosed in a vacuumed container whose inside surface is highly smooth and polished (see Fig.6.4.1). The photons that travel straight from the scintillator to the photodetector do not undergo any attenuation. Since a scintillator can emit photons in all directions therefore some of them will hit the walls of the container and get reflected. The smaller the angle of incidence the higher the probability that the photon will again hit the opposite wall of the container. After one or more such reflections the photons reach the photodetector and get counted. In principle, such a light guide should work quite efficiently. However, in reality these devices are quite inefficient. There are several reasons why simple reflection type light guides do not work well. First of all it is practically extremely difficult to ensure a very high degree of smoothness as any deviation would let the photons wander around more and even get absorbed by the material. Secondly the simple reflection is always associated with some degree of absorption by the material. Therefore the intensity of light emitted by the scintillator is bound to decrease as the light travels down the light guide.

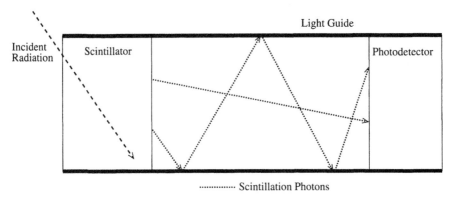

Figure 6.4.1: Transmission of scintillation photons through a simple reflection type light guide to a photodetector.

A.2 Total Internal Reflection Type

The light guides based on total internal reflection are the most popular amongst radiation detector technologists since they offer lower loss of light intensity and hence better sensitivity and superior resolution of the detector. The basic idea behind such a device is to ensure that the process of total internal reflection is guaranteed at all angles of incidence. The basic condition for total internal reflection to occur is

$$\sin \theta_i \geq \frac{n_m}{n}, \tag{6.4.1}$$

where θ_i is the angle of incidence (see Fig.6.4.2), n_m is the refractive index of the medium outside the waveguide (usually air), and n is the refractive index of the light guide. If the detector is kept in air then $n_m = 1$, which reduces the above condition to

$$\sin \theta_i \geq \frac{1}{n}. \tag{6.4.2}$$

For the generally used light guides having $n \approx 1.5$ for visible light, this gives

$$\theta_i \geq 41.8^0. \tag{6.4.3}$$

This shows that the photons striking the wall of the light guide at an angle less than 41.8^0 will suffer some loss. The light guide should then be constructed such that all probable incident angles are greater than this value. This implies that the geometry of a light guide plays a crucial role and should therefore be carefully designed.

To demonstrate the use of condition 6.4.2 in constructing a practical light guide, let us have a look at one of the commonly used light guide geometries, the so called *fish tail*. Fig.6.4.2 shows how a scintillator can be connected with a photodetector through such a light guide. The most widely used photodetectors are photomultiplier tubes, which usually have rounded entrance windows. Therefore one end of the fish-tail light guide is made round shaped. The other end that is connected to the scintillator is usually flat and thin to fit on the flat edge of the scintillator. The photons from the scintillator enter the light guide and travel outwards in straight lines. Obviously the ones that are in the *line of sight* of the photodetector, are captured by the detector most efficiently. However since the scintillation photons are emitted in all directions therefore some of them also hit the outer surface of the light guide. If the condition of total internal reflection is fulfilled, these photons get reflected from the surface. After one or more such reflections the photons eventually reach the photodetector.

Equation 6.4.3 implies that the light guide must be tapered at an angle such that the angle of incidence is always greater than 41.8^0. However the choice tapering angle θ_t does not depend only on this criterion. To minimize the loss of light due to reflections from the surface it is always desirable that the photons make minimum number of total internal reflections before reaching the photodetector. Also the angle at which the light enters the photodetector might be of significance for good photon collection efficiency. As shown in Fig.6.4.2, this angle can be determined by simple geometric considerations. Here we have used the fact that the angles of incidence and reflection are equal. Adding the angles of the triangle we get

$$\frac{\pi}{2} - \theta_p + \frac{\pi}{2} + \theta_t + \frac{\pi}{2} - \theta_i = \pi.$$

Rearrangement of this equation gives us the required dependence of θ_p on the angle of incidence and the tapering angle.

$$\theta_p = \frac{\pi}{2} - \theta_i + \theta_t \tag{6.4.4}$$

Let us now apply the condition of total internal reflection, that is equation 6.4.2, to the above relation. This gives

$$\sin\left(\frac{\pi}{2} + \theta_t - \theta_p\right) \geq \frac{1}{n}. \tag{6.4.5}$$

This simple relation can be used to determine the lower bound on the tapering angle required for a certain θ_p (see Example below).

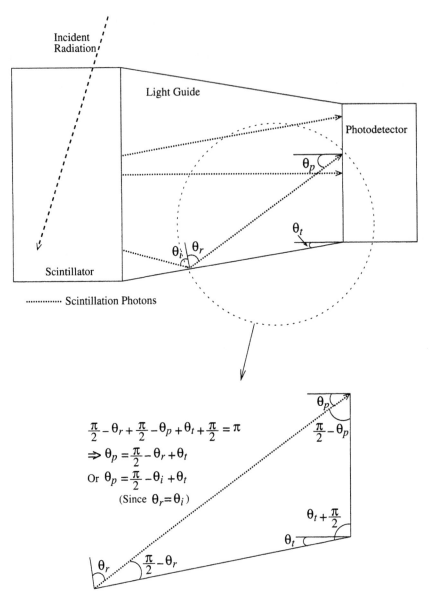

Figure 6.4.2: Transmission of scintillation photons through a fish-tail light guide to a photodetector. Also shown is the derivation of θ_p as a function of the angle of incidence θ_i and the tapering angle θ_t.

Example:
The scintillation material of a radiation detector is to be connected with a photomultiplier tube through a fish-tail type light guide. Assume that the light guide has a refractive index of 1.5 and the system will be used in air.

Determine the minimum tapering angle of the light guide if the maximum angle of light incident on photomultiplier surface should not exceed 60^0.

Solution:
For $n = 1.5$, equation 6.4.5 reduces to

$$\frac{\pi}{2} + \theta_t - \theta_p \geq 41.8^0$$
$$\Rightarrow \theta_t \geq \theta_p - 48.2.$$

Substituting $\theta_p = 60^0$ in the above equation gives

$$\theta_t \geq 18.2^0. \tag{6.4.6}$$

Hence the light guide must be tapered at least at an angle of 18.2^0.

The fish-tail is not the only geometry used in practical systems. The geometry of a light guide is in fact application dependent. for example if the photodetector can not be placed in line of sight of the scintillator then the light guide must have one or more bends. In some applications the distance between the scintillator and the photodetector is so large and the path is so complicated that it becomes extremely difficult to design and construct a stiff structure to guide the photons. In such cases, optical fibers offer the required flexibility. Optical fibers have become a standard in telecommunication where it is desired to transmit electrical signals at large distances with extremely low attenuation and high bandwidth. Optical fibers fulfill both of these requirements but for transmission of light signals. Therefore the electrical signal is first transformed into an optical signal and then transmitted. At the receiving end the signal is transformed back into the electrical signal and further processed. Optical fibers work on the principle of total internal reflection and suffer from very low light leakage.

A.3 Hybrid Light Guides

It is extremely difficult, if not impossible, to design and construct a light guide based on total internal reflection that ensures perfect transmission of photons without any loss. That is, there are always photons that impact the light guide at angles that do not favor total internal reflection and therefore refract out of the light guide. for example in the previous section we looked at the fish-tail type light guide and derived an expression to determine the tapering angle for total internal reflection. However the practical considerations of the system, such as surface areas of the scintillator and the photodetector and distance between them, may not favor implementation of a light guide that would ensure perfect total internal reflection and all angles of incidence. Depending on the application, the loss of light intensity in the light guide can be significant and even intolerable. Since it is extremely difficult to retain each and every scintillation photon inside a light guide therefore the only option left is to use some other means to decrease the loss. A simple reflecting surface around the light guide can serve the purpose. Fig.6.4.3 shows two types of hybrid structures with fish-tail light guide as one element and a reflecting surface as the other. The easiest way to construct the reflecting surface is to paint a highly reflecting paint on the outer surface of the fish-tail light guide as shown in Fig.6.4.3(b). However one could also opt for enclosing the fish-tail light guide in a separate reflecting enclosure as depicted in Fig. 6.4.3(a).

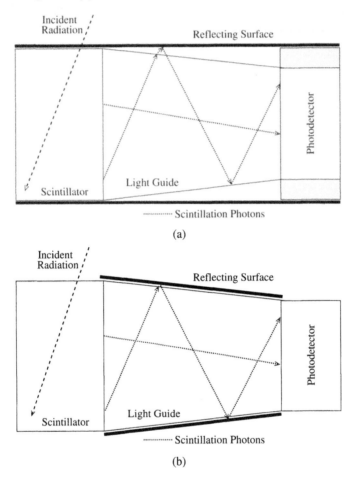

Figure 6.4.3: Two types of hybrid light guides. (a) A fish-tail light guide is enclosed in a container having highly polished reflecting inner surface. (b) The outer surface of a fish-tail light guide has been covered with a reflecting paint.

The surface of the reflecting part of a hybrid light guide must be extremely smooth and highly polished. Although the absorption of photons in such a surface can not be totally avoided but careful construction can help minimize the loss.

6.5 Photodetectors

The photons produced by scintillators can be detected by a number of means, the most notable of which are photomultiplier tubes (PMTs) and photodiodes (PDs).

Both of these detector types have their own pros and cons. for example photomultiplier tubes have sensitive mechanical structures that are prone to damage in mechanically unstable environments while the photodiodes are made of semiconductor materials and are therefore mechanically stable. On the other hand the response time of PMTs is much smaller than photodiodes and are therefore preferred in timing applications. Of course these are only two parameters and the choice of a photodetector is in reality a compromise between a number of factors including cost. We will learn about these factors as we discuss these two types of detectors in the following sections.

6.5.A Photomultiplier Tubes

Photomultiplier tubes are sensitive devices that are capable of converting light photons into a very large number of electrons. The basic building blocks of a complete PMT are a photocathode, an electron multiplication structure, and a readout electrode. The photons incident on the photocathode get converted into electrons through the process of photoelectric effect. The electron is then made to accelerate and strike a metallic structure called *dynode*, which results in the emission of more electrons. The newly produced electrons are again accelerated towards another dynode, where even more electrons are produced. This process of electron multiplication continues until the electrons reach the last dynode where the resulting current is measured by some electronic device. This process is graphically sketched in Fig.6.5.1.

Before we go on to the discussion of individual components of a PMT, let us have a look at the characteristics that make PMT based photodetectors desirable for radiation detection. The most important of these characteristics are listed below.

▶ High sensitivity.

▶ Good signal-to-noise ratio.

▶ Fast time response.

▶ Large photosensitive area.

The reader should bear in mind that here we are not comparing PMTs with photodiodes. With rapidly developing technology, new photodiode detectors are now being developed that match or even surpass these and other properties.

A.1 Photocathode

We learned about the process of photoelectric effect in Chapter 2 and saw that some particular materials emit electrons when they absorb photons of energy above a certain threshold. This threshold energy is called work function and is characteristic of the material. The material itself is called photocathode. If we assume that no energy is lost during this process then the energy of the emitted photon would simply be equal to the difference between the photon energy and the work function of the material, that is

$$E_e = \frac{hc}{\lambda} - \Phi, \tag{6.5.1}$$

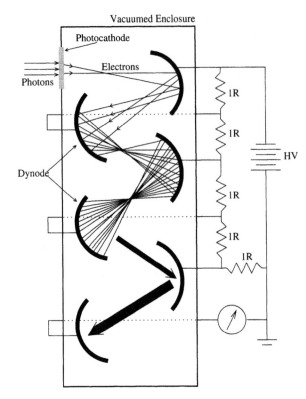

Figure 6.5.1: Working principle of a typical side-on type PMT. The light photons (e.g. coming from a scintillator) produce electrons through the photoemission process in the photocathode. These electrons are focused on to the first dynode where they produce secondary electrons, which then move towards the second dynode and produce more secondary electrons and so on. An actual PMT may contain 10 or more such dynodes. The amplified signal is measured at the final dynode or anode.

where λ is the wavelength of the incident light and Φ is the energy threshold or work function of the material. For a given material, this expression can be used to determine the maximum detectable wavelength (see example below).

Example:
A material having a work function of $2\ eV$ is used to convert photons into electrons, which are then detected by a photomultiplier tube. Compute the maximum wavelength of the photons it can convert into electrons.

Solution:
The maximum wavelength can be obtained by setting $E_e = 0$ in equation 6.5.1, which simply means that all of the incident energy has been used in the conversion process and the electron has not taken away any of the photon's

energy. Hence we have

$$0 = \frac{hc}{\lambda_{max}} - \Phi$$

$$\Rightarrow \lambda_{max} = \frac{hc}{\Phi}$$

$$= \frac{(6.63 \times 10^{-34})(2.99 \times 10^8)}{2 \times 1.602 \times 10^{-19}}$$

$$= 6.187 \times 10^{-7} \ m$$

$$= 618.7 \ nm.$$

Looking at equation 6.5.1 it is apparent that choosing a material with low work function is advantageous in terms of delivering energy to the outgoing electrons. However this is not the only criterion for selecting a photocathode for use in a PMT. for example if the PMT is to be used to detect scintillation photons then the most important factor is the conversion efficiency of the material at the most probably wavelength of the scintillation light. If the material does not have high enough efficiency at that wavelength then even a very low work function would not matter much. Since the realization of this problem, a number of extensive studies have been carried out to find the optimum photocathode materials at different scintillation wavelengths. Most of these efforts have gone into understanding the spectral response of the materials. By spectral response we mean the efficiency of the production of photoelectrons as a function of light wavelength. This efficiency is generally referred to as photocathode quantum efficiency and is simply defined as the ratio of the number of emitted photoelectrons to the number of incident photons, that is

$$QE = \frac{N_e}{N_\gamma}, \tag{6.5.2}$$

where N_e is the number of electrons emitted and N_γ is the number of incident photons. Quantum efficiency can also be expressed in terms of more convenient quantities, such as incident power and photoelectric current. To do that we first note that the incident power can be defined as

$$P_\gamma = n_\gamma h\nu, \tag{6.5.3}$$

where n_γ represents the number of photons of frequency ν incident on the detector per unit time. The expression for quantum efficiency can then be written as

$$QE = \frac{n_e}{P_\gamma / h\nu}$$

$$= \frac{I_{pe} h\nu}{e P_\gamma}. \tag{6.5.4}$$

Here n_e is the number of photoelectrons ejected per unit time and $I_{pe} = e n_e$ is the photoelectric current. This equation can be used to determine the photoelectric current for a particular value of the incident power (see example below).

Example:
A photocathode produces a current of 20 nA when exposed to light of wavelength 510 nm. Determine the photoelectric current if the wavelength of light is changed to 475 nm such that the incident power remains the same. Assume that the material has the same quantum efficiency at the two wavelengths.

Solution:
Equation 6.5.4 for photoelectric current can be written as

$$
\begin{aligned}
I_{pe} &= (QE)\frac{eP_\gamma}{hc}\lambda \\
&= K\lambda,
\end{aligned}
$$

where λ is the wavelength of the incident light and we have lumped together all the constant terms into one parameter K. This constant can be eliminated by writing the above equation for the two wavelengths and then dividing one with the other. Hence the required current is

$$
\begin{aligned}
I_{pe,2} &= I_{pe,1}\frac{\lambda_2}{\lambda_1} \\
&= 20\frac{475}{510} \\
&= 18.6 \ nA. \qquad\qquad (6.5.5)
\end{aligned}
$$

Fig.6.5.2 shows a typical spectral response curve of a photocathode. Here the quantum efficiency of the photocathode is plotted against wavelength of incident light. The interesting thing to note here is that the curve has a plateau where the variation in efficiency is not very large. As soon as one goes beyond this plateau on either side, the efficiency decreases rapidly. Therefore to build a good scintillation detector with a PMT, one should ensure that the spectrum of scintillation light has a peak somewhere in the middle of this plateau. There are some materials that have very short plateaus as well and using those in PMTs is generally not a good idea unless the scintillation spectrum is also narrow and has clearly defined peak that occurs at or near the peak of the spectral response curve.

As noted above the particular requirements of the system drive the choice of a photocathode in a PMT. Since there are a host of scintillation materials available having their unique scintillation spectra, therefore efficient detection of those photons with a PMT would require the use of photocathode materials with matching spectral response characteristics. Then there are also applications where the photons to be detected are not coming from a scintillator. That is, a PMT is a photodetector that can be used to detect photons no matter from where they are coming. for example the use of PMTs for detecting Cherenkov photons has recently gotten a lot of attention in highly sensitive large scale neutrino detectors. In essence, a PMT is a versatile photodetector that can be used to detect photons in virtually any environment provided it is equipped with a photocathode having the appropriate spectral response.

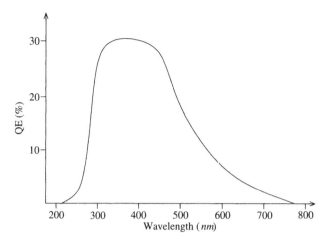

Figure 6.5.2: Quantum efficiency of a typical photocathode as a function of wavelength of incident photons.

Photocathodes can be used in essentially two different modes: transmission and reflection. In transmission mode the photocathode is semitransparent to allow transmission of photoelectrons (see Fig.6.5.3(a)). Such a photocathode is constructed by depositing a very thin layer of the material on the inside of the photon entrance window. Since most of the photoelectrons are emitted in the direction of travel of the incident photons therefore it is called transmission photocathode. Most PMTs are constructed with this type of photocathodes. There are also some photocathode materials that have high quantum efficiencies but very poor transmission properties. These *opaque* materials are used to construct the so called reflection photocathodes as shown in Fig.6.5.3(b). A reflection photocathode is made by depositing a thin layer of the material on a metal electrode inside the PMT.

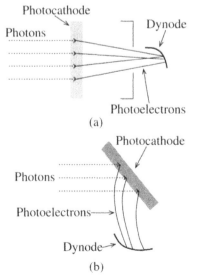

Figure 6.5.3: (a) Semitransparent photocathode used as a transmission photoemission device in a head-on type PMT. (b) Reflection type photocathode. Such photocathodes are generally used in circular type PMTs.

A host of photocathode materials have been identified with varying characteristics and spectral responses, some of which are listed below.

▶ *AgOCs*: This is one of the most widely used photocathode materials. It has a photoemission threshold of 1100 nm and a peak quantum efficiency at around 800 nm. The working range of this material is in the near infrared range of the photon spectrum. The main disadvantage of $AgOCs$ is its very low quantum efficiency, which has a peak of less than 1%. It is mainly used as a transmission photocathode.

▶ *GaAs(Cs)*: This material has a spectral response that ranges from ultraviolet to 930 nm. This broadband response makes it suitable for used with a wide range of scintillators without the need of a wavelength shifter. In most instances it is used in the transmission mode. Since the quantum efficiency of $GaAs(Cs)$ is temperature dependent with a peak at very low temperatures, it is sometimes operated at very low temperatures.

▶ *InGaAs(Cs)*: This material has greater sensitivity in the infrared range and higher signal-to-noise ratio in 900-1000 nm range than $GaAs(Cs)$.

▶ *SbCs₃*: This is one of the earliest used photocathode materials. It is still very popular amongst manufacturers as its spectral response ranges from ultraviolet to the visible region with a peak quantum efficiency of around 20%. The photoemission threshold of $SbCs_3$ lies at around 700 nm and has a peak at approximately 400 nm. Since it has very poor transmission capabilities therefore it is generally used as a reflection photocathode.

▶ *Bialkali Materials:* The bialkali materials such as $SbRbCs$ and $SbKCs$ are the most widely used of all photocathode materials due to their high sensitivities to blue light generated by NaI scintillators. The reader might recall that NaI is the most popular scintillator for radiation detection. The sensitivity to blue light is not the only reason for their popularity, though. These materials also have high quantum efficiencies with peaks of just less than 30%. Another advantage is their good stability at elevated temperatures. Some bialkali materials can be used at a temperature as high as 175^0C. A common bialkali material $SbKCs$ has a photoemission threshold of about 700 nm and maximum quantum efficiency of 28% at around 400 nm.

▶ *Multialkali Materials:* These materials have very wide spectral response ranging from ultraviolet to near infrared, making them highly suitable to be used with a number of different scintillators. Their main disadvantage is high thermionic emission of electrons even at room temperature and therefore external cooling is generally required. $NaKSbCs$ is a common multialkali.

▶ *CsTe, CsI*: These materials are sensitive to photons in ultraviolet region only and are therefore not very widely used.

A.2 Electron Focusing Structure

Since photocathodes have low quantum efficiencies therefore each electron produced is important and should be collected by the first dynode. This requires the use of an electron focusing structure to guide the photoelectrons to the dynode. The reason is that the photoelectrons are produced in all directions with varying energies and can easily go astray if they are not directed properly. An *ideal* electron focusing structure has the following two distinct characteristics.

▶ It directs the electrons from the photocathode to the first dynode such that all of them have the same transit time regardless of their initial energy. This is a very important requirement specially for tubes that are to be used in fast timing applications.

▶ It is able to focus all the photoelectrons produced in the photocathode.

Of course such strict requirements can be fulfilled only by an ideal focusing structure. However with a carefully designed structure one can achieve an electron collection efficiency of 80% or better. One thing that should be pointed out here is that not all PMTs have very stringent focusing requirements. Some PMT structures, as we will see later, have large first dynodes and are therefore able to collect most of the photoelectrons even without a focusing structure.

Focusing of electrons on to the first dynode depends on two factors: the geometry of the focusing structure and the electric field intensity in the space between the photocathode and the first dynode. The field intensity is generally made non-uniform and has a higher value near the photocathode to minimize the dependence of electron transit time on the initial electron velocity.

A.3   Electron Multiplication Structure

Earlier we saw that the quantum efficiency of a typical photocathode ranges between 10% and 30%. An efficiency of 10% means that for every 10 photons only one photoelectron is getting out of the photocathode. For low to moderate photon fluxes this is a problem since the resulting electron flux may not be sufficient to constitute a measurable current (see Example below). Hence to obtain a measurable signal and good signal-to-noise ratio the electron yield must somehow be increased. Now since the photocathode materials have their own physical and engineering limitations and can not be made more efficient, therefore the electron yield must be increased by some other process. This is accomplished in a PMT through the electron multiplication structure. The basic idea is to utilize the process of electron ejection from certain metals when they are bombarded by electrons. Such metals eject more electrons than the incident electrons provided the incident electrons have high enough energy. This implies that the electrons must first be accelerated as well. This may sound complicated but in actuality it can be done through a simple structure as shown in Fig.6.5.1. This PMT has 5 dynodes and a readout electrode. Each of these dynodes is kept at a higher potential than the preceding dynode to direct and accelerate the electrons towards itself. The result of these successive accelerations and secondary emissions is a cascade of electrons flowing down the dynode chain. Consequently the initial small number of photoelectrons gets multiplied into a very large number at the final electrode. To collect these electrons another metallic structure called *anode* is placed near the final dynode. The electron current at the anode is passed on to the associated electronics for further processing.

Example:
A NaI scintillator is bombarded by ionizing radiation that results in the deposition of 1.5 MeV of energy per second. The scintillation photons thus produced are detected by a PMT. Determine the current at the *first* dynode of this PMT. Assume that 15% of photons are lost before reaching the photocathode, which itself has a quantum efficiency of 20%. NaI produces about 40,000 photons per MeV.

Solution:
The number of electrons produced by the photocathode per unit time can be estimated from

$$N_e = (N_s)(\eta)(QE),$$

where N_s represents the number of scintillation photons produced per unit time, η is the efficiency of the process of transfer of photons to the photocathode, and QE is the quantum efficiency of the photocathode. The number of photons being produced per second in NaI is

$$N_s = (40,000)(1.5) = 60,000 \quad s^{-1}.$$

Hence the number of photoelectrons being produced per second is

$$
\begin{aligned}
N_e &= (60,000)(0.15)(0.2) \\
&= 1,800 \quad s^{-1}.
\end{aligned}
$$

Since each electron carries a unit electrical charge, the current at the first dynode is given by

$$
\begin{aligned}
I &= (1,800)\left(1.6 \times 10^{-19}\right) \quad Cs^{-1} \\
&= 2.9 \times 10^{-15} \quad A.
\end{aligned}
\tag{6.5.6}
$$

It is evident that it will be quite a difficult to measure this current of about 0.3 fA and also have a good signal-to-noise ratio. The reader should note that even when we started with a scintillator that had one of the highest light yields, the current at the first dynode turned out to be extremely small. Of course in this case the incident photon flux was also very low, which refers to the operation of the detector in low radiation fields. In high radiation fields the photoelectric current could be several tens of nanoseconds. In very low radiation fields, obtaining a measurable requires that the signal multiplication structure of the PMT has a high enough gain. for example a PMT having a gain of 10^4 would amplify the 0.3 fA signal to about 3 pA, which may still be too low for reliable measurements.

Every dynode in the electron multiplication structure does not have the same shape. for example the first dynode is specially designed to maximize the collection of photoelectrons that are already low in numbers. The structure of other dynodes depends on their particular locations. The focusing requirements for dynodes are also very stringent to avoid unnecessary loss of electrons. These requirements include not only proper designing of the dynodes but also their placement. Due to these issues the performance of a PMT is susceptible to mechanical jitters, which may lower the ability of one or more dynodes to focus or direct the electrons. Loss of electrons leads to signal deterioration and possibly non-linearity and therefore the sensitive mechanical structure of a PMT is one of its major disadvantages. However, as we will see later, some PMT designs are less prone to mechanical instabilities and should therefore be preferred if the environment is mechanically unstable.

The electron multiplication structure shown in Fig.6.5.1 is by no means a standard for PMTs. In fact, PMTs are manufactured in many different geometries. The choice of the geometry is mainly application dependent. In the following we will look at the most common of the PMT geometries and discuss their advantages and disadvantages.

▶ **Linear Focused Type:** This is perhaps the most famous type of electron multiplication structure (see Fig.6.5.4). Generally it is used in the head-on type of PMT. The distinguishing feature of this design is its extremely fast time response, which is achieved by positioning the dynodes to minimize the electron transit times. At each multiplication stage the electrons are focused by high electric field between the carefully placed dynodes, leading to a very fast response time. This design is susceptible to external magnetic field.

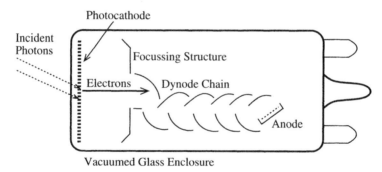

Figure 6.5.4:   Linear focused type PMT.

▶ **Box and Grid Type:** This is a somewhat simplified version of the linear focused type PMT. Such a structure is made of several quarter cylindrical dynodes installed in succession as shown in Fig.6.5.5. Since the collection area of the first dynode is larger than in the linear focused type therefore its focusing requirements are a bit relaxed. Although this simple design looks identical to the linear focused type but it lacks the progressive focusing of electrons between the dynodes. Hence it has slower response time and is therefore not a popular choice for most applications. Like linear focused type, this design also suffers from magnetic field dependent sensitivity.

▶ **Circular Cage Type:** This is a very compact structure with dynodes arranged in a circular fashion (see Fig.6.5.6). This type of PMT has a fast response time and is therefore suitable for use in timing applications. Although most of the circular cage type PMTs have side-on type structure but this geometry can be used to construct head-on type PMTs as well. The compactness of

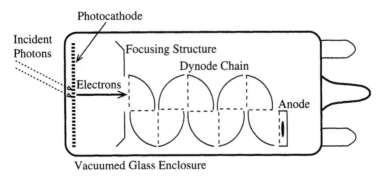

Figure 6.5.5: Box grid type PMT.

this structure demands high degree of precision in construction to achieve best performance.

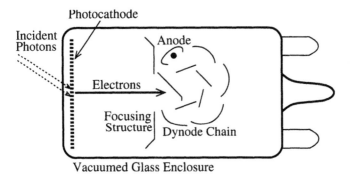

Figure 6.5.6: Circular cage type PMT.

▶ **Venetian Blind Type:** In the three types of PMTs we saw earlier the electrons were guided through a single path. This, of course, is not necessary as it is also possible to construct an electron multiplication structure where electrons can follow several different paths. Such a structure can minimize the loss of electrons and even simplify the design and construction. The venetian blind type of PMT is an example of such a design. As shown in Fig.6.5.6, the dynodes of this type of PMT are straight and make up an array that looks like a blind. In each row the dynodes are stacked parallel to each other and slanted at an appropriate angle with respect to the tube axis. A feature of this design is the larger area available to the electrons coming out of the photocathode. This simplifies the design considerably since it relieves the focusing requirements. The electrons can follow different paths as they travel down the dynode chain (see Fig.6.5.6). The biggest advantage of this design is the larger area it can cover as compared to the single-path type PMTs while its main disadvantage is the slow response time. The slow time response occurs due to the lower electric potential applied across the subsequent dynodes.

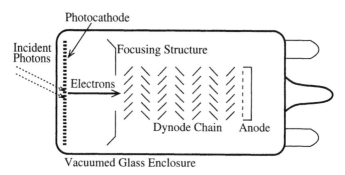

Figure 6.5.7: Venetian blind type PMT.

▶ **Mesh Type:** Just like the venetian blind type PMT, this structure also provides the electrons with different paths as they move down the dynode chain. In this structure, however, the dynodes are made of meshes that are stacked one after the other (see Fig.6.5.8). It has several advantages over other types such as position sensitivity. less susceptibility to external magnetic field, and good linearity.

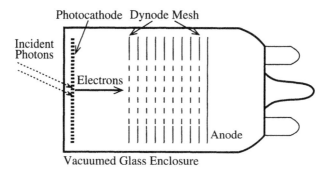

Figure 6.5.8: Mesh type PMT.

▶ **Metal Channel Type:** This structure is made of a large number of narrowly spaced dynodes providing fast electron transit capabilities and position sensitivity (see Fig.6.5.9).

▶ **Micro-Channel Plate (MCP) Type:** This structure consists of several million parallel electron multiplication structures in the form of micro glass tubes (see Fig.6.5.10(a)). The typical diameter of each of the glass tubes ranges from $5~\mu m$ to $25~\mu m$. As shown in Fig.6.5.10(b), each of the channels in this structure is capable of electron multiplication. The overall gain obtainable from a single micro-channel plate is on the order of 10^4. If higher gain is desired, two plates can be used in succession. The main feature of a PMT constructed with micro channel plates is its very fast time response and the possibility of position sensitivity. Another positive aspect of this design is low susceptibility to external magnetic fields.

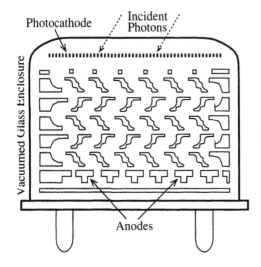

Figure 6.5.9: Metal channel type PMT.

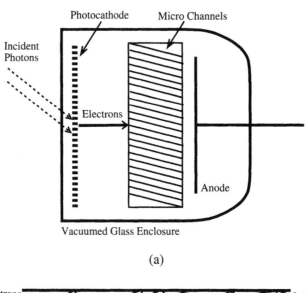

(a)

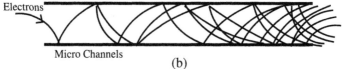

(b)

Figure 6.5.10:   (a) Micro channel type PMT. (b) Electron multiplication in a glass channel of MCP.

A.4 Voltage Divider Circuit

Proper focusing and acceleration of electrons at each stage of multiplication requires establishment of high potential gradient between the subsequent dynodes. We will see shortly that the electrons released by the last dynode are collected by the readout electrode called anode, which is kept at a higher potential than the last dynode. All the potentials to the dynodes and the anode can be provided from a single high voltage supply through a voltage divider circuit as shown in Fig.6.5.11.

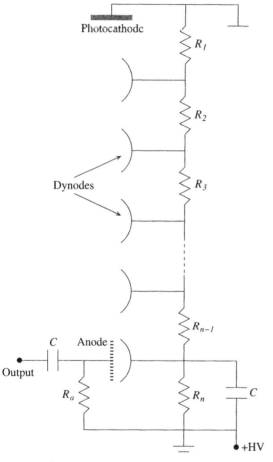

Figure 6.5.11: Typical voltage divider circuit of a PMT.

A.5 Electron Collection

The electrode that is used to measure the current is called anode. It is generally made up of a metallic grid and installed near the last dynode (see Fig.6.5.12). The spaces in the grid ensures that the electrons reach the last dynode without any appreciable collection by the anode. Since the secondary electrons from the last dynode are not accelerated towards another dynode therefore they form a space charge near the surface of the dynode. The gridded anode quickly collects these charges so that the linearity of the PMT response does not get compromised. One important factor to consider is that the geometry of the last dynode also plays an important part in efficient collection of electrons by the anode. There is also a potential gradient between the last dynode and the anode that directs the electrons towards the anode.

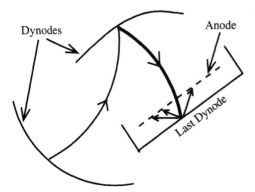

Figure 6.5.12: Sketch showing the collection of electrons by the gridded anode in a linear focused type PMT. The shape of the last dynode is different from the other dynodes to ensure efficient collection of charges by the anode.

A.6 Signal Readout

The output current at the anode of a PMT can be directly measured by a precision current measuring device. However the more convenient and generally used method is to convert this current into voltage and then pass this voltage through amplification stages to achieve good signal-to-noise ratio. The current to voltage conversion requires addition of a load resistance to the circuit as shown in Fig.6.5.14. In addition to this load resistance there is also some capacitance present at the output as represented by C_s in the figure. The value of the load resistance is generally chosen to be very large specially for low level current output. However, as we will see later when we discuss the frequency response of PMTs, its value can not be set arbitrarily high.

One practical aspect of the PMT readout is that, specially in high rate applications, the signal current at the last few dynodes may become much larger than the voltage divider current. This would force the last dynodes to draw the current from the voltage divider causing the interdynode voltage to change. Consequently the tube gain would change and the response of the tube would become nonlinear. The easiest way to overcome this problem is to provide a secondary source of current to the last one or two dynodes. Fig.6.5.14 shows such a circuit where the last two dynodes have been bypassed with a capacitor to ground.

A.7 Enclosure

The electron multiplication structure of a PMT is kept in a good vacuum to minimize the loss of electrons. The enclosure is generally made of glass but, in principle, can be made of any material. The entrance window, on the other hand, has strict requirements in terms of its transparency to the incident photons. The choice of the window material, therefore, is dependent on the type of photomultiplier tube, or more specifically, on the spectral range of detection of the tube. Different materials

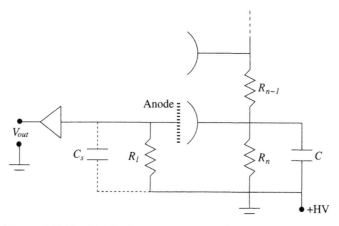

Figure 6.5.13: Typical output circuit of a PMT. The anode current is converted into voltage, which is measured across the load resistance R_l. C_s represents the combined stray capacitance due to the tube, output circuit, and the cables.

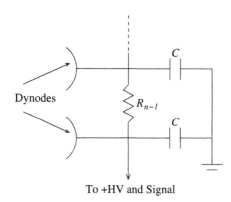

Figure 6.5.14: Bypass capacitors installed at the last two dynodes in a PMT act as a source of current and do not let the interdynode voltage to change.

have different cutoff wavelengths below which their transparency decreases to unacceptably low values. The definition of cutoff wavelength is somewhat arbitrary but generally a value at which the transmission falls to 10% is used. Fig.6.5.15 shows the typical transmission curves for fused silica, a material commonly used to construct PMT windows. The good thing about fused silica is that it has a very wide range in which the transmission remains almost constant. Another advantage is that it has low cutoff wavelength (160 nm).

Fused silica, though commonly used, is not the only material available with good transmission properties. Other materials of choice are lime glass, uv-glass, LiF, and sapphire.

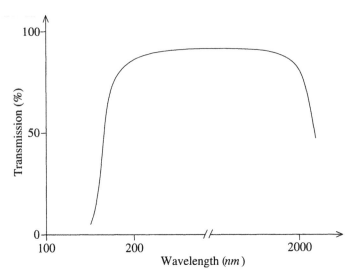

Figure 6.5.15: Typical transmission curve of a few mm thick fused silica.

A.8 Efficiency

There are two distinct processes that occur in a PMT before the electron multiplication starts. The first is the production of photoelectrons followed by their collection by the first dynode. The photoelectron production is independent of the transit of electrons towards the first dynode. Similarly the collection of charges by the dynode is independent of the physical processes taking place inside the photocathode. This implies that the efficiencies of these two processes are independent of each other. It was previously mentioned that the efficiency of the photocathode is generally termed as quantum efficiency or QE. The efficiency of collection of these electrons is called *collection efficiency* or simply CE. Although there are other processes involved that may decrease the overall efficiency of a PMT (such as collection of secondary electrons by subsequent dynodes and by the anode) but their effect is generally too small and the overall efficiency can be characterized by the product of QE and CE. These two efficiencies will be briefly discussed below.

Quantum Efficiency

Quantum efficiency defines the probability of production of photoelectrons in the photocathode. Practically it is measured by dividing the average number of photoelectrons emitted N_e by the average number of photons N_γ incident on the photocathode, that is

$$QE = \frac{N_e}{N_\gamma}. \tag{6.5.7}$$

The reader would recall that the number of photoelectrons produced by the incident photons depends on the energy of the photons. Or in other words the intensity or the flux of photoelectrons coming out of the photocathode depends on the wavelength of the incident photons. This implies that the quantum efficiency is also essentially dependent on the wavelength. This is a problem for the PMT manufactures who would very much like to assign a simple efficiency to the photocathode. Naturally it is very difficult, if not impossible, to measure the efficiency of the photocathode for each and every wavelength of interest. Therefore the practice is to either quote the quantum efficiency at a few wavelengths or use another measure instead. This other measure is the sensitivity of the PMT and will be discussed shortly.

Electron Collection Efficiency

The electron collection efficiency CE is a measure of how efficiently the photoelectrons are collected by the first dynode. Note that since the quantum efficiency of typical photocathodes is not very good, therefore the collection of each and every photoelectron is important. The collection efficiency is generally obtained by dividing the average number of photoelectrons collected by the first dynode N_e^{dy} by the average number of photoelectrons emitted by the photocathode N_e^{pc}, that is

$$CE = \frac{N_e^{dy}}{N_e^{pc}}. \tag{6.5.8}$$

Collection efficiency depends on many factors, such as shape of the photocathode, structure of the dynode, electric field profile in the space between the photocathode and the first dynode, and the orientation of the dynode with respect to the photocathode.

An interesting aspect of CE is that it can be improved by changing the mechanical structure and the electric field profile. In fact, the mechanical and electrical designs of the PMTs are optimized to ensure the best possible collection efficiency.

Overall Detection Efficiency

We just mentioned that the overall efficiency of a PMT can be approximated by multiplying the quantum efficiency with the collection efficiency. However some developers prefer to use a direct measure of the overall *detection* efficiency instead. The overall detection efficiency of a PMT is defined as the ratio of the average number of pulses counted by the measuring device to the average rate of incident photons.

$$\epsilon = \frac{N_c}{N_\gamma}, \tag{6.5.9}$$

where N_e represents the average count rate (average number of pulses counted per unit time) and N_γ is the average rate of incident photons. Note that this is a very useful quantity since it gives a direct measure of the efficiency of the whole system. The difficult part, however, is again the wavelength dependence of the photoelectric effect. Therefore, as with quantum efficiency, the overall detection efficiency is also quoted only at certain photon wavelengths. Another thing worth noting is the ease in determining this efficiency, since it can be done simply by illuminating the tube with a source of known strength and measuring the output response.

A.9 Sensitivity

Just like efficiency, sensitivity is also a measure to characterize the effectiveness of a PMT. In the following we will discuss different types of sensitivities that are generally quoted by PMT manufacturers.

Radiant Sensitivity

Radiant sensitivity is a measure of the sensitivity of the photocathode and is therefore closely related to the quantum efficiency we discussed earlier. Mathematically, radiant sensitivity is defined by the relation

$$S_r = \frac{I_{pe}}{P},\qquad(6.5.10)$$

where I_{pe} is the photoelectric current and P is the incident radiant power. Radiant sensitivity is also sometimes referred to as the PMT *responsivity* and is generally quoted in units of amperes per watt (A/W). Since the radiant sensitivity depends on the incident light wavelength therefore most PMT manufacturers prefer to quote it for the wavelength at which the sensitivity is maximum. However, the best method to compare different PMTs is to look at their spectral response curves. A spectral response curve shows the relationship between wavelength of the incident light and the quantum efficiency or spectral sensitivity of a tube. Unfortunately it is not practical for the manufactures to obtain such a curve for each tube they manufacture and therefore they normally quote the radiant sensitivity only at certain wavelengths together with some other sensitivity parameters that will be discussed shortly.

The reader would readily realize that the definition of radiation sensitivity is similar to the one of quantum efficiency. This is true since radiant sensitivity and quantum efficiency are related to each other through the relation

$$QE = \frac{1237 S_r}{\lambda}.\qquad(6.5.11)$$

where the wavelength λ is in units of nanometer (see the Example below for derivation).

Example:
Derive equation 6.5.11.

Solution:
We start with equation 6.5.10

$$S_r = \frac{I_{pe}}{P}.$$

The ionization current I_{pe} measured in time t is given by

$$I_{pe} = \frac{eN_e}{t},$$

where N_e is the number of photoelectrons and e is the unit electrical charge. The radiant power P absorbed by the photocathode in time t can be calculated from

$$P = N_\gamma \frac{hc}{\lambda}\frac{1}{t},$$

where N_γ represents the number of photons having wavelength λ. Substituting the expressions for I_{pe} and P in equation 6.5.10 yields

$$
\begin{aligned}
S_r &= \frac{eN_e/t}{N_\gamma hc/(\lambda t)} \\
&= \frac{N_e}{N_\gamma} \frac{e\lambda}{hc}
\end{aligned}
$$

Since N_e/N_γ is the quantum efficiency QE, we can write the above equation as

$$
\begin{aligned}
S_r &= QE\frac{e\lambda}{hc} \\
\Rightarrow QE &= \frac{S_r}{\lambda}\frac{hc}{e} \\
&= \frac{S_r}{\lambda} \frac{\left(6.626 \times 10^{-34}\right)\left(2.99 \times 10^8\right)}{1.602 \times 10^{-19}} \\
&\approx \frac{S_r}{\lambda}1237 \times 10^{-9}.
\end{aligned}
$$

If we take the wavelength λ in units of nanometer ($10^{-9}\ m$), the above equation can be written as

$$
QE = \frac{1237 S_r}{\lambda}.
$$

Both radiant sensitivity and quantum efficiency are interchangeably used to characterize the sensitivity of a photomultiplier tube. In practical photomultiplier tubes used in scintillation detectors, the incident light spectrum is very well known. Most scintillators produce light with a spectrum that peaks at either blue or green wavelength. Therefore radiant sensitivity can be effectively used to compare different photomultiplier tubes to be used with a particular scintillator.

Cathode Luminous Sensitivity

The cathode luminous sensitivity is defined as the average photoelectric current I_{pe} from the photocathode per incident photon flux Φ_γ from a tungsten filament lamp operated at a distribution temperature of 2856 K.

$$
S_{l,c} = \frac{I_{pe}}{\Phi_\gamma}. \tag{6.5.12}
$$

Anode Luminous Sensitivity

The definition of anode luminous sensitivity is similar to that of cathode luminous sensitivity. It represents the average anode current I_{anode} per incident photon flux Φ_γ from a tungsten filament lamp operated at a distribution temperature of $2856K$, that is

$$S_{l,a} = \frac{I_{anode}}{\Phi_\gamma}. \qquad (6.5.13)$$

Both cathode and anode luminous sensitivities are generally quoted in dimensions of amperes per lumen (A/lm).

Blue Sensitivity

Since most scintillators produce blue light therefore sometimes instead of the luminous sensitivity the manufacturers provide blue sensitivity. The blue sensitivity is defined as the average photoelectric current from the photocathode produced per unit flux of blue light. The blue light is conventionally produced by passing the light produced by a tungsten lamp operating at $2856K$ through a blue filter. Mathematically the blue sensitivity is given by

$$S_{blue} = \frac{I_{pe}}{\Phi_\gamma}. \qquad (6.5.14)$$

Blue sensitivity is usually expressed in amperes per lumen-blue (A/lm-b).

A.10 Gain

The gain of a PMT characterizes how effectively it amplifies the incident number of photons into a large enough signal at the output. It is generally described by the ratio of the anode current to the photoelectric current.

PMT gain mainly depends on two factors: the number of dynodes and the voltage applied to each of them. The gain of an individual dynode is described by the so called *secondary emission ratio*, δ. This ratio has been empirically found to be proportional to the potential at the dynode raised to a power between 0.7 and 0.8, that is

$$\begin{aligned}
\delta &\propto V_{dy}^\alpha \\
&= AV_{dy}^a, \qquad (6.5.15)
\end{aligned}$$

where V_{dy} represents the dynode voltage and α, having a usual value between 0.7 and 0.8, is a constant that depends on the geometric structure, orientation, and material of the dynode. A is the proportionality constant. The secondary emission ratio can be different for each of the dynodes in the multiplication structure of a PMT. However generally the difference is not very large and can be neglected. Assuming the δ to be the same for all n dynodes of the multiplication structure, the overall gain μ can be written as

$$\begin{aligned}
\mu &= \delta^n \\
&= \left[A V_{dy}^{\alpha} \right]^n \\
&= A^n \left[\frac{V}{n+1} \right]^{\alpha n} \\
&= \frac{A^n}{(n+1)^{\alpha n}} V^{\alpha n}.
\end{aligned} \qquad (6.5.16)$$

Here V is the supply voltage applied between the anode and the cathode. This voltage gets distributed to individual anodes through the voltage divider network (cf. Fig.6.5.1). Here we have assumed that all the resistors across the dynodes have the same value (hence we could use $V_{dy} = V/(n+1)$ in the above derivation). It is apparent from this equation that the overall gain is proportional to the voltage raised to the power αn, that is

$$\mu \propto V^{\alpha n},$$

For a tube having 10 dynodes, αn translates into a number between 7 and 8. The dependence of gain on voltage for a PMT having 10 dynodes as characterized by the above relation is shown in Fig.6.5.16. Typical photomultiplier tubes are operated between 1500 V and 2000 V with gains on the order of 10^5.

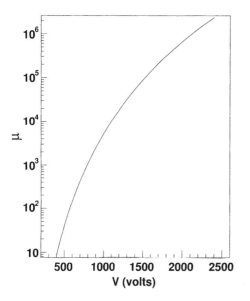

Figure 6.5.16: Variation of gain with applied voltage for a PMT having 10 dynodes.

Certainly the equation 6.5.16 is valid only for the case when all the dynodes have equal gains. In case, individual dynodes have different gains, something that is not uncommon in commercial PMTs, the overall gain should be computed from

$$\mu = \prod_{i=1}^{n} \delta_i, \qquad (6.5.17)$$

where δ_i denotes the gain of the ith dynode and $\prod$ stands for the product.

> **Example:**
> The multiplication structure of a PMT is composed of 10 dynodes. The gain of the first 4 dynodes is such that each of them produces 3 secondary electrons. The rest of the dynodes produce 4 secondary electrons. Compute the gain of the PMT.
>
> **Solution:**
> Given the gains of the first 4 dynodes as 3 and the remaining 6 dynodes as 4, the overall gain as calculated from equation 6.5.18 is
>
> $$\begin{aligned} \mu &= \left[\prod_{i=1}^{4}\delta_i\right]\left[\prod_{i=5}^{10}\delta_i\right] \\ &= \left(3^4\right)\left(4^6\right) \\ &= 3.3 \times 10^5. \end{aligned}$$

The strong dependence of gain on the applied voltage implies that the output of a PMT is highly susceptible to changes in the voltage. To quantify the variation of gain with change in applied voltage, let us differentiate both sides of equation 6.5.16. This gives

$$d\mu = \frac{A^n \alpha n}{(n+1)^{\alpha n}} V^{\alpha n - 1} dV. \tag{6.5.18}$$

The relation for the relative change in gain can be obtained by dividing this equation with equation 6.5.16. Hence we have

$$\frac{d\mu}{\mu} = \alpha n \frac{dV}{V}, \tag{6.5.19}$$

which can also be written as

$$\frac{\triangle \mu}{\mu} = \frac{\alpha n \triangle V}{V}. \tag{6.5.20}$$

For a PMT with 10 dynodes we have $\alpha n \approx 7$. For such a tube the relative change in gain would be about 7 times the relative change in applied voltage. Hence, if the applied voltage changes by 2%, the gain will change by about 14%. The effect is less dramatic for tubes constructed with lower number of dynodes. However, since the gain depends on the number of dynodes as well, the sensitivity of such a tube will also be lower. Fig.6.5.17 shows the dependence of the number of dynodes on gain for a PMT operated under the nominal voltage of 2000 volts.

A.11   Spatial Uniformity

In most applications it is desirable that the PMT is capable of covering as large an area as possible. This requires, above all, a large area photocathode. A uniformly built photocathode should have essentially same quantum efficiency throughout its entire coverage. Achieving a high degree of uniformity is, however, not an easy task

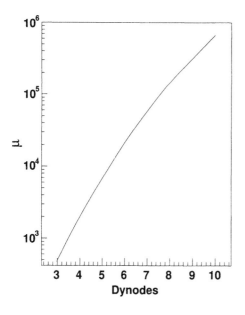

Figure 6.5.17: Dependence of gain on the number of dynodes in the multiplication structure of a PMT operated at 2000 volts.

specially at the ends of the photocathode. Furthermore the collection efficiency of the photoelectrons by the first dynode is usually not the same for the whole of the photocathode. The term *uniformity* is used to characterize the variations in the anode current when photons strike the photocathode at different positions. This uniformity is generally quantified by measuring the anode sensitivity to directional light incident on different points on the PMT entrance window. Fig.6.5.18 shows typical PMT sensitivity curves for head-on and side-on type PMTs. The response of head-on type PMT is spatially more uniform than the side-on type PMT. The sensitivity of side-on type PMT to light incident at and around the center of the photocathode is maximum and decreases to about 50% at the upper and lower corners. In contrast the head-on type PMT shows an abrupt decrease to almost zero near the corners of the upper face of the phototube. The anode sensitivity for the middle region, however, is fairly uniform. It is therefore recommended that the system is designed such that most of the incident photons strike the middle part of the PMT entrance window.

A.12 Time Response

In certain applications the time response of a PMT becomes extremely important. for example in Cherenkov detectors the timing information is used to construct the event vertex, that is the point of origination of the Cherenkov light. In such a situation a time uncertainty of less than a few nanoseconds may be desirable. Similarly in case of pulse shape measurements the reproduction of the incident waveform requires good PMT time response.

The response time of any PMT depends almost exclusively on the electron multiplication process. The reason is that the first stage of the photoelectron collection

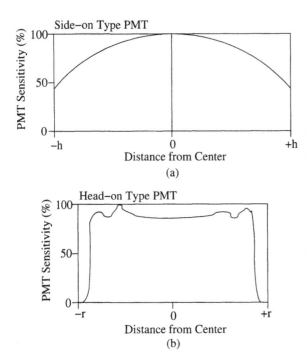

Figure 6.5.18: (a) Typical anode sensitivity curves depicting the response of PMTs to light incident on different photocathode positions normal to their entrance windows. (a) Side-on type PMT. (b) Head-on type PMT. The abrupt decrease of anode sensitivity at the edges requires care in tube positioning.

by the first dynode is almost independent of time. As the electrons move down the multiplication chain, their time distribution widens due to statistical nature of the process of electron emission from the dynodes. This implies that even if the initial photon pulse is a delta function, the output pulse at the anode will have a finite spread. The rise time of the output pulse is one of the two parameters used to characterize the time response of the PMT. The other parameter is the transit time of the electrons. The transit time is simply defined as the time between the peak of the anode pulse and the time of incidence of the photons (see Fig.6.5.19). The definition of this rise time is, however, somewhat arbitrary. Conventionally it is defined as the time it takes the anode pulse to rise from 10% to 90%.

If a train of very narrow width pulses are incident on a PMT, the transit time, rise time, and the decay time will show fluctuations about their mean values. For a well designed and constructed PMT, these fluctuations are due mostly to the statistical nature of the underlying physical processes. The spread in the transit time is what is used to characterize the time response of a PMT. The larger the spread the less time resolution the PMT has. The time spread can be obtained by taking the Fourier transform of the pulse transit times obtained by illuminating the PMT with very narrow width photon pulses. The FWHM of the frequency spectrum is generally taken as the transit time spread. The spread in the rise and decay times is, in most cases, very small. For the applications where the reproduction of the incident photon pulses is desired, these two times and their spread become very important.

One of the main reasons for the popularity of the linear focused type PMT is its superior time response characteristics compared to most other types of PMTs. The reason lies in the optimized dynode structure, placements, as well as the high

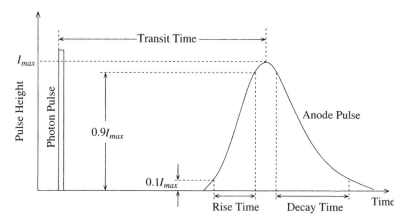

Figure 6.5.19: Time response of a PMT to a very narrow incident photon pulse. I_{max} is the highest measured anode current (or pulse height).

voltage distribution. There are also other types of PMTs, most notably the ones with multi channel plate structures, come close and sometimes surpass the timing performance of the linear focused type PMTs.

A.13 Frequency Response

The frequency response of PMTs is of significance for tubes working at high rates, specially the ones that are used for timing applications. In most applications one is interested in determining the *cutoff frequency*, that is, the frequency beyond which the tube's response falls below some acceptable value. In the previous section we studied the time response of PMTs. Although sometime the time response and frequency response are used interchangeably in literature but we will keep them separate even though they are closely related. The rationale behind this separation is simply the importance of the cutoff frequency that does not appear in the time response characteristics. Another factor that should be noted is that the time response of a PMT depends largely on its mechanical structure. The frequency response, on the other hand, depends not only on the mechanical structure but also on the external readout circuitry.

The dependence of the frequency response on the external circuitry can be appreciated from the following formula for the cutoff frequency.

$$f_c = \frac{1}{2\pi C_s R_{l,eff}} \tag{6.5.21}$$

Here C_s is the total capacitance at the output and $R_{l,eff}$ is the *effective* load resistance. The reader might recall that the load resistance is used to convert the output current into voltage and that the voltage is actually measured across it (cf. Fig.6.5.14). However it should be noted that the subsequent amplifier has its own input resistance too, as shown by R_a in Fig.6.5.20. This implies that the effective

effective resistance seen by the output signal will actually be

$$\frac{1}{R_{l,eff}} = \frac{1}{R_l} + \frac{1}{R_a},$$

$$\text{or} \quad R_{l,amp} = \frac{R_l R_a}{R_l + R_a}. \tag{6.5.22}$$

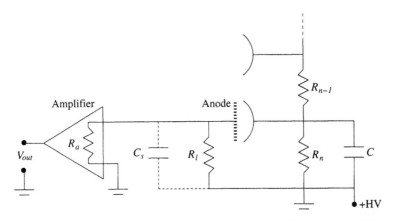

Figure 6.5.20: Typical output circuit of a PMT with amplifier having internal resistance R_a. The effective resistance seen by the output signal depends on R_l as well as R_a.

The product $R_{l,eff}C_s$ is generally referred to as the time constant of the output circuit, though it is also not uncommon to see $R_l C_s$ being referred to as the time constant instead. The capacitance C_s represents the combined capacitance of the tube and the stray capacitance of the cables and is generally very small. Since C_s depends on the mechanical structure of the tube and the cables therefore it can not be varied much. On the other hand, one is at liberty to choose a value of the load resistance R_l according to the requirements. Now, there are two competing requirements for this choice. One is the cutoff frequency, which should be as large as possible. This would require the load resistance to be as small as possible. The other requirement stems from the fact that the load resistance is used to convert the output current into voltage and hence, for reliable measurement of the output voltage, its value should be as large as possible. But we saw making it large has the downside of making the $R_{l,eff}$ large and cutoff frequency small. Another problem is that the larger R_l is the higher is the potential drop across it, which would make the potential drop between the last dynode and the anode small. The consequence of this would be the buildup of space charge at the last dynode and poor collection of charges leading to nonlinearity in the response, certainly not a very desirable effect. It is apparent that these conditions can not be met at the same time and one must choose an optimized solution based on the requirements of the particular application.

A.14 Energy Resolution

The pulse height at the anode of a PMT depends on the number of secondary electrons. Since we know the overall gain of the tube we might conclude that it would have a very good energy resolution. Unfortunately this is not really true since we have not considered the fact that the number of electrons produced at each step of the multiplication chain fluctuates around a mean value. We had noted earlier that the emission of secondary electrons is Poisson distributed, which implies that at each dynode the spread in the number of electrons can be written as

$$\frac{\sigma_{N_i}}{N_i} = \frac{\sqrt{N_i}}{N_i} = \frac{1}{\sqrt{N_i}}, \tag{6.5.23}$$

where N_i is the average number of secondary electrons produced by dynode i.

Since the spread depends inversely on the square root of the number of electrons therefore the first dynode, producing the least number of secondary electrons, introduces the largest uncertainty. As the multiplication progresses the overall uncertainty increases but at a slower rate after each step due to the increase in the number of electrons. The end result is a fairly broad peak at the anode and poor energy resolution. The only way out of this problem is to somehow increase the number of electrons at each stage. This can be done by using dynodes with a very large value of gain δ since it would guarantee emission of large number of electrons even at the first stage of multiplication. The PMTs designed to deliver good energy resolution are actually constructed with such dynodes. In general, however, PMTs have poor energy resolution.

If the PMT is used to detect the photons coming from a scintillator, the overall energy resolution will have a component due to scintillator alone as well. Generally, scintillators by themselves have good energy resolution and the largest uncertainty is introduced by the PMTs. This is one of the reasons why semiconductor photodetectors are now sometimes preferred over conventional PMTs.

A.15 Modes of Operation

A PMT can be operated in two distinct modes: digital and analog. Digital mode, also called the photon counting mode, is the one that actually utilizes the potential of PMT design to its fullest. In this mode the PMT is used to count the number of light photons that strike the photocathode, something that is extremely difficult, if not impossible, with other types of detectors. The caveat, of course, is that this does not work very well when the incoming photon intensity is very high and the associated electronics is not fast enough. This technique works best in situations when the incident photons are well separated in time.

When the photon intensity is low, a small number of photoelectrons are generated in the photocathode. Consequently the output pulses at the anode are well separated in time. The proportionality of the number of these pulses to the incident light forms the basis of the photon counting mode of operation. Fig.6.5.21 shows the pulse counting process in the digital operation mode of a PMT. It is evident that all of the photons incident on the photocathode do not generate photoelectrons. The ones that do, can reach the first dynode and initiate the electron multiplication process leading to a measurable pulse at the anode. Of course if the photoelectron goes astray and does not reach the first dynode, there will not be any pulse at the anode at all. It is therefore very important that, specially for measurement of low photon intensities, a PMT with a high electron collection efficiency is chosen.

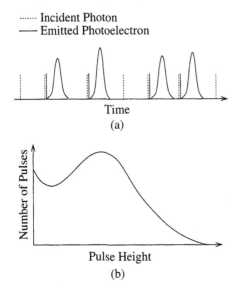

Figure 6.5.21: (a) Digital mode of PMT operation. The figure shows generation of pulses when the incident photon rate is low. (b) Pulse height spectrum of a typical PMT. X-axis is the pulse heights in a convenient bin size and y-axis is the number of pulses with heights lying within the corresponding bin.

As shown in Fig.6.5.21, although each pulse is generated by a single photon, still the height of the output pulse is not always the same. The major factor contributing to this variation is the stochastic nature of the electron multiplication process. Since the number of secondary electrons from a dynode depends on the incident electrons, therefore even a small variation in the electron population at early stages can potentially cause large variations in the charge collected by the anode. The fluctuation in the number of electrons can be due to two factors, one is the inherent statistical fluctuation in secondary electron emission process and the other is the collection efficiency of the dynode.

If the count rate is so high that the associated electronics can not differentiate between subsequent pules, the PMT output is a pulse-overlapped or time integrated signal (see Fig.6.5.22). In such a situation the counting of individual pulses is not possible and the output is an average current having typical shot noise characteristics. This is called the analog mode of PMT operation.

Example:
Estimate the peak output current and voltage of the output pulse of a PMT working in digital mode. Assume the tube gain and the output pulse width are 10^6 and 16 ns respectively. Take the input impedance of the amplifier to be 100 Ω.

Solution:
Since the PMT is operating in the photon counting mode therefore the output pulse is generated by a single photoelectron incident on the first dynode. With the tube gain of 2×10^6 and the elementary electronic charge of 1.6×10^{-19} C,

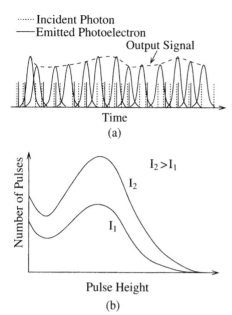

····· Incident Photon
——Emitted Photoelectron
            Output Signal

Time
(a)

Number of Pulses

$I_2 > I_1$

I_2

I_1

Pulse Height

(b)

Figure 6.5.22: (a) Analog mode of PMT operation. If the incident photon intensity is higher than the resolving power of the readout electronics, a pulse-overlapped (or time integrated) signal is obtained. In this mode the counting of individual pulses is not possible. (b) Pulse height spectrum of a typical PMT at two different incident photon intensities.

this electron multiplies into a total charge on anode of

$$Q_{anode} = (1.6 \times 10^{-19})(2 \times 10^6)$$
$$= 3.2 \times 10^{-13} \ C.$$

A pulse of width $t_w = 16ns$, will then produce a peak current of

$$I_{anode}^{peak} = \frac{Q_{anode}}{t_w}$$
$$= \frac{3.2 \times 10^{-13}}{16 \times 10^{-9}}$$
$$= 20 \ \mu A.$$

The corresponding voltage across the amplifier's input having impedance $R_{load} = 100 \ \Omega$ is then given by

$$V_{out} = I_{anode}^{peak} \times R_{load}$$
$$= (20 \times 10^{-6})(100)$$
$$= 2.0 \ mV.$$

This shows that the typical voltage output of a PMT working in digital mode is quite small and a good signal-to-noise ratio would require very sensitive and stable readout electronics.

A.16 Noise Considerations

Since PMTs are very sensitive instruments, their vulnerability to various types of noise should be carefully considered specially in situations where the count rate is very low. The most important PMT noise sources are described below.

▶ **Thermally Agitated Electron Emission:** We saw earlier that the photo-cathodes and dynodes generally used in PMTs have very low work functions. This implies that they are susceptible to thermal agitations and might sponta-neously emit electrons even at room temperature. There are two precautions that can be taken to minimize this effect: cooling and minimizing the size of the photocathode. Cooling the photocathode makes the thermal energy un-available to the electrons. Since the number of thermally emitted electrons also depends on the area of the photocathode therefore decreasing its size minimizes this effect.

▶ **Glass Scintillation:** The electron multiplication structure of a PMT is fitted inside a vacuumed glass enclosure, which itself can act as a scintillator. In normal operation, the deviation of trajectories of some of the electrons taking part in the multiplication process is inevitable. Some of these electrons can get attracted towards the outer glass enclosure, thus producing scintillation light.

▶ **Field Emission:** If a PMT is operated at a very high voltage (near its max-imum rating), it may suffer from small electrical discharges. The result is erroneous large random pulses. To avoid such a situation, the PMTs should be operated at a voltage well below the maximum rating.

▶ **Leakage Current:** No PMT structure can be made absolutely perfect. Small imperfections in insulations and contaminations are inevitable, causing a small leakage current to flow in the output circuit. This effect becomes more promi-nent at low operating voltages. To minimize this, all parts of the tube must be thoroughly cleaned.

▶ **Ringing:** Any impedance mismatch in the input of the preamplifier and the PMT can cause error in count rate. This so called *ringing* effect is a seri-ous problem for high rate situations. The best cure, of course, is to ensure impedance matching.

▶ **Mechanical Instability:** PMT is a sensitive mechanical structure and there-fore highly susceptible to mechanical vibrations. Such vibrations may cause the electrons to deviate from their usual trajectories. This may, for example lead to glass scintillation and consequently the signal-to-noise ratio may decrease considerably.

▶ **Inductive Noise:** Almost all of the PMTs are sensitive to external magnetic fields, which inhibit inductive noise in the system. Therefore the PMT struc-tures should either be electromagnetically shielded or the external field should be compensated. Some newer designs of PMTs, such as mesh or MCP type, are less susceptible to magnetic fields and are therefore preferred over other types if the time response and other requirements allow.

In the two subsequent sections we will look at the theoretical considerations of the signal-to-noise ratio computations in the analog and digital modes of PMT operation.

A.17 Noise in Analog Mode

We saw earlier that in analog mode of operation, an average photoelectric current is measured at the anode. The measured average photocurrent also has an AC shot noise component superimposed on its instantaneous value. There are several factors that contribute to this noise, the most important of which are described below.

▶ **Statistical Fluctuations:** The production of photoelectrons in the photo-cathode, like all other quantum mechanical processes, has certain statistical uncertainty. This implies that, even if we shine the PMT with monochromatic light having constant intensity, the production rate of photoelectrons will have some fluctuations characterized by the probabilistic nature of the process. Furthermore, the multiplication of these photoelectrons also has statistical uncertainty associated with it. These two fluctuations introduce uncertainty or *noise* in the system. The shot noise current σ_{st} corresponding to the overall statistical fluctuations in the photocathode and the dynodes can be written as

$$\sigma_{st} = \mu\sqrt{2eI_{pe}FB}. \tag{6.5.24}$$

Here μ is the PMT gain, I_{pe} is the photoelectric current, B is the bandwidth of the measuring electroncis, and F is the PMT noise figure, which can be calculated for a PMT having n dynodes with individual gains δ_i with $i = 1, ..., n$ by

$$F = 1 + \frac{1}{\delta_1} + \frac{1}{\delta_1.\delta_2} + ... + \frac{1}{\delta_1.......\delta_n}. \tag{6.5.25}$$

A point worth stressing here is that I_{pe} should not be confused with the anode current since it represents the photoelectric current, that is, the total charge being produced by the photocathode and collected by the first dynode per unit time. The terms F, B, and μ transform this current into the anode current. The photoelectric current can be obtained from the relation

$$I_{pe} = N_\gamma e(QE)(CE), \tag{6.5.26}$$

where N_γ is the number of photons incident on the photocathode per unit time, e is the electronic charge, QE is the quantum efficiency of the photocathode, and CE is the collection efficiency of the first dynode.

▶ **Background Light:** Photomultiplier tubes are very sensitive devices and therefore unless the system has been made extremely light tight, there are always some background photons contributing to the signal. The corresponding shot noise can be written as

$$\sigma_{bg} = \mu\sqrt{2eI_{bg}FB}. \tag{6.5.27}$$

Here I_{bg} is the background current, which is the average anode current measured without the incident light and can be computed from

$$I_{bg} = N_{\gamma,bg} e (QE)(CE),\qquad(6.5.28)$$

where $N_{\gamma,bg}$ is the number of background photons incident on the photocathode per unit time.

▸ **Dark Current:** This is the term used to describe the average noise current due to a number of sources. Dark current is dominated by thermionic emission of electrons from the photocathode. Other sources of dark current include current produced by ionization of residual gases inside the tube and leakage current between electrodes. The presence of dark current in any practical PMT is unavoidable but it can be minimized by proper designing and construction. A constant dark current in itself is not a problem as far as measurements are concerned since one can always subtract it out from the anode current. However the fluctuations of the dark current can introduce significant uncertainty in the final measurements. These fluctuations can be characterized by the shot noise current, which for the average dark current I_d can be estimated from

$$\sigma_d = \mu\sqrt{2eI_dFB}.\qquad(6.5.29)$$

where μ, F, and B represent the same parameters as described above. It is apparent from this equation that decreasing the dark current has the merit of decreasing the noise associated with it as well. Hence, keeping the dark current to the minimum possible value is highly desirable.

▸ **Johnson Noise:** Measurement of anode current requires that an amplifier be connected to the PMT load. The equivalent impedance of this circuit is subject to thermal variations causing injection of thermal or *Johnson* noise in the system. The Johnson noise for an amplifier having noise figure F_{amp} can be expressed as

$$\sigma_{amp} = \sqrt{\frac{4F_{amp}k_BTB}{R_{eqv}}}.\qquad(6.5.30)$$

Here k_B is the Botzmann's constant, T is the absolute temperature, and R_{eqv} is the equivalent circuit impedance.

As Johnson noise depends explicitly on temperature, it is instructive to see how it changes with changes in temperature. For that, let us differentiate equation 6.5.30 with respect to temperature. This gives

$$\frac{d\sigma_{amp}}{dT} = A\frac{1}{\sqrt{T}},\qquad(6.5.31)$$

$$\text{where}\quad A = \sqrt{\frac{FkB}{R}}.$$

This shows that the effect is more dramatic at lower temperatures. Note that here we are talking about the change in the Johnson noise and not the noise itself. The noise is larger at higher temperatures according to equation 6.5.30 but the variation in the noise level with respect to change in temperature is higher at lower temperatures.

Example:
If the temperature of a PMT based detector working at 200 K is elevated to the room temperature of 300 K, what would be the relative change in its Johnson noise.

Solution:
Since all other parameters are supposed to remain constant at the two temperatures, we can write equation 6.5.30 as

$$\sigma_{amp} = K\sqrt{T}$$

$$\text{where} \quad K = \sqrt{\frac{4F_{amp}k_B B}{R_{eqv}}}.$$

The percentage relative change in Johnson noise can be obtained by writing the above equation at the two temperatures and taking their relative difference.

$$
\begin{aligned}
\Delta\sigma_{amp} &= \frac{\sigma_{amp,1} - \sigma_{amp,2}}{\sigma_{amp,1}} \times 100 \\
&= \frac{\sqrt{T_1} - \sqrt{T_2}}{\sqrt{T_1}} \times 100 \\
&= \frac{\sqrt{200} - \sqrt{300}}{\sqrt{200}} \times 100 \\
&= 22.5\%.
\end{aligned}
\tag{6.5.32}
$$

All of the noise components described above contribute to the overall signal-to-noise ratio. If we know all the individual currents, we can compute the signal-to-noise ratio from

$$
\begin{aligned}
S/N &= \frac{I_\gamma}{\left[\sigma_{st}^2 + \sigma_{bk}^2 + \sigma_d^2 + \sigma_{amp}^2\right]^{1/2}} \\
&= \frac{I_\gamma}{\left[2e\mu^2 FB\{I_{pe} + I_{bg} + I_d\} + 4F_{amp}k_B TB/R_{eqv}\right]^{1/2}}.
\end{aligned}
\tag{6.5.33}
$$

However, practically it is difficult to separate the individual currents from the measurements. The easiest and most practical thing to do is to take two sets of measurements, one without and the other with incident light. The shot noise associated with the measurement without incident light is given by

$$\sigma_{bg,d,amp} = \sqrt{\sigma_{bk}^2 + \sigma_d^2 + \sigma_{amp}^2}, \tag{6.5.34}$$

while the total shot noise after shining the photocathode with incident light is

$$\sigma_{tot} = \sqrt{\sigma_{st}^2 + \sigma_{bk}^2 + \sigma_d^2 + \sigma_{amp}^2}. \tag{6.5.35}$$

Now, the currents corresponding to these noise levels can be subtracted to get the current due to light only, that is

$$I_{pe} = I_{tot} - I_{bg,d,amp}. \tag{6.5.36}$$

The overall signal-to-noise ratio from these measurements is then given by

$$S/N = \frac{I_{pe}}{\sigma_{tot}^2 + \sigma_{bg,d,amp}^2}$$

$$= \frac{I_{pe}}{\sigma_{st}^2 + 2\left(\sigma_{bk}^2 + \sigma_d^2 + \sigma_{amp}^2\right)}$$

$$= \frac{I_{pe}}{[2e\mu^2 FB\{I_{pe} + 2I_{bg} + 2I_d\} + 8F_{amp}k_B TB/R_{equ}]^{1/2}}. \qquad (6.5.37)$$

Except for the statistical fluctuations, all the noise sources described above are instrumental in nature and therefore can be reduced by proper design and construction. for example decreasing the bandwidth of the readout circuitry improves the signal-to-noise ratio. This can be done in practical system provided it does not affect the dynamic range of the system.

Example:
Determine the fluctuations in the dark current of 1 nA from a PMT having a gain of 10^4. The PMT is built with 6 dynodes and the system bandwidth is 50 Hz. Assume that the PMT noise figure is unity.

Solution:
Substituting the given values in equation 6.5.29, we get

$$\sigma_d = \mu\sqrt{2eI_d FB}$$

$$= 10^4 \left[(2)\left(1.602 \times 10^{-19}\right)\left(10^{-9}\right)(1)(30)\right]^{1/2}$$

$$= 9.8 \times 10^{-10} \quad A$$

$$= 0.98 \quad nA.$$

Apart from the signal-to-noise ratio, another parameter that is widely used to characterize the noise level of PMTs is the *equivalent noise input* or *ENI*. It represents the amount of light necessary to produce a signal-to-noise ratio of unity. *ENI* is mathematically defined as

$$ENI = \frac{1}{S_{l,a}}\left[2eI_d\mu B\right]^{1/2}, \qquad (6.5.38)$$

where $S_{l,a}$ is the anode luminous sensitivity in A/lm, I_d is the anode dark current in amperes, μ is the tube gain, and B is the bandwidth of the system. The dark current can easily be determined by measuring the tube output without any light input, not even the background light. For that the tube is generally kept in darkness for some time (a few minutes wait normally suffices) before measuring the output current.

A.18   Noise in Digital Mode

In digital mode the requirements on the noise are not as stringent as in analog mode. The reason is that in this mode the number of output pulses are counted in contrast with analog mode where actual pulse heights are measured. In digital mode one generally uses a discriminator at an early stage of the electronics, with the purpose of deciding which pulses are to be blocked. The discriminator generally has a lower and an upper threshold. The lower threshold is set to discriminate the noise pulses from the rest while the upper threshold has the purpose of eliminating spurious large pulses. The working principle is quite simple: only the pulses with heights within the lower and upper thresholds are allowed to pass through and get counted. All other pulses are simply rejected. In this way *most* of the noise is simply rejected at an early stage no matter what its origin is and *most* of the *good* pulses are allowed to pass through and counted. The reason why not all noise can be rejected in this manner lies in the uncertainty associated with the noise level. Complete rejection of noise would require the lower discriminator threshold to be set too high and upper discriminator threshold to be set too low, which would result in rejection of significant number of good pulses too. The settings of the discriminator are therefore based on a compromise between the acceptable signal-to-noise ratio and the count rate.

Except for the Johnson's noise, all the other noise components we discussed earlier for the analog mode can be defined for the digital mode as well. However now the shot noise values can be determined by simpler relations due to the digital nature of the readout.

▶ **Statistical Fluctuations:** Due to the statistical nature of photoelectric and electron multiplication processes, there is always some uncertainty in the number of counts detected. The shot noise attributable to this uncertainty can be obtained by noting that the process is Poisson in nature. Hence, the statistical noise counts are given by

$$\sigma_{st} = \sqrt{N_{st}}, \tag{6.5.39}$$

where N_{st} represents the number of counts due to incident light only.

▶ **Background Light:** If N_{bg} represents the counts due to background light only, the shot noise component due the uncertainty associated with it can be obtained from

$$\sigma_{bg} = \sqrt{N_{bg}}. \tag{6.5.40}$$

▶ **Dark Counts:** The shot noise component due to dark counts N_d is given by

$$\sigma_d = \sqrt{N_d}. \tag{6.5.41}$$

The total shot noise due to all sources in digital mode is then given by

$$
\begin{aligned}
\sigma_t &= \sqrt{\sigma_{st}^2 + \sigma_{bg}^2 + \sigma_d^2} \\
&= \sqrt{N_{st} + N_{bg} + N_d}.
\end{aligned} \tag{6.5.42}
$$

And the signal-to-noise ratio will be

$$S/N = \frac{N_{st}}{\sqrt{N_{st} + N_{bg} + N_d}}. \tag{6.5.43}$$

Though this is a simple relation, however using it requires one to know the counts due to all the three noise sources individually. Practically, it is extremely difficult to separate the data into the three components. What is done instead is to perform the measurements in two steps. In the first step the counts without any light source are accumulated for a certain period of time. This gives the counts due to background light and dark current only. That is

$$N_{bg,d} = N_{bg} + N_d. \tag{6.5.44}$$

In the second step, the PMT is illuminated with light and the counts are recorded for the same period of time. This gives the total counts due to all the noise sources and the incident light, that is

$$N_{total} = N_{st} + N_{bg} + N_d. \tag{6.5.45}$$

The counts due only to the incident light can then be obtained by subtracting $N_{bg,d}$ from N_{total}.

$$N_{st} = N_{total} - N_{bg,d} \tag{6.5.46}$$

Now, the two measurements performed have their own noise levels given respectively by

$$
\begin{aligned}
\sigma_{bg,d} &= \sqrt{\sigma_{bg}^2 + \sigma_d^2} \\
&= \sqrt{N_{bg} + N_d} \quad \text{and} \tag{6.5.47} \\
\sigma_{total} &= \sqrt{\sigma_{st}^2 + \sigma_{bg}^2 + \sigma_d^2} \\
&= \sqrt{N_{st} + N_{bg} + N_d}. \tag{6.5.48}
\end{aligned}
$$

The signal-to-noise ratio for the measurements will then be given by

$$
\begin{aligned}
S/N &= \frac{N_{st}}{\sqrt{\sigma_{tota}^2 + \sigma_{bg,d}^2}} \\
&= \frac{N_{st}}{\sqrt{N_{st} + N_{bg} + N_d + N_{bg} + N_d}} \\
&= \frac{N_{st}}{\sqrt{N_{st} + 2(N_{bg} + N_d)}}. \tag{6.5.49}
\end{aligned}
$$

Example:
A PMT working in the digital records dark and background counts of 5×10^4 and 1.5×10^2 respectively. Compute the total fluctuations due to these two noise sources.

Solution:
The shot noise due to the dark counts is

$$
\begin{aligned}
\sigma_d &= \sqrt{N_d} \\
&= \sqrt{5 \times 10^4} \\
&= 223.6.
\end{aligned}
$$

Similarly for the background counts we have

$$\sigma_{bg} = \sqrt{N_{bg}}$$
$$= \sqrt{1.5 \times 10^2}$$
$$= 12.2.$$

The total shot noise due to these two sources is given by

$$\sigma = \sqrt{\sigma_d^2 + \sigma_{bg}^2}$$
$$= \sqrt{223.6^2 + 12.2^2}$$
$$\approx 224.$$

A.19 Effect of Magnetic Field

One of the most problematic things about PMTs is their vulnerability to the external magnetic fields. The stem of this problem is the fact that some of the electrons traveling down the multiplication chain may change their paths due to the magnetic field. Such stray electrons do not contribute to the output signal and hence reduce not only PMT sensitivity but may also affect its linearity. This is a serious problem in situations where high magnetic field is present, such as particle physics experiments. In such a case there are three options to reduce this effect.

▶ **Less Susceptible PMTs:** Some PMT structures, such as multi channel type, are not much affected by external magnetic fields, and can therefore be preferred over the conventional PMTs.

▶ **External Shielding:** PMTs can be externally shielded, for example by constructing a Faraday cage around them.

▶ **Field Compensation:** The magnetic field can be compensated by another magnetic field near the PMTs such that the two cancel out each other. This method is more suitable for tubes working in low level magnetic fields, such as that of the Earth. The reason is that application of high compensating field may not be practically possible or even desirable due to its possible effect on other parts of the system.

6.5.B Photodiode Detectors

PMTs are not the only means of detecting light photons. New advancements in semiconductor technology have led to the development of the so called *photodiode* detectors having moderate to extremely high quantum efficiencies. In the previous Chapter we looked at some specific types of semiconductor detectors, such as PIN diode, heterojunction diode, and Schottky diode. Each of these detectors can be used as a photodetector since a photodiode detector is simply a semiconductor diode detector optimized to work in and around the visible spectrum of radiation.

Before we discuss the specific kinds of semiconductor photodetectors, let us have a look at the internal gain expected from a general photon sensor. Let us assume that there is a flux Φ_γ of photons having frequency ν uniformly illuminating the surface of a semiconductor detector of surface area A per unit time (Φ_γ has dimensions of photons per second.). The number of charge pairs generated per unit volume can then be written as

$$N = \eta \frac{\phi_\gamma}{h\nu},\tag{6.5.50}$$

where η is the quantum efficiency of the detector, which is simply the number of charge carriers generated per photon. The total current due to these charges, each having a charge q, is given by

$$\begin{aligned} i_\gamma &= qN \\ &= q\eta \frac{\phi_\gamma}{h\nu}. \end{aligned}\tag{6.5.51}$$

We saw in the previous Chapter that some of these charge pairs recombine with an average rate of $1/\tau$ (see equation 5.1.20), where τ is the average lifetime of the charges. Hence the above equation is an overestimation and in order to determine the actual current we must include the recombination rate into the equation as well. If n is the average number of charge pairs per unit volume, then their average generation rate can be written as

$$\begin{aligned} G &= \frac{n}{\tau} \\ &= \tau \frac{\eta(\phi_\gamma/h\nu)}{Ad}, \end{aligned}\tag{6.5.52}$$

where d is the detector thickness. This equation can be used to determine the average charge density n.

$$n = \frac{\eta(\phi_\gamma/h\nu)}{\tau Ad}.\tag{6.5.53}$$

To determine the current due to these charges we make use of Ohm's law, which states that the current density is proportional to the electric field, i.e.,

$$J = \sigma E$$

where σ is the conductivity of the material. Using $i = JA$, we find that the actual measurable current i_s is given by

$$\begin{aligned} i_s &= \sigma EA \\ &= q\mu n EA. \end{aligned}\tag{6.5.54}$$

Here μ is the charge mobility. We have assumed that the electrons and holes have same mobility, something that is not really true but is not expected to have a large effect on our derivation. Substituting n from equation 6.5.53 in the above relation gives

$$i_s = q \frac{\tau \eta \phi_\gamma \mu E}{dh\nu}\tag{6.5.55}$$

The photocurrent gain is then obtained by dividing i_s by i_γ. Hence we get

$$
\begin{aligned}
Gain &= \frac{i_s}{i_\gamma} \\
&= \frac{\tau \mu E}{d} \\
&= \frac{\tau}{t_d}.
\end{aligned}
\tag{6.5.56}
$$

Here we have assumed that the charges are produced at the surface and traverse the thickness d of the detector before being collected by the electrode in a time $t_d = d/v_d = d/\mu E$, with v_d being the average drift velocity of the charges. t_d is generally called transit or drift time and is an important parameter for estimating the gain of a photodetector. A well fabricated pn junction diode usually has a gain of 1, which signifies the fact that the recombination is minimal. Avalanche diode that we will discuss in the next section, may have a gain of 10,000 or more. The reason why one would want the gain to be so high is not difficult to understand if one considers the operation at very low incident photon fluxes. Even though most of the photodiodes have a quantum efficiency of more than 80% throughout the visible and near infrared regions, but since at nominal applied voltages there is no charge multiplication process therefore their signal-to-noise ratio at the single photon level is not large enough for obtaining meaningful results. Therefore, for low intensity measurements, the signal must be somehow amplified. In avalanche photodiodes, this is done internally in the bulk of the material through the process of charge multiplication. This emerging technology has already started to take the place of conventional PMTs in some applications and therefore deserves some attention. The next section is devoted to avalanche photodiodes.

6.5.C   Avalanche Photodiode Detectors (APD)

Avalanche photodiode detectors exploit the process known as *impact ionization*, in which the initial electrons create more free electrons by imparting energy to the molecules along their tracks. This process is very similar to the gas multiplication we discussed in chapter 3. The primary electrons produced by the incident radiation are made to attain high velocities under the influence of externally applied high electric field. If the energy attained by an electron is high enough, it can free one or more secondary electrons, thereby creating more charge pairs. It should be noted that theoretically such a process is only possible if the incident electron gains energy at least equal to the band gap energy of the material. However since an electron also looses energy through non-radiative scatterings, on the average energy of the electron should be much higher than the band gap energy. For most semiconductors an energy difference of a factor of 3 is normally required. The secondary electrons, being under the influence of the same electric field, produce tertiary charge pairs and so on. Once started, this process of charge multiplication grows and eventually causes avalanche multiplication of charge pairs. The large number of charge pairs thus produced create an electrical current that is much higher than obtained in conventional photodiode detectors. Since this current is proportional to the incident particle energy therefore the detector can be used for spectroscopic purposes.

Just like conventional photodiode detectors, the avalanche photodetectors can also be designed and built in different configurations. Fig.6.5.23(a) shows the conceptual design of a simple avalanche photodetector. The detector is made of an intrinsic (or lightly doped p type material) sandwiched between a heavy doped p side and a heavy doped n side. Another p type region is also established between the intrinsic material and the heavily doped n side. The contacts are made by metal deposition on two sides of the detector. A high reverse bias applied at the two ends creates an electric field profile similar to the one sketched in Fig.6.5.23(b). Based on this electric field profile two regions can be identified: an absorption region R_{abs} and a multiplication region R_{mult}. The incident radiation passing through the intrinsic or absorption region creates electron hole pairs along its track. The charges move in opposite directions under the influence of moderate electric field inside R_{abs}. The electrons move towards the p region, which is characterized by very high electric field intensity. This is where avalanche multiplication takes place. The large number of electron hole pairs generated here, as a result of charge multiplication, move in opposite directions and induce a voltage pulse that is measured by the readout electrode. Alternatively the current generated by the motion of charges can also be measured by the associated electronics.

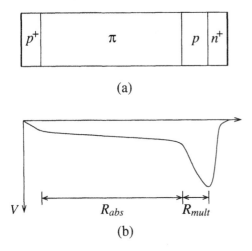

(a)

(b)

Figure 6.5.23: (a) Conceptual design of a simple avalanche photodetector. (b) The electric field profile of the structure shown in (a). π represents either a lightly doped p-material or intrinsic material while the superscript (+) refers to heavy doping.

The kind of structure shown in Fig.6.5.23 is generally known as *reach-through* structure. Fig.6.5.24 shows the practical design of a reach-through APD. The guard ring around the multiplication region is established to minimize the possibility of electrical breakdown.

C.1   Basic Desirable Characteristics

The desirable characteristics of an avalanche photodetector are not much different from the conventional photodiode detectors as can be deduced from the list below.

▶ **Small Leakage Current:** An idea APD does not have any leakage current. Real APDs have extremely small dark currents. The reason to strive for no leakage current is to avoid unwanted avalanches, something that may induce non-linearity in the detector response.

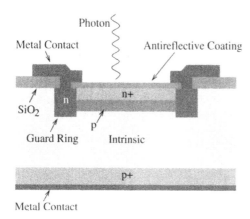

Figure 6.5.24: Typical structure of a silicon based reach-through type avalanche photodetector.

▶ **High Gain:** Since the sensitivity of the detector depends on its gain therefore the APDs are designed such that they have gains of up to 10^8. However depending on the application, one can opt for lower gain, which can be achieved by applying lower electric field.

▶ **Good Frequency Response:** Good frequency response is one of the most desirable characteristics of APDs, specially the ones that are to be used in high rate environments.

It should be noted that the last two characteristics are similar to the ones desired in photomultipliers, which of course is not a surprise since both detectors have essentially the same purpose. We will shortly see that modern APDs have characteristics near to their ideal values.

C.2 Multiplication Process and Gain Fluctuations

The process of electron multiplication in an APD is very much different from the one that occurs in a PMT. It resembles the avalanche process in gases that we studied in the Chapter on the gas filled detectors.

The process through which the charges multiply in an APD is known as *carrier impact ionization*. Intuitively thinking, it might seem a very simple process since it refers to the subsequent ionizations when high energy charge carriers make collisions with molecules along their tracks. However, the reality is that it is an extremely complicated process and is very difficult to evaluate analytically. The difficulty lies mainly in modeling the stochastically spread distribution of charges after impulse ionization. A detailed account of different existing models and the related analytical computations is beyond and scope of this book and therefore the interested reader is referred to other sources (see, for example (37; 24; 31) and the references therein).

The problems mentioned above lead to the difficulty in the determination of the probability density function of the electrons that make up the output signal. One might assume these electrons to have a Poisson distribution if the absorption of the incident light could be described by the Poisson process. This, however, is not really the case. In face, the distribution of electrons is fairly complicated and extremely difficult to evaluate numerically ((33)). Due to this difficulty a number of approximations to this probability density function have been proposed, including the one that assumes the shape of a Gaussian distribution, though with longer tails.

We will see later that even though the gain fluctuates around a mean value, still with proper attention to the noise sources one can build a detector based on APD that works at the limit of quantum fluctuations of the multiplication process. The mean value of the gain is an important parameter for design optimization, noise consideration, and calibrations. For the output signal due to electrons the expression for the mean gain can be written as (22)

$$\langle G_e \rangle = \frac{1-u}{e^{-\alpha d(1-u)} - u}, \tag{6.5.57}$$

where $u \neq 1$ is the ratio of the hole ionization rate to the electron ionization rate, α is the electron ionization rate, and d is the thickness of the depletion region. The ionization ratio u can have any value between 0 and 1 but is generally very small, on the order of 10^{-2} or 10^{-3}. Let us see what happens when we substitute $u = 0$ in the above expression. This gives

$$\langle G_e \rangle \approx e^{\alpha d}, \tag{6.5.58}$$

which, the reader might recall, looks remarkably similar to the expression that was obtained for gas multiplication in the chapter on gas filled detectors. From this expression one might conclude that higher gain can be obtained by simply increasing the depletion width. Although this, in principle, is true but is far away from the actual practice as it has the downside of increasing the fluctuations in the gain. We will return to this discussion in the next section. One thing to note is that this simplified expression for mean gain should not be used in actual computations as the effect of u on gain is not negligible (see example below).

Example:
An APD has a mean gain of 50 at hole-to-electron ionization rate ratio u of 0.1. Compute the relative change in its mean gain if u changes to 0.01. Assume that all other parameters remain the same.

Solution:
Using the given values we first compute the value of αd in equation 6.5.57.

$$\langle G_e \rangle = \frac{1-u}{e^{-\alpha d(1-u)} - u}$$

$$\Rightarrow \alpha d = -\frac{\ln\left[\frac{1-u}{\langle G \rangle} + u\right]}{1-u}$$

$$= -\frac{\ln\left[\frac{1-0.1}{50} + 0.1\right]}{1-0.1}$$

$$= 2.37$$

Now, using this value of αd we can calculate the gain at $u = 0.01$ as follows.

$$\langle G_e \rangle = \frac{1-u}{e^{-\alpha d(1-u)} - u}$$

$$= \frac{1 - 0.01}{e^{-2.37(1-0.01)} - 0.01}$$

$$= 11.5$$

The relative change in gain is

$$\Delta \langle G_e \rangle = \frac{50 - 11.5}{50} \times 100$$

$$= 77\%.$$

In general, the gain of an APD can be any value between 1 and 10^8. However the standard APDs are mostly operated between gains of 10 and 1000. APDs of gains higher than this are also used, though in situations where light level is extremely low, such as in single photon counting experiments.

The mean internal gain of an APD is a function of the applied voltage since the energy gained by the electrons between collisions depends on the electric field strength. The bias-gain curve of a typical APD is shown in Fig.6.5.25. It is apparent from this figure that an APD can, in principle, be operated in a range of gains, which makes it a versatile device that can be tuned according to the level of the incident light.

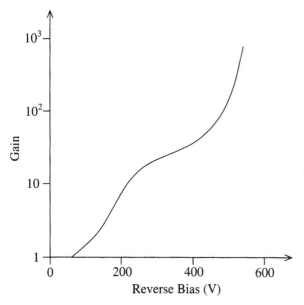

Figure 6.5.25: Variation of gain with applied reverse bias in a typical APD.

In most applications, APDs are used in a region of the voltage-gain curve where the relationship is approximately linear. In such a region the variation of gain with voltage can be written as

$$\frac{dG}{dV} = k_v G \quad \%V^{-1},$$
(6.5.59)

where V is the applied voltage and k_v is a constant that depends on the APD material and construction. Generally its value lies between 3 and 4.

Another parameter on which the gain depends heavily is the temperature. The gain increases with decrease in temperature according to the simple relation

$$\frac{dG}{dT} = -k_t G \quad \%^0C^{-1},$$
(6.5.60)

where T is the temperature in Celcius and k_t is a constant that depends on the type of APD. It normally assumes a value between 2 and 3. The negative sign simply signifies the inverse relationship between the gain and the temperature.

The above equations imply that operating an APD at a fixed temperature and voltage would yield a fixed gain. What these equations *do not* imply is that there is no uncertainty associated with the gain. As a matter of fact, the biggest problem with APD gain is actually its inherent uncertainty and not its variation with temperature and voltage. This uncertainty stems from the noisy nature of the carrier impact ionization process that multiplies the charges. Any measurement from an APD is therefore subject to these fluctuations as well, which of course can become undesirable for high precision measurements such as single photon counting. The fluctuations in the APD gain has thus become an active area of research and development.

Although some researchers prefer to work directly with the gain but most prefer the gain fluctuations that are characterized by the so called *excess noise factor* defined as

$$F = \frac{\langle G^2 \rangle}{\langle G \rangle^2},$$
(6.5.61)

where G is the APD gain and $\langle \rangle$ represents the mean value. A number of mathematical models exist that transform the above equation into a numerically computable form (see, for example (23)). We will not go into that discussion but it is worth mentioning here that the excess noise factor and hence fluctuations in the gain depend on

▶ the mean APD gain,

▶ the ionization coefficients of the charge carriers (electrons and holes), and

▶ the multiplication region.

The variation of the excess noise factor with the mean gain is the most profound one among the three. Though some APDs behave a bit differently, but generally there exists linear relationships between these two parameters, that is

$$F \approx AG + B,$$
(6.5.62)

where A and B are constants that depend on the APD material and construction. Typically A has a value on the order of 10^{-3}, while B assumes a value of around 2.

C.3   Quantum Efficiency and Responsivity

The quantum efficiency of a photodiode detector is used to characterize its efficiency of creating charges that contribute to the output signal. It is generally defined by the ratio

$$\xi = \frac{\text{rate of electron generation}}{\text{intensity of incident photons}}. \tag{6.5.63}$$

Note that here the *rate of electron generation* refers to the electrons that contribute to the output signal. Hence, if I_γ is the current in amperes that generates the output signal and is generated by the incident photons, then we can write

$$\text{Rate of electron generation} = \frac{I_\gamma}{e} \ \ s^{-1}, \tag{6.5.64}$$

with e being the electronic charge. I_γ is sometimes also referred to as the *photocurrent*. Since the photocurrent in an APD actually gets amplified by a factor G, therefore we can also write the above expression as

$$\text{Rate of electron generation} = \frac{I_{out}}{\langle G \rangle e} \ \ s^{-1}, \tag{6.5.65}$$

where $I_{out} = \langle G \rangle I_\gamma$ is the current measured at the APD output and $\langle G \rangle$ is the mean APD gain. The second term in the definition of the quantum efficiency is the *intensity of incident photons*. It can be calculated from the incident power P and the mean wavelength λ of the incident photons through the relation

$$\text{Intensity of incident photons} = \frac{P\lambda}{hc} \ \ s^{-1}. \tag{6.5.66}$$

Hence the expression for the quantum efficiency can be written as

$$
\begin{aligned}
\xi &= \frac{I_{out}/\langle G \rangle e}{P\lambda/hc} \\
&= \frac{I_{out}}{P\langle G \rangle} \frac{hc}{e\lambda} \\
&= \frac{R}{\langle G \rangle} \frac{hc}{e\lambda},
\end{aligned}
\tag{6.5.67}
$$

where $R = I_{out}/P$ is known as the *responsivity* of the photodetector. Responsivity is a standard parameter that is often quoted by the APD manufacturers and is used to compare the characteristics of different APDs. Its generally quoted units are amperes per watt. The expression for responsivity in terms of the efficiency can be deduced from the above expression as

$$R = \frac{\langle G \rangle e \xi \lambda}{hc}, \tag{6.5.68}$$

or in Standard units it can be written as

$$R = 8.07 \times 10^5 \langle G \rangle e \xi \lambda \ \ AW^{-1}. \tag{6.5.69}$$

Example:
An APD having a gain of 20 and responsivity of 8 A/W is subject to a photon beam of intensity $6.5 \times 10^9 \ s^{-1}$. Calculate the quantum efficiency of the APD at 1.6 μm. Also compute the average photocurrent.

Solution:
The efficiency can be calculated from equation 6.5.67.

$$\xi = \frac{R}{\langle G \rangle} \frac{hc}{e\lambda}$$

$$= \frac{8}{20} \frac{\left(6.63 \times 10^{-34}\right) \left(2.99 \times 10^8\right)}{\left(1.602 \times 10^{-19}\right) \left(1.6 \times 10^{-6}\right)}$$

$$= 0.31$$

To compute the average photocurrent I_γ we use the basic definition of responsivity, that is

$$R = \frac{I_{out}}{P}.$$

Using $I_{out} = \langle G \rangle I_\gamma$ this can be transformed into

$$I_\gamma = \frac{\langle P \rangle R}{G}$$

$$= \frac{R\phi hc}{\langle G \rangle \lambda}$$

where $\phi = P\lambda/hc$ is the incident photon intensity. Substituting the given values in the above equation yields

$$I_\gamma = \frac{(8) \left(6.5 \times 10^9\right) \left(6.63 \times 10^{-34}\right) \left(2.99 \times 10^8\right)}{(20) \left(1.6 \times 10^{-6}\right)}$$

$$= 3.2 \times 10^{-10} \quad A$$

$$= 0.32 \quad nA.$$

C.4 Modes of Operation

Just like PMTs, APDs can also be operated in analog or digital modes. However for APDs they are generally referred to as *linear* and *Geiger* modes.

We saw in the chapter on gas filled detectors that operating a detector in Geiger region means applying a bias voltage high enough to cause breakdown in the gas. This is also true for APDs operating in the Geiger mode. If a high enough reverse bias is applied to the APD, it will cause a large current pulse whenever a photon produces charge pairs in the bulk of the material. Hence in this way the detector can be used as a photon counting device just like a PMT. The problem, however, is the dark current in the bulk of the material, which is significant even at room temperatures. The trick, therefore, is to operate the detector at very low temperatures. Silicon APDs have been found to have very low leakage or dark currents and therefore high photon counting capabilities. With a pulse resolution of as low as $20ps$, a gain of as high as 10^8, and a quantum efficiency approaching 80%, the APDs are becoming more and more popular in single photon counting applications.

C.5 Noise Considerations

APDs are generally operated at low light levels due to their signal amplification characteristics. Their sensitivity is therefore limited by their inherent noise. There are different sources of noise in an APD, the most important of which are listed below.

▶ **Leakage Current:** An APD can have two types of leakage currents: surface leakage current and bulk leakage current. The bulk leakage current, which flows in the bulk of the material, is affected by the high electric field in a manner similar to the signal current, and hence gets amplified. We will represent the gain associated with the bulk leakage current by G_l to differentiate it from the multiplication factor for the signal. If I_{ls} and I_{lb} are the surface and bulk leakage currents respectively then the total leakage current I_l in an APD having a mean gain $\langle G \rangle$ can be expressed as

$$I_l = I_{ls} + \langle G_l \rangle I_{lb}. \tag{6.5.70}$$

The contribution of surface leakage current to the total leakage current is quite small and can therefore be safely ignored for most computations. Hence we can write

$$I_l \approx \langle G_l \rangle I_{lb}. \tag{6.5.71}$$

The fluctuations of this leakage or *dark* current is given by the shot noise formula

$$\sigma_l^2 = 2eI_l B \langle G_l \rangle^2 F_l, \tag{6.5.72}$$

where e is the unit electrical charge, B is the bandwidth of the system, and F_l is the excess noise factor for the leakage current. Note that here we have made use of the definition of the excess noise factor given earlier, that is

$$F_l = \frac{\langle G_l^2 \rangle}{\langle G_l \rangle^2}. \tag{6.5.73}$$

The excess noise factor in terms of electron ionization rate α, depletion width d, ionization rate ratio u, and gain factor G_l is given by (see, for example (21))

$$F_l = \frac{(1 + \alpha d \langle G_l \rangle)(1 + u \alpha d \langle G_l \rangle)}{\langle G_l \rangle}. \tag{6.5.74}$$

Determination of the noise factor and the fluctuations of the dark current requires the knowledge of the multiplication factor or gain, which can be computed from the relation (21)

$$\langle G_l \rangle = \frac{1 - e^{-\alpha d(1-u)}}{\alpha d \left[e^{-\alpha d(1-u)} - u \right]}. \tag{6.5.75}$$

It is instructive to compare equations 6.5.57 and 6.5.74, which represent the gains of signal and dark currents respectively. It can be shown that they are related through the relation

$$\langle G_l \rangle = \frac{1}{\alpha d} \left[\langle G_e \rangle - 1 \right]. \tag{6.5.76}$$

▶ **Excess Noise:** We have already introduced the term excess noise factor to characterize the fluctuations in the APD gain. This factor actually represents the so called excess noise that is related to the statistical nature of the underlying particle interactions that take place during the process of charge multiplication. Due to these fluctuations the distribution of amplitudes of the output current pulses assume a finite width even if the incident light can be described by an impulse or delta function.

We have already introduced the expression for the excess noise factor for the dark current (see equation 6.5.74. For the signal current due to electrons the noise factor can be written as (41)

$$F_e = u\langle G_e \rangle + \left[2 - \frac{1}{\langle G_e \rangle}(1 - u) \right], \tag{6.5.77}$$

where, as before, u represents the hole to electron ionization rate ratio and $\langle G_e \rangle$ is the mean signal gain.

▶ **Thermal Noise:** Thermal noise, though important, is not actually related to the APD itself. It refers to the noise generated in the output circuitry due to thermal agitations of the current carriers. During the discussion on PMT noise, we introduced the term *Johnson noise* to characterize the thermal noise generated in the preamplifier. The same argument can be applied for the case of APDs as well. Hence the variance of output current due to thermal noise is given by

$$\sigma_t = \sqrt{\frac{4F_t k_B T B}{R_{eqv}}}, \tag{6.5.78}$$

where F_t is the noise figure for the noise source (the output circuitry), k_B is the familiar Boltzmann's constant, T is the absolute temperature, B is the system bandwidth, and R_{eqv} is the equivalent impedance of the output circuit.

The signal-to-noise ratio can be determined by noting that the noise sources described above are independent of each other, since this allows us to represent the total fluctuations or variance of the output current by

$$\sigma_{out}^2 = \sigma_c^2 + \sigma_l^2 + \sigma_t^2, \tag{6.5.79}$$

where the subscripts *out*, *c*, *l*, and *t* refer to the variances corresponding to the output signal, charge carriers, leakage current, and thermal noise respectively. If the output signal is due to the electrons, as is usually the case, then the fluctuations in the electron current can be written as

$$\sigma_c^2 = 2e I_e B \langle G_e \rangle^2 F_e, \tag{6.5.80}$$

where B is the bandwidth of the system, $\langle G_e \rangle$ is the mean electron multiplication factor, I_e is the mean electron current, and F_e is the excess noise factor for the electron current. To compute the signal-to-noise ratio we also need the signal current, which for this case is simply given by $I_e \langle G_e \rangle$ since I_e is the signal current before multiplication. The expression for the signal-to-noise ratio can now be written as

$$
\begin{aligned}
S/N &= \frac{I_e \langle G_e \rangle}{\sigma_{out}} \\
&= \frac{I_e \langle G_e \rangle}{\sqrt{\sigma_c^2 + \sigma_l^2 + \sigma_t^2}} \\
&= \frac{I_e \langle G_e \rangle}{\sqrt{2eI_e B \langle G_e \rangle^2 F_e + 2eI_l B \langle G_l \rangle^2 F_l + 4F_t k_B T B R_{eqv}^{-1}}}.
\end{aligned} \tag{6.5.81}
$$

The signal-to-noise ration depends on the mean gain which in turn has a dependence on the ratio of hole and electron ionization rates u. In fact S/N is highly sensitive to the value of u as can be deduced from Fig.6.5.26. This implicit dependence on u puts a tough constraint on the value of gain that would yield high signal-to-noise ratio. This is specially true for higher values of u. For small u, the curve is more or less flat after a certain gain and therefore gives some flexibility in terms of choosing the gain according to the particular application.

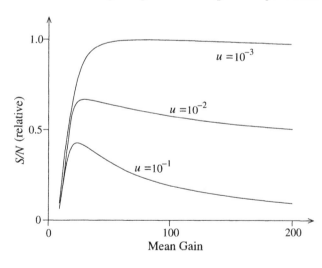

Figure 6.5.26: Dependence of signal-to-noise ratio of an APD on the mean gain at different values of the ratio of the hole ionization rate to the electrion ionization rate.

C.6 Radiation Damage

The radiation damage to semiconductor detectors has already been discussed in the previous Chapter. The damage mechanisms in the APDs are the same as in conventional semiconductor detectors. The overall effect of the damage is, however, more dramatic due to the higher sensitivity of these detectors. In an APD being operated in a hostile radiation environment, the following two distinct types of radiation induced damages can occur.

▶ **Bulk Damage:** This occurs when the lattice atoms in the bulk of the material get displaced from their usual sites through incident particle impacts. An atom displaced in such a way creates a defect site that can trap electrons and thus contribute to the dark current.

▶ **Surface Damage:** The incident radiation can also create similar defect on the surface of the APD's active volume, decreasing the quantum efficiency of the detector.

Problems

1. If the light yield of a scintillation material is 25,000 photons per MeV, estimate its efficiency of producing green scintillation light.

2. Estimate the number of luminous centers in a $NaI : Tl$ scintillator if it emits 4×10^4 photons after absorbing 2 MeV of energy. Assume its quantum efficiency of luminous centers to be 0.96.

3. Compute the efficiency of a $CsI : Tl$ scintillator if its quantum efficiency of luminous centers is 0.92 and the efficiency of luminous center activation is 0.85 (note that these two numbers have been arbitrarily assigned and therefore should not be assumed to represent a real crystal).

4. Calculate the minimum tapering angle of a fish-tail type light guide used in air for a maximum angle of incidence of 75^0. Assume the refractive index of the light guide to be 1.55.

5. Assume that the light guide of previous exercise is immersed in light water. Calculate the angle above which the total internal reflection inside the guide will be guaranteed.

6. You want to design a photomultiplier tube that could detect photons having wavelength up to 520 nm. Determine the work function of the photocathode that would be suitable for the PMT.

7. A PMT having a gain of 10^5 is exposed to a photon beam of intensity 1.5×10^5 s^{-1} at a wavelength of 470 nm. Compute the maximum anode current and the voltage at the PMT output assuming that the quantum efficiency of the photocathode is 60%.

8. A photomultiplier detection system having a bandwidth of 100 Hz has a dark current of 200 pA. Determine the relative change in the fluctuations of the dark current if the system bandwidth is changed to 1 kHz.

9. In an effort to determine the shot noise due to all sources the output current of a PMT is first measured in the dark and then by exposing it to a photon flux. The first measurement yielded a current of 1.6 nA while the second one gave a current of 20.5 nA. Determine the total shot noise due to all sources.

10. Determine the minimum power that can be detected from a PMT having a background current of 300 pA and a dark current of 1.7 nA.

11. Determine the relative change in the Johnson noise for a PMT when it is cooled down from room temperature (300 K) to 250 K.

12. A PMT has 12 dynodes, the last 8 of which have individual gains of 5 each. Assume that the first 4 dynodes produce the same number of secondary electrons, what should be their individual gains if the overall gain of the PMT is to remain between 10^7 and 10^8.

13. A PMT having 10 dynodes with a gain of 6×10^6 is used to detect 540 nm photons. Assuming that the incident power is 100 nW, compute the photoelectric

current and the anode current. How much voltage this current corresponds to if the equivalent resistance of the output circuit is 100 Ω. Assume the quantum and collection efficiencies to be 40% and 80% respectively.

14. An APD has a gain of 100 and is used to measure photon intensity from a laser light having mean wavelength of 2.2 μm. If the quantum efficiency of the APD is 0.56, compute its responsivity. Also calculate the photocurrent if the incident photon flux is 2.5×10^{11} s^{-1}.

15. Compute the relative change in the mean gain of an APD if the hole-to-electron ionization rate ratio changes from 0.001 to 0.1. The APD has a gain of 100 at $u = 0.001$.

16. Compute the average photocurrent flowing through an APD having a gain of 50 and responsivity of 10 if it is exposed to a 4.8 eV photon beam of intensity 1.6×10^8 s^{-1}.

Bibliography

[1] Alimonti, G. et al., A **Large Scale Low Background Liquid Scintillation Detector: The Counting Test Facility at Gran Sasso**, Nucl. Instrum. Meth. A414, 1998.

[2] Amsler, C. et al., **Temperature Dependence of Pure** CsI **: Scintillation Light Yield and Decay Time**, Nucl. Instrum. Meth. A, 480, 2002.

[3] Arnfield, M.R. et al., **Radiation Induced Light in Optical Fibers and Plastic Scintillators: Application to Brachytherapy Dosimetry**, IEEE Trans. Nucl. Sci., 43, 1996.

[4] Baldini, A. et al., **Liquid Xe Scintillation Calorimetry and Xe Optical Properties**, oai:arXiv.org:physics/0401072, 2004. bibitembeddar1 Beddar, A.S., A **New Scintillator Detector System for the Quality Assurance of Co-60 and High Energy Therapy Machines**, Phys. Med. Biol., 39, 1994.

[5] Beddar, A.S., **Radiation Response of Optical Fibers to Clinical Radiotherapy Photon and Electron Beam**, OPTO-88, ESI Publications, Paris, 1988. bibitembeddar1 Beddar, A.S. et al., **Water-Equivalent Plastic Scintillation Detectors for High-Energy Beam Dosimetry**, Phys. Med. Biol., 37, 1992.

[6] Birks, J.B., Proc. Phys. Soc. A64, 847, 1951.

[7] Bohnet, I. et al., **Radiation Hardness of Different Calorimeters Studied with a Movable CO-60 Source at the ZEUS Detector**, Calorimetry in High Energy Physics, Lisbon, 1999.

[8] Brown, G.J., **Photodetectors: Materials and Devices V: 26-28 January 2000, San Jose, California**, SPIE-International Society for Optical Engine, 2000.

[9] Busenitz, J. et al., **The KAMLAND Experiment**, Int. J. Mod. Phys. A 16S 1B, 2001.

[10] Carel, van E., W.E., **Inorganic Scintillators in Medical Imaging**, Phys. Med. Biol. 47, 2002.

[11] Clarke, H.B. et al., **The Scintillation Phenomenon in Anthracene I. Radiation Damage**, Proc. Phys. Soc., Vol.79, 1962.

[12] Cohen, M.J., **Semiconductor Photodetectors**, Society of Photo Optical, 2005.

[13] Cova, S. et al., **Avalanche Photodiodes for Near-Infrared Photon counting**, SPIE Proceedings, Vol.2388, 1995.

[14] Dautet, H.I. et al., **Photon-Counting Techniques with Silicon Avalanche Photodiode**, Applied Optics, 32(21), 1993.

[15] Dennis, P.N.J., **Photodetectors**, Springer, 1986.

[16] Donati, S., **Photodetectors: Devices, Circuits and Applications**, Prentice Hall PTR, 1999.

[17] Dorenbos, P. et al., **Afterglow and Thermoluminescence Properties of** $Lu_2SiO_5 : Ce$ **Scintillation Crystals**, J. Phys.: Condens. Matter 6, 1994.

[18] Dorenbos, P., van Eijk, C.W., **Inorganic Scintillators & Their Applications: Proceedings of the International Conference, Scint 95**, Coronet Books. 1996.

[19] Frlez, E. et al., **Radiation Hardness of the PIBETA Detector Components**, Fizika B, No.12, 2003.

[20] Golovkin, S.V. et al., **New Liquid Scintillators for Particle Detectors Based on Capillary Fibers**, IHEP 96-13, Protvino, 1996.

[21] Hakim, N.Z. et al., **Generalized Excess Noise Factor for Avalanche Photodiodes of Arbitrary Structure**, IEEE Trans. Electron Devices, Vol.37, 1990.

[22] Hakim, N.Z., **Signal-to-Noise Ratio for Lightwave Systems Using Avalanche Photodiodes**, J. Lightwave Technology, Vol.9, No.3, March 1991.

[23] Hayat, M.M., **An Analytical Approximation for the Excess Noise Factor of Avalanche Photodiodes with Dead Space**, IEEE Trans. Electron Device Lett., Vol.20, No.7, July 1999.

[24] Hayat, M.M. et al., **Gain-Bandwidth Characteristics of Thin Avalanche Photodiodes**, IEEE Trans. Electron Devices, Vol.49, No.5, May 2002.

[25] Horrocks, D., **Organic Scintillators**, Gordon and Breach, 1968.

[26] Huhtinen, M., **High-Energy Proton Induced Damage in** $PbWO_4$ **Calorimeter Crystals**, arXiv:physics/0412085 v2, 2004.

[27] Ikhlef, A. et al., **X-Ray Imaging and Detection Using Plastic Scintillating Fibers**, Nucl. Instr. and Meth. A, 442, 2000.

[28] Kappers, L.A. et al., **Afterglow, Low-Temperature Radioluminescence and Thermoluminescence of** $Lu_2O_3 : Eu$ **Ceramic Scintillators**, Nucl. Instrum. Meth. A, 537, 2005.

[29] Lecoq, P. et al., **Inorganic Scintillators for Detector Systems : Physical Principles and Crystal Engineering**, Springer, 2006.

[30] Letourneau, D, Pouliot, J., Roy, R., **Miniature Scintillating Detector for Small Field Radiation Therapy**, Med. Phys., 12, 1999.

[31] McIntyre, R.J., **A New Look at Impact Ionization–Part I: A Theory of Gain, Noise, Breakdown Probability and Frequency Response**, IEEE Trans. Electron Devices, Vol.46, Aug. 1999.

[32] McIntyre, R.J., **Multiplication Noise in Uniform Avalanche Diodes**, IEEE Trans. Electron Devices, Vol.ED-13, 1966.

[33] McIntyre, R.J., **The Distribution of Gains in Uniformly Multiplying Avalanche Photodiodes: Experimental**, IEEE Transactions on Electron Devices, ED-19(6):713-718, June 1972.

[34] Moses, W.W., **Current Trends in Scintillator Detectors and Materials**, Nucl. Instrum. Meth. A, 487, 2002.

[35] Rodnyi, P.A., **Physical Processes in Inorganic Scintillators**, CRC, 1997.

[36] Ross, H. et al., **Liquid Scintillation Counting and Organic Scintillators**, CRC, 1991.

[37] Saleh, M.A. et al., **Impact-Ionization and Noise Characteristics of Thin III-V Avalanche Photodiodes**, IEEE Trans. Electron Devices, Vol.48, Dec. 2001.

[38] Stanton, N.R., **A Monte Carlo Program for Calculating Neutron Detection Efficiencies in Plastic Scintillator**, S.N., 1971.

[39] Swank, R., **Characterization of Scintillators**, Annu. Rev. Nucl. Sci., Vol. 4, 1954.

[40] Taylor, C.J. et al., **Response of Some Scintillation Crystals to Charged Particles**, Phys. Rev., Vol.84, No.5, 1951.

[41] Teich, M.C. et al., **Excess Noise Factors for Conventional and Superlattice Avalanche Photodiodes and Photomultiplier Tubes**, IEEE J. Quantum Electron., Vol.QE-22, 1986.

[42] Trishenkov, M.A., **Detection of Low-Level Optical Signals : Photodetectors, Focal Plane Arrays and Systems**, Springer, 1999.

[43] Weber, M.J., **Selected Papers on Phosphors, Light Emitting Diodes, and Scintillators: Applications of Photoluminescence, Cathodoluminescence, Electroluminescence, and Radioluminescence**, SPIE-International Society for Optical Engineering, 1998.

Position Sensitive Detection and Imaging

The spatial variation in energy and intensity of radiation has been found to be a very useful tool in many disciplines. For example medical x-ray diagnostics utilize the change in radiation intensity due to relative attenuation coefficients of different organs and bones inside the body to create a two-dimensional image. In fact, with the advent of electronic detectors, a lot of progress has been made in building two- and three-dimensional imaging systems for medical diagnostic purposes. However the development has not been limited to medical applications. Other fields, such as particle physics research, environmental radiation monitoring, and non-destructive testing of materials have equally benefited from these developments. Latest advancements, such as charged coupled devices, multistrip silicon detectors, and microstrip gas chambers have taken this field of research and development to a higher level of sophistication and utility.

Position sensitive detection and imaging are two fields that have some overlapping areas in terms of technology. For example the detector technologies used in both systems are very similar but the processing of data for imaging systems is much more involved than for position sensitive detectors. By position sensitive detector we mean a system with which one could find the position of an interacting particle up to a certain resolution. In such a system the radiation is allowed to interact directly with the detector. An imaging system, on the other hand, is concerned about creating the image of an object by exposing it to radiation and then detecting the reflected or transmitted radiation.

In this chapter we will first discuss the position sensitive detection and then move on to the imaging devices. But before we do that it is worthwhile to introduce the reader to some terminologies and quantities that are used to characterize the operation of these devices.

7.1 Some Important Terminologies and Quantities

Since position sensitive detectors and imaging devices are essentially based on similar concepts of particle detection, therefore most of the terminologies and quantities that will be introduced here are common to both of them.

7.1.A Spatial Resolution

A basic requirement of all position sensitive and imaging detectors is that they must be able to distinguish between intensities at two closely spaced points in space. The spacing of these points and how well the detector is able to separately record radiation intensities at the points determine the spatial resolution of the system. Position resolution is a critical criterion used to characterize the usefulness of the detection system for a particular application and can therefore be said to be application dependent. For position sensitive detectors, the requirement for spatial resolution are generally much more stringent than for imaging systems. For example in detectors used in high energy experiments, a position resolution of a few tens of micrometers is not uncommon. On the other hand, in a CT scanner, which is an imaging system, a spatial resolution of a few millimeters is considered practical for diagnostic purposes.

Most modern position sensitive detectors consist of a large number of very thin but long and closely spaced detectors. For example a multiwire proportional counter comprises of a large number of very closely spaced thin anode wires. Similarly a multistrip silicon detector consists of very large number of a few micrometer wide and a few centimeters long silicon detectors doped on a single crystal sheet.The spatial resolution of such detectors is dependent on the individual detector elements and the free space between them.

Let us now have a look at the imaging systems. In principle, a position sensitive detector can be used for imaging applications and vice versa. However, due to different requirements of spatial resolution and frequency response the general practice is to design the systems for either position resolution or imaging. A drawback of the imaging systems is that no matter how precise they are, there is always some discrepancy between an object and its image. The reason for this discrepancy lies in the way imaging detectors are constructed. Most of the modern imaging systems consist of a number of discrete elements, generally known as *pixels*, each having some shape and all lumped close together to form an array of detectors. Each of these pixels is able to record the intensity of the incident radiation. Ideally there should not be any image deterioration if the pixels are small enough and are kept very close to each other. However there are two distinct effects that can cause degradation of the spatial resolution. One is the crosstalk and the other is aliasing. The crosstalk can affect the position sensitive detectors as well while aliasing is specific to imaging systems. These two effects are discussed below.

A.1   Crosstalk

The term crosstalk has actually been borrowed from electronics where it refers to the pickup of electrical signal by nearby signal lines and devices. Since position sensitive detectors and imaging devices are made up of large number of individual detector elements therefore it is possible that a part of the signal received by one of the channels (that is, one of the detectors) gets transfered to the neighboring channels. The detectors based on a single bulk of the material are most affected by crosstalk. The reason is that such detectors provide alternative paths to the signal charge carriers after they get scattered. The charges that go astray are thus picked up by the neighboring channels.

A.2   Aliasing and Antialiasing

Aliasing is an effect that causes blurring of the image. The image appears blurred when it acquires extra information that the original object did not contain. This unwanted information is artificially produced either by wrong sampling or by wrong image reconstruction. Let us discuss these two causes of aliasing in some detail.

Aliasing due to Sampling Frequency

Most of the modern imaging systems generate the so called *digital image* of the object by *sampling* the analog information. The term sampling is used to characterize the process of gathering spatial information at regular time intervals and passing it on to the analog to digital converters. Of course there is no reason to exclude the possibility of sampling at irregular time intervals but the general practice is to do it at well defined regular intervals since it simplifies the further processing of the information. There are also imaging systems that perform only the *spatial sampling*, that is, the image is sampled at only one point in time. An obvious example would be the digital cameras that sample and store information through charged coupled photon detection devices. In most medical imaging systems both spatial and temporal sampling is done at the same time to preserve and reconstruct as many details of the object as possible.

Let us concentrate now on the process of spatial sampling since it can cause aliasing of the final image. To understand this we first note that the object, whose image is to be captured, has some spatial frequencies. In other words there are shades in the image occurring at different spatial intervals. In general, the spatial variation of shades in an image is fairly irregular and therefore contains a number of frequencies. The spatial frequency content of the object can be obtained by taking the Fourier transform of variation of its shades. In general the Fourier spectrum shows a number of frequencies, the maximum of which corresponds to the bandwidth of the system (this is true only for baseband systems but most images can be characterized to be baseband). Now, if we want to recreate the image of this object through an array detector then the natural question to ask is how many pixels (or individual detection elements) per unit distance should be enough to create an acceptable image. In other words, what should be the sampling frequency of the detector array be? Intuitively we would think that the sampling frequency must be higher than the maximum frequency of the shades in the actual image. In fact, in reality the sampling frequency must be greater than twice the maximum frequency. This is formally stated by the following *sampling theorem.*

An image can be faithfully reconstructed if and only if the sampling frequency is greater than or equal to twice the bandwidth of the input signal.

Mathematically, this can be written as

$$f_s \geq 2B, \tag{7.1.1}$$

where F_s is the sampling frequency and B is the bandwidth of the detector. In the specific case of a baseband system, for which the highest frequency f_{max} coincides with the bandwidth, the sampling condition can be written as

$$f_s \geq 2f_{max}, \tag{7.1.2}$$

where $2f_{max}$ is sometimes also called the *Nyquist frequency*, a term that was originally introduced for sampling in time domain. Hence we can say that an image will be aliased or will not be faithfully reconstructed if

$$f_s < f_{Nyq}, \tag{7.1.3}$$

where $f_{Nyq} = 2f_{max}$ is the Nyquist frequency for a baseband system. An array detector having pixels at a distance of d from each other we have $f_{max} = 1/d$ and the Nyquist frequency is given by

$$f_{Nyq} = \frac{2}{d}, \tag{7.1.4}$$

and the condition that the image does not get aliased or gets faithfully reconstructed is

$$f_s \geq \frac{2}{d}. \tag{7.1.5}$$

In most instances the Nyquist frequency is reported in units of line pairs per millimeter (lp/mm) or cycles/mm. Only the frequencies below this value faithfully pass through the system and therefore the device acts as a spatial low pass filter. The frequency spectrum beyond the Nyquist value is folded about that frequency and added to the spectrum of lower frequencies causing aliasing of the reconstructed image. It is obvious that aliasing results is an artificial increase in the spectral content of the signal and therefore should be minimized for a faithful reconstruction of the actual image.

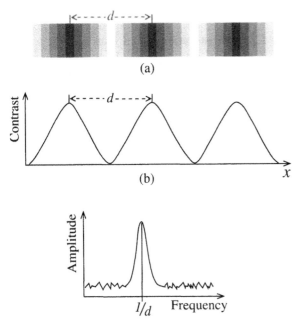

Figure 7.1.1: (a) The shades in an object's image that can be described by a sine function as shown in (b). Contrary to the figure shown, assume that the shades are continuously distributed and do not have defined boundaries. (c) The Fourier transform of the sine function is a single peak at a frequency that is equal to the inverse of the wavelength.

Let us now try to intuitively understand how sampling at a frequency lower than the Nyquist frequency can cause aliasing. Fig.7.1.1(a) shows a one-dimensional image having periodically occurring shades that can be described by a sine function as shown in Fig.7.1.1(b). The Fourier transform of the function as sketched in Fig.7.1.1(c) shows a single peak at a frequency that corresponds to the *wavelength* of the sine function.

Let us see what happens if we use an array detector having a spatial frequency equal to the frequency of the sine wave or half the Nyquist frequency. Such a detector would consist of pixels or individual detection elements arranged at a distance d from each other (see Fig.7.1.2). Of course the detector can record information only at the pixel locations. This means that, for a periodic function, the recorded contrast will be the same at all points. The reconstructed image, as shown in Fig.7.1.2 will then simply be a single shade. The wrong sampling frequency has thus added information in the reconstructed image that was not there in the actual object. In other words the image has been aliased due to sampling at a frequency less than the Nyquist frequency.

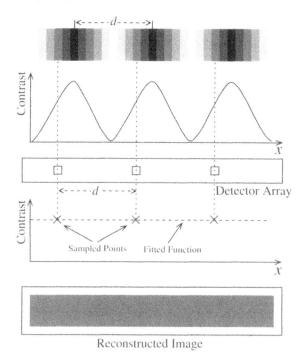

Figure 7.1.2: Sampling of the image having the Nyquist frequency of $2/d$ by a detector array having spatial frequency of $1/d$. Since the acquired data points lie on a single straight line, the reconstructed image is severely aliased. That is, the final image contains information that was not there in the actual image. The contrast in the final image depends on the position of detector pixels relative to the image being sampled.

Another example of aliasing is shown in Fig.7.1.3, where the detector has a regular array structure with a sampling frequency of $f_s = 5/4d$. Note that here also the detector does not meet the condition set by the sampling theorem (that is, $f_s \geq 2/d$). As shown in the figure, the image is sampled only at the pixel locations. The reconstructed image how has more information as compared to the case of Fig.7.1.2 where the sampling frequency was even lower. Hence increasing the sampling frequency has increased the image quality. However still the reconstructed image contains a lot of information not contained in the actual image. Or in other words, it is still aliased.

Let us now increase the sampling frequency further such that it becomes equal to the Nyquist frequency. Since now it satisfies the sampling condition, we would

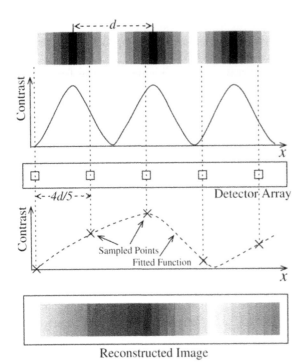

Figure 7.1.3: Sampling of the image having the Nyquist frequency of $2/d$ by a detector array having spatial frequency of $5/4d$. Since the sampling condition is not met, the reconstructed image is aliased.

expect that the reconstructed image would not get aliased. For such a system, the detector array elements or pixels must be a distance $d/2$ apart from each other as shown in Fig.7.1.4. It is clear from the figure that sampling at this frequency helps in reconstructing the image fairly well. A point to note here is that although the reconstruction of the image at Nyquist frequency is shown to be perfect, but in reality since the images are not always so regular therefore sampling at a higher frequency is generally preferred. This process is sometimes referred to as *oversampling* and is a common practice.

Up until now we have talked about sampling in spatial domain. However aliasing is best understood if we look at the sampling in the frequency domain. It should be noted that sampling in spatial domain is equivalent to multiplying by a spike function at each spatial point (see Fig.7.1.5(a)), a process generally known as *multiplication*. Now, if we wanted to perform the same operation in frequency domain, we will have to *convolute* the frequency spectrum of the image by a spike function. This process of convolution is mathematically represented by

$$f(\omega) \otimes g(\omega) = \int f(u)g(\omega - u)du. \qquad (7.1.6)$$

where f and g are the two functions (for example frequency spectrum and spike function) in frequency domain. This implies that convolving in frequency domain is equivalent to multiplying in spatial domain. The opposite is also true, that is, convolution in spatial domain is like multiplication in frequency domain. Mathematically, these equivalences for two functions f and g in frequency ω and spatial x

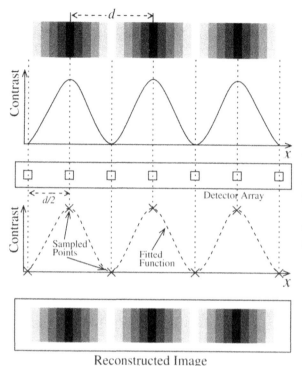

Figure 7.1.4: Sampling of the image having the Nyquist frequency of $2/d$ by a detector array having spatial frequency of $2/d$. Since the sampling condition is satisfied, the reconstructed image is not aliased.

domains can be represented by

$$f(\omega) \otimes g(\omega) \quad \Leftrightarrow \quad f(x) \times g(x) \quad \text{and} \tag{7.1.7}$$
$$f(\omega) \times g(\omega) \quad \Leftrightarrow \quad f(x) \otimes g(x), \tag{7.1.8}$$

where $\otimes$ and $\times$ represent convolution and multiplication respectively.

Going back to our discussion, we first examine the shape of the convoluted spectrum as shown in Fig.7.1.5(b). It is apparent that convolution has introduced copies of the replicated image. This is an unwanted byproduct of the process and must somehow be filtered out. This is most conveniently done by *multiplying* the function by a box function (see Fig.7.1.5(c)). If the spectra were well separated, as they should, the result is good reconstruction of the original spectrum as shown in the figure. The original image can then be obtained by taking the inverse Fourier transform of this spectrum. However, if the copies of the convoluted power spectrum were overlapped due to sampling at lower than Nyquist frequency, then the high and low frequencies will get overlapped at end positions as shown in Fig.7.1.5(d). This would result in an effective frequency that is different from the actual one by an amount determined by the overlap. The next step in the reconstruction process, that is, multiplication by a box function is shown in Fig.7.1.5(e). Clearly the result would be a spectrum different from the actual one. The inverse Fourier transform of this *distorted* frequency spectrum will not be able to faithfully reproduce the original image. The resultant image would then be aliased, that is, it would contain spatial frequencies that were not present in the actual image.

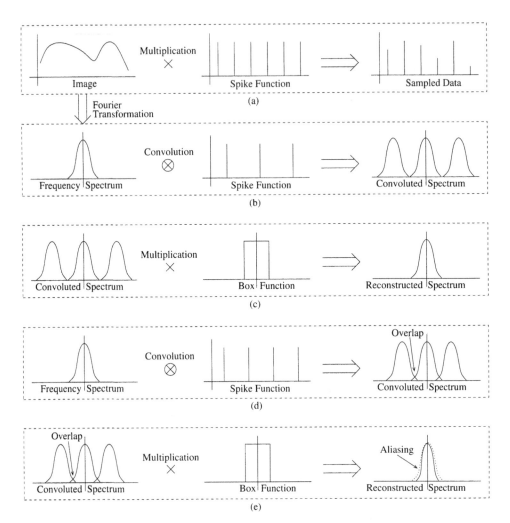

Figure 7.1.5: (a) Sampling in spatial domain is equivalent to multiplication of the image by a spike function. (b) Sampling in frequency domain is equivalent to convolution of the frequency spectrum with a spike function. (c) Multiplication of the convoluted frequency spectrum by a sinc function leads to reconstruction of the original frequency spectrum provided the sampling condition has been satisfied. (d) Convolution of the frequency spectrum with spike function leading to sampling at a frequency lower than the Nyquist frequency. The resulting copies of the frequency spectrum overlap at the two ends. (e) Multiplication of the frequency spectrum obtained in the last step with a box function to determine the original frequency spectrum leads to aliasing.

The above examples clearly demonstrate the importance of Nyquist condition ($f_s \geq N_{Nyq}$) in image sampling and reconstruction. However, even though this is a necessary condition but is in no way the sufficient condition for producing an image that is not aliased. Aliasing can also creep into the image at the reconstruction stage, which is the topic of our next discussion.

Aliasing due to Reconstruction

Aliasing can also occur due to poor reconstruction. We saw earlier that reconstruction in the frequency domain can be done by multiplying the convoluted frequency spectrum by a box function. Now, suppose instead of this standard practice, one resorts to using a sinc function instead. The sinc function actually represents the Fourier transform of the box function. In other words, convolution of a function in the spatial domain by a box function is equivalent to multiplying it by a sinc function in spatial domain. For any variable x, the sinc function is defined by

$$\text{sinc}(x) = \frac{\sin(\pi x)}{\pi x}. \tag{7.1.9}$$

Fig.7.1.6 shows the process of multiplication of the convoluted frequency spectrum by a sinc function. It is apparent that even when there is no overlap between the neighboring copies of the spectrum, the final spectrum has unwanted frequencies. The inverse Fourier transform of this spectrum will contain a high degree of aliasing. Note that this kind of aliasing has nothing to do with the sampling frequency and is solely due to poor reconstruction.

A.3 Point Spread Function (PSF)

PSF determines how well a detector is able to distinguish between two perfect points separated in space. Let us suppose there is a perfect point whose image is recorded by an imaging system consisting of a large number of pixels. These pixels can be regarded as individual detectors separated by some small distance or pitch. If the radiation being used to create the image of the spot on these pixels is monochromatic and perpendicular to both surfaces, the image will consist of a round shape having highest intensity in the middle and rapidly decreasing intensity with distance away from the center (see Fig.7.1.7(a)). If one now plots this intensity with respect to any one coordinate (the image will be radially symmetrical for a perfect spot), a Gaussian-like distribution can be obtained. This one-dimensional distribution is our point spread function. The distribution, of course, is never perfectly Gaussian and can even be skewed, but, for most practical purposes, it does not really matter how it looks like since we can still define a quantity called *Full Width at Half Maximum* or *FWHM* as shown in Fig.7.1.7(b). The reason for defining this quantity can be understood from Fig.7.1.8, which shows the distributions of two closely spaced spots. It is quite clear that if the distance between the two spots is equal to or larger than *FWHM*, the detector will be able to distinguish between them. The spatial resolution of the system therefore depends on the value of *FWHM*.

The *PSF* for most imaging systems can be characterized by a Gaussian distribution, given by

$$f(x) = \frac{1}{\sqrt{2\pi\sigma^2}} \exp\left[\frac{-(x - x_0)^2}{2\sigma^2}\right], \tag{7.1.10}$$

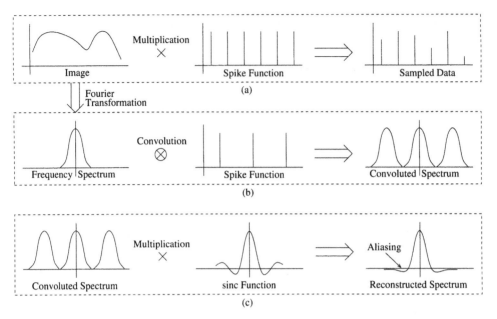

Figure 7.1.6: (a) Sampling in spatial domain is equivalent to multiplication of the image by a spike function. (b) Sampling in frequency domain is equivalent to convolution of the frequency spectrum with a spike function. (c) Multiplication of the convoluted frequency spectrum by a sinc function leads to aliasing even if the copies of the frequency spectrum are well apart. This type of aliasing is therefore solely due to reconstruction.

where x_0 represents the center of the function and σ is the standard deviation (see Fig.7.1.9). The $FWHM$ of this function is given by

$$
\begin{aligned}
FWHM &= 2\sqrt{2\ln 2}\sigma \\
&\approx 2.36\sigma.
\end{aligned}
\tag{7.1.11}
$$

A.4   Line Spread Function (LSF)

Line spread function is similar to PSF in the sense that it uses the same technique to quantify the spatial resolution. However, in this case, the image of an ideal line is used instead of a point. The LSF is then drawn perpendicularly through this image. Since a line can be thought to consist of a large number of closely spaced points, therefore the LSF produces a whole profile of $FWHM$s instead of a single value as for the PSF. The resolution of the system at any point can then be deduced from this profile. Although LSF is a good measure of the system resolution but it is very difficult to determine experimentally. The main difficulty lies in manufacturing an ideal line object.

(a)

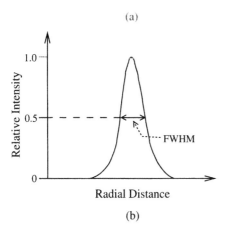

Radial Distance

(b)

Figure 7.1.7: (a) Image of a point made by a pixel detector. (b) Point spread function of image in (a). $FWHM$ is the width of this function at exactly half of the full amplitude. A real PSF is not necessarily Gaussian-like as shown in this figure. The shades shown in the figure should not be taken as having sharp boundaries, as the figure might suggest.

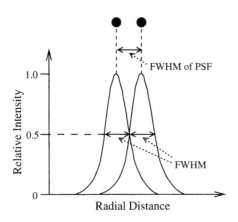

Radial Distance

Figure 7.1.8: $PSFs$ of images of two points close together taken by a pixel detector. Since the points are $FWHM$ distance apart, they are distinguishable. Hence PSF can be used to quantify the spatial resolution of an imaging detector.

A.5 Edge Spread Function (ESF)

The LSF does not offer a very practical means of determining the system resolution due to the engineering problem of manufacturing an ideal line object. ESF in this sense is more convenient since it considers the image of an ideal step function, which can easily be obtained from a rectangular object as shown in Fig.7.1.11.

The edge spread function can be defined as the integral of one dimensional point spread functions. That is

$$ESF(x) = \int LSF(x)dx \qquad (7.1.12)$$

$$\text{or} \quad LSF(x) = \frac{d}{dx}[ESF(x)]. \qquad (7.1.13)$$

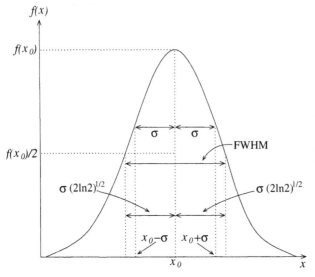

Figure 7.1.9: Definition of various terms in a the expression 7.1.10 for Gaussian distribution. The *PSF* of most imaging systems has Gaussian shape.

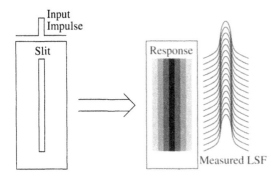

Figure 7.1.10: The *LSF* from a thin line can be thought to consist of a large number of *PSF*s at closely spaced points on the line. The shades shown in the figure should not be taken as having sharp boundaries, as the figure might suggest.

A.6   Modulation Transfer Function (MTF)

The modulation transfer function determines how much contrast of the original object is maintained by the detector. In other words it characterizes how faithfully the spatial frequency content of the object gets transferred to the image.

To understand the MTF, we must first understand what is meant by the modulation function. Earlier we introduced the concept of spatial frequency and saw that the spatial variation of contrast in an image can be represented by waves having different frequencies. Fig.7.1.12 shows the graphical representation of such a wave. Since contrast can be defined in terms of how much light is transmitted, therefore such a graph will always remain positive valued at all spatial points. The modulation

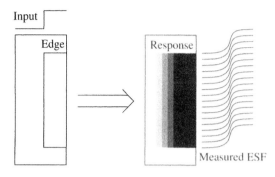

Figure 7.1.11: Edge spread function can be obtained by placing a rectangular object having sharply defined edges between the source and the detector. The shades shown in the figure should not be taken as having sharp boundaries, as the figure might suggest.

of the graph is defined by

$$
\begin{aligned}
M &= \frac{y_1}{y_2} \\
&= \frac{(N_{max} - N_{min})/2}{(N_{max} + N_{min})/2} \\
&= \frac{N_{max} - N_{min}}{N_{max} + N_{min}},
\end{aligned}
\tag{7.1.14}
$$

where y_1 and y_2 are as defined in Fig.7.1.12. N_{max} and N_{min} represent the maximum and minimum values of the function used to quantify the contrast. The function could, for example be the transmittance determined by the pixel readout. Modulation, in principle, can have any value between 0 and 1, although a value of 1 is very difficult, if not impossible, to achieve.

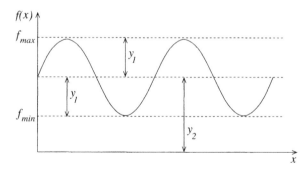

Figure 7.1.12: Definition of the parameters in equation 7.1.15. $f(x)$ represents any convenient function for quantifying the contrast.

Now, the modulation is present not only in the object but also in the image. The ratio of their respective modulations is called modulation transfer ratio, that is

$$
MT = \frac{M_{im}}{M_{ob}},
\tag{7.1.15}
$$

where M_{im} and M_{obj} represent respectively the modulations in the image and the object. The modulation transfer ratio defined here is a function of spatial frequency and therefore can not be used to characterize the response of the system. For that one must determine the modulation transfer ratio at each spatial frequency. The dependence on the modulation transfer ratio on the spatial frequency is called the modulation transfer function or MTF. Fig.7.1.13 shows the shape of a typical MTF.

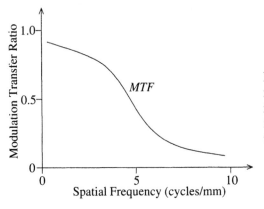

Figure 7.1.13: Typical dependence of modulation transfer ratio on spatial frequency. The parameterized function is called the modulation transfer function.

Since MTF characterizes how well the contrast in the object gets transferred to the image, therefore it is inherently related to the various spread functions we discussed earlier. In fact, MTF can be defined as the modulus of the Fourier transform of the line spread function, that is

$$
\begin{aligned}
MTF(\nu) &= |\mathcal{F}\{LSF(x)\}| \\
&= \frac{1}{\sqrt{2\pi}} \int_{-\infty}^{\infty} LSF(x)e^{\imath 2\pi\nu x}dx,
\end{aligned} \qquad (7.1.16)
$$

where ν is the spatial frequency, x is the spatial distance, and $\{$ represents the Fourier transform. In case of digital sampling, the above integral should be changed to a summation. The problem with the above formalism is that the line spread function is not easy to determine directly. Instead, as mentioned earlier, the edge spread function is normally determined. Since line spread function can be obtained by differentiating the edge spread function therefore the MTF formula given above can be used.

A big advantage of using MTF to characterize the contrast resolution is that if the system consists of different components having their own MTFs, then the overall modulation transfer function MTF_{total} can be obtained by simply multiplying them together. That is

$$
MTF_{total} = \prod_{i} MTF_{i} \qquad (7.1.17)
$$

where MTF_{i} represents the modulation transfer function of ith component of the system.

Most modern imaging systems, such as CCD cameras, are based on regularly spaced pixel detectors. The maximum MTF of such a pixel is well known and is given by the sinc function (see example below)

$$
\begin{aligned}
MTF(\nu)_{pix} &= \text{sinc}(d\nu) \\
&= \frac{\sin(d\nu\pi)}{d\nu\pi},
\end{aligned} \qquad (7.1.18)
$$

where ν is the spatial frequency and d is the pixel size. This expression does not take into account the variation in image quality due to pixel to pixel spacing. The MTF due only to these spacings can also be represented by a similar sinc function given by

$$MTF(\nu)_{sp} = \frac{\sin(p\nu\pi)}{p\nu\pi}. \tag{7.1.19}$$

Here p represents the pixel spacing. We can now used equation 7.1.18 to determine the overall MTF of the detector. This gives

$$
\begin{aligned}
MTF(\nu)_{total} &= MTF(\nu)_{pix} MTF(\nu)_{sp} \\
&= \frac{\sin(d\nu\pi)\sin(p\nu\pi)}{pd\nu^2\pi^2}. \tag{7.1.20}
\end{aligned}
$$

This shows that the MTF of a pixel detector can be improved by decreasing the size of the pixels and the pixel-to-pixel spacings. In the limit that these parameters tend to zero, the MTF tends to unity. That is

$$\lim_{p,d\to 0} MTF(\nu)_{total} = 1. \tag{7.1.21}$$

Of course, this does not represent a *practical* physical limit on MTF since it is almost impossible to design a detector that satisfies this condition. The physical limit is still governed by the actual pixel width and the pixel-to-pixel spacing according to the relation 7.1.20.

7.1.B Efficiency

Efficiency of any detection system characterizes its usefulness for a particular task. Any position sensitive or imaging system consists of a number of individual elements working together. For example a nuclear imaging system consists of a radioactive source, collimation setup, imaging detector array (generally a charged coupled device), data acquisition and storage system, and image processing software. Each of these elements has its own efficiency, which must be taken into account to obtain a realistic measure of the signal-to-noise ratio. Since a complete discussion of all such efficiencies is out of the scope of this book, we will restrict ourselves to the quantum efficiency, which is related directly to the process of detection.

B.1 Quantum Efficiency

Quantum efficiency is traditionally used to quantify the efficiency of x-ray based systems. However it can, in principle, be used for any photon based system. Since most imaging systems use photons as incident radiation therefore it deserves some attention. Before we go on to its mathematical formalism, a point worth noting is that the actual definition of quantum efficiency is dependent on the underlying physical processes. For example during our discussion on photomultiplier tubes we saw that quantum efficiency of photocathode is described in terms of number of incident photons and the number of generated photoelectrons. In this chapter we will define quantum efficiency to characterize how efficiently x-ray photons get absorbed in the material through which they pass. That is how efficiently the photons are absorbed in the detection material. The rationale behind this definition is that the absorption efficiency is directly related to the efficiency of generation of charge pairs, which in turn determines the efficiency of signal generation.

We saw in chapter 3 that the passage of photons through matter can be fairly accurately described through an exponential of the form

$$I = I_0 e^{-\mu x},$$

where I is the photon intensity at a depth x and I_0 is the incident intensity. μ is the attenuation coefficient, which depends on the photon energy. Let us write this equation in a slightly different form.

$$\frac{I_0 - I}{I_0} = 1 - e^{-\mu x}$$

Here the numerator $(I_0 - I)$ is simply the number of photons absorbed per unit time in the material as the beam traverses the distance x. The ratio of this quantity with the incident photon intensity I_0 is called *quantum efficiency (QE)* of the material. Hence we can write

$$QE = 1 - e^{-\mu x}. \tag{7.1.22}$$

From this definition it is clear that the quantum efficiency actually represents the efficiency of the material to absorb the incoming particles. Whether this absorption is useful or not for generating a detectable signal in a detector is another story. In a semiconductor detector, for example a part of the incident radiation gets consumed in increasing the lattice vibrations, something that can be termed as *parasitic absorption* for radiation detection purposes. This is also true for gas filled detectors where some of the incident radiation may be absorbed by the walls of the chamber or by some other elements in the gas without creating charge pairs. However, generally speaking most of the absorbed radiation does actually generate detectable signal and with proper calibration this signal can be fairly accurately related to the energy of the incident radiation. Since in a radiation detector the height of the output signal depends on the energy absorbed in the material therefore the efficiency of absorption is directly related to the efficiency of the detector. Same is true for an imaging system where it is desired that the system has as high a quantum efficiency in the energy range of interest as possible. Looking at the above equation it is apparent that this efficiency can be increased in two ways: by increasing the thickness of the active detection material and by using material with a higher attenuation coefficient.

The energy dependence of attenuation coefficient shows more or less similar behavior for most of the materials of interest. The coefficient is high at lower energies and decreases with energy. If the energy range of interest contains the atomic absorption edge of the material, then the coefficient will show a sharp dip. Consequently there will be a peak in quantum efficiency at that level. Such an energy region can be utilized to create images of high contrast if the region of interest is filled with such a material. In fact this has been tried in experiments of non-invasive angiography where the patient receives potassium intravenously and synchrotron radiation around the K-edge of potassium is used to create the images of the arteries.

B.2 Spatial Detective Quantum Efficiency ($DQE(f)$)

The detective quantum efficiency is perhaps the most widely used parameter for characterizing the effectiveness of imaging systems. The reason is that it determines the overall effect of not only quantum efficiency but also of resolution and noise on the system. In essence, it gives a measure of the overall system performance and how well it produces a quality image.

For imaging systems, DQE is a function of spatial frequency and is therefore sometimes referred to as *spatial* detective quantum efficiency. Mathematically, it is defined as

$$DQE(f) = \frac{SNR_{out}^2(f)}{SNR_{in}^2(f)}. \tag{7.1.23}$$

where SNR_{out} is the signal-to-noise ratio in the output image and SNR_{in} represents the signal-to-noise ratio of the incident radiation, which can be written as

$$
\begin{aligned}
SNR_{in} &= \frac{N_0}{\sigma_N} \\
&= \frac{N_0}{\sqrt{N_0}} \\
&= \sqrt{N_0} \\
\Rightarrow SNR_{in}^2 &= N_0.
\end{aligned}
\tag{7.1.24}
$$

Hence the detective quantum efficiency can be evaluated from

$$DQE(f) = \frac{SNR_{out}^2(f)}{N_0}. \tag{7.1.25}$$

$DQE(f)$ is a frequency dependent parameter and is perhaps the most widely used one for imaging detectors.

7.1.C   Sensitivity

Sensitivity of a position sensitive detector or an imaging device characterizes the minimum signal that can be faithfully detected and measured.

7.1.D   Dynamic Range

Dynamic range gives an idea of the range of the signal that the detector is able to cover. It is defined by

$$D = \frac{N_{max}}{N_{min}}, \tag{7.1.26}$$

where N_{max} and N_{min} represent the maximum and minimum detectable signals respectively.

7.1.E   Uniformity

Most imaging systems are two dimensional devices consisting of an array of pixels. Each of these pixels can be thought of as an individual detector having its own sensitivity, gain, and quantum efficiency. However if the material of all these pixels is the same, as is normally the case, then the variations in these parameters is not very large and the offsets can be corrected in the data acquisition and analysis software. Large variations, on the other hand, are not acceptable in practical systems as they may lead to false results specially for low radiation environments. Uniformity is a term used to describe the fact that all of the pixels show uniform response when illuminated by an spatially uniform beam of radiation. To ensure this, a mask of the response of each pixel under uniform and no illumination are obtained and then subsequently used to correct the actual images.

7.1.F Temporal Linearity

Temporal linearity is of course one of the most desirable properties of any imaging system. However due to aging, radiation damage, and other effects, the detector response changes with time. This requires periodic calibration of the system, something that is painstaking but absolutely necessary to ensure proper functioning.

7.1.G Noise and Signal-to-Noise Ratio (S/N)

Modern position sensitive detectors and imaging systems consist of not only high resolution detection material but also of highly sensitive analog and digital electronics. All the subsystems of such a device contain their own sources of noise. For systems that are supposed to work at very high resolutions, characterization of all of these noise sources is extremely important. If the spread in the signal due to individual components is known then the total noise can be obtained by taking the square root of the sum of their squared values, that is

$$\sigma_{total} = \left[\sum_{i=1}^{N} \sigma_i \right]^{1/2} , \qquad (7.1.27)$$

where σ_i corresponds to the noise of the ith component of the total of N components.

Although one would ideally want to be able to analytically model all the noise sources but in practice such a task is extremely difficult, if not impossible. The general practice, therefore, is to determine the noise experimentally. One point worth noting here is that there are a few noise sources that are always there irrespective of the complexity of the system. This makes life a bit easier since one can then determine at least the physical limit of the overall system resolution. Following is the list of such known noise sources.

▶ Statistical fluctuations of the incident radiation flux.

▶ Statistical fluctuations of the radiation absorption and charge pair production.

▶ Thermal or Johnson noise.

We will look at these and other noise sources later in the chapter.

The term signal-to-noise ratio, as the name suggests, is the ratio of the signal to the total noise. It plays a very important role in determining the usefulness of a particular detection system for a certain application. It should however be noted that it is not always necessary to maximize this ratio. For example in systems where resolution is not of much concern, other factors, such as sampling frequency, may be more relevant than S/N.

7.2 Position Sensitive Detection

7.2.A Types of Position Sensitive Detectors

Broadly speaking, position sensitive detectors can be grouped into three categories: array devices, scanning devices, and timing devices.

A.1 Array Devices

The position sensitive detectors that are made up of a large number of individual detectors arranged in the form of an array fall into this category. The individual detectors of the array can be of any shape, such as linear, circular, or square. The example of linear arrays are multiwire proportional counters and silicon microstrip detectors, both of which were originally developed for high energy experiments. We will discuss these and other types of array devices later in the chapter.

A.2 Scanning Devices

It is in principle possible to perform position measurement by scanning a single detector over the object. However such devices are generally used for imaging and therefore their discussion is deferred to the section on imaging devices.

A.3 Timing Devices

The array devices we discussed earlier can not always be made very large. To understand this point let us suppose that the detector is to be used to find the position of a particle within a precision of a few centimeters along a linear dimension that stretches for several meters. Furthermore suppose that the incident particle can come from any direction. Building an array detector for this purpose would require enclosing the whole volume with small detectors, which may not be practically possible. The solution to this problem is to build the so called timing device in which the position is measured by determining the time of arrival of the pulse generated by the incident particle. This concept is graphically depicted in Fig.7.2.1.

Let us now discuss some of the most commonly used position sensitive detectors.

7.2.B Multiwire Proportional Chambers (MWPCs)

A multiwire proportional counter is simply a gas filled wire chamber consisting of a large number of anode wires stretched in a plane for position measurement. The original motivation of developing such detectors came from the need to track particles in high energy physics experiments. The wire chambers thus developed worked amazingly well and helped scientists make great discoveries at particle accelerators. Although these chambers are still used in some laboratories but since the advent of high resolution silicon detectors their utility has somewhat diminished.

Fig.7.2.2 shows the sketch of a typical multiwire proportional chamber. The closely spaced thin anode wires are stretched in a plane that is midway between two cathode planes. The cathodes are kept at negative potential with respect to anode wires that are kept at ground potential. Each of the anode wires acts as an independent proportional counter and must therefore be read out independently.

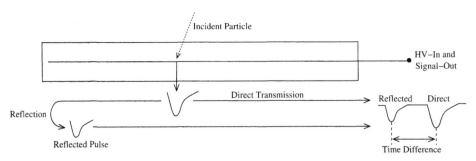

Figure 7.2.1:   Position sensitive detection from a timing device. The signal generated in the central anode wire is transmitted in both directions. The direct and reflected pulses arrive at the processing electronics (such as an oscilloscope) at different times. The time difference is then used to determine the position of the pulse generation provided the system has been calibrated.

The spacing between the anode wires, also called pitch, determines its position resolution. A pitch of 2 mm to 3 mm is typical of most wire chambers. It should, however, be remembered that the position resolution is generally better than this spacing. The cathode planes are generally separated by a distance of approximately six times the pitch. A point worth noting here is that the cathode plane can also consist of closely spaced wires or strips to enhance the position resolution. However it complicates the design and is typically not done.

Typical anode wire diameter is about $20 \mu m$, which produces a high potential gradient with respect to the cathode plane. The potential at any point in the active volume of the chamber can be calculated from

$$V(x,y) = -\frac{q}{4\pi\epsilon_0} ln \left[4\sin^2 \left(\frac{\pi x}{s} \right) + 4\sinh^2 \left(\frac{\pi y}{s} \right) \right], \qquad (7.2.1)$$

which represents the distribution of potential around an array of parallel line charges q measured in C/m along z-axis and located at $(x,y) = (i \times s, i \times s)$ with $i = 0, \pm 1, \pm 2 \ldots$. Naturally the potential near the anode wires should be high enough to cause avalanche multiplication required by proportional counters. The electric field intensity in a typical multiwire chamber is shown in Fig.7.2.3. It is evident that the field intensity is very high near the anode wire and that is where the avalanche begins. The details of the multiplication process that leads to an output pulse proportional to deposited energy has already been discussed in the chapter on gas filled detectors and therefore will not be repeated here. The interested reader is encouraged to go through the section on proportional counters in that chapter. Here we will concentrate on the aspects particular to the multiwire operation of the chamber.

As with single wire proportional counters, the radiation passing through the chamber volume produces electron ion pairs. The ions move towards the cathode while the electrons rush toward the anode wires. The high electric potential at the anode causes the electrons to gain enough energy between collisions to produce secondary ionization and ultimately avalanche. A major concern in multiwire proportional counters is the spreading of this avalanche since it can potentially cause deterioration of the position resolution. The reader might recall, that in single wire

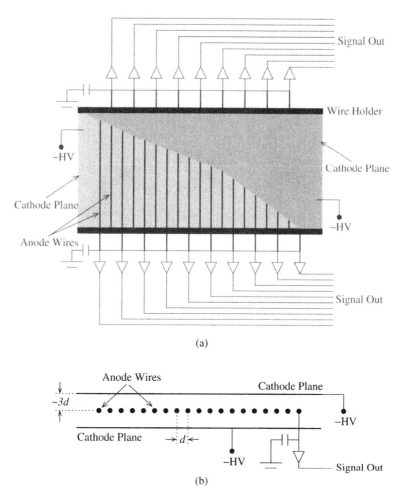

Figure 7.2.2: Top (a) and cross sectional (b) views of a simple multiwire proportional chamber. The spacing between a cathode plane and the plane of anode wires is approximately three times the anode-to-anode distance or pitch, which is typically 2 mm to 3 mm.

proportional counter, avalanche spreading is not of much concern. The best way to deal with this problem is to use a quenching gas. The basic idea behind this is to suppress the electron population through absorption by the quencher molecules. Again, the quenching mechanisms and quenching gases have already been discussed in the chapter on gas filled detectors.

The output of a multiwire proportional chamber depends on the associated electronics, which in turn depends on its mode of operation. In digital mode, in which only counting is performed and the information is binary, the electronics might consist of simple level crossing discriminators and counters. However if the pulse height is also to be measured then the circuitry would become more complicated and involved. Since a multiwire proportional counter can be thought of as a combination

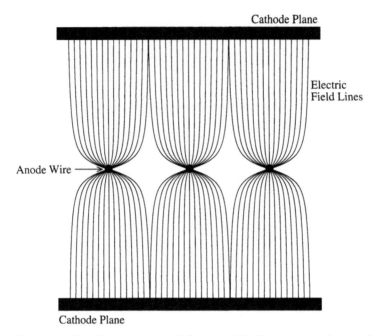

Figure 7.2.3: Electric lines of force inside the active volume of a multiwire proportional chamber.

of many single proportional counters therefore the pulse profile recorded from each of the wires is similar to the one obtained from a conventional single wire cylindrical proportional counter. The time profile of the pulse is mainly dictated by the time constant of the circuit as shown in Fig.7.2.4.

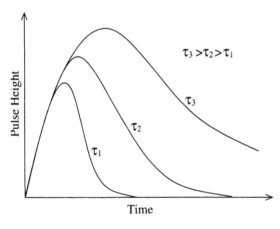

Figure 7.2.4: Typical pulse time profile recorded from a single anode wire of a multiwire proportional counter. Note that the shape of the pulse depends on the time constant τ of the circuit.

The MWPC described above can give only one dimensional position resolution. To obtain two dimensional event identification, one generally uses two such detectors with wires perpendicular to each other. A time coincidence unit can then tag the events in the two chambers as being from the same particle. An even higher resolution can be obtained by using several chambers in succession such that the wires of each of them are tilted at a different angle.

7.2.C   Multiwire Drift Chamber

A multiwire drift chamber is a tracking device that was built to track particles in high energy physics collisions. It uses the timing information to determine the position of an ionizing particle. In principle it is similar to MWPC we just discussed but uses another detector to make the relative time measurements. A drift chamber has two other features as well that distinguish it from a conventional multiwire proportional chamber. One is the anode wire spacing, which in this case is generally a few centimeters as opposed to a few millimeters in case of a MWPC. The other is the shape of the cathode. In a drift chamber the cathode is made up of either closely spaced wires or strips that are kept at distributed potentials. The advantage of this design is that the electrons produced by the incident radiation are directed towards the nearest anode wire with a constant drift velocity.

The design and principle of operation of a multiwire drift chamber are shown in Fig.7.2.5. The incident radiation produces electron ion pairs inside the active volume that drift towards the nearest anode and cathode. After passing through the chamber, the radiation is detected by a fast detector, such as a scintillation counter, which produces a pulse. This pulse initiate a logic-enable pulse that starts a timing counter as shown in Fig.7.2.5(b). The counter keeps on counting until it gets a stop signal that is generated when the electrons eventually reach the anode, produce avalanche, and produce an anode pulse. As shown in the figure, a logic enable pulse is initiated that stops the timer. The number of ticks of the counter determines the time it has taken the electrons to reach the anode wire. Now, if the drift velocity of the electrons in the chamber gas is known, one can determine the distance traveled by the electron and hence position at which the electron was produced by the radiation.

7.2.D   Microstrip Gas Chambers

Microstrip gas chambers are essentially based on the concept of MWPC. Here instead of anode wires stretched between cathode planes, alternating anode and cathode strips are realized on a suitable substrate. On top of this whole structure a drift plane is provided and the chamber is filled with gas. In most modern MSGCs the anode and cathode strips are formed through the process of photo-lithography. The electrodes are generally made of gold of chromium, though in principle one can use any high conductivity material. Although with this technique one can form strips as narrow as a few microns, but the typical widths of anode and cathode strips are 10 μm and 100 μm respectively. The electrode to electrode distance depends on the position resolution desired and can range from a few tens of micormeters to more than a hundred micrometer. The thickness of the electrodes is generally less than 2 μm. A typical MSGC is shown in Fig.7.2.6.

MSGCs are known to show high radiation hardness, which makes them suitable for use in hostile radiation environments. Together with this advantage they also offer very good position resolution. Detectors with 20 μm to 30 μm resolution have been developed and operated at high energy physics research facilities. Although

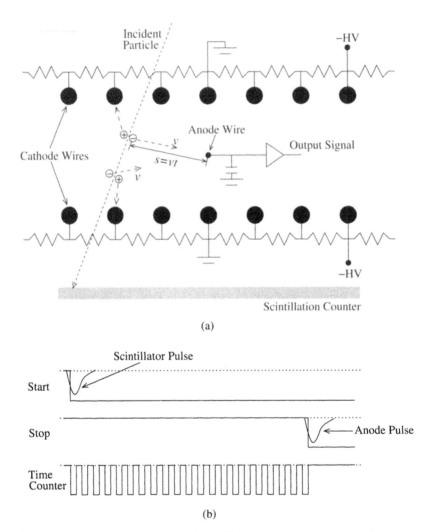

Figure 7.2.5: Sketch of a typical multiwire drift chamber and the timing diagram of its operation. See text for explanations.

MSGCs have seen extensive applications in high energy physics experiments but their utility is in no way limited to this field only. For example they have found applications in medical imaging too, though with a modified structure. We will see later one such design that employs this concept but with inclusion of another layer of electrodes perpendicular to the anodes and cathodes for two dimensional position sensitivity.

7.2.E   Semiconductor Microstrip Detectors (SMSDs)

Semiconductor microstrip detectors are based on the same principle employed to build MWPCs and MWGCs with the exception that the detection medium in this

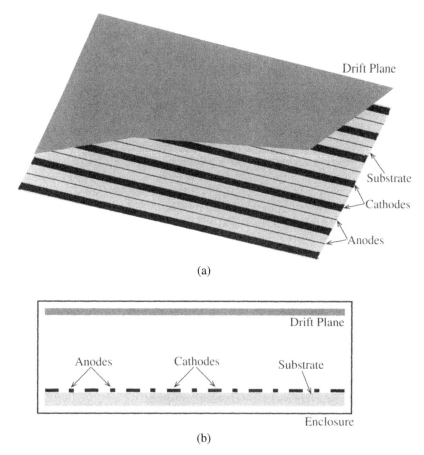

Figure 7.2.6: (a) Top and (b) cross-sectional view of a typical MSGC. The drift plane is kept at a distance of a few millimeters from the substrate.

case is a semiconductor material. The commonly used materials to build SMSDs are silicon and germanium, with silicon being the most popular. Silicon microstrip detectors are now seen as standard tracking devices used in high energy physics experiments. On the other hand, germanium based detectors have also been built and employed for versatile applications such as x-ray spectroscopy and medical imaging. The biggest problem with microstrip semiconductor detectors is that their dimensions are limited by the dimensions of the semiconductor wafer, which can not be made larger than a few tens of a centimeter. Therefore to built a large area detector one must use several small size detectors placed close together.

Two variants of the basic structure of a simple one-sided silicon microstrip detector are shown in Fig.7.2.7. Here the base material or substrate is an n-type semiconductor, which has been implanted with strips of n+ material. A reverse bias is established between these strips and the aluminum implantation on the other side of the silicon wafer. This creates a depletion region, which for best performance should extend up to the end of both sides. This ensures that all the charge pairs created in the bulk of the silicon constitute the output signal. The detector in such a configuration is said to operate in fully- or over-depleted mode, which is the usual mode of operation of SMSDs. The output signal is read out through aluminum strips implanted over the n+ strips. The two are however separated by an insulation strip usually made of SiO_2. The insulation creates a capacitance between the two strips and therefore the readout is actually capacitively coupled. In other words, mirror charges on the aluminum strips are actually seen by the readout circuitry. This design eliminates the need to install external coupling capacitors, something that can be hard to implement considering the available space. Fig.7.2.7(a) shows another layer of p-spray material. The reason for this layer is to discourage oxide buildup between the strips that could lead to shorts and consequent impairment of position resolution. Another design that can achieve this is shown in Fig.7.2.7(b), where instead of a spray, strips of p+ material are implanted between the n+ strips. At present p-stop design is more popular than p-spray design mainly due to engineering difficulties.

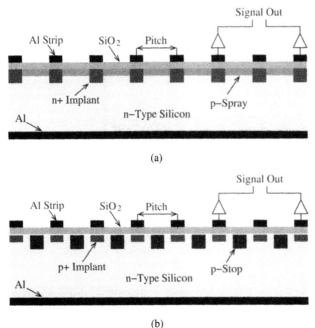

(a)

(b)

Figure 7.2.7: Simple one sided silicon microstrip detector of (a) p-spray and (b) p-stop types .

For proper functioning each of the strips must be biased such that the detector gets fully depleted. Supplying bias is, however, not a trivial matter as it involves distributing the output of a power supply to all the strips through biasing resistors. These bias resistors take space and therefore must be made as small as possible. The available structures include implanted resistors, polysilicon resistors, and punch-through resistors; each having their own pros and cons. The particular choice depends on the type of detector, the radiation environment, and the cost.

The main dimension that determines the spatial resolution of a microstrip detector is its pitch p, which is simply the distance between two consecutive readout strips. In digital mode of operation, where only binary information is recorded, the resolution is given by

$$R \approx \frac{p}{\sqrt{12}}. \tag{7.2.2}$$

This implies that the resolution can be improved by decreasing the pitch. However the pitch can not be decreased to very low values due to engineering constraints and therefore a compromise has to be made. Most silicon detectors are built with a pitch of 40 μm. The resolution can also be increased by operating the detector in analog mode. The reason is that the relative distribution of charges amongst neighboring strips can be used to reconstruct the spatial profile of the charge with the peak that determines the actual position of particle interaction.

Reading out each and every strip of a microstrip detector is not always possible or even desired. The reader can visualize how difficult it would be to make connections on pads that are only 40 μm apart. Of course the soldering is done through specialized machines but even then the task is extremely difficult. Also diagnostics and repair of such a structure would be very labor intensive and involved. Therefore, in general, the SMSD based systems have intermediate strips that are not read out.

Example:
Estimate the improvement in spatial resolution of a microstrip detector operating in digital mode if the pitch is decreased from 100 μm to 40 μm.

Solution:
The resolution with the pitch of 100 μm is given by

$$
\begin{aligned}
R_1 &= \frac{p_1}{\sqrt{12}} \\
&= \frac{100}{\sqrt{12}} \\
&= 28.8 \ \mu m.
\end{aligned}
$$

Similarly the resolution with 40 μm pitch is

$$
\begin{aligned}
R_2 &= \frac{40}{\sqrt{12}} \\
&= 11.5 \ \mu m.
\end{aligned}
$$

The improvement in resolution is then given by

$$
\begin{aligned}
\delta R &= \frac{R_1 - R_2}{R_1} \times 100 \\
&= \frac{28.8 - 11.5}{28.8} \times 100 \\
&= 60\%.
\end{aligned}
\tag{7.2.3}
$$

Perhaps the biggest advantage of semiconductor microstrip detectors over multiwire proportional chambers is that they can be constructed as double sided detectors. In such a detector the strip implantation is done on both sides. Fig.7.2.8 shows a simple double sided silicon detector. Here, the strips on both sides are orthogonal to each other giving a two-dimensional position resolution. Depending on the application it may as well be desired to keep the two sides at another angle with respect to each other. The purpose of the guard rings shown in Fig.7.2.8 is to reduce the possibility of electrical breakdown. Generally several guard rings are implanted on a detector module such that the high voltage gradually decreases instead of a large abrupt change.

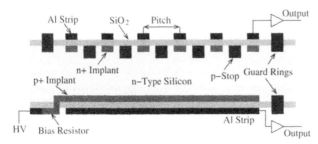

Figure 7.2.8: Design of a simple double sided silicon microstrip detector.

7.3 Imaging Devices

At the most basic level, creating an image of an object requires exposing it to a radiation field, such as x-rays or light photons, and then detecting the reflected, refracted, and transmitted particles on a two dimensional position sensitive detector. Common examples of imaging devices are the familiar medical x-ray system and the digital camera. Broadly speaking, imaging devices can be divided into two categories: conventional and electronic. The film based x-ray system is a conventional device while digital camera is an electronic imaging system. In the following sections we will look at some of the important devices belonging to both of these categories.

7.3.A Conventional Imaging

As stated earlier, by conventional imaging we mean the devices that do not acquire, process, and store the signals electronically.

A.1 X-ray Photographic Films

X-rays are the most widely used means of producing images of objects hidden behind other objects that are opaque to visible light. For example the images of internal organs of human body are routinely taken for medical diagnostic purposes. To make such an image, x-rays are made to fog a photographic film after passing through the sample. If the sample has many different materials (such as tissues and bones in human body), then the x-rays will be variably attenuated depending on the attenuation constant of the materials. Since the photographic film is sensitive to the intensity of the x-rays therefore a two dimensional image of the object is formed. In this way, the photographic film acts as a position sensitive x-ray detector.

The base material of photographic films is generally polyester or celluloid. The film is coated with an emulsion made of some light sensitive material, such as silver nitrate or silver halide salts. After the film has been exposed to x-rays it must go through a series of chemical processes to make the image visible.

The main advantage of photographic films is that they can cover large areas. On the other hand, they have several disadvantages, some of which are listed below.

▶ **Low Sensitivity:** The sensitivity of photographic films to x-rays is so poor that a typical x-ray film absorbs only about 1% of the radiation.

▶ **Low Flux Detection Limit:** Typical photographic films can not detect x-ray flux of less than 0.1 photons per μm^2.

▶ **Logarithmic Response:** Photographic film emulsions respond logarithmically to the incident x-ray flux. Consequently the image contrast needs very careful interpretation.

▶ **Film Development:** The photographic films have to be chemically developed to view the image. Although with modern technology the process is not any more time consuming or labor intensive but still it can not match the real time data acquisition and processing offered by electronic imaging systems.

▶ **Electronic Archiving:** Saving films electronically on a computer memory device for archiving and electronic transfers is difficult, if not impossible.

A.2 Thermoluminescent Detector Arrays

We have already discussed the detectors based on thermoluminescence phenomenon in the chapter on solid state detectors. As a reminder, thermoluminescent materials have the ability to store energy delivered by radiation and then release it after being exposed to heat. It is, in principle, possible to develop an imaging system based on an array of these devices. This is routinely done at high energy physics experiments. An number of closely spaced TLDs are installed at the desired location, for example near a beam pipe. The TLDs keep on accumulating the dose until they are retrieved and then read out. An image reconstruction software can then be used to construct the spatial radiation profile of that area. Use of TLDs in such a way at particle detector experiments is not uncommon. The main advantage of such a system is that it does not require any electronic circuitry and is therefore fairly easy to install. The downside is that it can only provide a spatially coarse radiation profile.

7.3.B Electronics Imaging

Electronic imaging is one of the fastest growing fields mainly due to its applicability in consumer electronics. Digital cameras have now literally replaced the conventional film cameras due to their cost effective availability and high resolution. These cameras use the so called charged coupled devices (CCDs) to register photons coming from the object. The resulting signals are then used by a processor to reconstruct the object's image. The CCDs also have applications in medical diagnostics. In other areas, such as astronomy and particle physics, the electronic imaging devices have seen enormous progress. The fast paced development of such devices has also added a fair bit of complexity in not only their hardware but also the software.

7.3.C   Charged Coupled Devices

Most of the disadvantages of photographic films have been overcome through the development of different kinds of electronic imaging devices. One such detection system is the so called Charged Coupled Device or CCD. This semiconductor based multichannel imaging system is used for two dimensional imaging and is now a standard in many applications. The basic working principle of a CCD is the same as any other semiconductor detector, that is conversion of incident radiation into electron-hole pairs and then measuring the resulting signal. There are different CCD design variants that have been developed over the years based on the specific applications. However we can safely divide these devices into two broad categories: direct imaging systems and indirect imaging systems.

It should be noted that CCDs are in no way limited to the detection of photons. They can be easily configured to measure any type of radiation capable of creating electron-hole pairs in their active detection media. In fact, as we will see later, CCDs can also distinguish between different kinds of incident radiation. These qualities make these devices so versatile that they are currently being successfully used in very different applications ranging from digital cameras to medical radiation scans to radiation monitoring in earth's upper atmosphere.

7.3.D   Direct Imaging

A direct imaging CCD is used to convert the incident radiation directly into electron-hole pairs by letting it interact with the semiconductor material. As in other semiconductor detectors that we visited in the previous Chapter, this conversion takes place inside the depletion region (see Fig.7.3.1). The free electrons and holes move towards opposite electrodes under the influence of the externally applied electric field and induce charges on the readout electrodes. The resulting signal pulse is processed by the electronics attached to the readout electrodes.

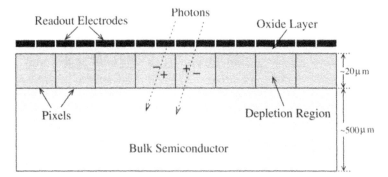

Figure 7.3.1: Schematic of a typical CCD working in direct detection mode with front illumination. The active medium in such devices is generally silicon. An incident particle produces charge pairs in the depletion region, which induce charges on respective readout electrodes that are read out through the connected circuitry.

D.1 Properties of a Direct Imaging CCD

The performance of a CCD is characterized by the following properties.

▶ **Spatial Resolution:** The spatial resolution of a direct imaging device is equal to or better than the dimensions of its pixels. Hence, the smaller the pixel the better the resolution.

▶ **Quantum Efficiency:** The quantum efficiency of a CCD determines the efficiency of *useful absorption* of incident particles in the depletion region. By useful absorption we mean the absorption that leads to generation of at least one electron-hole pair inside the depletion region. In a conventional direct imaging device, the incident particles have to first pass through the readout electrodes and the oxide layer before reaching the depletion region. These layers can significantly attenuate the particle beam especially at lower energies, thus decreasing the overall quantum efficiency of the detector. At higher energies, however, the particles may pass through the depletion region without depositing any or very little of their energy, again resulting in a decreased quantum efficiency. The adverse effect of absorption in front layers can be minimized if the CCD is illuminated from the back side, though it has its own drawback of parasitic absorption in the thick diffusion region. This absorption can be minimized by making this region as thin as possible (see Fig.7.3.2). In such a configuration the back-illumination technique is quite effective in increasing the quantum efficiency at low energy end and is therefore commonly employed in practical systems.

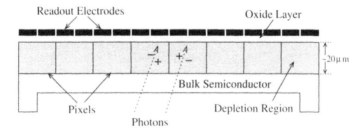

Figure 7.3.2: Schematic of a back-illuminated CCD working in direct detection mode. The semiconductor (generally silicon) layer is made as thin as $10\mu m$ to minimize the low energy photon absorption in the undepleted region of the detector. The acid etching technique that is generally used for this purpose ensures uniformity of the layer.

A point to note here is that there are two different parameters that are sometimes mistakenly used in place of each other. The QE as defined here refers to the ratio of the number of incident photons usefully absorbed $N_{\gamma,abs}$ inside the depletion region to the total number of incident photons $N_{\gamma,inc}$, that is

$$QE = \frac{N_{\gamma,abs}}{N_{\gamma,inc}}. \tag{7.3.1}$$

There is another quantity, which is also sometimes referred to as the quantum efficiency, that actually determines the number of electrons generated $N_{e,gen}$ per incident photon $N_{\gamma,inc}$. To avoid confusion, we will call it conversion efficiency CE.

$$CE = \frac{N_{e,gen}}{N_{\gamma,inc}} \tag{7.3.2}$$

The ratio of these two quantities gives the effective quantum yield η of the CCD, that is, the number of charge pairs generated per absorbed photon.

$$\eta = \frac{CE}{QE} = \frac{N_{e,gen}}{N_{\gamma,abs}} \tag{7.3.3}$$

▶ **Dynamic Range:** The dynamic range of any detection system determines the smallest and the largest signals detectable by the device. It is conventionally determined by dividing the largest detectable signal by the smallest. For a CCD, the dynamic range depends on three factors: resolution of the readout electronics, system gain and the energy of the incident radiation. The electronics resolution is determined by the resolution of the analog to digital converter. The sytem gain G is calculated by taking the ratio of the number of electron-hole pairs generated N_{eh} to the number of ADC counts N_c

$$G = \frac{N_{eh}}{N_c}. \tag{7.3.4}$$

The numerator in the above equation is simply determined by dividing the energy of the incident radiation by the w value of the semiconductor material, that is, $N_{eh} = E/w$. Hence above equation can also be written as

$$G = \frac{E}{wN_c}$$
$$\text{or} \quad N_c = \frac{E}{wG}. \tag{7.3.5}$$

If G and E are known, we can easily determine the minimum detectable signal N_c from the above equation. The dynamic range D would then be simply given by

$$D = \frac{N_{max}}{N_c}, \tag{7.3.6}$$

where N_{max} is the maximum number of measurable counts and is determined by the ADC resolution (see example below).

Example:
Plot the dynamic range of a silicon based CCD having a gain of 10 photoelectrons per ADC count in the photon energy range of 50 eV to 20 keV. Assume that the ADC has 15 bit resolution (It is generally safe to assume that a system having a commercially available 16-bit ADC has a net resolution of 15 bits.).

Solution:
The maximum counts that can be recorded by a 15-bit ADC are $N_{max} = 2^{15} =$ 32768 counts. Hence according to equation 7.3.6, the dynamic range is

$$D = \frac{32768}{N_c}$$
$$= \frac{(32768)(wG)}{E}$$
$$= (32768)(3.62)(10)\frac{1}{E},$$

where $w = 3.62 \ eV$ is the ionization energy for silicon. This variation of dynamic range with energy has been plotted in a semilogarithmic graph (see Fig.7.3.3).

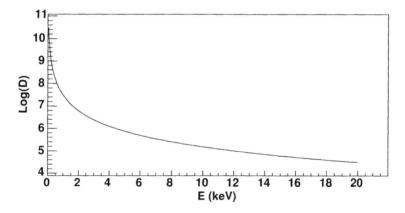

Figure 7.3.3: Semilogarithmic graph of variation of dynamic range with incident photon energy in a silicon based CCD having a gain of 10 photoelectrons per ADC count and 15-bit ADC resolution.

D.2 Disadvantages of Direct Imaging

This direct detection method works quite well for x-rays having high interaction probability in the depletion region. However it suffers from several disadvantages, which make it less desirable for x-ray imaging, some of which are stated below.

▶ Production of electron-hole pairs in the depletion region depends on the energy of incident x-ray photons. At energies higher than $20 keV$, the interaction probability becomes so low that direct detection method does not produce reasonable signal-to-noise ratio.

▶ Like all semiconductor detectors, CCDs also get damaged by instantaneous and integrated radiation doses. Direct detection exposes the CCD to radiation and thus degrades its performance over time due to damage caused by the radiation.

▶ The active area of direct detection CCDs is small and is not suitable for most imaging requirements.

7.3.E Indirect Imaging

The disadvantages of direct detection can be overcome by converting x-ray photons into visible light photons, for which the standard CCD sensors have the highest sensitivity. The trick, of course, is to ensure linearity of the conversion process and make it as efficient as possible. Different types of phosphors and scintillators have been identified as efficient photon converters. Phosphors are granular solids while scintillators are found in polycrystalline powder form. Generally phosphors are used in CCD cameras. The efficiency of the conversion process can be optimized by changing the chemical composition and grain size of the phosphors. Sometimes additional mirror coating is also provided to increase the x-ray interaction probability in the material.

Fig.7.3.4 shows the sketch of the main components of a typical imaging device based on CCD sensors. The x-rays get absorbed in the phosphor, which promptly emit light. These light photons are guided through an optical coupling, such as fiber optic, to the photocathode of an image intensifier. The role of this intensifier is to amplify the light coming from the initial phosphor. The photocathode converts the light photons into electrons, which are multiplied by passing them through multiple stages of a photomultiplier tube. These electrons are then converted into light photons by another layer of phosphor or scintillator. These light photons are made to pass through a fiber optic guide before entering the CCD depletion region to produce electron-hole pairs. Such a multi-stage process ensures that the x-rays do not reach the sensitive CCD sensors, prolonging the life of the detector. Also such a device can be used for a broad range of x-ray energies as compared to the direct imaging systems, which become essentially insensitive even at moderate energies.

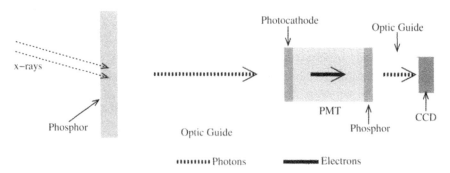

Figure 7.3.4: Sketch of an indirect CCD for x-ray imaging.

7.3.F   Microstrip and Multiwire Detectors

The microstrip and multiwire detectors we discussed earlier in the chapter can be used as imaging detectors. However since these devices provide only a one dimensional spatial information therefore in order to use them as imaging devices one should do either of the following.

▶ The design can be modified such that there are two planes of readout detectors that are kept orthogonal to each other. Both MSGCs or SMSDs can be modified in this way. The reason is that they are produced by implanting strips on a substrate, which can be done on the other side of the substrate as well. We have seen earlier that double sided silicon detectors, where readout strips on one side are normally implanted orthogonal to the other side, are a reality and constitute one of the most commonly used position sensitive detectors. On the other hand, modifying the multiwire proportional or drift chambers in this way carries a number of engineering and operational difficulties.

▶ The other method is to use two or more one-dimensional detectors in succession such that the readout strips or wires are orthogonal or at some other angle with others. Multiwire proportional chambers are operated in this configuration, though to obtain good position resolution and not as imaging devices.

7.3.G   Scintillating Fiber Detectors

In the chapter on scintillation detectors we saw that certain materials produce light when exposed to radiation. One of the problems with these systems is that, in order to avoid absorption of the light produced by them, one must use other phosphors to change the light wavelength. This process has its own efficiency, which decreases the overall efficiency of the detector. Scintillating fibers, on the other hand, are specially designed solids that can not only produce scintillation light but also guide the photons with minimal self absorption. Imaging detectors based on such fibers are a reality now as their variations have been shown to perform amazingly well in particle physics research as well as medical imaging.

Fig.7.3.5 shows a typical scintillation fiber having two components: a scintillation core and an optical cladding. Radiation passing through the scintillation core produces light photons. The optical cladding ensures that most of these photons travel down the fiber through the process of total internal reflection. Most commercially available scintillation fibers produce blue or green light and are therefore suitable for detection by commonly available photomultiplier tubes. The diameter of a typical fiber is around 1 mm^2 but fibers of 100 μm or less in diameter are also available. The fibers can be made in any cross sectional shape but round and square shaped fibers are the most popular.

Most imaging devices use photomultiplier tubes to detect scintillation photons produced in the fibers. Since a typical imaging device contains several tens of thousands of scintillation fibers therefore a single channel PMT is not suitable for the purpose. Other PMT structures such as microchannel PMTs are therefore often used. The scintillation light is transported through an optical guide, such as clear fibers, to the photocathode of the PMT. Fig.7.3.6 shows the conceptual design of an imaging system based on scintillating fibers. Here two sets of fiber bundles are

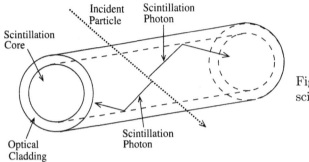

Figure 7.3.5: Sketch of a typical scintillation fiber.

shown orthogonal to each other. The spatial resolution of this system would depend on the fiber to PMT channel mapping and PMT spatial resolution.

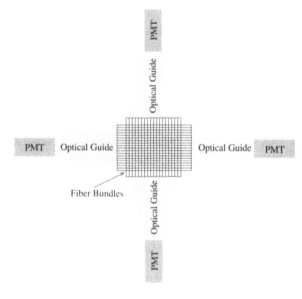

Figure 7.3.6: Conceptual design of an imaging system based on scintillating fibers.

Problems

1. Calculate the spatial resolution of a position sensitive detector having a pitch of 40 μm and working in digital mode.

2. Estimate the change in spatial resolution of a semiconductor microstrip detector if the readout scheme is changed from all strips to every third strip.

3. Calculate the effective quantum yield of a CCD camera having quantum efficiency of 40% and conversion efficiency of 87%.

4. Determine the change in the dynamic range of a CCD if its ADC resolution is changed from 12 bits to 16 bits. Assume that all other parameters remain the same.

5. Compute the dynamic range of a silicon based CCD system at an energy of 1.5 keV. The system has a gain of 20 photoelectrons per ADC count and 16-bit ADC resolution.

6. What would be the dynamic range of the CCD camera described in the previous exercise for an absorbed energy of 10 MeV?

Bibliography

[1] Antonuk, I. et al., **Signal, Noise, and Readout Considerations in the Development of Amorphous Silicon Photodiode Arrays for Radiotherapy and Diagnostic Imaging**, Proc. SPIE, 1443, 1991.

[2] Barrett, H, Swindell, W., **Radiological Imaging: The Theory of Image Formation, Detection, and Processing**, Academic Press, 1996.

[3] Bogucki, T.M. et al., **Characteristics of a Storage Phosphor System for Medical Imaging**, Technical and Scientific Monograph No.6, Estman Kodak Health Sciences Division, 1995.

[4] Bushberg, J.T. et al., **The Essential Physics of Medical Imaging**, Lippincott Williams & Wilkins, 2001.

[5] Bushong, S.C., **Radiation Protection: Essentials of Medical Imaging Series**, McGraw-Hill Medical, 1998.

[6] Carpenter, J.M. et al., **Neutrons, X Rays, and Gamma Rays: Imaging Detectors, Material Characterization Techniques, and Applications**, Society of Photo Optical, 1993.

[7] Charpak, G., **Research on Particle Imaging Detectors: Localization of Ionizing Radiators**, World Scientific Publishing Company, 1995.

[8] Chugg, A.M., Hopkinson, G.R., **A New Approach to Modelling Radiation Noise in CCD's**, IEEE Trans. on Nucl. Science, Vol.45, No.3, 1998.

[9] Dereniak, E.L., **Imaging Detector Arrays**, SPIE, 1992.

[10] Dereniak, E.L., **Infrared Detectors, Focal Plane Arrays, and Imaging Sensors**, SPIE, 1989.

[11] Epstein, C.L., **Mathematics of Medical Imaging**, Prentice Hall, 2003.

[12] Fahng, R. et al., **X-Ray Imaging with Amorphous Selenium: Detective Quantum Efficiency of Photoconductive Receptors for Digital Mammography**, Med. Phys., 22, 1995.

[13] Fraser, R.G. et al., **Digital Imaging of the Chest**, Radiology, 171, 1989.

[14] Fujieda, I. et al., **High Sensitivity Readout of 2D - Si Image Sensors**, Japan. J. Appl. Phys., 32, 1993.

[15] Grove, A.S., **Physics and Technology of Semiconductor Devices**, John Wiley and Sons, 1967.

[16] Guerra, A.D., **Ionizing Radiation Detectors for Medical Imaging**, World Scientific Publishing Company, 2004.

[17] Hopkinson, G.R, **Proton Damage Effects on P-Channel CCD's**, IEEE Trans. on Nucl. Science, Vol.46, No.6, 1999.

[18] Hussain, E.M., **Handbook on Radiation Probing, Gauging, Imaging and Analysis**, Springer, 1999.

[19] Janesick, J.R., **Scientific Charge-Coupled Devices**, SPIE Press, 2001.

[20] Johnson, C.B., **Photoelectronic Detectors, Cameras and Systems**, SPIE, 1995.

[21] Kaplan, D.M., **Introduction to Subatomic-Particle Spectrometers**, arXiv:physics/9805026 v2, 12 June, 2002.

[22] Kouris, K., **Imaging With Ionizing Radiations**, Sheridan House, 1982.

[23] Lomheim, T.S. et al., **Imaging Charged-Coupled Device (CCD) Transient Response to 17 and 50 MeV Proton and Heavy-Ion Irradiation**, IEEE Trans. on Nucl. Science, Vol.37, No.6, 1990.

[24] Mandelkern, M., **Nuclear Techniques for Medical Imaging: Positron Emission Tomography**, Ann. Rev. Nucl. Part. Sci., 45, 1995.

[25] McLean, I.S., **Electronic Imaging in Astronomy: Detectors and Instrumentation**, John Wiley & Sons, 1997.

[26] Miyaguchi, K. et al., **CCD Developed for Scientific Applications by Hamamatsu**, Nucl. Instr. and Meth. A, 436, 1999.

[27] Murphy, D.B., **Fundamentals of Light Microscopy and Electronic Imaging**, Wiley-Liss, 2001.

[28] Murray, S.S., **Development of High Resolution Imaging Detectors for X-Ray Astronomy**, National Aeronautics and Space Administration National Technical Information Service, 1985.

[29] Nier, M.C., **Standards for Electronic Imaging Technologies, Devices, and Systems: Proceedings of a Conference Held 1-2 February 1996, San Jose, California**, SPIE, 1996.

[30] Oppelt, A., **Imaging Systems for Medical Diagnostics**, Wiley-VCH, 2006.

[31] Park, J., **Imaging Detectors in High Energy, Astroparticle and Medical Physics: Proceedings of the UCLA International Conference**, World Scientific Publishing Company, 1996.

[32] Pope, J., **Medical Physics: Imaging**, Heinemann Educational Books, 1999.

[33] Rabbani, M. et al., **Detective Quantum Efficiency of Imaging Systems with Amplifying and Scattering Mechanisms**, J. Opt. Soc. Am. A, 4, 1987.

[34] Sandrik, J.M., Wagner, R.F., **Absolute Measures of Physical Image Quality: Measurement and Application to Radiographic Magnification**, Med. Phys., 9, 1982.

[35] Saxby, G., **The Science of Imaging: An Introduction**, Taylor & Francis, 2001.

[36] Schreiber, W.F., **Fundamentals of Electronic Imaging Systems: Some Aspects of Image Processing**, Springer, 1993.

[37] Siegmund, O., **High Spatial Resolution Investigations of Microchannel Plate Imaging Properties for UV Detectors**, National Aeronautics and Space Administration National Technical Information Service, 1996.

[38] Suetens, P., **Fundamentals of Medical Imaging**, Cambridge University Press, 2002.

[39] Sulham, C.V., **Special Nuclear Material Imaging Using a High Purity Germanium Double Sided Strip Detector**, Storming Media, 2004.

[40] Webb, S., **The Physics of Medical Imaging**, Taylor & Francis, 1988.

[41] Westbrook, E.M, Naday, I., **Charged-Coupled Device-Based Area Detectors**, Methods in Enzymology, 276, 1997.

[42] Wolken, J.J., **Light Detectors, Photoreceptors, and Imaging Systems in Nature**, Oxford University Press, USA, 1995.

[43] Yaffe, M.J., Rowlands, J.A., **X-Ray Detectors for Digital Radiography**, Phys. Med. Biol., 42, 1997.

Chapter 8

Signal Processing

Signal processing plays an extremely important role in extracting useful information from detectors. Generally speaking two pieces of information are important with respect to detection and measurement of radiation: the amplitude and the timing of the output pulse. Amplitude information is important with respect to applications, such as energy spectroscopy, in which the measurement of the energy deposited by the incoming radiation is desired. On the other hand there are applications, such as particle tracking, in which precise timing of pulses is required. In order to extract such information from the narrow width and low amplitude detector pulses, a number of analog and digital signal processing steps are required. Broadly speaking the signal can be either processed entirely through a chain of analog circuitry or it can be converted to digital form for analysis. We will refer to the former in this book by analog signal processing. With the advent of cost effective computing, digital signal processing (DSP) is now gaining a lot of popularity. In fact it is now a method of choice wherever possible. It should, however, be noted that digital signal processing does not eliminate the need for analog circuitry from the electronics chain. Some analog units are always needed to amplify the signal and make it usable for the analog to digital converters (ADC). For example although the availability of very fast ADCs (called flash ADCs) has made it possible to replace the pulse shaping step by a digital process, preamplifiers are still needed. Fig.8.0.1 shows typical analog and digital signal processing steps adapted in detection systems.

Whether the processing is analog or digital, the detector pulse must almost always be first preamplified. The reason is that generally the detector pulse has a very low amplitude and very short duration, making the direct measurements of amplitude and time difficult and prone to large systematic errors. The preamplification step not only amplifies the pulse but also increases the pulse width, thus making it suitable for processing by the next electronic circuitry. The subsequent processing steps highly depend on the application and data analysis requirements. If the requirements dictate the use of analog processing techniques then the pulse is further shaped, amplified, and fed into a pulse height analyzer. On the other hand, if digital signal processing is desired, then the signal is amplified and then converted to digital counts through an ADC. It should be noted that the type of processing is highly dependent on the application and therefore there is no general electronics chain that could be universally adopted. In this chapter we will visit the most important of the signal processing methods used in modern radiation detection systems.

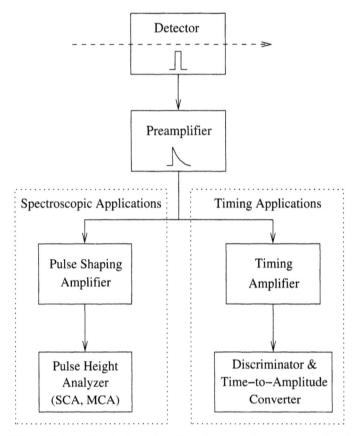

Figure 8.0.1: Typical analog signal processing steps involved in timing and spectroscopic applications.

8.1 Preamplification

The analog signal pulse produced as a result of passage of radiation through a detector usually has very narrow width and amplitude and therefore can not be directly digitized or even counted. Unless the detector signal has enough strength (i.e., the height of the pulse is large enough), it must first be preamplified before transporting it to other processing units. A preamplifier is a simple but efficient amplifier that is directly connected to the detector output.

Different kinds of preamplifiers can be constructed to suit the specific detector and processing requirements. Several parameters are considered when designing a preamplification circuitry, most of which belong to competing requirements. Hence the design process is actually optimization of some of these parameters to suit the specific needs. Some of the important design specifications of any preamplier are

▶ signal to noise ratio (S/N),

▶ range of input signal.

▶ response time,

▶ power consumption,

▶ dynamic range,

▶ pulse pile-up, and

▶ common mode.

Broadly speaking the preamplifiers used in radiation detection systems can be divided into the following categories.

▶ Voltage Sensitive Preamplifier

▶ Current Sensitive Preamplifier

▶ Charge Sensitive Preamplifier

8.1.A   Voltage Sensitive Preamplifier

Voltage sensitive preamplifier is the most basic type of preamplifier that can be used in radiation detection systems. Its function is to simply amplify the potential at its input stage by some gain factor defined by its components. Fig.8.1.1(a) shows the principle of working of such an instrument. Here we have represented the detector as a voltage source. For such a circuit the voltage at the input stage of the amplifier V_a is related to the signal voltage V_s through the relation

$$V_a = \frac{R_a}{R_s + R_a} V_s, \tag{8.1.1}$$

where R represents resistance with subscripts a for amplifier input and s for signal respectively. For such an amplifier to work properly it is necessary that it does not draw any current from the source, since any current drawn by it would decrease the potential drop across R_s. This would require its input resistance (more accurately, impedance[1]) to be infinite, something that is not practical and can only be achieved up to a good approximation. In fact, looking at the above relation it can be inferred that for V_a to approach V_s the input impedance of the preamplifier should be very large *as compared to* the source impedance, that is

$$R_a \gg R_s \implies V_a \approx V_s.$$

Since the output of a voltage sensitive linear amplifier should be proportional to the voltage at its input stage, we can write

$$\begin{aligned} V_{out} &= AV_a \\ &\approx AV_s. \end{aligned} \tag{8.1.2}$$

Here A is the gain of the amplifier, which depends on the type of amplifier and the external components.

---

[1]The terms impedance and resistance can be used interchangeably when the current is not changing with respect to time. If the current changes with time, the effects of circuit and cable capacitance and inductance also come into play. It is a good idea to always use the term impedance since it characterizes the system better than the simple resistance.

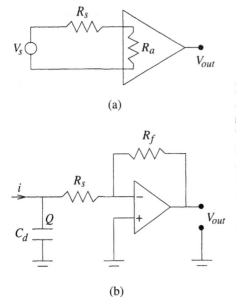

(a)

(b)

Figure 8.1.1: (a) Principle design of a voltage sensitive preamplifier connected to a
source, such as a radiation detector output.
The input impedance R_a of the preamplifier
is kept very large so that it draws minimal
current from the source. (b) A simplified
but realistic voltage sensitive preamplifier
with feedback resistor R_f. C_d is the combined detector and stray capacitance and R_s
is the combined impedance.

This simplistic picture does not include a feature found in all real detectors, that
they possess some capacitance as well. In fact the potential difference V_s actually
appears across the combined capacitance of the detector and other components such
as cable. This potential is approximately given by

$$V_s \approx \frac{Q}{C_d}, \tag{8.1.3}$$

where Q is the charge collected by the readout electrode and C_d is the *combined*
detector and stray capacitance. Hence we can conclude that the voltage at the
output of a realistic amplifier, to a good approximation, is given by

$$V_{out} \approx A\frac{Q}{C_d}. \tag{8.1.4}$$

The reader should readily realize the problem with this configuration, i.e., the
output voltage depends on the detector capacitance. For a perfectly linear operation,
therefore, any change in detector capacitance must be accounted for in the conversion
constants through recalibration. This, of course, is not practical for systems, such
as semiconductor detectors, whose capacitance may change during operation due to
small fluctuations in temperature. For other systems, such as ionization chambers,
this type of amplifier can be effectively used.

An advantage of voltage sensitive amplifiers is that they provide low impedance
output signal, which is ideal for signal transportation at large distances. But alone
this advantage can not justify its preference over other types of preamplifiers. A
realistic but simplified voltage sensitive preamplifier is shown in Fig.8.1.1(b).

Up until now we have not included the detector's instantaneous current in the
discussion. We know that the charges produced in the active volume of the detector
by incident radiation actually constitute a current, which can be measured. In

fact most of the radiation detectors can be modeled as current sources. So how do we relate this current to the detector's voltage that we amplify by a voltage sensitive preamplifier? This can be done if we take into account the fact that this instantaneous current actually gets integrated on detector's capacitance. If at any instant in time t a current i_s flows through the detector, then the total charge Q accumulated on the detector's capacitance C_d from $t = 0$ to $t = t_0$ will be given by

$$Q = \int_0^{t_0} i_s(t)dt. \tag{8.1.5}$$

Of course C_d here is the *combined* detector and stray capacitance. Hence the preamplifier's output voltage can be expressed in terms of detector's instantaneous current as

$$V_{out} = \frac{A}{C_d} \int_0^{t_0} i_s(t)dt. \tag{8.1.6}$$

This expression implies that the output voltage has a direct dependence on capacitive load of the amplifier. This capacitive load C_d together with the amplifier's input impedance R_a determines how fast the capacitor discharges through the time constant of the circuit $R_a C_d$. The integrated current and the time constant determine the shape of the voltage pulse.

Since we are integrating the current to convert it into voltage, C_d should discharge slower than the charge collection time t_d. The criterion

$$t_d \ll R_a C_d, \tag{8.1.7}$$

is therefore sometimes used to decide if a voltage sensitive preamplifier is suitable for a detector.

8.1.B Current Sensitive Preamplifier

In certain applications it is desirable to measure the instantaneous current flowing through the detector. This can be done through a current sensitive preamplifier, which converts the detector's instantaneous current into a measurable voltage. Therefore, this device can also be called a current-to-voltage converter. A current sensitive amplifier can be constructed in the same way as a voltage sensitive amplifier with the exception that the input impedance in this case must be kept at the minimum to allow the current to flow through the amplifier (see Fig.8.1.2(a)).

The current i_a flowing into the preamplifier of Fig.8.1.2(a) is related to the source current i_s by

$$i_a = \frac{R_s}{R_s + R_a} i_s. \tag{8.1.8}$$

This equation implies that the requirement, that the current flowing into the preamplifier is approximately equal to the source current, can be fulfilled by making preamplifier's input impedance very small as compared to source impedance, i.e.,

$$R_a \ll R_s \implies i_a \approx i_s$$

Since a linear current sensitive preamplifier is simply a current-to-voltage converter, therefore its output voltage is proportional to the source current.

$$V_{out} \propto i_s \tag{8.1.9}$$

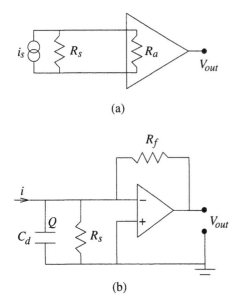

(a)

(b)

Figure 8.1.2: (a) Working principle of a current sensitive preamplifier connected to a current source, which could be a radiation detector. A current sensitive amplifier is actually a current to voltage converter. In order for it to be able to measure instantaneous current, its input impedance must be very small as compared to the detector's output impedance. (b) A simplified but realistic current sensitive preamplifier with feedback resistor R_f. C_d is the combined detector and stray capacitance and R_s is the combined impedance.

Conversion of instantaneous current into voltage requires that the detector capacitance C_d discharges much quicker than the charge collection time of the detector. This implies that we must have

$$\tau_d \gg R_s C_d.$$

Here τ_d is the charge collection time and we have neglected the amplifier resistance R_a since $R_a \ll R_s$.

8.1.C Charge Sensitive Preamplifier

The dependence of voltage sensitive preamplifier on the input capacitance is a serious problem for many detection systems. In fact this very problem prompted the detector designers to develop the so called *charge sensitive* preamplifiers. In this device, instead of directly amplifying the voltage or converting the current to voltage, the charge accumulated on the detector capacitance is *integrated* on another capacitor. The resulting potential on that capacitor is then directly proportional to the original charge on the detector. Since it essentially accumulates the charge on detector capacitance on another capacitor, a charge sensitive preamplifier acts as an integrator of charge. Fig.8.1.3 shows the principle of working of this device. The feedback capacitor C_f accumulates the charge on combined detector and stray capacitance C_d. The output voltage is then simply given by

$$V_{out} \propto \frac{Q_f}{C_f}$$

$$\text{or} \quad V_{out} \propto \frac{Q_d}{C_f}, \tag{8.1.10}$$

where Q_d is the charge accumulated on C_d, which is proportional to the charge Q_f integrated on the feedback capacitor. The condition that $Q_f \approx Q_d$ can only be achieved if no current flows into the preamplifier's input. This implies that the amplifier's input impedance should be very large ($R_a \approx \infty$).

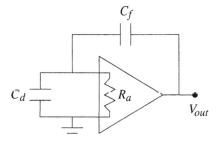

Figure 8.1.3: Basic principle of a charge sensitive preamplifier. The charge accumulated on the detector capacitance C_d is allowed to integrated on a feedback capacitor C_f. The voltage at the output is then proportional to the input charge.

Fig.8.1.4 shows a detector, whose output is connected to the *inverting* input of a charge integrating preamplifier. Without the feedback capacitor, this circuit would act as an inverting voltage sensitive preamplifier with the output voltage of

$$V_{out} = -AV_d. \tag{8.1.11}$$

Here V_d is the input voltage and A is the preamplifier gain. Typical operational amplifiers have very large gains ($A \gg 1$). If we now connect a feedback capacitor to the circuit, the input voltage would be given by the sum of voltage across this capacitor V_f and the output voltage, i.e.,

$$\begin{aligned} V_d &= V_f + V_{out} \\ &= V_f - AV_d \\ \Rightarrow V_f &= (A+1)V_d. \end{aligned} \tag{8.1.12}$$

This of course is valid only if the preamplifier has infinite impedance, which is generally true to a good approximation. Using $Q = CV$, we can write the above relation in terms of accumulated charges on input and feedback capacitances.

$$\begin{aligned} \frac{Q_f}{C_f} &= (A+1)\frac{Q_{in}}{C_{in}} \\ \Rightarrow C_{in} &= (A+1)C_f\frac{Q_{in}}{Q_f} \\ &\approx (A+1)C_f \quad \text{(since } Q_{in} = Q_f) \end{aligned} \tag{8.1.13}$$

Here Q_{in} and C_{in} are the charge and capacitance at the input of the amplifier respectively. This equation shows that the input capacitance is a function of feedback capacitance and gain of the amplifier. C_{in} is sometimes referred to as the *dynamic capacitance* of the preamplifier. Note that C_{in} is not detector's capacitance, which we will include in the equations later. A quantity that is often quoted is the *charge gain*, which is obtained by taking the ratio of the output voltage to the corresponding input charge. Unlike the conventional open loop gain A, this gain is not a dimensionless quantity. Charge gain A_Q can be computed from

$$
\begin{aligned}
A_Q &= \frac{V_{out}}{Q_{in}} = \frac{AV_{in}}{C_{in}V_{in}} = \frac{A}{C_{in}} \\
&= \frac{A}{A+1}\frac{1}{C_f} \\
&\approx \frac{1}{C_f} \quad (\text{since } A \gg 1).
\end{aligned}
\tag{8.1.14}
$$

This shows that the charge gain can be increased by choosing C_f of lower capacity.

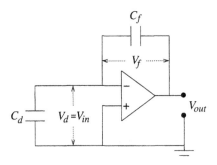

Figure 8.1.4: Schematic of a simple charge sensitive preamplifier. The detector is shown as a capacitive load and is connected to the inverting input of the preamplifier. Without the feedback capacitor, the circuit will work as an inverting voltage sensitive preamplifier since the input impedance of the preamplifier is very large.

Another quantity of interest is the *charge transfer efficiency* η_{in} of the preamplifier. It characterizes the efficiency of transfer of charge on *detector's capacitance* to the feedback capacitance. Only in ideal case it is possible to have a 100% charge transfer efficiency but it can be approached up to a good approximation by choosing appropriate capacitances. η_{in} is defined as

$$
\eta_{in} = \frac{Q_f}{Q_t},
\tag{8.1.15}
$$

where Q_t is the total charge on detector and dynamic capacitance. If we represent detector capacitance by C_d, then $Q_t = Q_d + Q_{in}$. Using this and the fact that $Q_f = Q_{in} = C_{in}V_{in}$, we can write the charge transfer efficiency as

$$
\begin{aligned}
\eta_{in} &= \frac{Q_{in}}{Q_{in} + Q_d} \\
&= \frac{1}{1 + Q_d/Q_{in}} \\
&= \frac{1}{1 + C_d V_d/C_{in}V_{in}} \\
&= \frac{1}{1 + C_d/C_{in}}.
\end{aligned}
\tag{8.1.16}
$$

Here we have made use of the fact that the voltage across the detector capacitance is equal to the voltage at amplifier's input stage, i.e., $V_d = V_{in}$. This is a true statement as can be verified by inspecting Fig.8.1.4. It is apparent from this expression that a

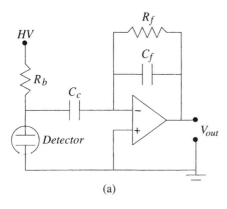

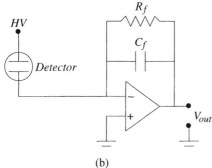

Figure 8.1.5: Schematics of (a) an AC coupled and (b) a DC coupled resistive feedback charge sensitive preamplifiers. The charge accumulated on the feedback capacitor C_f decays through the feedback resistor R_f with a time constant of $R_f C_f$.

100% charge transfer efficiency would require the detector capacitance to be much larger than the input capacitance, that is for $\eta_{in} \approx 1$ we must have

$$C_d \ll C_{in}$$
$$\text{or} \quad C_d \ll (A+1)C_f. \tag{8.1.17}$$

Up until now we have just considered how the feedback capacitor charges. For any subsequent measurement it must be discharged or *reset* to allow the next charge accumulation. Two methods are commonly used for this purpose: resistive feedback and pulsed reset.

C.1   Resistive Feedback Mechanism

In a resistive feedback charge sensitive preamplifier, a feedback resistor is provided so that the charge integrated on the feedback capacitor could decay with a predefined time constant (see Fig.8.1.5). The result is a sharply rising pulse that decays exponentially with a time constant $R_f C_f$.

The time variation of the output pulse from an *ideal* resistive feedback charge sensitive preamplifier is given by

$$v_{out}(t) = \frac{Q_f}{C_f} e^{-t/R_f C_f}$$
$$= \frac{\eta_{in} Q t}{C_f} e^{-t/R_f C_f}. \tag{8.1.18}$$

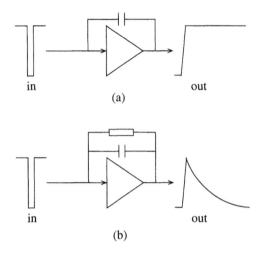

Figure 8.1.6: (a) Typical output voltage pulse shapes of a simple charge sensitive preamplifier (a) and a resistive feedback charge sensitive preamplifier (b). In both cases the amplitude of the output pulse is proportional to the original detector pulse, which in turn is proportional to the energy deposited in the detector. Without a feedback resistor the feedback resistor does not discharge and the output voltage does not change with time after reaching the maximum value. In a resistive feedback preamplifier, on the other hand, the pulse decays exponentially with a time constant equal to $R_f C_f$.

where, as before, Q_f is the charge accumulated on the feedback capacitor C_f, Q_t is the total input charge, and η_{in} is the charge transfer efficiency. It is apparent form these equations that the amplitude of the signal is given by

$$V_{out} = \frac{Q_f}{C_f} = \frac{\eta_{in} Q_t}{C_f}. \qquad (8.1.19)$$

This maximum value of the voltage occurs at $t = 0$, which implies a step response of the circuit to the detector output. However, in reality, it always takes some finite amount of time to reach its maximum value. In most of the cases we are not really concerned about the rise time of the pulse since its shape is always conditioned by subsequent analog processing units. However in some timing applications it is desired to know this quantity. V_{out} is proportional to the detector pulse and is therefore a measure of the energy deposited by the incident radiation. After reaching this maximum value, the pulse decays exponentially through the feedback resistor R_f (see Fig.8.1.6). The time for this decay depends on the time constant of the feedback circuit $\tau_f = R_f C_f$ and can therefore be chosen according to the requirements.

A nice feature of resistive feedback preamplifiers is that they do not exhibit any dead time, though at the expense of pulse pile up at high count rates. That is, even though such a preamplifier will keep on amplifying even if the next pulse arrives during the pulse decay time but the offset of the previous pulse will be added to the new one (see Fig.8.1.7). It is due to this reason that for a high rate situation such a preamplifier is not recommended. At low count rates, however, the resistive feedback mechanism gives reasonable performance; *reasonable* because of the noise added to the system by the feedback resistor itself.

C.2   Pulsed Reset Mechanism

The feedback resistor of the resistive feedback charge sensitive preamplifier is itself a source of noise and is therefore not suitable for high resolution systems. Pulsed

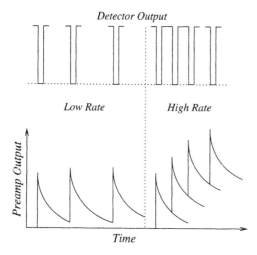

Figure 8.1.7: Low rate and high rate measurements from a resistive feedback charge sensitive preamplifier. Although the preamplifier does not exhibit a dead time even at high rates, but the pulse pileup can be a serious problem.

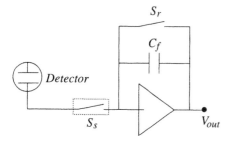

Figure 8.1.8: Conceptual design of a pulsed reset charge sensitive preamplifier. S_r is the reset switch that is used to drain the charge accumulated on the feedback capacitor. Some preamplifiers are also equipped with a set switch S_s to minimize dead time (see text).

reset preamplifiers provide an alternate mechanism to reset the feedback capacitor after the pulse has reached its maximum value. The idea is to replace the feedback resistor with an electronic switch providing a current drain path to discharge the feedback capacitor *on demand*. This switch is toggled on and off through an additional timing circuitry, which is sometimes connected to the preamplifier output to generate a trigger signal for the switch. Fig.8.1.8 shows the basic principle of such a device. A typical timing diagram of the circuit is shown in Fig.8.1.9(a). The reader should readily realize that the pulses received by the preamplifier during the reset time are lost. This can of course be minimized by keeping the reset time at the minimum. A better way, however, is to introduce another switch at the input of the preamplifier. If this switch is kept open during the reset time (i.e. S_s open and S_r close), then the detector pulses would keep on integrating charges on the combined detector and stray capacitance. Closing S_s after the feedback capacitor had completely discharged would transfer these charges to the feedback capacitor, hence eliminating the information loss. However, there is one caveat to this scheme, namely the time jitter of the switching pulses, which introduces some uncertainty in the rising and falling edges of the pules. To overcome this problem, normally a small sacrifice is made by keeping the S_r-open pulse in the middle of the S_r-close pulse (see Fig.8.1.9(b)).

Most of the electronic switches used in commercially available preamplifiers are transistor based. Timing of the reset operation in these devices is generally done by

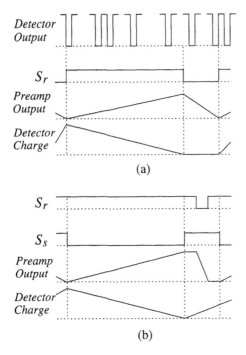

Detector Output

S_r

Preamp Output

Detector Charge

(a)

S_r

S_s

Preamp Output

Detector Charge

(b)

Figure 8.1.9: Timing diagrams for operation of a pulsed reset charged sensitive preamplifier (a) without and (b) with a switch at the preamplifier input (see also Fig.8.1.8). The diagram in part (b) is generally said to represent *dynamic recording*, since it does not permit loss of information during feedback capacitor reset. All timings shown are *active-low*, i.e., the level 0 corresponds to the active state.

built-in circuitry that sends a reset pulse to the transistor switch if the preamplifier output reaches a predefined value (normally the upper limit of preamplifier's range). Fig.8.1.10 shows the working principle of such a preamplifier. It is also possible to toggle the switch through an external circuit that does not take any input from the preamplifier. Such a system is generally configured to time the operation at regular intervals of time and therefore provides more flexibility in terms of further data readout and processing.

The transistor based reset switches may suffer from time jitters, logic level edge excursions, and digital charge injections from the logic unit. These problems can be minimized, if not eliminated, by using the so called *optical reset trigger*. In this scheme the resetting is done through light instead of electrical current. Certainly such a device does not eliminate the need for timing circuitry but, by electrical isolation from the preamplifier input, it does not cause any charge injection in the circuit. The result is a much cleaner toggling of the switch as compared to the electronically triggered switch. Fig.8.1.10(b) shows a typical optical feedback pulsed preamplifier. Here a light emitting diode in the logic circuit inhibits the preamplifier switch. The switch itself is a field effect transistor (FET).

8.2 Signal Transport

It may at first seem quite trivial to transport signal from the preamplifier, if the detector has one, to other pulse processing units. One would think that all is needed is a cable that takes the signal from one end to the other or a simple wireless transmitter and receiver. However, as it turns out, there are a few points that must

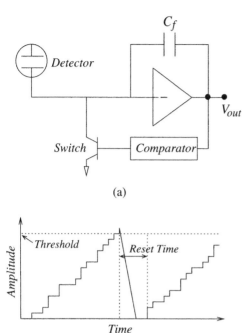

(a)

(b)

Figure 8.1.10: (a) Sketch of a transistor based pulsed reset charge sensitive preamplifier. The feedback capacitor discharge starts automatically when the preamplifier output crosses a predefined threshold. The switch (transistor) closes during the reset and drains the current to ground. (b) Amplification of detector pulses for the circuit shown in (a). During the reset time, the amplifier can not amplify the detector output. This is referred to as the *dead time* of the preamplifier.

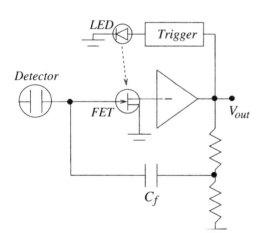

Figure 8.1.11: A typical optical pulsed feedback charge integrating preamplifier. The trigger is issued when the output level crosses some predefined threshold. This lits the LED, which shines light on the FET, which in this condition creates a closed circuit to ground, thus discharging the feedback capacitor (see also Fig.8.1.10(b)).

be carefully considered while deciding on the signal transport methodology. In the following we will discuss the most important of these points in relation to typical radiation detection systems. But before we go on we should note on important point that, although using a wireless data transport mechanism might at first seem promising, but in reality it has many pitfalls, the most important of which is the electromagnetic interference. Therefore wherever possible, physical cables are used to transport signal. Here also we will concentrate on signal transport through cables.

8.2.A Type of Cable

A typical signal cable consists of a central metallic wire with a metal shield around it. No matter how good the wire of the cable is, it still has some resistance given by

$$R = \rho\frac{L_c}{A_c}, \tag{8.2.1}$$

where ρ is the resistivity of the metal, L_c is its total length, and A_c is its cross sectional area. The voltage drop across the cable due to this resistance can simply be calculated by Ohm's law, which gives

$$V = IR = \rho I\frac{L_c}{A_c}. \tag{8.2.2}$$

Here I is the current flowing through the wire. This voltage drop must be kept as low as possible as compared to the minimum signal voltage. It is apparent from this equation that this can be done in essentially three ways:

▶ **Low Resistivity Material:** Resistivity ρ is material and temperature dependent. Silver is a very good conductor with a resistivity of 1.62×10^{-8} Ωcm but is quite expensive as compared to other materials, such as copper or aluminum. Copper has a resistivity of 1.69×10^{-8} Ωcm, which is slightly higher than that of silver but with two obvious advantages: it is much cheaper and can be easily shaped into conductors of different thicknesses. Another good material is aluminum with a resistivity of 2.75×10^{-8} Ωcm. This can be cheaper than copper but has a slightly higher resistivity.

▶ **Small Cable Length:** Keeping the length small is the most simplistic way to decrease signal attenuation due to cable. However there are normally engineering constraints to the minimum possible length. For example in a detector working in high radiation field, the data processing units are generally located a few hundred feet away from the detector and proper routing of cables puts a minimum limit on the cable length.

▶ **Large Cross Sectional Area:** The technical term describing the cross sectional area of a wire is its *gage*. The lower the gage, the thicker the wire. Resistivity and hence the voltage drop in low gage wires is definitely small, but there may be practical considerations, such as cost and cable routing (the lower the gage the stiffer the cable), that may put a limit on the wire thickness for a particular system.

It is a general practice to use standard cables in radiation detection systems, even though a custom made cable is expected to make signal transportation a bit better. There are two reasons for that. Firstly the cost of custom made cables is significantly higher than the standard cables, and secondly a custom made cable has to be available in large quantities to replace the damaged cables. We will discuss here some of the important types of standard cables available.

Example:
A 50 m long single wire copper cable is used to transport signal current. The wire has a cross sectional area of 0.5 mm^2. Assume that the resistivity of

copper is 1.69×10^{-8} Ωcm. Compute the resistance the cable offers to the signal current.

Solution:
We can use equation 8.2.1 to compute the resistance.

$$R = \rho \frac{L_c}{A_c}$$
$$= \left(1.69 \times 10^{-8}\right) \frac{50}{0.5 \times 10^{-6}}$$
$$= 1.69 \ \Omega$$

A.1 Coaxial Cable

As shown in Fig.8.2.1(a), a typical coaxial cable has the following components.

▶ A central conductor, mostly made of copper.

▶ Two insulators surrounding the central wire. One thin and other thick.

▶ Metallic shield surrounding the outer insulator.

▶ Outermost insulator surrounding the metal shield.

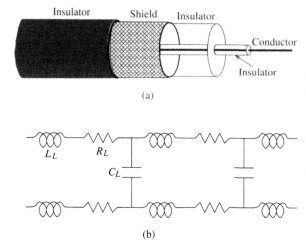

(a)

(b)

Figure 8.2.1:   (a) Sketch of a coaxial cable. This type of cable is commonly used for signal transport due to its low attenuation and high shielding capabilities. (b) Model of a typical coaxial cable having inductance, resistance, and capacitance per unit length of the cable.

The purpose of the shield is to alienate the central conductor from the outside electromagnetic field. However there is no coaxial cable that provides perfect shielding and the decision to either use it or to go for some other type of cable is based on the relative strengths of the signal and the background radiation.

The insulators in between the central conductor and the shield provide a low distortion path or *waveguide* to the signals. This ensures minimal dispersion and hence small attenuation of signals specially the short duration pulses.

All coaxial cables have characteristic resistance, capacitance and inductance which can be modeled as shown in Fig.8.2.1(b). It can be shown that the capacitance and inductance are given by

$$C = \frac{2\pi\epsilon}{\ln(b/a)L_c} \tag{8.2.3}$$

$$L = \frac{\mu}{2\pi}\ln\left(\frac{b}{a}\right)L_c \tag{8.2.4}$$

where ϵ and μ are the effective permittivity and permeability of the dielectric (i.e. insulators between central wire and outer shield) respectively. a and b are the radii of the central wire and the shield respectively. L_c, as before, represents the total length of the cable. These relations can be used to determine the characteristic impedance of the cable as

$$Z_c = \left(\frac{L}{C}\right)^{1/2}$$

$$= \frac{\sqrt{\mu}}{2\pi\sqrt{\epsilon}}\ln\left(\frac{b}{a}\right). \tag{8.2.5}$$

The characteristic impedance of most of the commercially available standard coaxial cables lies between 50 Ω and 200 Ω. For example Z_c for RG58 is 50 Ω.

Another quantity of interest is the velocity of signal propagation, which becomes important in timing applications. It can be approximated from the relation

$$v = \frac{L_c}{\sqrt{LC}}$$

$$= \frac{1}{\sqrt{\mu\epsilon}}. \tag{8.2.6}$$

Typically signal velocities are on the order of 80% to 99% of the velocity of light in vacuum. This implies that the signal in a cable actually propagates as an electromagnetic wave. If this wave encounters an imperfection on its way, it may get distorted or reflect back. These imperfections could be due to cable joints or terminations. If the end of the cable is not connected to anything, then the wave will see an infinite load impedance and will completely reflect back. This can also be seen from the relation of reflection coefficient Γ defined as

$$\Gamma = \frac{Z_l - Z_c}{Z_l + Z_c}, \tag{8.2.7}$$

where Z is the characteristic impedance and the subscripts l and c stand for the load and cable respectively. If the end of the conductor is open the reflection coefficient will be 1 ($Z_l = \infty \Rightarrow \Gamma = 2$). In this situation the signal will be reflected back with the same *phase*. The other extreme is to short the other end of the cable. In this case the $Z_l = 0$ and $\Gamma = -1$. Hence the reflected pulse will be *inverted in phase*. Based on this discussion it be concluded that any impedance mismatch between two cables connected together or the cable and the load may produce reflected pulses and deteriorate the signal transmission. It is therefore imperative that care is taken to ensure impedance matching at any point of cable discontinuity.

A.2   Twisted Pair Cable

The coaxial cable is a well shielded cable but unfortunately never provides 100% shielding specially against radio frequency background. An alternative to this is the twisted pair cable, which is used to transport the so called *differential signals*. Before we discuss the geometry of the twisted pair cable, we should first understand the difference between single and differential outputs.

Most of the general purpose commercial preamplifiers produce the so called *single output*. Such a signal is highly susceptible to external noise sources such as background radio-frequency emitters that are generally present in environments containing different electronics devices. The main mechanism is the pickup of energy by the signal cable through electromagnetic interactions.

The shielding of the signal cable (such as in coaxial cable) is the most widely used method because of its simplicity. However, as we stated earlier, this does not always yield perfect results as most of the commercially available shielded signal cables are not 100% effective.

The other method is to use differential output. Such an output is carried through two cables instead of one such that the information is contained in the difference of potentials on the cables instead of the absolute value of the potential as in the case of single output. The noise in this case is picked up equally by both cables, keeping the difference unchanged. However, since the electromagnetic interference is proximity dependent, the cables must be twisted together in the shape of a spiral to ensure equal pickup at each point. Such cables are generally referred to as *twisted pair cables* and are available in unshielded form.

To further process this signal, for example by an amplifier/shaper, the subsequent circuitry must be able to accept differential input. It should however be noted that the notion of *exactly same* pickup by the two cables is somewhat idealized. The pickup is actually dependent on the gains of each input and therefore modern amplifiers are built with a *differential gain balancing* circuitry to match the two gains for proper pickup noise cancellation.

A.3   Flat Ribbon Cable

Flat ribbon cables are extensively used to transport digital and analog signals. Such a cable is in the form of a ribbon with several individually isolated wires bonded together. Ribbon cables come mostly in unshielded form but their shielded versions are also available.

Ribbon cables should not be used to transport detector signals because even in their shielded versions the cables are not individually shielded making the inter-cable cross talk a possibility.

Such cables are generally used to transport power and digital signals. Digital signals are basically logic signals and therefore a small attenuation in the signal does not cause any problems. However if an unshielded ribbon cable is used to transport high frequency (such as few hundred Mhz) signals, the same cable becomes a source of radio frequency emissions. Consequently the environment becomes hostile for analog signal cables and even good coaxial cables may pick up substantial noise. The result of this pickup is generally appearance of high frequency oscillatory components in the data.

8.3 Pulse Shaping

The raw signal coming out of a detector is usually very narrow and therefore not very useful for extracting information such as height of the pulse. This narrow signal must first be shaped into a broader pulse with a rounded maximum. The broader pulse is necessary to reduce noise and the rounded peak is needed to measure the amplitude with precision. In general, there are two objectives for transforming the detector signal into a well defined pulse,

▶ Increase signal to noise ratio.

▶ Increase pulse pair resolution.

As we will see, these two requirements are somewhat contradictory and therefore an optimized solution is sought based on the particular application of the detector.

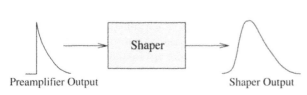

Preamplifier Output   Shaper Output

Figure 8.3.1: A pulse shaping amplifier can transform a preamplifier pulse having sharp peak and small width into a well-shaped pulse with rounded maximum and appropriate rise and decay times. Shapers can produce uni-polar as well as bi-polar pulses.

For detectors used in spectroscopy and precision measurements it is desired that the signal to noise ratio (S/N) is improved. This can be done by increasing the pulse width. The shapers designed for this task not only increase the pulse width but also produce a rounded peak of the pulse (see Fig.8.3.1). The rounded maximum facilitates the measurement of the pulse height. If the signal level has no effect on the shape of the pulse, the pulse height is essentially the measure of the energy absorbed by the detection medium. This proportionality ensures that the energy spectrum of the incident radiation can be inferred from the pulse height spectrum.

If the rate of incident radiation is very high then increasing the pulse width is not a good option because consecutive pulses can pile up over one another (see Fig.8.3.2). In such a situation, the shaping time (or the pulse width) is decreased such that the signal to noise ratio does not become unacceptable.

A number of pulse shaping methods are available each with its own pros and cons. The choice of the shaping methodology is therefore highly application specific. In the following we will look at some commonly used pulse shaping strategies.

8.3.A Delay Line Pulse Shaping

Delay line pulse shaping is used for detectors with high internal gains, such as, scintillation detectors. For detectors with low internal gains, the signal-to-noise ratio with delay line pulse shaping is smaller than the other available methods.

The basic idea behind this kind of pulse shaping is to combine the propagation delay of distributed delay lines to produce an essentially rectangular output pulse

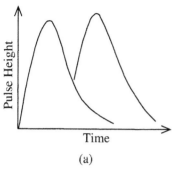

(a)

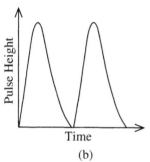

Time

(b)

Figure 8.3.2: (a) Pulse pile-up in high rate applications. (b) Decreasing pulse width eliminates pile-up.

from each preamplifier output pulse. The rectangular output pulse has fast rise and
fall times and can therefore be very conveniently used for pulse shape discrimination
and timing applications. Fig. 8.3.3 shows a typical delay line pulse shaping circuitry.
The step pulse from the preamplifier is inverted, delayed and then added back to
the original pulse to obtain a rectangular pulse having width equal to the delay time
of the delay line.

8.3.B   CR-RC Pulse Shaping

The CR-RC pulse shaping is perhaps the simplest and the most widely used method
of shaping preamplified detector pulses. The shaper consists of two parts: a CR
differentiator and an RC integrator (see Fig. 8.3.4).

The CR differentiator acts like a high pass filter with pulse amplitude given by

$$V_{out} = V_{in} - \frac{Q}{C_d}. \tag{8.3.1}$$

Differentiating this with respect to time gives a differential equation

$$\frac{dV_{out}}{dt} - \frac{dV_{in}}{dt} + \frac{V_{out}}{R_d C_d} = 0, \tag{8.3.2}$$

where we have used $V_{out} = R_d I = R_d dQ/dt$.

$R_d C_d$ in the above equation is the time constant of the differentiator which mainly
determines the decay time of the output signal. In detection systems the preamplifier
shaping time is generally larger than this time constant. This implies that the

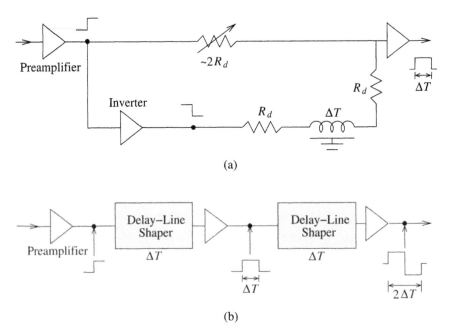

(a)

(b)

Figure 8.3.3: (a) Typical single delay line pulse shaping circuitry. (b) Double delay line shaping in which the output of the first shaper is fed into the input of the second single line shaper.

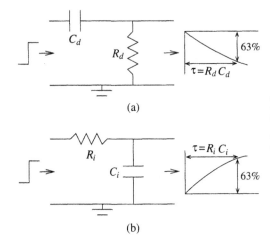

(a)

(b)

Figure 8.3.4: Response of simple (a) CR differentiator or high pass filter and (b) RC integrator or low pass filter to a step input pulse.

preamplifier output going into the CR differentiator can be considered a unit step function. In such a case the output can be approximated by

$$V_{out} = e^{-t/\tau_d}, \tag{8.3.3}$$

where $\tau_d = R_d C_d$ is the time constant.

The above equation implies that the output pulse decays exponentially with time with a width that can be controlled by the time constant τ_d. In radiation measurement systems the main objective is to measure the height of the pulse, which is proportional to the energy deposited by the radiation in the active volume of the detector. However, as can be seen in Fig.8.3.5, the output pulse from a CR differentiator has a sharp peak, which makes it difficult to measure the pulse height. Another disadvantage of the sharp peak is that it is most affected by the high frequency noise in the system. Obviously the solution to the problem is to somehow make this peak rounded. To solve this problem the output of the CR differentiator can be passed through an RC integrator.

An RC integrator with a time constant $\tau_i = R_i C_i$ is shown in Fig.8.3.4. The figure also depicts the behavior of the circuit in response to a step input pulse. In such a case the time profile of the output pulse can be approximated by

$$V_{out} = 1 - e^{-t/\tau_i}. \tag{8.3.4}$$

If now the fast rising output of CR differentiator is passed through this RC integrator circuit, the result will be a well shaped pulse with rounded maximum (see Fig.8.3.6 and Fig.8.3.5).

If we assume that the preamplifier output is a step function then the pulse profile after passing through subsequent CR and RC stages can be approximated by

$$V_{out} = \frac{\tau_d(\tau_d e^{-t/\tau_d} + \tau_i e^{-t/\tau_i})}{\tau_d \tau_i (\tau_d - \tau_i)}. \tag{8.3.5}$$

As we saw above the differentiator and integrator act like high-pass and low-pass filters respectively and can be combined together to form a band pass filter or pulse shaper where the shaping takes place in two steps. At first step the preamplifier pulse goes through the high pass filter, which attenuates the noisy low frequency components from the signal. The resulting signal is then fed into the low pass filter, which allows only the low frequency clean signals to pass through. The high frequency components consisting mainly of noise are attenuated at this stage.

According to equation 8.3.5 the rise and decay times of the final pulse depend on the time constants of the CR and RC shapers. Generally the time constants are chosen such that a pulse with rounded peak, short rise time, and long decay time is obtained. This can, for example be ensured by setting the two decay constants to be equal ($\tau_d = \tau_i = \tau$). In this case equation 8.3.5 becomes

$$V_{out} = \frac{t}{\tau} e^{-t/\tau}. \tag{8.3.6}$$

However the practical shapers used in radiation detection systems do not produce output pulses that can be approximated by a step function. The reason is that the shaping time of the preamplifier and the shaper do not differ by several orders of magnitude as required by the step function approximation. Realistic preamplifiers produce exponentially decaying pulses, which resemble the output of a CR filter in response to a step input pulse (cf. Fig. 8.3.5). If this pulse is then input into a CR-RC shaper, it inhibits a significant undershoot in the decaying part of the output pulse (see Fig. 8.3.6). This undershoot can be a serious problem for high rate systems where a new pulse can arrive before the previous one has fully decayed.

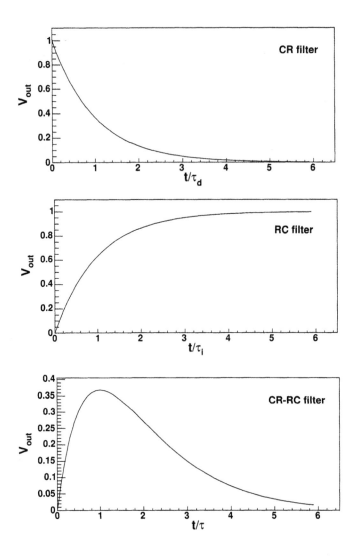

Figure 8.3.5: Response of CR (upper), RC (middle), and CR-RC (bottom) filters to step input pulse. The step input pulse approximation is valid only for preamplifier-shaper combinations in which the decay time of the amplifier pulse is much larger than the shaper's time constant. In real systems this approximation is generally not valid and consequently the CR-RC shaper suffers from small undershoot in the decaying part of the output pulse.

The new pulse will then ride on the decaying pulse whose undershoot will lower its effective height.

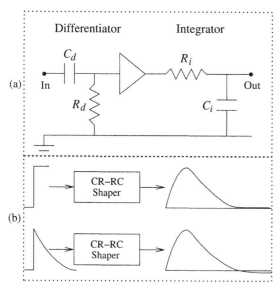

Figure 8.3.6: (a) A simple CR-RC shaper and (b) its response to two different input pulses: a step input and a more realistic preamplifier output. A perfect step input does not produce any undershoot while a significant undershoot can be expected when the output of a practical preamplifier is fed into the shaper. This undershoot can result in underestimation of the height of a subsequent pulse.

The choice of time constant of a CR-RC shaper depends on the particular requirements of the detection system. The measurement resolution and high rate capability are two competing factors that must be considered to find an optimized solution. Good resolution demands that the time constant be large enough to ensure complete integration of the detector signal. For example for scintillation detectors the time constant is chosen to be at least three times the decay constant of the scintillator. However such a long pulse duration might be problematic in high rate situations particularly due to the pulse undershoot problem mentioned above. In such a case the time constant is shortened at the expense of resolution.

Whatever time constant we choose, there is always a possibility for a pulse to arrive before the previous one has reached the baseline. If a large number of pulses arrive during the undershoot times the measured pulse amplitudes will be far less than the actual amplitudes. It may lead, for example to significant broadening of the measured energy distributions and consequently the system resolution will be seriously affected. Fortunately this problem can be solved by the so called pole-zero cancellation circuitry.

B.1   Pole-Zero Cancellation

As we saw above, the output of a realistic preamplifier can not be approximated by a step input pulse. In fact the exponentially decaying pulse from the preamplifier output looks very similar to the step input response of a CR filter. This implies that the response of a CR-RC shaper to such a pulse can be studied by assuming a CR-CR-RC circuit with a step input pulse. To study such a circuit we should take a look at its *transfer function*, which is simply a mathematical representation of its behavior under different conditions. For computational purposes the transfer function is mostly transformed into a complex space with a continuous or a discrete domain (Laplace domain for continuous function or Z domain for discrete function).

Detailed discussion about transfer functions and their derivations is beyond the scope of this book and the reader is referred to standard texts in electronics (11; 6). The Laplace transform of the transfer function of the CR-CR-RC circuit with a step input is given by

$$H(s) = \frac{\tau_p}{(\tau_p s + 1)} \frac{\tau_d s}{(\tau_d s + 1)(\tau_i s + 1)}, \tag{8.3.7}$$

where s is a complex variable of the Laplace transform. The subscripts d, i, and p in τ refer to CR differentiator, RC integrator, and preamplifier (which we have taken to be a CR differentiator with step input) respectively.

It is apparent that the above function has singularities (points at which the function becomes infinite) at $s = -1/\tau_p, -1/\tau_d, -1/\tau_i$. These are called *poles* of the function. One of the effects of these poles is the appearance of significant undershoot in the decaying part of the pulse (see Fig.8.3.7).

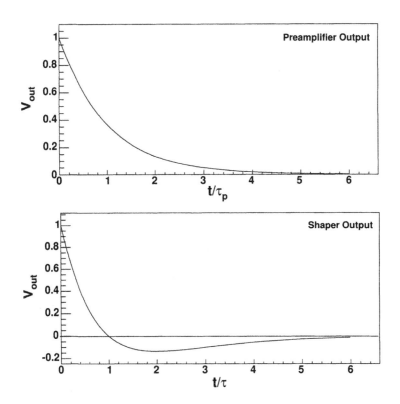

Figure 8.3.7: Typical output of a preamplifier having shaping time τ_p (upper plot). The response of a CR-RC shaper to this preamplifier output is shown in the lower plot. The graph has been generated using $\tau_d = \tau_i = \tau$. The significant undershoot in the decaying part of the pulse reduces system resolution and must therefore be either reduced or accounted for in the data.

Elimination of such undershoots can be accomplished if the responsible pole is somehow removed from the function. This can be done by introducing similar quantities in the numerator of the function such that they cancel the pole out. Since the quantity used to cancer the pole has the effect of vanishing the function in the absence of the pole therefore it is termed as *zero* of the function. Hence the term pole-zero cancellation. To determine which pole we should concentrate on removing from the function 8.3.7, we note that the undershoot becomes evident only after the shaping time of the preamplifier becomes comparable with that of the shaper. Hence we conclude that $s = -1/\tau_p$ is the pole that has to be removed.

Let us have a look at a simple but practical method of removing this pole. Fig. 8.3.8 shows a CR-RC shaper with a variable resistor R_{pz} installed in parallel to R_d. Inclusion of this resistor modifies the transfer function 8.3.7 as

$$H(s) = \frac{\tau_p s}{(\tau_p s + 1)} \frac{\tau_d (R_{pz} C_d s + 1)}{(R_{pz} C_d \tau_d s + R_{pz} C_d + \tau_d)} \frac{1}{(\tau_i s + 1)}. \tag{8.3.8}$$

It is apparent from this function that a properly chosen value of R_{pz} can eliminate the undershoot altogether (see Example at the end of this section). However in practical systems where the parameters related to physical circuit elements do not remain constant over time, the approach is to minimize the undershoot. Manually this can be done by changing R_{pz} while looking at the output signal on an oscilloscope. Some newer developments utilize additional circuitry to automatically adjust this resistance and are therefore more desirable in systems with randomly drifting electronic parameters and having large number of channels.

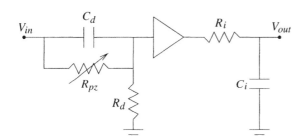

Figure 8.3.8: Simple pole zero cancellation circuit. The variable resistor can be used to minimize the undershoot.

Theoretically pole-zero cancellation strategy we saw above should work perfectly. However we should remember that addition of a zero involves adding an additional piece of circuitry with its own limitations and irregularities. Minor physical changes, for example a small shift in temperature may slightly shift the pole and zero. Consequently the pole may not get canceled at all. The variable resistor R_{pz} can compensate for some drifts and modern systems are designed to accommodate small changes. However such circuits have their own limitations and do not always work in all possible scenarios. The wide spread implementation of pole zero cancellation circuitry in detection systems is therefore not due to its effectiveness over other techniques, such as control theory, rather due to the ease and simplicity in its application.

Example:
Determine the optimal value of R_{pz} to eliminate undershoot from the output

of circuit 8.3.8.

Solution:
Examination of transfer function 8.3.8 of circuit 8.3.8 reveals that if we choose $R_{pz} = \tau_p/C_d$, it will transform the numerator $(R_{pz}C_d s + 1)$ into $(\tau_p s + 1)$. This will cancel out the identical term in the denominator thus eliminating the undershoot from the output. The transfer function in this case will become

$$H(s) = \frac{\tau_p \tau_d s}{\tau_p(\tau_d s + 1) + \tau_d} \frac{1}{(\tau_i s + 1)}. \tag{8.3.9}$$

It should be noted that even a small drift in C_d will make the undershoot reappear and the pole will not be totally eliminated. Modern systems employ additional circuitry to adjust for such changes automatically to keep the undershoot at minimum.

B.2 Baseline Shift Minimization

In many applications it is desired that the electronic circuitry following the amplifiers is AC-coupled. This might induce baseline shifts in the signal with varying count rate. In order to actively minimize such base line shifts a baseline restorer circuitry is introduced. One such circuit is shown in figure 8.3.9 Here the addition of another CR differentiator after the RC integrator produces a bipolar pulse instead of a unipolar one. Although such a circuit minimizes the baseline shifts but the longer duration of its bipolar output makes it unsuitable for high rate applications. Its signal to noise ratio is also worse than the simple CR-RC shaper.

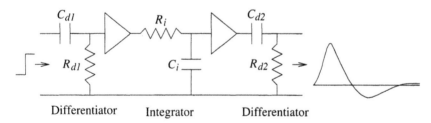

Figure 8.3.9: CR-RC-CR shaper. The output of this shaper is a bipolar pulse which minimizes the baseline shifts.

8.3.C Semi-Gaussian Pulse Shaping

The simple CR-RC pulse shapers we visited earlier work well for applications where high resolution is not required and signal-to-noise ratio is not a very important consideration. For systems where improvement in signal-to-noise ratio is of prime importance, the simple RC integration circuitry must be replaced with a complicated active integrator. Such networks can improve not only the signal-to-noise ratio but are also capable of decreasing the width of the output pulse. Fig. 8.3.10 shows a typical 2-stage active integrator circuit.

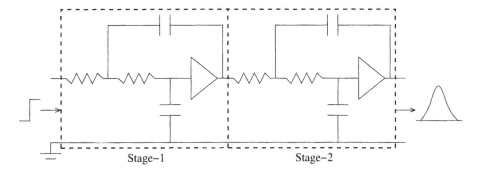

Figure 8.3.10:   Step input response of a typical two-stage semi-Gaussian pulse shaper.

Such a shaper is called semi-Gaussian because the output pulse has a Gaussian-like shape (see Fig.8.3.11). The shaper could be designed to have unipolar or bipolar outputs. Generally unipolar output is preferred due to a much better signal to noise ratio.

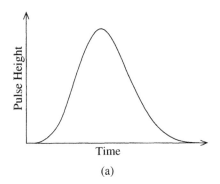

(a)

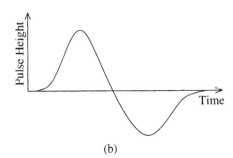

(b)

Figure 8.3.11:   (a) Unipolar and (b) bipolar outputs of a semi-Gaussian pulse shaper.

8.3.D   Semi-Triangular Pulse Shaping

The semi-Gaussian pulse shaper we discussed earlier is one of the most popular circuits used in radiation detection systems. If a number of such circuits are connected in succession, as shown in Fig.8.3.12, it results in an output pulse with approximately linear rising and falling edges. The output pulse then has a shape that looks like a triangle (see Fig.8.3.13). Such a shaper is referred to as semi-triangular pulse shaper.

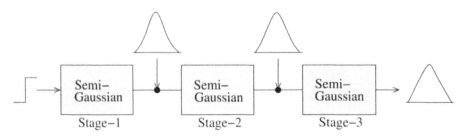

Figure 8.3.12: Step input response at each stage of a three-stage semi-triangle pulse shaper.

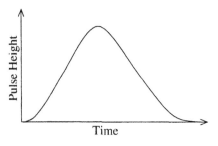

Figure 8.3.13:   Output of a semi-triangular pulse shaper.

8.4   Filtering

Due to pickup from external and internal noise sources, it is often desirable to filter the pulse such that only certain frequencies are allowed to pass through to the next electronic stage. Though it was not specifically mentioned earlier but the pulse shapers we just discussed actually perform the filtration function.

There are essentially three classes of filters depending on the frequency range that needs to be blocked. The low pass filters are perhaps the most commonly used as they block all the high frequency components from the signal. The high pass filters, on the other hand, block low frequency components, which may be desirable in some applications. The combination of these two is called the band pass filter, which allows only a range of frequencies to pass through. Fig.?? shows schematics of different types of filters.

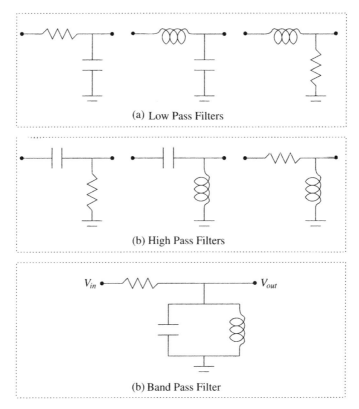

Figure 8.4.1: (a) (b) Schematics of different types of low pass and high pass filters. (c) A simple band pass filter.

8.4.A   Low Pass Filter

The RC integrator we discussed earlier (see Fig.8.3.4(b)) is actually a low pass filter as can be seen from the following expression for its output voltage.

$$V_{out} = V_{in} \frac{1}{\sqrt{1 + (2\pi f RC)^2}} \tag{8.4.1}$$

Here V_{in} is the input voltage and f is the frequency of the input signal. It is obvious that the response of the circuit is weaker at higher frequencies. The frequency at which the response becomes too weak is called the cutoff frequency of the filter. Its value depends not only on the circuit components but also on the definition of the *too weak output*. The most commonly agreed upon value of the cutoff frequency is the value at which the response becomes -3 *dB*. To translate this value in terms of resistance and capacitance of the circuit, let us take the base-10 logarithm of both sides of the above equation and multiply by 2. We are doing this to convert the ratio V_{out}/V_{in} into decibel (*dB*) units.

$$2\log\left(\frac{V_{out}}{V_{in}}\right) = 2\log\left[\frac{1}{\sqrt{1 + (2\pi f RC)^2}}\right] \tag{8.4.2}$$

Let us now substitute $f = 1/2\pi RC$ in this relation. This gives

$$2\log\left(\frac{V_{out}}{V_{in}}\right) = -3\ dB. \tag{8.4.3}$$

Hence, according to our definition, the frequency

$$f_{cutoff} = \frac{1}{2\pi RC}, \tag{8.4.4}$$

is the cutoff frequency for this simple RC filter. This frequency is sometimes also referred to as the *breakpoint* since above this value the attenuation of the circuit becomes too large. Quantitatively speaking, above the breakpoint the response of the circuit varies by 6 dB per octave of frequency increase (see Example below). On the other hand at frequencies much lower than this cutoff frequency the attenuation becomes negligibly small. This can also be seen from equation 8.4.2 where we can neglect the second term in the denominator on the right hand side for $f >> f_{cutoff}$. This gives

$$2\log\left(\frac{V_{out}}{V_{in}}\right) \approx 2\log(1) = 0$$
$$\Rightarrow V_{out} \approx V_{in},$$

which simply means that at frequencies much lower than the cutoff value the output voltage (or power) is nearly equal to the input voltage (or power).

Example:
Quantify the change in the response of a simple RC low pass filter with change in frequency above the breakpoint of $f_{cutoff} = 1/2\pi RC$.

Solution:
We start with equation 8.4.2 and note that at frequency higher than the cutoff frequency the first term in the denominator on the right hand side can be neglected.

$$2\log\left(\frac{V_{out}}{V_{in}}\right) = 2\log\left[\frac{1}{\sqrt{1 + (2\pi f RC)^2}}\right]$$
$$\approx 2\log\left[\frac{1}{\sqrt{(2\pi f RC)^2}}\right]$$
$$= -2\log(2\pi f RC)$$
$$\propto -2\log(f)$$

The above proportionality means that an 8 fold increase in frequency would correspond to a 6 dB change in the response. Hence we can say that the response of a low pass RC filter changes by 6 dB per octave of frequency increase.

The simple RC low pass filter we just discussed is seldom used in modern radiation detection systems, which are equipped with more complicated active circuits.

8.4.B   High Pass Filter

The CR differentiator circuit shown in Fig.8.3.4(a) is a high pass filter. Its output voltage is given by

$$V_{out} = V_{in} \frac{2\pi RC}{\sqrt{1 + (2\pi f RC)^2}}, \tag{8.4.5}$$

where, as before, V_{in} is the input voltage and f is the input signal frequency. Let us write this relation in terms of the decibel notation by first taking base-10 logarithm of both sides and multiplying by 2. This gives

$$
\begin{aligned}
2\log\left(\frac{V_{out}}{V_{in}}\right) &= 2\log\left[\frac{2\pi f RC}{\sqrt{1 + (2\pi f RC)^2}}\right] \\
&= 2\log\left[\frac{1}{\sqrt{1 + 1/(2\pi f RC)^2}}\right].
\end{aligned}
\tag{8.4.6}
$$

This relation shows that the response of the system is inversely proportional to the logarithm of the frequency. As in the case of the low pass filter, substituting $f = 1/2\pi RC$ in this relation gives

$$2\log\left(\frac{V_{out}}{V_{in}}\right) = -3 \, dB. \tag{8.4.7}$$

This frequency at which the response is -3 dB is called the cutoff frequency or the breakpoint. Below this frequency the attenuation becomes high and increases rapidly at a rate of 6 dB per octave of frequency change. The circuit can therefore be said to effectively block the low frequencies. At frequencies much higher than the cutoff value the second term in the denominator of the right hand side of equation 8.4.6 can be neglected, giving

$$
\begin{aligned}
2\log\left(\frac{V_{out}}{V_{in}}\right) &\approx 2\log(1) = 0 \\
\Rightarrow V_{out} &\approx V_{in}.
\end{aligned}
$$

This shows that for frequencies much higher than the breakpoint, the output voltage of the CR filter is approximately equal the voltage at its input.

8.4.C   Band Pass Filter

A band pass filter can be constructed by connecting a low pass filter with a high pass filter in series as shown in Fig.8.4.2(a). A realistic RC-CR band pass filter is shown in Fig.8.4.2(b). The values of the individual resistances and capacitances can be chosen according to the cutoff frequency requirements. It should be noted that here there will be two cutoff frequencies or breakpoints corresponding to the low and high pass filters.

8.5   Amplification

The two main purposes of the preamplifiers we discussed earlier are to amplify the low level signal as it comes out of the detector and to match the detector and

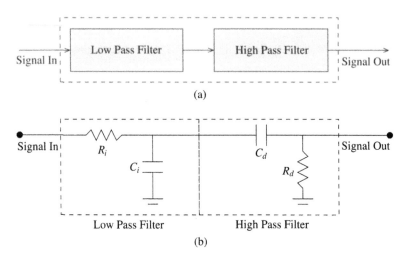

Figure 8.4.2: (a) Conceptual design of a band pass filter. (b) A simple but realistic band pass filter consisting of a RC integrator and a CR differentiator.

external circuit impedances. Therefore even if the preamplification is not required, it is a good idea to connect the detector output directly to a preamplifier with unity gain. It should however be pointed that in principle it is possible to avoid using a preamplifier if the output signal is large enough to be transported and processed.

Whether a preamplifier is used or not, a main amplifier at a later stage is almost always present. Generally one uses a shaper that amplifies the signal as well. In fact almost all active shapers can also amplify the signal. The basic working principle of such a combination is the same as we have already discussed in the sections on preamplifiers.

8.6 Discrimination

Often it is desired to check whether the analog output signal is above a predefined threshold or not. This is done through a *discriminator* module, which accepts an analog input and produces a digital output. It has a comparator circuitry that compares the input voltage to a preset voltage. The preset voltage, also called the discriminator *threshold*, can normally be adjusted through a potentiometer. If the input pulse amplitude is greater than the preset threshold, the output logic level becomes active, otherwise it remains in its previous state. Now it is also possible to design a discriminator that has a high threshold level as well. The output of such a module would become high only if the input lies between two predefined thresholds. This concept is graphically depicted in Fig.8.6.1.

The output of a discriminator is binary. That is, it can be either high or low. Definition of low and high is arbitrary because one can define any voltage level to represent high and any other to represent low. For a system that is designed to work on a certain logic definition, this arbitrariness does not pose a problem. However in expandable systems, where it is desired that modules made by other

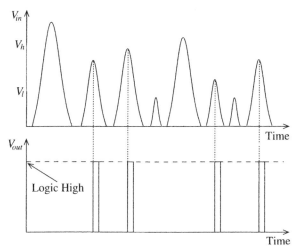

Figure 8.6.1: Pulse discrimination by a two-level discriminator. Only pulses that are between the two preset thresholds are counted by the subsequent circuity.

developers could be integrated into the system, interconversion of logic levels could pose an engineering problem. To avoid this problem, certain logic standards have been developed. It is highly recommended that the electronics designers try their best to use one of these standards in their circuitry. The following are the most commonly used logic standards.

▶ **NIM Logic:** NIM stands for Nuclear Instrumentation Modules, a set of modules that were developed for nuclear instrumentation. A logic was developed specifically for these modules. In this logic $0\ V$ is defined as logic 0. Logic 1, however, is defined in terms of current and not voltage. A current of -16 mA into a 50 Ω resistor corresponds to logic high, which is equivalent to about -800 mV.

▶ **TTL Logic:** TTL is an acronym of Transistor Transistor Logic. It is perhaps the most widely used logic standard. As opposed to NIM, it is defined solely in terms of voltage: a voltage of $0\ V$ corresponds to logic 0 while the logic is said to be high if the level is between $2\ V$ and $5\ V$.

▶ **ECL Logic:** The Emitter Coupled Logic or ECL is based on differential amplification of the digital signals and adjustment of the dc voltage levels through emitter followers. The ECL logic levels are conventionally defined as -1.6 V for low and -0.75 V for high.

8.6.A Pulse Counting

In the previous chapters we saw that in most detection systems the height of the output pulse is proportional to the energy delivered by the incident radiation. This implies that the energy spectrum of the radiation can be determined by counting pulses of different amplitudes. This is essentially what is done in spectroscopic systems. However counting pulses is not limited to radiation spectroscopy since one might even desire to simply count all the pulses to determine the total radiation intensity. Such measurements are performed using two devices called single channel and multi channel analyzers, which will be discussed in some detail now.

A.1 Single Channel Analyzer (SCA)

In a large number of applications one is interested in counting the number of *interesting* pulses. The definition of an interesting pulse depends on what one intends to look at. For example one might be interested in looking at all the pulses generated in the detector by the radiation. Or, since the height of a pulse depends on the energy deposited by the radiation, one might want to count pulses corresponding to some energy range of interest. No matter which pulses are to be analyzed, the remaining pulses must be filtered out or blocked so that a *clean* dataset could be obtained. This is necessary even if one intends to look at all radiation induced pulses since there are always low level noise and high level spurious pulses. This filtration requires a two-level discriminator, which we have already discussed in the previous section. An instrument that discriminates the pulses based on set thresholds and produces logic outputs is called a single channel analyzer. Here *single* signifies the fact that all the pulses are discriminated using only one set of thresholds. The block diagram of a typical SCA is shown in Fig.8.6.2. Here LLD and ULD stand for upper level discriminator and lower level discriminator respectively. In most modern commercial single channel analyzers the thresholds can either be set through potentiometer knobs available on the front panel or through an external port.

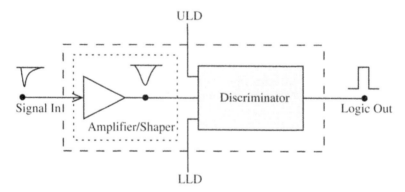

Figure 8.6.2: Block diagram of a simple single channel analyzer.

After the pulse has been discriminated and a logical output has been generated, it must be counted. Most commercially available SCAs do not have counting functionality and therefore one must use a separate counting module for the purpose.

Single channel analyzers are most suitable for discriminating pulses within a certain height range. A common application is measuring intensity of an x-ray peak from an x-ray machine. Recall that the x-ray peak from an x-ray machine is superimposed on a bremsstrahlung continuum. Hence if one knows the energy of the peak, it can be easily discriminated by setting SCA thresholds around it.

Now, suppose we are interested in determining the whole x-ray spectrum of an x-ray machine. This can be done counting pulses at different threshold windows such that the whole energy range is spanned. However it is apparent that this is quite a laborious and time consuming task, not to mention inefficient and prone to uncertainties related to time variation of radiation flux. Therefore for such measurements single channel analyzers are not generally used.

A.2 Multi Channel Analyzer (MCA)

A multi channel analyzer can be thought of as a modified version of single channel analyzer and a counter since it essentially performs the same task albeit with a number of threshold windows. Thus it eliminates the need to count pulses at each threshold window individually, making the process faster and less vulnerable to uncertainties related to time variations in radiation flux. In terms of internal functioning, multi channel analyzers are quite different from single channel analyzers since they actually digitize the input signal for analysis as opposed to the all-analog processing by single channel analyzers. Fig.8.6.3 shows a simple block diagram of a multi channel analyzer. The analog detector pulse is first amplified and shaped. The height of the amplified pulse is then digitized. Since the ADC output is linearly related to the input analog voltage, therefore the counts thus obtained correspond to the energy deposited by the radiation. After digitization, a count in the corresponding memory bin is incremented.

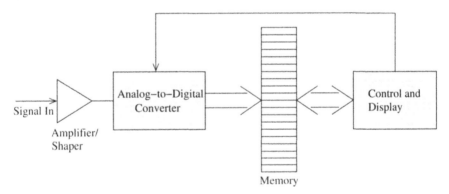

Figure 8.6.3: Block diagram of a simple multi channel analyzer designed for pulse height analysis.

Generally, multi channel analyzers have a number of channels, such as 512, 1024, or more, giving the user some choice in selecting the appropriate resolution. Each of these channels actually corresponds to memory locations that can be addressed by the corresponding ADC counts and thus incremented.. The system can be tuned such that the whole dynamic range is covered.

Most multi channel analyzers can be operated in two different modes. One is the usual pulse height analysis mode that we just discussed and the other is the multi channel scalar mode. In this mode the pulses are counted with respect to some other parameter. The process starts with a logic pulse that starts incrementing the counts in the first memory location. This counting continues until another logic pulse arrives that move the counting to the next memory location. The logic pulses can correspond to any parameter of interest. It could even be time, in which case the system simply measures the total intensity.

8.7   Analog to Digital Conversion

Converting an analog pulse into digital form is very convenient in terms of analyzing and storing the relevant information contained in the signal. The so called *analog to digital converters* (ADC) are extensively used in detection systems to convert the information contained in the analog pulse into an equivalent digital form. Before we go on to discuss different types of ADCs, let us have a look at some factors that need to be considered while deciding on the method of conversion.

8.7.A   A/D-Conversion Related Parameters

A.1   Conversion Time

This is the time it takes the ADC to perform one complete conversion. It depends not only on the A/D conversion method but also on the pulse height.

A.2   Dead Time

The total time it takes the ADC to acquire a signal, complete the conversion, and become available for the next acquisition is called dead time because during this time it can not accept a new signal. The dead time typically consists of:

▶ signal acquisition time,

▶ conversion time,

▶ data transfer to buffers, and

▶ reset time.

In well designed systems, the transfer of converted signal to the memory is performed during the reset time. How that memory is subsequently read out and handled is not part of ADC performance. However since such a memory is generally of limited capacity (such as 1kB FIFOs) , it must be read out continuously to avoid overflow, which leads to loss of information.

Additionally, the operations on ADCs are controlled through digital signals, which have their own response times and uncertainties. Such uncertainties are referred to as *time jitters* and should be given proper consideration in high resolution systems.

If the event rate is such that the pulses arrive during dead time of ADC, the information gets lost unless the charge is dynamically integrated on some capacitor for later acquisition by the ADC. Another possibility is to determine the average dead time of the system and then correct for it in the final analysis. This strategy works well for random signals where all events experience the same dead time. Determination of the dead time is a straight forward process in which the ADC is fed with a known stream of pulses and the output is recorded. The comparison of output to the input gives quantitative measure of the dead time.

A.3   Resolution

The ADC output is in the form of the so called *digital word*, which is simply a number. This number should be directly proportional to the analog input at the full dynamic range of the system. The lowest number that the ADC can assign to a meaningful analog input determines its resolution.

The resolution of an ADC is generally represented in *bits*. A n-bit ADC ideally has a resolution of

$$\frac{\triangle V}{V} = \frac{1}{2^n}. \tag{8.7.1}$$

Example:
Determine the resolutions of an 8-bit and a 16-bit ADC.

Solution:
The 8-bit ADC as a resolution of

$$\begin{aligned} \frac{\triangle V}{V} &= \frac{1}{2^n} \\ &= \frac{1}{2^8} \\ &= 3.90 \times 10^{-3}. \end{aligned}$$

Similarly the resolution of the 16-bit ADC is

$$\begin{aligned} \frac{\triangle V}{V} &= \frac{1}{2^{16}} \\ &= 1.52 \times 10^{-5}. \end{aligned}$$

A.4 Nonlinearity

Any nonlinearity in the analog to digital conversion process can be a serious problem at least for high resolution detection systems. It is therefore extremely important that the linearity of the ADC used in the system is ensured. Unfortunately the linearity of the analog to digital conversion process does not depend just on the ADC design and circuitry but also on the shape of the analog pulse and its duration.

There are two types of nonlinearities that must be eliminated or at least minimized in a practical ADC. These are generally referred to as *differential* and *integral* nonlinearity.

▶ **Differential Nonlinearity:** Differential nonlinearity is a measure of the uniformity of the increments through which an ADC goes during the conversion process.

▶ **Integral Nonlinearity:** If the ADC counts are not linearly proportional to the analog pulse amplitude then it is said to have integral nonlinearity.

A.5 Stability

It is generally desired that the physical parameters, on which the analog to digital conversions depend, do not significantly change with time. Therefore stability is one of the important criteria for choosing an ADC.

8.7.B   A/D Conversion Methods

B.1   Digital Ramp ADC

This simple technique involves a digital to analog converter (DAC), which is used to convert the output of a binary counter. A comparator compares this output to the analog input signal height to decide whether the counter should continue with counting or stop. The counter is stopped when the two voltages are within the set tolerance level. The counter is provided with a clock pulse through either a built-in or an external oscillator. Fig.8.7.1 shows a schematic diagram of such an ADC.

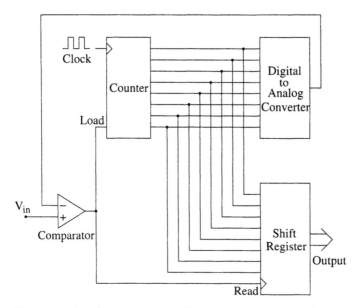

Figure 8.7.1:   A digital ramp ADC having 8-bit resolution.

At the start of the conversion cycle the DAC output is lower than the input voltage and consequently the comparator output goes high. This forces the counter to start counting up with each clock pulse (see Fig.8.7.2). The output is then fed directly to the DAC, which outputs a slightly higher voltage. This voltage is then again compared with the input voltage. If the input voltage is still higher than the DAC output, the comparator output will remain high and the counter will continue with counting. This counting process stops as soon as the DAC output exceed the input voltage, since at that point the comparator's output goes low. To read out counter's value at that point, a shift register is provided which loads the binary count as soon as the comparator's output becomes low. This low output of the comparator also causes the counter to reset to zero and become available for the next cycle.

This technique of digital ramping suffers from the following two major problems that make it undesirable for some applications.

▶ **Slow Sampling.** The fact that the counter has to count from zero at each conversion cycle makes the process very slow and unsuitable for high rate applications.

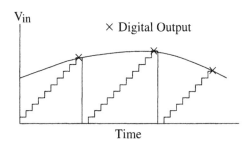

Figure 8.7.2: Counting cycles in a digital ramp ADC.

▶ **Irregular Sampling.** Since the counter keeps on counting until the DAC output does not exceed the input voltage, the conversion time depends on the height of the signal. This implies that the sampling is irregular for time varying input signal, something which makes the readout timing and analysis difficult.

B.2 Successive Approximation ADC

As we saw in the previous section, the digital ramp ADC suffers from slow speed since the counter always starts from zero at the start of the conversion cycle. To overcome this disadvantage the counting process must be modified. This can be accomplished by replacing the binary counter by a successive approximation register (SAR) (see Fig.8.7.3). This register is still a counter but it does not count up in binary sequence, rather it increments the counts by trying all bits and comparing the result of DAC output to the analog input at each step. It starts with the most significant bit and ends its cycle with the least significant bit. At each step it sets the bits according to whether the difference between DAC output and analog input is less than or greater than zero. The advantage of this method is that the DAC output converges to the analog input faster than the simple binary counter. Fig.8.7.4 shows typical conversion iterations of a SAR.

The SAR provides an additional bit for the shift register to notify of its completion of the conversion. This can also be a predefined regularly spaced clock cycle, in which case the output is not irregularly spaced in time as in the digital ramp ADC.

The main drawback of successive approximation ADC is its high differential non-linearity, which could go as high as 20

B.3 Tracking ADC

Tracking ADC uses an up/down counter as opposed to a regular counter in digital ramp ADC (see Fig.8.7.5). The counter counts continuously on a clock and counts up or down according to whether the comparator output is high or low. In this way the counter continuously tracks the input voltage and never gets reset unless the input voltage itself drops to zero. The net effect is the fast conversion time and simplicity in design as in this case the shift register is not needed. Fig.8.7.6 shows how such an ADC tracks the input analog voltage.

The main disadvantage of this design is that the counter continuously counts up or down and consequently the digitized output is never stable. This variation can, however, be minimized by latching the counter output to a shift register only when the output changes by some predefined value.

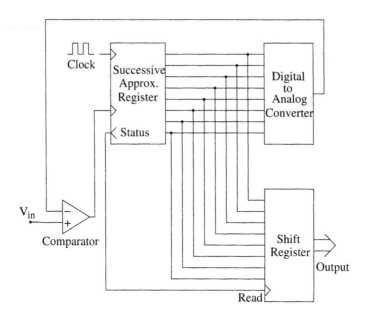

Figure 8.7.3: Schematic of an 8-bit successive approximation ADC.

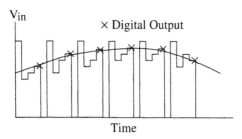

Figure 8.7.4: Conversion cycles in a successive approximation ADC.

B.4 Wilkinson ADC

Wilkinson ADC belongs to a class of *slope* or *integrating* ADCs. As opposed to other ADCs we visited earlier, slope ADCs do not employ DAC, thus eliminating the major source of differential nonlinearity.

The main idea behind this type of ADC is to save the input pulse on an analog memory capacitor and then allow it to discharge slowly. A counter keeps on counting during the linear discharge of the capacitor. The final digital word is proportional to the analog input. The following are the steps that a typical Wilkinson ADC takes during a conversion cycle (see also Fig.8.7.7(a)).

1. The input pulse amplitude is stretched to a wide pulse using a pulse stretcher circuitry. Commonly used method is the so called *sample and hold* circuit, in which the analog voltage is sampled on a capacitor through a FET switch (see Fig.8.7.7(b)).

2. The stretched pulse is transferred to a memory capacitor.

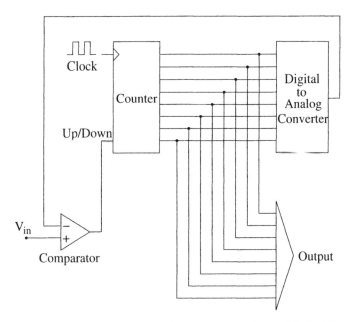

Figure 8.7.5: Schematic of an 8-bit tracking ADC. The counter starts counting up (down) whenever the comparator's output is high (low). Shift register is not required in such a design.

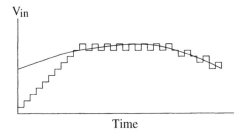

Figure 8.7.6: Conversions in a tracking ADC. The counter never gets reset once it starts counting unless the analog input drops to zero.

3. The capacitor is disconnected from the pulse stretcher circuit.

4. A current source is connected to the capacitor, which starts discharging it linearly.

5. At the same time, a counter is started to count the number of clock pulses it takes the voltage on the capacitor to reach the baseline voltage. If the source provides a constant current i_s throughout the discharge then it will take time

$$t_d = V_{in}\frac{C}{i_s} \qquad (8.7.2)$$

to completely discharge the capacitor having capacitance C. Here V_{in} is the input voltage. Since this time is proportional to the digital counter therefore the final digitized word is proportional to the input voltage (see also Fig.8.7.8).

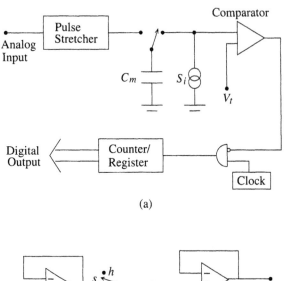

(a)

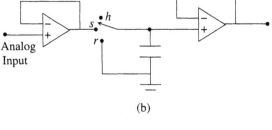

(b)

Figure 8.7.7: (a) Schematic of a Wilkinson ADC. C_m is a memory capacitor, S_i is a current source, and V_t is the reference voltage used by the comparator to make the conversion start/stop decision. Normally V_t is kept at the ground potential, in which case the comparator is called a zero-crossing comparator. (b) Sketch of a simple sample and hold circuit that can be used as a pulse stretcher in Wilkinson ADC.

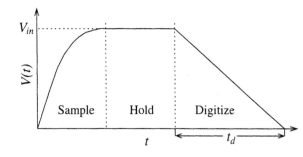

Figure 8.7.8: Conversion cycle of a typical Wilkinson ADC. The input pulse amplitude is sampled and held on a capacitor and then digitized. The digitization time t_d is proportional to the input voltage V_{in}.

A big disadvantage of Wilkinson A/D technique is its long conversion time, which can be calculated from

$$T_{conv} = 2^n T_{clock} = \frac{2^n}{f_{clock}}, \qquad (8.7.3)$$

where n is the ADC resolution in bits and T_{clock} is the period of counter's clock having frequency f_{clock}. Hence a 12 bit ($n = 12$) Wilkinson ADC having a clock frequency of 100 MHz will complete one conversion in about 41 μs, which is a very long time for high rate applications. Increasing the resolution to 16 bits at the same clock frequency would take conversion time to about 655 μs. Of course the conversion time can be shortened by increasing the clock frequency but it has its own engineering difficulties. Nevertheless Wilkinson ADCs working at 400-500 MHz have been manufactured and successfully operated in detection systems.

B.5 Flash ADC

As the name suggests, a flash ADC performs the conversion in a *flash*, which simply means that it is extremely fast. Such ADCs are becoming more and more popular due to the conversion speed they offer as compared to conventional ADCs. The downside is their resolution, which is difficult to increase beyond the 8 bit mark.

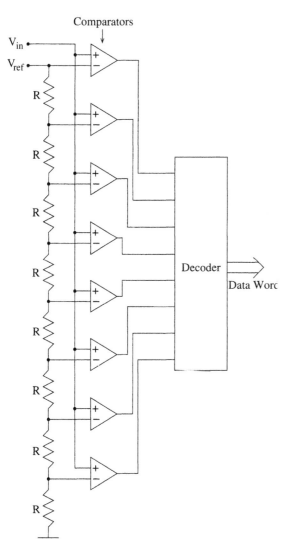

Figure 8.7.9: A 4-bit flash ADC having 8 comparators. The decoder could, for example be a priority encoder that generates a binary number based on the highest order active input. V_{ref} is a stable reference voltage, which is generally provided by a precision voltage regulator that is a part of the converter circuit.

Fig.8.7.9 depicts the working principle of a typical flash ADC. As shown, the ADC consists of a number of comparators in parallel, each of which compares the input signal to a unique reference voltage. The outputs of the comparators are fed to a decoder that produces a binary output. A flash ADC with n-bit resolution requires 2^n comparators. This is a big disadvantage if high resolution is required. Generally in radiation detection systems a compromised 8-bit circuitry composed of 256 comparators is employed. The component intensity of flash ADCs can be appreciated by noting that even a moderate 12-bit resolution system would require 4096 comparators, a task extremely difficult to accomplish, if not impossible, without jeopardizing the small size and simplicity of the design. The disadvantages of flash ADCs include

▶ lower resolution (generally 8 bit),

▶ high power consumption,

▶ high input capacitance,

▶ differential non-linearity, and

▶ large number of components required (an 8 bit flash ADC requires 256 comparators).

8.7.C Hybrid ADCs

As we saw in the previous sections, different A/D conversion methods have different pros and cons. Designing a particular ADC is an optimization process of the two competing parameters: resolution and conversion time. A conventional ADC working at high resolution suffers from longer conversion time and is therefore unsuitable for high rate situations. Fortunately there is a practical way to achieve both of these traits, i.e. high resolution and short conversion time, in a single ADC. Such an ADC is called *hybrid ADC*, since it is a combination of two conventional ADCs.

Suppose we require a 14 bit ADC that could perform a single conversion in less than $1\mu s$. It will be quite a challenge to build this ADC using one of the techniques we visited earlier. However using hybrid technique, we can divide the resolution in two parts: coarse (say 8 bits) and fine (the remaining 6 bits). For coarse conversion we can use a flash ADC, which can perform the conversion in $10ns$ or less. For fine conversion we can use a Wilkinson ADC, which will take about $0.6\mu s$ for a conversion if working at a clock frequency of $100MHz$. Hence one full 14 bit conversion will be completed in less than $1\mu s$ with very good integral and differential linearities.

Because of their superior performance over conventional ADCs, the hybrid ADCs are becoming more and more popular in high resolution radiation detection systems designed to work in high rate environments.

8.8 Digital Signal Processing

Digital signal processing is an alternative to the analog signal processing we have been discussing up until now. The basic idea behind it is to directly digitize the preamplified signal and then process the data digitally. Since here the shaping and

subsequent filtration are done digitally therefore the process totally eliminates the
need to shape the signal before its digitization.

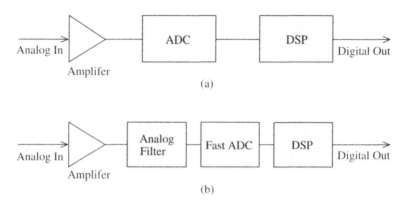

(a)

(b)

Figure 8.8.1:   (a) Basic steps involved in digital signal processing.
(b) Fast ADCs and pre-filtration is required in most applications to
avoid aliasing of data.

Fig.8.8.1(a) shows the basic steps needed to digitally process the signal. Before
the signal could be digitally processed, it must first be preamplified and converted to
digital form. The reader would recall that the pulse from a preamplifier generally has
a fast rise time. This implies that the ADC must be able to perform the conversion
so quickly that the reconstructed pulse does not get aliased. Aliasing is a term that
was introduced in the chapter on imaging detectors. There, we were concerned with
spatial aliasing and saw that its effect is to introduce additional data and/or miss
actual data points due to under-sampling of the image. The same is true for temporal
aliasing. If the signal has high frequency components (such as the ones embedded
in the fast rise time of the pulse) and the ADC sampling frequency is lower than
the Nyquist frequency then the data will be aliased. As described in the chapter on
imaging detectors, Nyquist frequency is twice the highest frequency component in
the signal. That is, if one Fourier transforms the analog pulse, the Nyquist frequency
would be twice the highest frequency in the Fourier spectrum. Fig.8.8.2 shows the
effect of sampling frequency on the *reconstructed* analog pulse. It is apparent that
high frequency components are most affected by undersampling. In Fig.8.8.2(a)
the data is sampled at a frequency much lower than the Nyquist condition and is
therefore severely aliased. On the other hand, as shown in Fig.8.8.2(b), sampling at a
frequency higher than the Nyquist value leads to an almost perfect replication of the
analog pulse. Now, practical ADCs have sampling and conversion rate limitations
and therefore it is not always possible to satisfy the Nyquist condition for all types
of input analog signals. A way around this problems is to pre-filter the data before
digitization. This is actually what is done in most practical systems (see Fig.8.8.1).
  After digitization by the analog to digital converter, the data are fed into the
digital signal processing unit. This is where the data are digitally shaped, filtered,
and further processed using complex mathematical algorithms. Here the processing
is done solely through software (exceptions are the digital processing units consisting
of hardware components designed to perform specific operations). This is where the

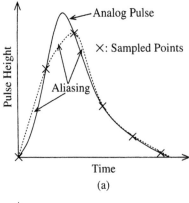

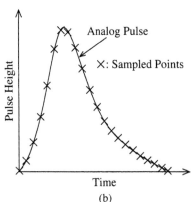

(a)

Figure 8.8.2: (a) Undersamling of data leads to aliasing. (b) Aliasing can be avoided by sampling at higher than Nyquist frequency.

(b)

digital signal processing becomes much superior and powerful than its analog counterpart. The tasks of complicated pulse manipulations, which would otherwise take extremely complex analog circuitry, can be handled using computer codes. Furthermore the algorithms and the computer codes can be modified and changed according to the particular requirements without changing any hardware components.

Although digital signal processing has many advantages, its design and implementation are not very easy. First of all, as we just discussed, aliasing is one of the biggest problems in DSP design as it requires use of very fast ADCs. But this is not the whole story, since the resolution and dynamic range of the ADC should also be good enough to properly sample and digitize the signal. However, as we saw earlier in the chapter, the resolution and speed have conflicting requirements on ADC design. Another problem is that the ADCs themselves introduce noise in the system, thus deteriorating the overall signal to noise ratio.

8.8.A Digital Filters

Discussion of digital filters is a very broad subject in itself and is out of the scope of this book. However due to its importance we will introduce here the reader to the basic concepts and designs of such filters.

The basic building blocks of a digital filter are

▶ adders,

▶ delay function, and

▶ multipliers.

Since typical digital filters consist of very large number of such building blocks or functions, therefore they are prone to errors introduced by the associated mathematical operations. The errors introduced by additions and multiplications can pile up and become significant at the output. Therefore designing a digital filter not only warrants care in designing the algorithms but also in its implementation.

Most of the modern digital filters are based on floating point mathematics with feedback. That is, the functions are recursively calculated until some predefined condition is met. The technical name of such filters is *infinite impulse response filters*, since here the transfer functions are actually represented by infinite recursive series. The foremost advantage of this technique is that it can be used to design and implement filters that are not realizable in conventional analog signal processors. That is why most modern multi channel analyzers have built in digital filters and can therefore be called digital signal processors.

8.9 Electronics Noise

In the real world, there are no electronic components that behave ideally. Their deviation from the expected ideal behavior can be random or systematic, both of which are collectively called *electronics noise*. Depending on the particular application, this noise may or may not be a significant source of degradation of the signal to noise ratio in radiation measurement systems. It is therefore important to first estimate the contribution of the electronics noise to the overall noise and make a judgment on whether to invest in electronics noise reduction or not. Low resolution systems working at high rates, generally do not require low noise electronics. On the other hand for high resolution systems, reduction in electronics noise is a major challenge.

One of the highly desired traits of a good system is that its electronics noise should not depend on the signal itself. That is, the output fluctuations with no input (often called pedestal fluctuations) should not change when an actual signal enters the electronics chain. Fortunately, with modern electronics components, deviation from this behavior is only seldom observed. Therefore, in most situations, we are left with quantifying the pedestal noise. Comparing this noise with the expected signal would then tell us whether it is worthwhile to put efforts to decrease the noise further or not. Let us elaborate this with an example.

Suppose we want to measure the energy of 90 *keV* photons using a silicon detector. The mean number of electron hole pairs generated by photons depositing energy E_{dep} can be calculated from

$$N = \frac{E_\gamma}{W}$$

where W is the energy needed to create a charge pair. For silicon we have $W = 3.6\ eV$. We suppose that the incident photon is completely stopped in the active medium, implying that the deposited energy is 90 keV. The number of electron hole pairs generated by this photon will be

$$
\begin{aligned}
N &= \frac{90 \times 10^3}{3.6} \\
&= 2.5 \times 10^4 \text{ electron hole pairs}
\end{aligned}
$$

The statistical fluctuation in this number with the Fano factor $F = 0.1$ will be

$$
\begin{aligned}
\sigma_{stat} &= \sqrt{FN} \\
&= \sqrt{0.1 \times 25000} \\
&= 50 \text{ electron hole pairs}
\end{aligned}
$$

The presently available electronics for semiconductor detectors can have a noise level of as low as 10 electron hole pairs. However such a noise level is difficult and expensive to achieve. Most of the systems have an inherent noise of several hundred electrons. Even if we assume that the electronics noise level σ_{elec} is 100 electrons, it has a significant effect on the signal to noise ratio as shown below.

$$
\begin{aligned}
S/N &= \frac{N}{\sigma_{stat}} \qquad\qquad \text{(with only statistical fluctuations)} \\
&= \frac{2.5 \times 10^4}{50} = 500 \\
S/N &= \frac{N}{\left(\sigma_{stat}^2 + \sigma_{elec}^2\right)^{1/2}} \qquad \text{(with statistical and electronics fluctuation)} \\
&= \frac{2.5 \times 10^4}{(50^2 + 100^2)^{1/2}} \\
&\approx 224
\end{aligned}
$$

$$(8.9.1)$$

This clearly shows how large an impact does the electronics noise have on the signal to noise ratio. In this particular case, therefore, the strategy should be to decrease the electronics noise as much as possible.

Example:
A silicon detector is used to measure the intensity of a 150 keV photon beam. The signal to noise ratio is found to be 300. Determine the noise introduced by the associated electronics.

Solution:
We first compute the statistical noise level. To do that we must first estimate the number of charge pairs generated. Assuming that all of the incident energy is absorbed in the active volume of the detector, the number of generated

electron hole pairs is

$$N = \frac{E_{dep}}{W}$$

$$= \frac{150 \times 10^3}{3.6} \quad \text{(since } W = 3.6 \ eV/\text{charge pair for Si)}$$

$$= 4.17 \times 10^4 \text{ charge pairs.}$$

The statistical fluctuations associated with the generation of charge pairs is given by

$$\sigma_{stat} = \sqrt{FN}$$

$$= \sqrt{(0.1)(4.17 \times 10^4)} \quad \text{(since } F \approx 0.1 \text{ for Si)}$$

$$= 64.5 \text{ charge pairs.}$$

The electronics noise can be estimated from the expression for the signal to noise ratio as follows.

$$S/N = \frac{N}{\left(\sigma_{stat}^2 + \sigma_{elec}^2\right)^{1/2}}$$

$$\Rightarrow \sigma_{elec} = \left[\left(\frac{N}{S/N}\right)^2 - \sigma_{stat}^2\right]^{1/2}$$

$$= \left[\left(\frac{4.17 \times 10^4}{300}\right)^2 - 64.5^2\right]^{1/2}$$

$$= 123 \text{ charge pairs}$$

8.9.A   Types of Electronics Noise

Broadly speaking, electronics noise can be divided into two categories: random and non-random. The former follows Gaussian statistics while the latter generally has $1/f$ spectral distribution (f is the frequency). In the following we will discuss some of the most commonly encountered types of noise belonging to these two categories.

A.1   Johnson Noise

The free charges in a conductor possess kinetic energy, which is a function of their temperature. Due to this energy they constantly remain in random thermal motion. Since motion of a charge carrier constitutes an electrical current therefore because of this motion the electrical state of the system fluctuates even in the absence of any current driving force. These fluctuations are referred to as Johnson, thermal, or Nyquist noise. To elaborate on this further, let us suppose we have an electrical conductor having contacts at both ends but not connected to any voltage source. If we measure the voltage across these contacts using a voltmeter or an oscilloscope, the average voltage V will be zero. However we will see fluctuations about this average value (see Fig.8.9.1(a)). These fluctuations are due to the thermal noise of the conductor. The question is, if the average voltage is zero, then why we should care about these fluctuations. The answer lies in the fact that it is actually the signal power that determines the information content of the signal. Since power is proportional to the square of the voltage (V^2), therefore in this case it will have a value greater than zero whenever the actual voltage fluctuates from its average value of zero (see Fig.8.9.1(b)). Let us now see how this noise can be quantified.

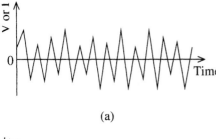

(a)

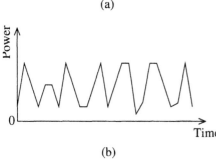

(b)

Figure 8.9.1: (a) Typical variation of voltage or current in a conductor not connected to any current source. The average voltage or current is zero. (b) Power, which is proportional to square of voltage or current is non-zero even in the absence of a current source.

Johnson noise is usually represented by spectral noise power density p_j^2, which is simply the noise power per unit frequency bandwidth. This quantity has been found to be proportional to the absolute temperature T.

$$p_j^2 \equiv \frac{dP_{noise}}{df} = 4k_BT \tag{8.9.2}$$

Here k_B is the Boltzmann's constant. This can also be represented in terms of current and voltage by noting that the power carried by a current I in the presence of a voltage V is given by

$$P = VI = I^2R = \frac{V^2}{R}, \tag{8.9.3}$$

where we have used the Ohm's law $V = IR$ for a conductor having resistance R. Hence the spectral noise current density can be obtained by substituting this in the above definition of p_j as

$$i_j^2 \equiv \frac{dI_{noise}^2}{df} = \frac{4k_BT}{R}. \tag{8.9.4}$$

Similarly the spectral noise voltage density is given by

$$v_j^2 \equiv \frac{dV_{noise}^2}{df} = 4k_BTR. \tag{8.9.5}$$

Note that these expressions do not represent total noise, only the noise density or noise per unit bandwidth. The noise voltage and current can be obtained by multiplying the above relations by the system bandwidth $\triangle f$, that is

$$I_j^2 = \frac{4k_B T}{R}\triangle f \tag{8.9.6}$$

$$\text{and} \quad V_j^2 = 4k_B T R \triangle f. \tag{8.9.7}$$

A point worth mentioning here is that the Johnson noise associated with a resistor in series can be modeled by a voltage noise source. A parallel resistor, on the other hand, can be modeled by a current noise source.

The expressions above reveal an important fact about the Johnson noise. That is, the noise power is proportional to the bandwidth of the system. This implies that the noise power for a bandwidth of 0 to 2 Hz would be the same as for a bandwidth of 40000 to 40002 Hz. In other words, the noise power is independent of the frequency. This kind of noise is generally referred to as *white noise*.

Example:
Determine the thermal noise voltage in a 200 Ω resistor at 27 ^0C for a system having a bandwidth of 0 to 500 MHz.

Solution:
The noise voltage can be calculated from equation 8.9.7.

$$\begin{aligned} V_j &= 4k_B T R \triangle f \\ &= \left[(4)\left(1.38 \times 10^{-23}\right)(300)(200)\left(500 \times 10^6\right)\right]^{1/2} \\ &= 40.7 \ \mu V. \end{aligned} \tag{8.9.8}$$

A.2   Shot Noise

The current in a conductor is always carried by discrete charges. In most cases these charges are injected into the system in such a way that their behavior is *stochastically* independent, meaning that creation or arrival of a new charge is independent of its predecessors. Due to the inherent statistical nature of the underlying phenomena, the numbers of these charges at any time fluctuates from a mean value. This fluctuation causes the electrical current to randomly fluctuate around its mean value. Shot noise corresponds to these random fluctuations. It has Gaussian characteristics both in time and frequency domains and belongs to the category or white noise.

The power density of shot noise at a particular frequency is given by

$$p_s^2 \equiv \frac{dP_{noise}}{df} = 2eV. \tag{8.9.9}$$

The power density can be used to define the spectral shot noise current density i_s^2 and spectral shot noise voltage density v_s^2 as follows.

$$i_s^2 \equiv \frac{dI_{noise}^2}{df} = 2eI \tag{8.9.10}$$

$$v_s^2 \equiv \frac{dV_{noise}^2}{df} = 2eRV \tag{8.9.11}$$

As with the Johnson noise, here also the above expressions must be multiplied by the system bandwidth to determine the absolute noise current or voltage, that is

$$I_s = \sqrt{2eI\triangle f} \tag{8.9.12}$$

$$\text{and} \quad V_s = \sqrt{2eRV\triangle f}. \tag{8.9.13}$$

These expressions reveal why shot noise is called a white noise: the noise power is dependent on the bandwidth and not on the frequency itself. Or in other words the noise power at a bandwidth of 2 to 3 Hz would be the same as at a bandwidth of 200002 to 200003 Hz.

Example:
Compute the total noise current in a semiconductor based detection system having a bandwidth from 0 to 500 kHz if a current of 0.5 μA flows through it.

Solution:
The detector can be assumed to be a current noise source. the magnitude of the current can be computed from the equation 8.9.12 as follows.

$$
\begin{aligned}
I_s &= \sqrt{2eI\triangle f} \\
&= \left[(2)\left(1.602 \times 10^{-19}\right)\left(0.5 \times 10^{-6}\right)\left(500 \times 10^{3}\right)\right]^{1/2} \\
&= 0.28\ nA
\end{aligned}
\tag{8.9.14}
$$

A.3   1/f Noise

One over f noise corresponds to a number of *non-random* noise sources in a detection system. Its name derives from the fact that its power spectrum has an approximate $1/f$ dependence, that is

$$p_f \equiv \frac{dP_{noise}}{df} \propto \frac{1}{f^\alpha}, \tag{8.9.15}$$

where the exponent α is approximately equal to unity for most systems. Since power is proportional to the square of the voltage, therefore the noise voltage density can be written as

$$v_f^2 = \frac{A}{f^\alpha}, \tag{8.9.16}$$

where A is a system dependent constant of proportionality. The above expression when multiplied by the system bandwidth $\triangle f$ gives the total noise voltage, that is

$$v_f = \sqrt{\frac{A}{f^\alpha}\triangle f}. \qquad (8.9.17)$$

Unfortunately even though $1/f$ noise is exhibited by most natural systems, its sources are not fully understood. The electronic components based on MOS technology are most affected by this noise. On the other hand, it is negligibly small for usual metallic conductors. Experimental determination of $1/f$ noise in detection systems is of prime importance specially for the ones that are supposed to work at high resolution. This involves determining the total noise power density at different frequencies using a spectrum analyzer. The spectrum analyzer is needed to perform the Fourier transformation of the signal and averaging the noise power density. A typical result of such an experiment is shown in Fig.8.9.2.

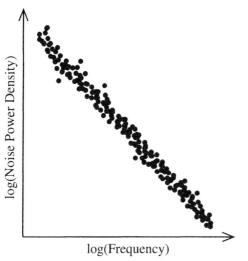

Figure 8.9.2: Typical $1/f$ noise spectrum.

The good thing about $1/f$ noise in electronic systems is that it can be decreased by improving the design and quality of the electronic components constituting the system. Of course another approach is to work at higher frequencies, if other factors allow.

A.4 Quantization Noise

If the process of analog to digital conversion involves step-wise or discrete sampling of the analog signal, the variation in signal at each step has sharp edges. This introduces high frequency noise in the system. The usual method to eliminate this noise is to block the high frequencies by using a low pass filter.

8.9.B   Noise in Specific Components

B.1   Noise in Amplifiers

All amplifiers exhibit inherent internal noise, the level of which depends on the way they have been designed and constructed. However the noise at an amplifier input is particularly troublesome because it gets amplified by the same amplification factor as the input signal. Determination of the amplifier noise requires consideration of both internal and external noise sources. The noise sources at the input of an amplifier in a typical detection system include

▶ amplifier's internal noise sources.

▶ shot noise due to detector bias,

▶ Johnson noise due to series resistance, and

▶ Johnson noise due to parallel resistance.

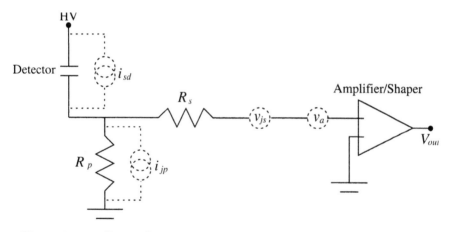

Figure 8.9.3: Circuit for equivalent noise determination of an amplifier.

The equivalent noise at the input of an amplifier therefore consists of these four main components as shown in Fig.8.9.3. Here the amplifier's internal noise source has been represented as an input voltage noise source having spectral voltage density v_a^2. The thermal voltage noise power density of the series resistor is represented by v_{js}. The noise current power density through the parallel resistor has been represented by i_{jp}^2. The bias voltage applied to the detector acts as a source of shot noise, with an spectral noise current density i_{sb}^2. The last noise source is the shunt or parallel resistor with a spectral noise current density of i_{sp}. The amplifier's equivalent internal noise voltage spectral density can be assumed to be composed to two components: a random or white noise v_{as}^2 and the $1/f$ noise characteristic of electronic systems. Hence we can write

$$v_a^2 = v_{as}^2 + \frac{A}{f}, \tag{8.9.18}$$

where A is a proportionality constant (recall that spectral power density for $1/f$ noise is proportional to the inverse of the frequency).

The shot noise due to detector bias voltage can be very well represented by a current noise source as shown in Fig.8.9.3. Using equation 8.9.10, we write the spectral current density for detector bias as

$$i_{sd}^2 = 2eI_d, \tag{8.9.19}$$

where I_d is the current flowing through the detector. Of course here we have assumed that no current flows through the amplifier/shaper, which is a valid assumption since for most practical purposes the input impedance of an amplifier can be taken to be infinite. Also, no current should be flowing through the parallel resistance, which is again a valid assumption since typical shunt resistances are very large.

We now move to the noise induced by the parallel resistance. Since the parallel resistance acts as a current source of Johnson noise, therefore we can use equation 8.9.4 to represent its power density.

$$i_{jp}^2 = \frac{4k_BT}{R_p}. \tag{8.9.20}$$

Similarly the series resistance induces noise at the amplifier's input that can be represented by a voltage noise source. The Johnson noise voltage power density for this resistance can be written as (cf. equation 8.9.5)

$$v_{js}^2 = 4k_BTR_s. \tag{8.9.21}$$

Now we are left with the task of determining the total noise power density at the input of the amplifier/shaper. Since some of the noise components are in terms of voltage while others are in terms of current therefore we can not simply add them together. If the voltage amplification factor of the amplifier is knows, it is easiest to transform all the noise currents into noise voltages using Ohm's law

$$v = iR. \tag{8.9.22}$$

Hence the total noise voltage at the input of the amplifier is given by

$$V_{n,in} = \left[\left(v_{as}^2 + \frac{A}{f}\right)\triangle f + 2eI_d\tau R_s^2\triangle f + \frac{4k_BT}{R_p}\triangle f + 4k_BTR_s\triangle f\right]^{1/2}. \tag{8.9.23}$$

Here we have multiplied all the noise densities by the bandwidth $\triangle f$ to determine the total noise voltage. If the amplifier's amplification factor is A, then the output noise voltage will be given by

$$\begin{aligned} V_{n,out} &= AV_{n,in} \\ &= A\left[\left(v_{as}^2 + \frac{A}{f}\right)\triangle f + 2eI_d\tau R_s^2\triangle f + \frac{4k_BT}{R_p}\triangle f + 4k_BTR_s\triangle f\right]^{1/2} \end{aligned} \tag{8.9.24}$$

Now, if the input signal voltage is $V_{s,in}$, the signal output voltage will be given by

$$V_{s,out} = AV_{s,in}. \tag{8.9.25}$$

The signal to noise ratio for the system is then

$$S/N = \left[\frac{V_{s,out}}{V_{n,out}}\right]^2$$

$$= \frac{A^2 V_{s,in}^2}{A^2 \left[\left(v_{as}^2 + \frac{A}{f}\right)\Delta f + 2eI_d \tau R_s^2 \Delta f + \frac{4k_B T}{R_p}\Delta f + 4k_B T R_s \Delta f\right]}$$

$$= \frac{V_{s,in}^2}{\left(v_{as}^2 + \frac{A}{f}\right)\Delta f + 2eI_d \tau R_s^2 \Delta f + \frac{4k_B T}{R_p}\Delta f + 4k_B T R_s \Delta f}. \qquad (8.9.26)$$

This expression shows that the signal to noise ratio is independent of the amplifier's amplification factor. The only frequency dependent component in this expression is the amplifier's $1/f$ noise. All other factors have bandwidth dependence.

B.2 Noise in ADCs

The most significant ADC noise source is the process of quantization itself. The corresponding quantization noise is given by

$$\sigma_q = \frac{\Delta V}{\sqrt{12}}, \qquad (8.9.27)$$

where ΔV is the change in analog voltage corresponding to one bit. In other words it is given by the ADC resolution multiplied by its dynamic range. Recall that the resolution of a n-bit ADC is given by

$$R = \frac{\Delta V}{V} = \frac{1}{2^n}, \qquad (8.9.28)$$

This implies that the analog voltage increment can be written as

$$\Delta V = \frac{V}{2^n}. \qquad (8.9.29)$$

Substituting this in the above expression for ADC noise gives

$$\sigma_q = \frac{V}{2^n \sqrt{12}} \qquad (8.9.30)$$

Example:
Determine the quantization noise corresponding to a 12-bit ADC having an analog range of 0 to 10 volts.

Solution:
The voltage range of the ADC is 10 volts and $n = 12$. Hence, according to equation 8.9.30, the quantization noise is given by

$$\sigma_q = \frac{V}{2^n \sqrt{12}}$$

$$= \frac{10}{2^{12}\sqrt{12}}$$

$$= 0.7 \, mV.$$

8.9.C   Measuring System Noise

As mentioned earlier, the particular detector application dictates the design of signal processing circuitry. However, after the design and development process, it is imperative that the noise of the system is carefully measured to determine the equivalent noise charge (ENC). The equivalent noise charge is defined as

$$ENC = \frac{Q_{signal}}{S/N} \tag{8.9.31}$$

$$= \frac{V_{noise}}{V_{signal}} Q_{signal} \tag{8.9.32}$$

where V_{noise} and V_{signal} are the output noise and signal pulse heights and Q_{signal} is the input signal charge. These three parameters must be experimentally determined to compute the ENC. The best method to to this is to inject a known signal into the circuitry and measure the relevant parameters of the output pulse. Since in such a situation the input charge Q_{signal} is known therefore one needs to determine only the other two parameters. Of course the output depends on the input as well as the circuitry and therefore if the input signal itself has significant noise then it would be highly unlikely that the ENC of the system is accurately determined. Therefore in general one tends to use high quality pulse generators for this purpose. There are different ways to analyze the signal output, with use of spectrum analyzers being the most popular choice. Most modern spectrum analyzers have the capability to determine the spectral distribution through Fourier transformation of the input signal. They also provide a host of mathematical manipulation functions to determine other related parameters, such as peak widths and $FWHM$.

The heights of the signal and the noise voltages can thus be determined from the spectrum analyzer. These values can be substituted in the above expression to determine the equivalent noise charge.

A point worth mentioning here is that one could, in principle, use a known radiation source instead of a pulse generator. This would be advantageous if one is interested in determining the ENC for the whole system including the detector. When a pulse generator is used, the measured equivalent noise charge corresponds to the electronic components only.

8.9.D   Noise Reduction Techniques

Any radiation detection system contains a number of noise sources. Elimination or reduction of these noises is, however, dependent on the *acceptable* signal to noise ratio. The acceptable S/N is of course application dependent. It is therefore a good practice to first estimate the maximum noise level that could be afforded in the system before deciding on the signal processing and transport modules.

The most bothersome of the noise sources in a detection system are related to electronics components. We visited some of these sources in the previous section. Now we will discuss how we can reduce their effect on the overall signal to noise ratio of the system.

D.1   Detector Signal

Enhancing the detector signal is one way of increasing the signal to noise ratio. For example in a PMT the primary photoelectron is multiplied several thousand times resulting in a large pulse at the anode. This certainly increases the statistical noise as well but by a lower factor since the statistical noise increases by the square root of the signal. Neglecting any other noise source, the signal to noise ratio for a signal of height N is given by

$$S/N = \frac{N}{\sqrt{N}} = \sqrt{N}.$$

D.2   Frequency Filters

All detection systems contain more than one sources of low frequency noise. Let us see what we can do to increase the signal to noise ratio in such systems.

▶ **Using Band Pass Electronic Filters:** Consider a system in which the most dominant noise has the $1/f$ behavior. Eliminating sources of such a noise can be quite challenging since some of them may depend on the design of the electronics components. In such a situation, a preferable solution is to work at higher frequencies, if of course the physical detection processes allow that. Working at higher frequencies requires that the low frequencies are somehow blocked at some point in the system. This can be done by using a *high pass filter*, which is an electronic circuit that allows only low frequencies to pass through. Similarly one can also block higher frequencies using a *low pass filter*. In most of the systems, however, it is desired to filter both low and high frequencies, something that can be achieved by the *bandpass filters*.

▶ **Using Software Filters:** It is also possible to block certain frequencies in a system through software. Modern systems are generally hooked to a powerful central processing unit (CPU), that can perform millions of mathematical operations in a second. It is therefore possible to perform the filtering operations *online* provided the CPU is able to handle it. If online processing is not possible then of course the data can also be filtered offline later on. A commonly used method of online filtering is through the Fourier Transform of the data. The steps generally taken are outlines below.

1. **Step-1:** Take Fast Fourier Transform (FFT) of a block of data. Most of the FFT algorithms require data points to be exactly 2^n in number (n being an integer) and at least 1024.

2. **Step-2:** Find peaks in the Fourier spectrum. This can be done by any peak finding algorithm. These peaks correspond to the dominant frequency components in the system. For example in a poorly grounded system the $60Hz$ power line noise creeps into the signal and one should see a peak at $60Hz$ in the Fourier spectrum of the data.

3. **Step-3:** Eliminate the peaks by using some interpolation scheme. Generally it suffices to use the mean baseline value of the output to replace the peaks.

4. **Step-4:** Take inverse FFT to obtain the filtered data values.

Problems

1. A semiconductor detector having a total capacitance of 1 pF is connected to a voltage sensitive preamplifier having a gain of 10. When the detector is placed in a constant radiation field, it produces pulses of height 500 mV at the preamplifier's output. Estimate the charge accumulated on the detector's capacitance corresponding to this voltage.

2. Suppose the temperature of the detector described in the previous exercise can not be maintained at a fixed value and its variation can induce up to 10% change in detector's capacitance. Estimate the width of the charge distribution corresponding to these fluctuations.

3. Compute the voltage drop across a 20 m long copper wire having a cross sectional area of 0.8 mm^2 if a current of 0.5 mA flows through it.

4. Quantify the change in the response of a simple CR high pass filter with change in frequency above the breakpoint of $f_{cutoff} = 1/2\pi RC$.

5. A silicon based detection system has an inherent equivalent electronics noise of 150 electrons. Compute the signal to noise ratio if the detector is placed in a beam of 200 keV photons such that 90% of the photons get absorbed in its active volume.

6. Determine the thermal noise currents of a 1 $M\Omega$ resistor connected in parallel at room temperature and at -10 ^0C. Assume the system bandwidth to be 200 MHz.

7. Determine the Johnson noise voltage at 27 ^0C if a 100 Ω resistor is connected in series. The system bandwidth is from 100 Hz to 500 kHz.

8. Calculate the shot noise in a silicon detector if a current of 15 μA flows through it. Assume the system bandwidth to be 1 MHz.

9. Compute the quantization noise of an 8-bit ADC having a conversion range from -10 V to +10 V.

Bibliography

[1] Bouquet, F.L., **Radiation Effects on Electronics**, Systems Co, 1994.

[2] Bozic, S.M., Chance, R.J., **Digital Filters and Signal Processing in Electronic Engineering: Theory, Applications, Architecture, Code**, Albion/Horwood Publishing Ltd., 1999.

[3] Crecraft, D., Gergely, S., **Analogue Electronics and Signal Processing**, Butterworth-Heinemann, 2001.

[4] Delaney, C.F.G., DeLaney, C., **Electronics for the Physicist with Applications**, Ellis Horwood Ltd., 1980.

[5] Donati, S., **Photodetectors, Devices, Circuits, and Applications**, Prentice Hall Inc., 1999.

[6] Faissler, W.L., **Introduction to Modern Electronics**, Wiley, 1991.

[7] Friel, J.J., **X-Ray and Image Analysis in Electron Microscopy**, Princeton Gamma-Tech, Princeton, NJ, 1995.

[8] Goodnick, S.M. et al., **Materials and Electronics for High-Speed and Infrared Detectors: Proceedings of Spie 19-20 and 23 July 1999 Denver, Colorado**, SPIE, 1999.

[9] Jayadev, T.S.J., **Infrared Sensors: Detectors, Electronics, and Signal Processing : 24-26 July, 1991, San Diego, California**, Society of Photo Optical, 1991.

[10] Jones, M.H., **A Practical Introduction to Electronic Circuits**, Cambridge University Press, 1995.

[11] Horowitz, P, Hill, W., **The Art of Electronics**, Cambridge University Press, New York, 1989.

[12] Knoll, G.F., **Radiation Detection and Measurement**, Wiley, 2000.

[13] Leo, W.R., **Techniques for Nuclear and Particle Physics Experiments**, Springer, 1994.

[14] Leven, A., **Applied Physics for Electronic Technology, A Problem Solving Approach**, Butterworth-Heinemann, 1998.

[15] Lyons, R.G., **Understanding Digital Signal Processing**, Prentice Hall, 2004.

[16] Meddins, R., **Introduction to Digital Signal Processing**, Newnes, 2000.

[17] Messenger, G.C., **The Effects of Radiation on Electronic Systems**, Van Nostrand Reinhold Co., 1986.

[18] Moon, T.K., Stirling, W.C., **Mathematical Methods and Algorithms for Signal Processing**, Prentice Hall, 1999.

[19] Motchenbacher, C.D., Connelly, J.A., **Low-Noise Electronic System Design**, John Wiley & Sons Inc., 1995.

[20] Oppenheim, A.V. et al., **Signals and Systems**, Prentice Hall, 1996.

[21] Orfanidis, S., **Introduction to Signal Processing**, Prentice Hall, 1995.

[22] Ott, H., **Noise Reduction Techniques in Electronic Systems**, John Wiley & Sons Inc., 1988.

[23] Proakis, J.G., Manolakis, D.K., **Digital Signal Processing: Principles, Algorithms and Applications**, Prentice Hall, 1995.

[24] Rogalski, A., Bielecki, Z., **Detection of Optical Radiation**, Bull. Polish Acad. Sci., Technol. Sci., Vol.52, No.1, 2004.

[25] Shani, G., **Electronics for Radiation Measurements**, CRC Press, 1995.

[26] Strong, J.A., **Basic Digital Electronics**, Chapman & Hall, 1991.

[27] Zichichi, A.L., **Gas Detectors & Electronics for High Energy Physics**, World Scientific Publishing Company, 1992.

Chapter 9

Essential Statistics for Data Analysis

Statistics is perhaps the most powerful available technique for analyzing experimental data. However its proper use requires careful attention not only to the techniques used but also to the system being analyzed. The advent of modern but very complicated radiation detection and measurement systems has shifted the experimenter's intention from merely taking averages of data to the more complicated tasks of full scale statistical analysis. Most of the modern statistical techniques are labor intensive and are almost impossible to perform without using computers. A modern experimenter, therefore, uses statistical analysis software to analyze the data, a task that is easy to perform but prone to errors if not carefully done. Unfortunately these software work as black boxes; they take data in and produce the final results. No matter what is fed in, something comes out as the result. Whether it makes sense or is correct is up to the person doing the analysis to decide. Therefore it is imperative that an experimenter fully understands the statistical techniques and their underlying theories before using statistical software.

Statistics is a vast field and it should not be expected that the reader becomes familiar with all of its intricacies after going through this chapter alone. However it will provide enough information that would enable the reader to analyze the data more efficiently.

Before we begin, let us see what our main objective should be in terms of drawing reasonable inferences from the outcome of an experiment. Suppose we want to measure the half life of a radioactive isotope. To do this, we use a suitable detector and measure the activity of the sample at several time intervals. However, there is a problem with this scheme; each of these measurements has some uncertainty associated with it. This uncertainty could be a combination of several effects such as the random nature of the radioactive phenomenon, the randomness in the conversion process of incident radiation into charge pairs, and the errors in the measurement system. Now we have a huge problem here; we have a number, namely the measurement at a point in time, but do not know how much faith we should put into it. The best and the easiest solution to this problem is to take several measurements instead of one and then report the average of these with a range within which any subsequent measurement is *expected* to lie, that is

$$A = \bar{A} \pm \triangle A.$$

The calculation of the average $\bar{A}$ and the dispersion $\triangle A$ and their interpretation is the task from where we start this chapter.

9.1   Measures of Centrality

Whenever we talk about *data* we generally mean some numbers obtained through
some experiment and corresponding to measurable quantities. An example would
be the nuclear scan of a patient obtained by a CCD camera. The output of such a
scan would consist of ADC counts observed by each pixel of the camera at regular
intervals of time. After the data has been obtained there must be some algorithm to
analyze it. Such algorithms are of course application dependent and are developed
according to the requirements. However there are certain quantities that are almost
always sought after in every analysis. One such quantity is a suitable measure
of centrality. For our example, this might be the *average* number of ADC counts
received during the scan period by each pixel, or the *average* total counts received
by each pixel.

Now, what is this average and why is it so important? This is not very hard
to understand if we keep in mind that one of the purposes of any experiment is to
determine how the system *normally* behaves and what value should be *expected* if
another measurement is taken. This normal or expected value is actually what is
referred to as the average or measure of central tendency.

There are different ways in which the measure of central tendency can be obtained. The three most commonly used measures are

▶ Mean,

▶ Median, and

▶ Mode.

The true meaning of these measures will become clear when we discuss the probability density functions later in the chapter. However at this point it is worthwhile
to see what we *normally* mean by these quantities. For this discussion we will assume that we have taken several measurements of a quantity, such as activity of
a radioactive sample, at regular intervals of time. Even if this quantity is not expected to change with time, we will still see fluctuations in the measurements. These
fluctuations will mainly be due to two effects: statistical nature of the process (radioactivity, in this case) and measurement uncertainty (of the detection system).
After we are done with the measurements, we can just add all the numbers and
divide the result by the number of data points. This is called *mean* of the data.
Mathematically we can write it as

$$\bar{x} = \frac{1}{N} \sum_{i}^{N} x_i. \tag{9.1.1}$$

where N represents the number of measurements and x_i is the number value of the
parameter being measured at each point.

Sometimes such a computation of mean is not very meaningful. For example, if
we know that each of these data points has different importance with respect to all
the other measurements, then we must *weigh* each of the data point accordingly. In
this case the expression for the mean will be

$$\bar{x} = \frac{\sum_{i}^{N} w_i x_i}{\sum_{i}^{N} w_i}, \tag{9.1.2}$$

where w_i is the weight or importance of each data point. We will see later that this weighting mean is the most commonly used method of computing the average.

Mean is easy to calculate but also very easy to get misleading results. If, for example the output of a series of measurements contains large excursions due to any reason related to the measurement process, the average of all the data will not be a faithful representation of the parameter unless those excursions are excluded from the calculations. This may or may not be possible, depending on the volume of data and the available computing time and power. In such a situation, there is another quantity that can be used instead of the mean, that is the *median*. Median is simply the middle number of the sample. To determine median, data is arranged in ascending or descending order and the middle value is picked. This eliminates the erroneous data points from the calculation of average (see example below). If there are two values in the middle, a simple mean of the two values is taken as the median.

Mode is the most frequently occurring value in the data. It is rarely used in data analysis.

Example:
A parallel plate ionization chamber is used to measure the intensity of x-rays coming from an x-ray machine with constant output. The data is amplified, shaped, digitized by an 8-bit ADC, and stored in the computer memory. The following is a sample of the ADC counts recorded.

34, 30, 28, 33, 29, 30, 31, 255, 27, 35, 29, 255, 33, 32, 28,30

Compute the measures of central tendency from the data.

Solution:
The out-of-bound values (255) at two points should be excluded from the measurement of mean. However to see how these two values would affect the computation of all the three measures of central tendency, let us compute these quantities with and without these erroneous data points. Using equation 9.1.1

$$\bar{x} = 58.7 \quad \text{with all values}$$
$$\bar{x} = 30.6 \quad \text{with two erroneous data points excluded}$$

For median we write the data in ascending order.

27, 28, 28, 29, 29, 30, 30, 30, 31, 32, 33, 33, 34, 35, 255, 255

Since there are two central values therefore the median will be their mean.

$$\text{Median} = \frac{30 + 31}{2} = 30.5$$

If we exclude the two erroneous data points then the median will be

$$\text{Median} = \frac{30 + 33}{2} = 30.$$

Mode is the highest occurring value, which in both cases is

$$\text{Mode} = 30$$

It is apparent that mean is the measure of central tendency that is most affected by bad data points. Therefore computation of mean should follow proper filtration of data to obtain meaningful results.

9.2   Measure of Dispersion

The advantage of computing the measure of central tendency, such as mean, is that it tells us what to expect if another measurement is taken. Measures of dispersion tell us how much fluctuation around the central value we should expect. The most commonly used measure of dispersion is the standard deviation defined by

$$\sigma = \frac{1}{N-1} \sum_{i=1}^{N} (x_i - \bar{x})^2 , \tag{9.2.1}$$

for a sample of N measurements having a mean of $\bar{x}$.

We will learn more about this measure when we discuss the probability density functions.

9.3   Probability

Probability gives a quantitative way to define the chance of occurrence of a certain event from a class of other possible events. For example, the chance of getting a tail when we toss a coin or the chance that an incident photon on a photocathode will cause a photoelectron to emit. Numerically the value of probability lies between 0 and 1. A probability of 0 means there is absolutely no chance that the particular event would occur while a probability of 1 guarantees with absolute certainty that it will occur. Our common sense might tell us that only these two extremes should have any physical significance. For example, we would find it very difficult to associate an element of chance to whether an event occurs or not. However in the microscopic world, which is mainly governed by quantum mechanical phenomena, this is exactly what happens. When an incident α-particle enters a gaseous detector, it *may* or *may not* interact with the atoms of the gas. It is impossible, according to quantum mechanics, to say with absolute certainty whether an interaction will take place or not. However, fortunately enough, we can associate statistical quantities to a large number of incident particles and talk in probabilistic terms. We have seen one such quantity, the interaction cross section, in earlier chapters. This approach to predicting events at the microscopic level has been found to be extremely successful and is therefore extensively used.

Mathematically speaking, probability can be defined by considering a sample set S and its possible subsets A, B, ... and so on. The probability P is a real valued function defined by

1. For every subset A in S, $P(A) \geq 0$,

2. For disjoint subsets (that is, $A \cap B = \emptyset$), $P(A \cup B) = P(A) + P(B)$,

3. $P(S) = 1$

The subsets A, B etc. of the sample space S can be interpreted in different ways. Two most commonly used interpretations use the so called frequentist and Bayesian approaches. Each of these approaches has its own pros and cons in terms of usability and ease of application. Which approach to use is largely dependent on the application and the personal bias of the experimenter. Since these two approaches have real world significance and are not of just academic interest, we will spend some time to understand them in practical terms.

9.3.A   Frequentist Approach

This is the most common approach toward defining the subsets of the sample space. Here the outcomes of a repeatable experiment are taken as the subsets. The limiting frequency of occurrence of an event A is then assigned to the probability $P(A)$. In simple terms it means that in this approach if we perform a repeatable experiment then the probability will be the frequency of the outcome.

The problem with this approach is that it does not provide a platform to include subjective information into the process, such as experimenter's prior belief about the behavior of the system. Consequently in the frequentist approach sometimes it becomes harder to treat systematic uncertainties. In such cases the Bayesian approach provides a more natural way to draw meaningful inferences.

9.3.B   Bayesian Approach

In this approach the subsets of the sample space are interpreted as hypotheses, which are simply true or false statements. This approach actually defines a degree of certainty to the hypothesis as opposed to the frequentist statistics in which there are only two degrees (True implies $P(A) = 1$ and False implies $P(A) = 0$).

Let us assume that we perform an experiment to examine the validity of a mathematical model (or theory). The theory gives us a degree of certainty or probability $P(T)$ about the outcome of the experiment. If we represent the probability of the outcome given the theory by $P(D|T)$, then Baye's theorem states that

$$P(T|D) \propto P(D|T)P(T). \tag{9.3.1}$$

Determination of $P(T)$ is one of the fundamental concerns in Bayesian statistics, which does not provide any fundamental rules for that. Another problem is that there could be more than one possible hypotheses and hence the right hand side of the above equation must be summed over all the possibilities to normalize the equation.

Both of these interpretations of probability yield almost same answers for large data sets and therefore the choice largely depends on the personal bias of the experimenter.

9.3.C   Probability Density Function

The outcome of a repeatable experiment, usually referred to as a *random variable*, is not always discrete and usually can take any value within a continuous range. If x is continuous then the probability that the outcome lies between x and $x + dx$ can be written as

$$P(x) = f(x; \theta)dx.$$

Here $f(x, \theta)$ is called the *probability density function* or simply p.d.f. Some physicists prefer to call it *distribution function*. A p.d.f. may depend on several parameters, which we have collectively represented by θ. Generally θ is unknown and its value is determined through measurements of x.The function $f(x; \theta)$ need not be continuous, though. It can take discrete values, in which case it will itself represent the probability. The advantage of using a p.d.f. is that it enables us to predict a number of variables related to the outcome of an experiment, such as the mean and the frequency with which any random data will take on some particular value or lie within a range of values.

Any p.d.f. must first be normalized before use. This can be done by noting that the maximum probability of any event occurring is always 1. Hence we can integrate the p.d.f. over all space to get the normalization constant N, that is

$$N \int_{-\infty}^{\infty} f(x, \theta)dx = 1.$$

C.1   Quantities Derivable from a P.D.F

A probability density function is very convenient in terms of evaluating different quantities related to its independent parameter.

▶ **Cumulative Distribution Function:** Sometimes we are interested in finding the probability of the occurrence of an event up to a certain value of the independent parameter. This can be done by the so called cumulative distribution function $F(a)$, which gives the probability that the variable x can take any value up to a value a. It is defined by

$$F(a) \equiv F(-\infty < x \leq a) = \int_{-\infty}^{a} f(x)dx. \tag{9.3.2}$$

If the function $f(x)$ is normalized then the probability that x can take any value from a onwards can be obtained from

$$F(a \leq x < \infty) = 1 - F(a). \tag{9.3.3}$$

▶ **Expectation Value:** The expectation value of any function $g(x)$ is obtained by taking its weighted mean with the distribution function of its random variable. For any general distribution function $f(x)$, it is given by

$$E(g(x)) = \frac{\int_{-\infty}^{\infty} g(x)f(x)dx}{\int_{-\infty}^{\infty} f(x)dx}. \tag{9.3.4}$$

If the function $f(x)$ has already been normalized, that is

$$\int_{-\infty}^{\infty} f(x)dx = 1, \tag{9.3.5}$$

then the expression for the expectation value becomes

$$E(g(x)) = \int_{-\infty}^{\infty} g(x)f(x)dx. \tag{9.3.6}$$

In this book, unless otherwise stated, we will assume that the p.d.f has already been normalized.

▶ **Moments:** Using the definition of the expectation value given above, we can compute the the expectation value of x^n. This quantity is called the nth moment of x and is defined as

$$\alpha_n = \int_{-\infty}^{\infty} x^n f(x)dx. \tag{9.3.7}$$

The most commonly used moment is the first moment (generally represented by μ), which simply represents the weighted mean of x. For a normalized distribution function $f(x)$, it is given by

$$\mu \equiv \alpha_1 = \int_{-\infty}^{\infty} x f(x)dx. \tag{9.3.8}$$

α_1 is also called the expectation value of x (generally represented by $E(x)$), as this is the value that we should expect if we perform another measurement. How confident we are about this *expectation*, depends on the distribution function.

▶ **Central Moments:** The nth central moment of any variable x about its mean μ is defined as the expectation value of $(x - \mu)^n$. If x follows a probability distribution function $f(x)$, its nth central moment can be calculated from

$$m_n = \int_{-\infty}^{\infty} (x - \mu)^n f(x)dx. \tag{9.3.9}$$

The variance $\sigma \equiv m_2$ is the most commonly used central moment. According to the above definition, it is given by

$$\sigma = \int_{-\infty}^{\infty} (x - \mu)^2 f(x)dx. \tag{9.3.10}$$

Variance quantifies the spread of the values around their mean and is used to characterize the level of uncertainty in a measurement.

It can be shown that

$$\begin{aligned} \sigma^2 &= E(x^2) - (E(x))^2 \\ &= \alpha_2 - \mu^2. \end{aligned} \tag{9.3.11}$$

▶ **Characteristic Function:** Up until now we have assumed that the distribution function can be explicitly written and easily manipulated to determine the moments. Unfortunately this is not always the case. Sometimes we come across situations where the data can not be represented by a distribution for which the integral of $g(x)^n f(x)$ can be easily evaluated to find the moments. There is way out of this situation, however. Even though the moments are not calculable for some functions but their Fourier transforms can be determined by evaluating the integral

$$\psi(\nu) = \int_{-\infty}^{\infty} e^{i\nu x} f(x) dx. \tag{9.3.12}$$

Where i is the complex number and ν is a parameter having dimensions that are inverse of the parameter x. For example, if x represents time then ν is the frequency. This transformation of the distribution function dependent on x into a *characteristic function* dependent on ν is equivalent to taking the expectation value of $e^{i\nu x}$. The good thing about this characteristic function is that it can be used to determine the moments of $f(x)$ through the relation

$$\alpha_n = i^{-n} \frac{d^n \psi}{d\nu^n} \bigg|_{\nu=0}. \tag{9.3.13}$$

▶ **Skewness:** Not all real distribution functions are symmetric. In fact, we seldom find a parameter that can be represented by a distribution function having no skewness whatsoever (see Fig.9.3.1). Unless this skewness is negligibly small, it must be quantified to extract useful information from the distribution. The best and the most commonly used way for this quantification is the computation of the so called *coefficient of skewness* defined as

$$\gamma_1 = \frac{m_3}{\sigma^3}. \tag{9.3.14}$$

The higher the value of γ_1, the more skewed is the distribution. Although γ_1 is extensively used in analyses, but in fact any odd moment about the mean can be used as a measure of skewness.

▶ **Kurtosis:** Besides skewness, the tail of a distribution also contains useful information and should not be neglected in the analysis. A commonly used measure of this tail is known as kurtosis of the distribution and is defined as

$$\gamma_2 = \frac{m_4}{\sigma^4} - 3. \tag{9.3.15}$$

The reason for defining kurtosis in this way lies in the use of the Gaussian distribution as a standard for comparison. As we will see later in this chapter, most of the natural processes can be described by the Gaussian distribution and therefore using it as a standard is justified. A perfect Gaussian distribution has a kurtosis of 0 since it always satisfies $m_4 = 3\sigma^4$. Note that $\gamma_2 = 0$ means that the tail of the distribution *falls off* according to the Gaussian distribution, not that it does not have any tail. If the distribution has a long tail, that is, if kurtosis is negative, it is called a *platykurtic* distribution. A *leptokurtic* distribution, on the other hand, has positive γ_2 and its tail falls off quicker than the Gaussian distribution.

▶ **Median:** In the beginning of this chapter we introduced *median* as a measure
of central tendency. Since we now know about probability density functions, we
are ready to appreciate the true meaning of this parameter. Median actually
represents the value at which the probability is 1/2, that is

$$P(x_{med}) = \int_{-\infty}^{x_{med}} f(x)dx = \int_{x_{med}}^{\infty} f(x)dx = \frac{1}{2}. \qquad (9.3.16)$$

Fig.9.3.1 represents this concept in graphical form. The computation of median becomes meaningful if the distribution is too skewed or has high negative
kurtosis, in which case the mean will not be a faithful representation of the
data.

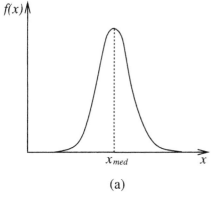

(a)

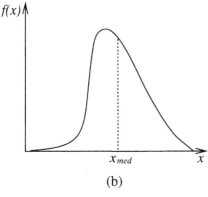

(b)

Figure 9.3.1: Mode of a symmetric (a) and
a skewed (b) distribution. The mode is the
value of x at which the probability is 1/2, that
is, the areas under the curve on the left and
right hand sides of the dotted line are equal.

C.2   Maximum Likelihood Method

The Bayesian approach we discussed earlier allows an experimenter to analyze the
data against some hypotheses. This can be done by the so called *maximul likelihood
method*. To elaborate on this methodology, we will look at a simplified example of
two hypotheses.

   Let us suppose that we have two hypotheses about the outcome of an experiment.
With each of these hypotheses, we can associate a probability distribution function.

If we now take N measurements, the joint probabilities associated with a particular result of each of these hypotheses will be given by

$$dp_1 = \prod_{i=1}^{N} f_1(x_i)dx_i \quad \text{and}$$

$$dp_2 = \prod_{i=1}^{N} f_2(x_i)dx_i. \tag{9.3.17}$$

However we are not interested in these individual probabilities since our aim is to judge the first hypothesis against the other. In other words we want to know the odds that hypothesis 1 is true against hypothesis 2 (or vice versa). This can be quantified by using the so called *likelihood ratio* defined as

$$R = \prod_{i=1}^{N} \frac{f_1(x_i)}{f_2(x_i)} \tag{9.3.18}$$

This ratio tells us how much faith we should put into a hypothesis against the other but does not in any way rule out the possibility of other hypotheses being more correct than these two. This is especially true for situations where a large number of hypotheses can be associated with the experiment. Let us suppose that we have an infinite number of hypotheses, which can be represented by a continuous variable h of the normalized probability density function $f(h; x)$. The joint probability that a particular hypothesis is true can then be obtained by taking the product of all the individual distributions $f(h; x_i)$ associated with each of the experimental results $x_1, x_2, ..., x_N$. This is called the *likelihood function* and is represented by

$$L(h) = \prod_{i=1}^{N} f(h; x_i). \tag{9.3.19}$$

The likelihood function $L(h)$ is a distribution function of h and can assume any shape depending on the probability density functions from which it has been derived. If we plot this function with respect to h, the most probable value of h (generally represented by h^*) will be the value at which $L(h)$ is maximum (see Fig.9.3.2). However in terms of computing this mathematically, since most of the probability density functions are exponential in nature therefore generally the natural logarithm of this function is used used instead. This function defined as

$$l(h) = \ln(L(h)), \tag{9.3.20}$$

is commonly known as *log-likelihood function*. h^* can then be found by simply equating the derivative of this function with respect to h to zero, that is

$$\frac{\partial l(h)}{\partial h} = 0 \tag{9.3.21}$$

Although the most probable value as obtained from the expression above is very useful, still it alone is not a faithful representation of the function since it does not tell us anything about how the other values of h are spread out. This spread, also known as the rms or root-mean-squared value, of h about h^* can be calculated from

$$\triangle h = \left| \frac{\int (h - h^*)^2 L dh}{\int L dh} \right|^{1/2} \quad \text{or}$$

$$\triangle h = \left[\frac{-\partial^2 l(h)}{\partial h^2} \right]^{-1/2}. \tag{9.3.22}$$

The second expression is extensively used to compute errors and due to its importance its derivation will be provided when we discuss the distribution functions later in the chapter.

Although $L(h)$ can assume any shape but it can be shown that for large values of N (that is $N \to \infty$), it approaches a Gaussian distribution as shown in Fig.9.3.2(b). We will learn about this particular distribution in the next section when we take a look at some of the commonly used distribution functions.

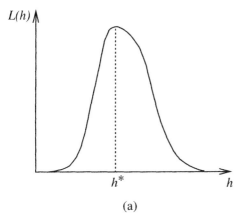

(a)

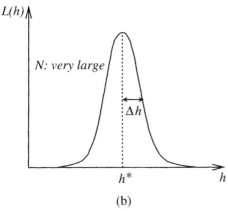

(b)

Figure 9.3.2: Typical likelihood functions for relatively small (a) and very large (b) number of data points. For large N the function approaches Gaussian distribution.

A point worth noting here is that the case of low statistics needs careful attention. The reason is that, as seen in Fig.9.3.2(b), the distribution can look fairly broad and asymmetric if the available number of data points are low. In such a case, merely quoting h and $\triangle h$ might not be sufficient and one should present the distribution function plot as well.

Up until now we have assumed that the likelihood function is described by a single variable h. If we have k number of parameters instead, we will have to solve the following k simultaneous equations to find the maximum likelihood solution.

$$\frac{\partial \ln L(h_1, h_2, ..., h_k)}{\partial h_i}\bigg|_{h_i=h_i^*} = 0 \qquad (9.3.23)$$

Now that we know what maximum likelihood function is, what do we do with it? Well, to do any Maximum likelihood analysis we first need a probability distribution function. In the next section we will look at some commonly used distribution functions and employ the Maximum likelihood methodology to draw inferences about them.

9.3.D Some Common Distribution Functions

D.1 Binomial Distribution

The binomial distribution can be used to determine the probability of r successes out of N outcomes of an experiment and is defined by

$$f(r; N, p) = \frac{N!}{r!(N - r)!}p^r(1 - p)^{N-r}, \qquad (9.3.24)$$

with $r = 0, 1, ..., N$ and $0 \leq p \leq 1$.

The events must be random, mutually exclusive, and independent, which in simple terms essentially means that the occurrence of one event should not influence the outcome of the next. The outcome of an experiment describable by binomial distribution has only two possible outcomes, such as getting head or tails when a coin is flipped or detecting or failing to detect a particle when it passes through the active medium of a detector. This means that if the probability of getting an event is p then probability of not seeing the event would simply be $(1 - p)$.

Let us now write the likelihood function for the occurrence of an event and then try to calculate its most probable value and the corresponding error. The likelihood function for a continuous variable p that follows binomial distribution can be written as

$$L(p) = \frac{N!}{r!(N - r)!}p^r(1 - p)^{N-r}. \qquad (9.3.25)$$

To compute the most probable value p^* of p, we take the derivative of its natural logarithm with respect to p and then equate it to zero (see equation 9.3.21). First we take the logarithm of the function keeping in view that we are only interested in evaluating terms that explicitly contain p.

$$\begin{aligned} \ln L(p) &= \ln\left[\frac{N!}{r!(N - r)!}p^r(1 - p)^{N-r}\right] \\ &= r\ln(p) + (N - r)\ln(1 - p) + \ln\left[\frac{N!}{r!(N - r)!}\right] \end{aligned} \qquad (9.3.26)$$

The derivative of this with respect to p is

$$\frac{\partial \ln(L)}{\partial p} = \frac{r}{p} - \frac{N - r}{1 - p}. \qquad (9.3.27)$$

Hence maximum of $\ln(L)$ at p^* is

$$\frac{r}{p^*} + \frac{N-r}{1-p^*} = 0$$

$$\Rightarrow p^* = \frac{r}{N}. \qquad (9.3.28)$$

Now in order to evaluate the error in p^* we differentiate again equation 9.3.27 with respect to p to get

$$\frac{\partial^2 \ln(L)}{\partial p^2} = \frac{r}{p^2} - \frac{N-r}{(1-p)^2} \qquad (9.3.29)$$

According to equation 9.3.22, the error in p^* is then given by

$$\triangle p = \left[\frac{-\partial^2 \ln(L)}{\partial p^2} \right]^{-1/2}$$

$$= \left[\frac{r}{p^{*2}} + \frac{N-r}{(1-p^*)^2} \right]^{-1/2}$$

$$= \left[\frac{p^*(1-p^*)}{N} \right]^{1/2}, \qquad (9.3.30)$$

where we have used $r = p^*N$.

D.2 Poisson Distribution

Poisson distribution represents the distribution of Poisson processes and is in fact a limiting case of the Binomial distribution. By Poisson processes we mean the processes that are discrete, independent and mutually exclusive.

The p.d.f. of a Poisson distribution is defined as

$$f(x; \mu) = \frac{\mu^x e^{-\mu}}{x!}, \qquad (9.3.31)$$

with $x = 0, 1, ...$ represents the *discrete* random variable, such as ADC counts obtained from a detection system and $\mu > 0$ is the mean. Fig.9.3.3 depicts this distribution for different values of μ. It is apparent that the width of the distribution increases with μ, which indicates that the uncertainty in measurement increases with an increase in the value of x.

Let us now apply the maximum likelihood method to determine the best estimate of mean of a set of n measurements assuming that the underlying process is Poisson in nature. The best way to do this is to use the maximum likelihood method we outlined earlier and applied in the previous section while discussing the Binomial distribution. Since Poisson distribution is a discrete probability distribution therefore its likelihood function for a set of n measurements can be written as

$$L(\mu) = \prod_{i=1}^{n} f(x_i, \mu)$$

$$= \prod_{i=1}^{n} \left[\frac{\mu^{x_i} e^{-\mu}}{x_i!} \right]$$

$$= \frac{\mu^{\sum x_i} e^{-n\mu}}{x_1! x_2!...x_n!} \qquad (9.3.32)$$

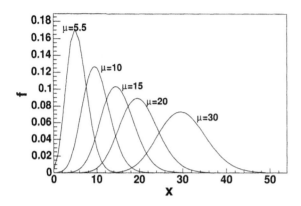

Figure 9.3.3: Poisson probability density for different values of μ. The width of the distribution, a reflection of the uncertainty in the measurement, increases with increase in μ.

The log likelihood function of $L(\mu)$ is

$$l \equiv \ln(L) = \left(\sum_{i=1}^{n} x_i\right) \ln(\mu) - n\mu - \ln(x_1! x_2! ... x_n!). \qquad (9.3.33)$$

Following the maximum likelihood method ($\partial l / \partial \mu = 0$) we get

$$\frac{\partial}{\partial \mu}\left[\left(\sum_{i=1}^{n} x_i\right) \ln(\mu) - n\mu - \ln(x_1! x_2! ... x_n!)\right] = 0$$

$$\frac{1}{\mu^*}\sum_{i=1}^{n} x_i - n = 0$$

$$\mu^* = \frac{1}{n}\sum_{i=1}^{n} x_i. \qquad (9.3.34)$$

This shows that the simple mean is the most probable value of a Poisson distributed variable. To determine the error in μ, we fist take second derivative of the log likelihood function and then substitute it in equation 9.3.22.

$$\frac{\partial^2 l}{\partial \mu^2} = -\frac{1}{\mu^2}\sum_{i=1}^{n} x_i$$

$$\triangle\mu = \left[-\frac{\partial^2 l}{\partial \mu^2}\right]^{-1/2}$$

$$= \left[\frac{\mu^{*2}}{\sum_{i=1}^{n} x_i}\right]^{1/2}$$

$$= \frac{1}{n}\left[\sum_{i=1}^{n} x_i\right]^{1/2} \qquad (9.3.35)$$

This is one of the most useful results of the Poisson distribution. It implies that if we make one measurement, the statistical error we should expect in it would simply be the square root of the measured quantity. For example, if we count the number of γ-ray photons coming from a radioactive source using a GM-tube and get a number N, the statistical error we should expect will simply be $\sqrt{N}$. Fortunately, most of the processes we encounter in the field of radiation detection and measurement, such as activity of a radioisotope, photoelectric effect, and electron multiplication in a PMT tube, can all very well be described by Poisson statistics.

D.3 Normal or Gaussian Distribution

The normal distribution was originally developed as an approximation to the binomial distribution. Its usefulness was soon recognized by scientists and soon it became one of the most commonly used probability distributions in not only these fields but also in other sciences. The utility of normal distribution can be appreciated by noting an amazing property of most of the physical processes that their random variables can be safely approximated to be distributed normally. Therefore a common practice is to assume that a random variable having unknown distribution can be defined by a normal distribution. This property of random variables is actually the result of the so called *central limit theorem*, which states that the mean of any set of variables with any distribution tend to the normal distribution provided their mean and variance are finite. Although the term normal distribution is very commonly used, still some scientists prefer to call it Gaussian distribution.

Gaussian distribution has a bell-shaped curve (see Fig.9.3.4) and is defined for a variable x in the domain $x \in (-\infty, \infty)$ by

$$f(\mu; x) = \frac{1}{\sigma\sqrt{2\pi}}e^{-(x-\mu)^2/2\sigma^2}, \qquad (9.3.36)$$

where μ and σ are the mean and standard deviation of the distribution respectively. Both μ and σ are finite for a normally distributed variable.

If we substitute $\mu = 0$ and $\sigma^2 = 1$ in the above equation, we obtain the so called *standard normal distribution* with a probability density function given by

$$P(x) = \frac{1}{\sqrt{2\pi}}e^{-x^2/2}. \qquad (9.3.37)$$

This is simply a special kind of Gaussian distribution having a symmetric bell shaped curve centered at $x = 0$ (see Fig.9.3.4). In fact by changing the variables, any normal distribution can be easily converted into a standard normal distribution (see Example below).

Let us now apply our maximum likelihood method to compute the most probable value and its accuracy assuming the variable to be Gaussian distributed. Suppose we make N measurements of a variable and represent the result by x_i. Each of these measurements will have its own error σ_i. Then according to equation 9.3.19, the likelihood function is given by

$$L(\mu) = \prod_{i=1}^{N} \frac{1}{\sigma_i\sqrt{2\pi}}e^{-(x_i-\mu)^2/2\sigma_i^2} \qquad (9.3.38)$$

In order to apply the condition 9.3.23, we rewrite the above equation in the form

$$L(\mu) = \prod_{i=1}^{N}(e^{-(x_i-\mu)^2/2\sigma_i^2})(\sigma_i^{-1})(2\pi)^{-1/2} \qquad (9.3.39)$$

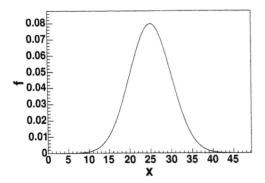

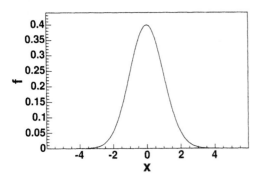

Figure 9.3.4: (a) Gaussian distribution for $\mu = 25$ and $\sigma = 5$. (b) Standard normal distribution having $\mu = 0$ and $\sigma = 1$. With proper change of scale, any Gaussian distribution can be transformed into a standard normal distribution.

Taking the natural logarithm of both sides of this equation gives

$$\ln(L) = \sum_{i=1}^{N} \left[-\frac{(x_i - \mu)^2}{2\sigma_i^2} - \ln(\sigma_i) - \frac{\ln(2\pi)}{2} \right]. \qquad (9.3.40)$$

The maximum likelihood solution is then obtained by differentiating this equation with respect to μ and equating the result to zero. Hence we get

$$\frac{\partial \ln(L)}{\partial \mu^*} = \sum_{i=1}^{N} \frac{x_i - \mu^*}{\sigma_i^2} = 0 \qquad (9.3.41)$$

$$\Rightarrow \sum_{i=1}^{N} \frac{\mu^*}{\sigma_i^2} = \sum_{i=1}^{N} \frac{x_i}{\sigma_i^2}$$

$$\Rightarrow \mu^* = \frac{\sum_{i=1}^{N} w_i x_i}{\sum_{i=1}^{N} w_i}, \qquad (9.3.42)$$

where $w_i = 1/\sigma_i^2$. Hence the most probable value is simply the weighted mean with respective inverse variances or *errors* as weights. If we assume that each measurement has the same amount of uncertainty or error then using $\sigma_i = \sigma$ the maximum likelihood solution will become

$$\mu^* = \frac{\sum_{i=1}^{N} x_i/\sigma}{\sum_{i=1}^{N}(1/\sigma)}$$

$$= \frac{1}{N}\sum_{i=1}^{N} x_i, \tag{9.3.43}$$

which is nothing but the expression for calculating simple mean. Hence we have found that calculating mean by this method require that the variable is distributed normally and that each measurement has the same error associated with it. Measuring activity of a radioactive source falls into this category provided all the conditions including the state of the detector does not change with time.

Let us now try to calculate the error in the calculation of the solution we just obtained. Note that we are interested in finding out the spread of μ about μ^* and not the errors in individual measurements. To do this we make use of the argument that for large number of measurements ($N \to \infty$), $L(\mu)$ approaches a normal distribution. Hence we can write

$$L(\mu) = \frac{1}{\sigma_t\sqrt{2\pi}}e^{-(\mu-\mu^*)^2/2\sigma_t^2} \tag{9.3.44}$$

.

Here the subscript t in σ_t is meant to differentiate the standard deviation of μ from that of x. Again the condition 9.3.23 can be used to obtain the maximum likelihood solution for this distribution. We first take the natural logarithm of both sides of the above equation to obtain

$$\ln(L) = \ln\left[\left(e^{-(\mu-\mu^*)^2/2\sigma_t^2}\right)(\sigma_t)^{-1}(2\pi)^{-1/2}\right]$$

$$= -\frac{(\mu-\mu^*)^2}{2\sigma_t^2} - \ln(\sigma_t) - \frac{\ln(2\pi)}{2}. \tag{9.3.45}$$

Differentiating this twice with respect to μ^* gives

$$\frac{\partial \ln(L)}{\partial \mu^*} = \frac{\mu-\mu^*}{\sigma_t^2}$$

$$\Rightarrow \frac{\partial^2 \ln(L)}{\partial \mu^{*2}} = -\frac{1}{\sigma_t^2}. \tag{9.3.46}$$

Hence the error in calculation of μ, is given by

$$\triangle\mu = \sigma_t = \left[-\frac{\partial^2 \ln(L)}{\partial \mu^{*2}}\right]^{-1/2}. \tag{9.3.47}$$

This general expression for computing errors is of central importance in likelihood method and is extensively used in data analysis. Let us now use this expression to derive the expression for the total error in measurements when each measurement is characterized by its own error σ_i. This can be done by differentiating equation

9.3.41 again with respect to μ^* and substituting the result in the above expression.

$$\frac{\partial^2 L}{\partial \mu^2} = -\sum_{i=1}^{N} \frac{1}{\sigma_i^2}$$

$$\Rightarrow \frac{1}{\sigma_t^2} = \sum_{i=1}^{N} \frac{1}{\sigma_i^2} \tag{9.3.48}$$

This expression represents the *law of combination of errors*, which states that for repeated measurements of a normally distributed variable having errors σ_i, the inverse of the total error in the calculation of mean is equal to the sum of inverse of individual measurement errors.

Unfortunately, in a number of practical problems, an analytic determination of $\Delta\mu$ is not possible. In such cases one tries to find the value of the likelihood function at each point by iterating μ (or more accurately, by trying different values of μ). The points thus obtained are then plotted and the likelihood function is obtained by performing the best fit through the points. In most cases with large number of data points the likelihood function is Gaussian like. If it isn't, one must perform a weighted average to determine the error function, that is

$$\left\langle \frac{\partial^2 \ln(L)}{\partial \mu^2} \right\rangle = \frac{\int \frac{\partial^2 \ln(L)}{\partial \mu^2} L d\mu}{\int L d\mu}. \tag{9.3.49}$$

This is an important relation since it can be used to show that (see problems at the end of the chapter) the maximum likelihood error in μ can be evaluated from

$$\Delta\mu = \left[\frac{1}{N} \int \frac{1}{L} \left(\frac{\partial L}{\partial \mu} \right)^2 dx \right]^{1/2}, \tag{9.3.50}$$

where N is the number of measurements. An interesting aspect of this result is that it allows one to determine the number of measurements necessary to obtain a particular value of the parameter μ with a certain accuracy, that is

$$N = \frac{1}{(\Delta\mu)^2} \int \frac{1}{L} \left(\frac{\partial L}{\partial \mu} \right)^2 dx. \tag{9.3.51}$$

D.4 Chi-Square (χ^2) Distribution

χ^2-distribution is one of the most extensively used probability distributions to perform goodness-of-fit tests, which we will discuss later in the chapter. It is defined as

$$f(x; n) = \frac{x^{n/2-1} e^{-x/2}}{2^{n/2} \Gamma(n/2)}, \tag{9.3.52}$$

where in order to avoid confusion due to the exponent 2 of χ^2 we have represented it by x. $\Gamma()$ is the *gamma function* and $x \geq 0$. The tables as well as analytical forms of gamma functions can be found in standard texts of statistics and mathematics. The parameter n in the above definition is called the *degrees of freedom* of the system. The meaning of this term can be understood by looking at the definition of x or χ^2.

Suppose we have m independent normally distributed random variables u_i having theoretical means μ_i and variances σ_i^2. χ^2 is then defined as

$$\chi^2 \equiv x = \sum_{i=1}^{m} \frac{(u_i - \mu_i)^2}{\sigma_i^2}. \qquad (9.3.53)$$

The parameter n in the definition of the χ^2 probability distribution is then related to the number of independent variables in this equation. For large n the χ^2 distribution reduces to the Gaussian distribution with mean $\mu = n$ and variance $\sigma^2 = 2n$. Fig.9.3.5 shows the shapes of the chi-square distribution for different degrees of freedom. As n increases, the distribution assumes a shape that becomes more and more like a Gaussian or normal distribution.

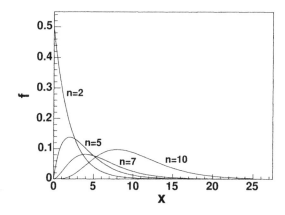

Figure 9.3.5: χ^2 probability density functions for different degrees of freedom n. The shape of the distribution approaches that of a Gaussian distribution with increasing n.

To understand the utility of this distribution function, let us have a closer look at the definition 9.3.53. The numerator in this equation represents the deviations of the normally distributed variable u_i from its theoretical means at each data point while the denominator represents its expected standard deviations. In other words the numerator and denominator are the actual and expected deviations respectively. If the data u_i is really Gaussian distributed, then ideally the actual deviation should be equal to the expected one. However in real data there are always fluctuations and consequently these deviations are not equal. The χ^2 probability density function 9.3.52 actually tells us how the probability of this deviation is distributed. Hence this distribution function can be used to judge the data against a hypothetical mean. We will learn more about this when we discuss the goodness-of-fit tests later in the chapter.

D.5 Student's t Distribution

Student's t distribution is a widely used probability distribution. It forms the basis of Student's t-test, which we will discuss later in the chapter. To define this distribution, let us first write

$$z = \sum_{i=1}^{n} x_i^2$$

$$\text{and} \quad t = \frac{x}{\sqrt{z/n}},$$

for n independent Gaussian variables having 0 mean and 1 variance. The variable z in this expression follows the χ^2-distribution we defined above and the variable t follows Student's t distribution with n degrees of freedom defined by

$$f(t; n) = \frac{1}{\sqrt{n\pi}} \frac{\Gamma[(n+1)/2]}{\Gamma(n/2)} \left[1 + \frac{t^2}{n}\right]^{-(n+1)/2}, \qquad (9.3.54)$$

where Γ is the familiar gamma function, the variable t can take any value ($-\infty < t < \infty$), and n can be a non-integer.

The Student's t distribution looks very similar to Gaussian distribution. For small n, however it has wider tails, which approach that of a Gaussian distribution with increasing n.

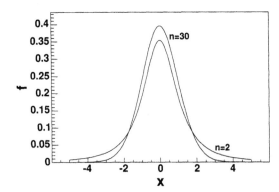

Figure 9.3.6: Student's t distribution for two values of degrees of freedom n. As n increases the tails of the distribution approaches that of a Gaussian distribution.

D.6 Gamma Distribution

For a Poisson process the distance in x from any starting point to the kth event follows Gamma distribution given by

$$f(x; \lambda, k) = \frac{x^{k-1} \lambda^k e^{-\lambda x}}{\Gamma(k)}, \qquad (9.3.55)$$

with $0 < t < \infty$ and k can be a noninteger.

For $\lambda = 1/2$ and $k = n/2$ it reduces to the χ^2-distribution we defined above.

Using Maximum Likelihood Method

Now that we have learned all the basics of maximum likelihood methodology, we are ready to use it in practical situations. By practical situations we mean the real cases corresponding to distributions that are not perfectly described by the standard distribution functions we just studied. Let us suppose we have a variable x, whose probability distribution function can be written as

$$p(x, k) = ke^{-kt},$$

where k is a constant. Suppose we take 4 measurements of the parameter t: $t = 12, 14, 11, 12$. What we want to do is to use the maximum likelihood method to compute the value of the constant k. To do this we first need to determine the maximum likelihood function. According to equation 9.3.19, this is give by

$$\begin{aligned} L(k) &= \prod_{i=1}^{4} ke^{-kt_i} \\ &= e^{-k^4(t_1+t_2+t_3+t_4)} \\ &= k^4 e^{-49k}. \end{aligned}$$

Now we take the natural logarithm of this function.

$$\ln(L) = 4\ln(k) - 49k$$

According to the maximum likelihood method, differentiating the above function with respect to k and equation the result to zero gives the required maximum likelihood estimate of k.

$$\begin{aligned} \frac{\partial \ln(L)}{\partial k} &= 0 \\ \Rightarrow \frac{4}{k} - 49 &= 0 \\ \Rightarrow k &= \frac{4}{49}. \end{aligned}$$

Let us now look at another example. This time we want to know how many measurements we must make so that the parameter $k = 0.21$ of the distribution

$$f(x, k) = kx \quad ; x \in (0, 1),$$

can be determined with an accuracy of 5%. That is, the relative error in $k = 0.21$ is

$$\frac{\triangle k}{k} = 0.05. \tag{9.3.56}$$

This can easily be done by using equation 9.3.51, which for our case becomes

$$N = \frac{1}{(\triangle k)^2} \int_0^1 \frac{1}{f} \left(\frac{\partial f}{\partial k}\right)^2 dx.$$

With $f = kx$, we get

$$\left(\frac{\partial f}{\partial k}\right)^2 = x^2$$

$$\Rightarrow \frac{1}{f}\left(\frac{\partial f}{\partial k}\right)^2 = \frac{x}{k}$$

$$\Rightarrow \int_0^1 \frac{1}{f}\left(\frac{\partial f}{\partial k}\right)^2 dx = \frac{1}{2k}.$$

Substituting this in the above relation for N gives

$$N = \frac{1}{(\triangle k)^2}\frac{1}{2k}.$$

But we have $k = 0.21$ and $\triangle k = 0.05k = 0.0105$, which gives

$$N = \frac{1}{(0.0105)^2}\frac{1}{(2)(0.21)}$$

$$= 2.1 \times 10^4.$$

9.4   Confidence Intervals

Suppose we have a rough idea of the activity level in a radiation environment and we use this information to estimate the dose that we expect a radiation worker to receive while working there for some specific period of time. The problem with this scheme is that there are a number of uncertainties involved in the computations, such as level of actual activity, its space dependence (the radiation sources might not be isotropic), and its variation with time. In such a situation what can be done is to define a confidence interval within which the value is *expected* to lie with a certain *probability*. For example, we can say that there is 90% probability that the person will receive a dose of somewhere between 10-20 *mrem*. Here we have two parameters that we are reporting: the confidence interval and its associated probability. The choice of a confidence interval is more or less arbitrary, although generally it is based on some rationale, such as our rough estimation based on some known parameters (in our example, we might have gotten a value of 15 *mrem* and then decided to give ourselves a leverage of ±5 *mrem* to compensate for any uncertainty in the calculations.). The probability, on the other hand, depends on the confidence interval and the probability distribution.

If the probability distribution of a variable x (such as dose) is given by $L(x)$, then the probability that x lies between x_1 and x_2 is given by

$$P(x_1 < x < x_2) = \frac{\int_{x_1}^{x_2} L(x)dx}{\int_{-\infty}^{\infty} L(x)dx}. \tag{9.4.1}$$

If the function $L(x)$ is normalized, then the denominator becomes 1 and the probability is simply given by

$$P = \int_{x_1}^{x_2} L(x)dx. \tag{9.4.2}$$

This probability is actually the area under the curve of $L(x)$ versus x between the points x_1 and x_2 and therefore depends on the choice of the confidence interval (see Fig.9.4.1).

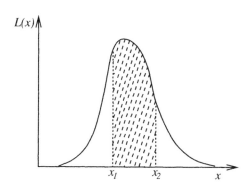

Figure 9.4.1: The probability that a value x lies within a confidence interval of (x_1, x_2) is obtained by dividing the area under the curve (shaded section) by the total area. If the distribution is normalized then the denominator will be 1 and the shaded area will simply be the required probability. This probability, therefore, depends on the choice of confidence interval. In practice, the probability is first selected and then the confidence interval is obtained from the cumulative distribution function of the probability distribution.

We saw above that the choice of confidence interval is arbitrary while the probability depends on it. Therefore one would assume that the interval is first chosen and then the probability is calculated. However the general practice is quite the opposite. If it is known that the process under consideration has a certain probability distribution then the probability is first chosen and then the confidence interval is deduced from some available tables or curves. For example, for Gaussian distribution, which is the most commonly used distribution, the tables of probability integrals are used to find the confidence intervals.

Let us now take a look at the example of a normally distributed variable x having mean μ and variance σ^2. We are interested in finding the probability that the measured value lies between $\mu - \delta x$ and $\mu + \delta x$. This probability, according to the definition above, can be evaluated from

$$\begin{aligned} P &= \frac{1}{\sigma\sqrt{2\pi}} \int_{\mu-\delta x}^{\mu+\delta x} e^{-(x-\mu)^2/2\sigma^2} dx \\ &= erf\left(\frac{\delta x}{\sigma\sqrt{2}}\right). \end{aligned} \tag{9.4.3}$$

Here $erf(u)$ is the *error function* of u, whose values are available in tabulated form in standard texts. To get a feeling of what different values of P would mean with respect to σ, we look at some typical values.

$$P(\mu - \sigma < x < \mu + \sigma) = 0.6827$$
$$P(\mu - 2\sigma < x < \mu + 2\sigma) = 0.9545$$
$$P(\mu - 3\sigma < x < \mu + 3\sigma) = 0.9973$$

What these values essentially show is that if the data can be represented by a perfect Gaussian distribution, then we can be only 68.27% sure that the next measurement will lie within the range $\mu \pm \sigma$. However if we wanted to be more than 99% sure about this we will have to stretch the range to around 3σ on both sides of the distribution. Fig.9.4.2 explains this concept in graphical form.

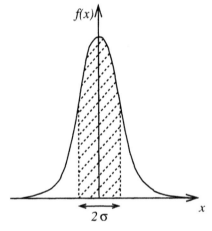

Figure 9.4.2: Confidence interval of a standard Gaussian distribution. The shaded area represents the probability that the next measurement of x will lie within the interval $\mu - \sigma < x < \mu + \sigma$. For a perfect Gaussian distribution this turns out to be 0.6827 meaning that one could be up to 68.27% sure that the value will not be out of these bounds.

9.5 Measurement Uncertainty

There is always some uncertainty associated with a measurement no matter how good our measuring device is and how carefully we perform the experiment. There are different types of uncertainties associated with any measurement but they can be broadly divided into two categories: systematic and random.

9.5.A Systematic Errors

All measurements, direct or indirect, are done through some type of measuring device. Since there is no such thing as a perfect device, therefore one should expect some error associated with the measurement. This type of error falls into the category of systematic errors, which refer to the uncertainties in the measurement due to the measurement procedures and devices. Unfortunately it is not always easy to characterize systematic errors. Repeating the measurements does not have any effect on them since they are not random. In other words, systematic errors are not statistical in nature and therefore can not be determined by statistical methods.

The good thing is that the systematic uncertainties can be minimized by modifying the procedures and using better devices. For example, one can use a detector having better accuracy or, in case of a gas filled detector, improve on its accuracy by using a more efficient gas mixture. Similarly easy steps can be taken to decrease the systematic uncertainties associated with readout electronics. An obvious example of that would be the use of an ADC having better resolution. Another way to decrease the systematic uncertainty is to properly calibrate the system.

Systematic uncertainties are system specific and therefore there is no general formula that could be used for their characterization. It is up to the experimenter to carefully determine these errors and faithfully report them in the final results.

9.5.B Random Errors

Random errors refer to the errors that are statistical in nature. For example, the radioactive decay is a random process. Even though we know the *average* rate of decay of a sample, we can not predict when the next decay will happen. This implies that there is an *inherent* time uncertainty associated with the process. Similarly the production of charge pairs in a radiation detector by passing radiation is also a random process (see chapter 2). We can say that *on the average* how many charge pairs will be produced by a certain amount of deposited energy but we can not associate an absolute number to it. Such uncertainties that are inherent to the process and are statistical in nature are categorized as random uncertainties.

Fortunately most physical processes are Poisson in nature. This makes is fairly easy to estimate the random error associated with a measurement. The random error associated with a measurement during which N counts were recorded, is given by

$$\delta_{stat} = \sqrt{N}. \tag{9.5.1}$$

For example, let us suppose that we measure the activity of a radioactive sample by taking three consecutive readings by a single channel analyzer/counter: 2452, 2367, 2398. The absolute random errors associated with these measurements will be:

$$\delta_{stat,1} = \sqrt{2452} = 49.52$$
$$\delta_{stat,2} = \sqrt{2367} = 48.65$$
$$\delta_{stat,3} = \sqrt{2398} = 48.97$$

9.5.C Error Propagation

Let us suppose we perform an experiment and make N independent measurements x_i each having uncertainty δx_i and standard deviation $sigma_{x_i}$. We then use these measurements to evaluate some function $u = f(x_1, x_2, ..., x_N)$. The question is: how can we estimate the standard deviation and error in the quantity we thus determine? This is where the error propagation formulae come into play, according to which the combined variance and standard error in the function u can be evaluated from

$$\sigma_u^2 = \left[\frac{\partial f}{\partial x_1}\right]^2 \sigma_{x_1}^2 + \left[\frac{\partial f}{\partial x_2}\right]^2 \sigma_{x_2}^2 + ... + \left[\frac{\partial f}{\partial x_N}\right]^2 \sigma_{x_N}^2 \tag{9.5.2}$$

$$\text{and} \quad \delta_u = \left[\left[\frac{\partial f}{\partial x_1}\right]^2 \delta x_1^2 + \left[\frac{\partial f}{\partial x_2}\right]^2 \delta x_2^2 + ... + \left[\frac{\partial f}{\partial x_N}\right]^2 \delta x_N^2\right]^{1/2}. \tag{9.5.3}$$

These general relations can be used to derive formulae for specific functions as shown below.

C.1   Addition of Parameters

Suppose we have $u = x_1 + x_2 + ... + x_N$. In this case the derivatives of $u = f(x)$ will be given by

$$\frac{\partial f}{\partial x_1} = \frac{\partial f}{\partial x_2} = ... = \frac{\partial f}{\partial x_N} = 1. \tag{9.5.4}$$

Equation 9.5.3, then reduces to

$$\delta_u = \left[\delta x_1^2 + \delta x_2^2 + ... + \delta x_N^2\right]^{1/2}, \tag{9.5.5}$$

which states that the total error in the measurement will simply be equal to the square root of the sum of individual errors squared.

Note that the above formula also holds if the some or all of the parameters have negative signs. In other words, the formula remains the same whether the parameters are added or subtracted in the function.

C.2   Multiplication of Parameters

Let us now see how the errors propagate if the function has the multiplicative form. For simplicity we will restrict ourselves to two variables, that is, we will assume that $u = x_1 x_2$. In this case the derivatives of $u = f(x)$ will be given by

$$\frac{\partial f}{\partial x_1} = x_2 \tag{9.5.6}$$

$$\text{and} \quad \frac{\partial f}{\partial x_2} = x_1. \tag{9.5.7}$$

Substituting these values into equation 9.5.3 gives

$$\frac{\delta_u}{u} = \left[\left(\frac{\delta x_1}{x_1}\right)^2 + \left(\frac{\delta x_2}{x_2}\right)^2\right]^{1/2}. \tag{9.5.8}$$

The generalized form of this equation for N parameters is given by

$$\frac{\delta_u}{u} = \left[\left(\frac{\delta x_1}{x_1}\right)^2 + \left(\frac{\delta x_2}{x_2}\right)^2 + ... + \left(\frac{\delta x_N}{x_N}\right)^2\right]^{1/2}. \tag{9.5.9}$$

Note that here δ_u/u refers to relative error. For absolute error this must be multiplied by u. What the above formula tells us is that the relative error in the measurement of N independent measurements is simply the square root of the sum of the individual relative errors squared.

The reader can verify that the above formula does not change its form in case of division. For example, the error in $u = x_1 x_2/x_3$ can be determined from the above formula without any modifications.

9.5.D   Presentation of Results

Now that we know how to calculate errors associated with parameters by using errors in individual measurements, we should discuss how to present our final results. We saw earlier that there are basically two classes of errors: systematic and random. Though it is a common practice to combine both errors together in the final result but a better approach, as adopted by many careful experimenters, is to explicitly state them separately. For example, the result of an experiment might be represented at 1σ confidence as

$$\xi = 205.43 \pm 6.13^{syst} \pm 14.36^{rand},$$

where the superscripts *syst* and *rand* stand for systematic and random errors respectively.

A word of caution here. By looking at the above numbers, one might naively conclude that all the values would lie between $205 - 6.13 - 14.36$ and $205 + 6.13 + 14.36$. This is not really true. Earlier in the chapter we discussed the confidence intervals and we saw that, for normally distributed data, a 1σ uncertainty guarantees with only about 68% confidence that the result lies within the given values (that is, between $\bar{\xi} - \sigma$ and $\bar{\xi} + \sigma$). For higher confidence, one must increase the σ-*level*. For example, for a 99% confidence, the above result will have to be written as

$$\xi = 205.43 \pm 6.13^{syst} \pm 43.08^{rand},$$

where we have multiplied the random error of 1σ by a factor of 3. Note that, since systematic uncertainty does not depend on statistical fluctuations, there is no need to multiply it by any factor. Now we can say with 99% confidence that the value of the parameter lies between $205 - 6.13 - 43.08$ and $205 + 6.13 + 43.08$.

 ## 9.6   Confidence Tests

Computing different quantities from a data set obtained from an experiment is helpful in understanding the characteristics of the system but if we have a certain bias about the behavior of the system we might also want to judge the data against our hypothesis. This *judgment* can be qualitative, such as just a visual sense of how the data looks like with respect to the expectation, or quantitative, which is the subject of the discussion here.

To judge a data sample quantitatively against a hypothesis we perform the so called *confidence* or *goodness-of-fit* test. For this we first define a goodness-of-fit statistic by taking into account both the data and the hypothesis. The idea is to have a quantity whose probability of occurrence could tell us about the level of agreement between the data and the hypothesis. Of course the choice of this statistic is arbitrary but several standard functions have been generated that can be applied in most of the cases. Before we look at some of these functions, let us first see how the general procedure works.

Let us represent the goodness-of-fit statistic by t such that its large values correspond to poor agreement with the hypothesis h. Then the p.d.f $g(t|h)$ can be used to determine the probability p of finding t in a region starting from the experimentally obtained value t_0 up to the maximum. This is equivalent to evaluating the cumulative distribution function

$$p \equiv 1 - P(t_o) = 1 - \int_{-\infty}^{t_o} g(t|h)dt \quad \text{or}$$

$$p = \int_{t_o}^{\infty} g(t|h)dt \tag{9.6.1}$$

A single value of p, however, does not tell us much about the agreement between data and hypothesis because at each data point the level of agreement could be different. The trick is to see how the value of p is distributed throughout its range, that is, between 0 and 1. Of course if there is perfect agreement, the distribution will be uniform.

Let us now take a look at some of the commonly used goodness-of-fit statistics.

9.6.A Chi-Square (χ^2) Test

This is perhaps one of the most widely used goodness-of-fit statistic. In the following we outline the steps needed to perform the test.

1. The foremost thing to do is to construct a hypothesis, which has to be tested. This hypothesis should include a set of values μ_i that we *expect* to get if we perform measurements and obtain the values u_i. These set of values may have been derived from a known distribution that the system is supposed to follow.

2. Decide on the number of degrees of freedom. If we take N measurements, the degrees of freedom are not necessarily equal to N because there may be one or more relations connecting the measured values u_i. If the number of such relations are r, then the degrees of freedom will be given by $\nu = N - r$.

3. Using the measured values u_i, compute a sample value of χ^2 from the relation 9.3.53
$$\chi^2 = \sum_{i=1}^{N} \frac{(u_i - \mu_i)^2}{\sigma_i^2}.$$

4. Compute the *normalized* χ^2, that is, χ^2/ν.

5. Decide on the acceptable significance level p, which represents the probability that the data is in agreement with the hypothesis or not. A commonly chosen value of p is 0.05, which gives a confidence of 95%.

6. Determine the value of $\chi_{\nu,p}^2$ at which p is equal to the chosen value. This means evaluating the integral
$$p = \int_{\chi_{\nu,p}^2}^{\infty} f(x)dx \tag{9.6.2}$$

for $\chi_{\nu,p}^2$ (see also equation 9.6.1). $f(x)$ is of course the χ^2 probability density function. The solution to this equation requires numerical manipulations, which can be done, for example by employing the Monte Carlo integration technique. However this is not generally done since there are tables and graphs available that can be used to deduce the values of $\chi_{\nu,p}^2$ with respect to p and ν.

7. Compare χ^2/ν with $\chi^2_{\nu,\alpha}/\nu$.

Let us now see what we can infer from this comparison.

▶ **Case-1**, $\chi^2/\nu \simeq \chi^2_{\nu,\alpha}/\nu$: We are up to $\alpha \times 100\%$ confident that our hypothesis was correct.

▶ **Case-2**, $\chi^2/\nu > \chi^2_{\nu,\alpha}/\nu$: This may mean one of the following.

1. The model we have chosen to represent the system is not adequate.

2. The model is adequate but there are some bad data points in the sample. It takes only a few large excursions in the data that are far away from the mean to yield a large value of chi-square. Care should therefore be taken to ensure that proper filtration of the data is performed to eliminate such data points.

3. The data values are not uniformly distributed about their means. The is the most troubling scenario, since it would mean that this goodness-of-fit method is not really applicable and we should either resort to some other method or look closely at the data to find out if just a few values are causing this deviation from the normal distribution. Generally, discarding a few data points does the trick.

▶ **Case-3**, $\chi^2/\nu < \chi^2_{\nu,\alpha}/\nu$: This means that the squares of the random normal deviates are less than expected, a situation that demands as much attention as the previous one. The following possibilities exist for this case.

1. The expected means were overestimated. This does not mean that the model was wrong.

2. There are a few data points that have caused the chi-square value to become too small.

9.6.B Student's t Test

Student's t test is the most commonly used method of comparing the means of two low statistics data samples. To perform the test, first the following quantity is evaluated.

$$t = \frac{|\bar{x}_1 - \bar{x}_2|}{\sigma_{12}} \tag{9.6.3}$$

Here $\bar{x}_1$ and $\bar{x}_2$ represent the means of first and second datasets and σ_{12} is the standard deviation of the difference between the two means. It can be computed from

$$\sigma_{12} = \left[\frac{\sigma_1^2}{N_1} + \frac{\sigma_2^2}{N_2} \right]^{1/2}, \tag{9.6.4}$$

where σ_1 and σ_2 are the standard deviations of the two datasets having N_1 and N_2 number of data points. Note that here what we have done is to simply taken the square root of the sum of the standard errors associated with each dataset.

The next step is to compare the calculated t-value with the tabulated one. The tabulated values, derived from the Student's t distribution we presented earlier, are generally given for different degrees of freedom and levels of significance. The total degrees of freedom for the dataset are given by

$$
\begin{aligned}
\nu &= (N_1 - 1) + (N_2 - 1) \\
&= N_1 + N_2 - 2.
\end{aligned}
\tag{9.6.5}
$$

The choice of level of significance depends on the level of confidence one intends to have on the analysis. If one chooses a value of 0.05 and the calculated t value turns out to be less than the tabulated one, then one could say with 95% confidence that the means are not significantly different.

Example:
An ionization chamber is used to measure the intensity of x-rays from an x-ray machine. The experiment is performed at two different times and yield the following values (arbitrary units).

Measurement-1: 380, 398, 420, 405, 378
Measurement-2: 370, 385, 400, 419, 415, 375

Perform Student's t test at 95% and 99% confidence levels to see if the means of the two measurements are significantly different from each other.

Solution:
First we compute the means of the two datasets.

$$
\begin{aligned}
\bar{x}_1 &= \sum_{i=1}^{N_1} \frac{x_{1,i}}{N_1} \\
&= 396.2 \\
\bar{x}_2 &= \sum_{i=1}^{N_2} \frac{x_{2,i}}{N_2} \\
&= 394
\end{aligned}
$$

Next we determine the standard deviations of the two means.

$$
\begin{aligned}
\sigma_1 &= \frac{1}{N_1 - 1} \sum_{i=1}^{N_1} (x_{1,i} - \bar{x}_1)^2 \\
&= 17.61 \\
\sigma_2 &= \frac{1}{N_2 - 1} \sum_{i=1}^{N_2} (x_{2,i} - \bar{x}_2)^2 \\
&= 20.59
\end{aligned}
$$

The standard deviation of the mean is given by

$$
\begin{aligned}
\sigma_{12} &= \left[\frac{\sigma_1^2}{N_1} + \frac{\sigma_2^2}{N_2}\right]^{1/2} \\
&= \left[\frac{17.61^2}{5} + \frac{20.59^2}{6}\right]^{1/2} \\
&= 11.52.
\end{aligned}
$$

Now we are ready to compute the t value.

$$
\begin{aligned}
t &= \frac{|396.2 - 394|}{11.52} \\
&= 0.191
\end{aligned}
$$

To compare this t value with the tabulated values we must first determine the degrees of freedom of the dataset. This is given by

$$
\begin{aligned}
\nu &= N_1 + N_2 - 2 \\
&= 5 + 6 - 2 = 9.
\end{aligned}
$$

For a 95% confidence level and 9 degrees of freedom the tabulated t value is 2.26. And for a 99% confidence level the tablulated t value is 1.83. Since both of these values are greater than the calculated t value of 0.19, therefore we can say with at least 99% confidence that the two dataset means are not significantly different.

9.7 Regression

Regression analysis is perhaps the most widely used technique to draw inferences from experimental data. The basic idea behind it is to fit a function that closely represents the trend in the data. The function can then be used to make predictions about the variables involved.

Fitting a function to the data through regression analysis is not always a very pleasant experience, specially if the data shows variations that can not be characterized by standard functions, such as polynomial, exponential, or logarithmic. The easiest form of regression analysis is the simple linear regression, which we will discuss in some detail now. Later on we will look at other kinds of regression analysis.

9.7.A Simple Linear Regression

Simple linear regression refers to fitting a straight line to the data. The fitting is mostly done using a technique called *least square fitting*. To understand this technique, let us start with the equation of a straight line

$$y = mx + c, \tag{9.7.1}$$

where m is the slope of the line and c is its y-intercept. Since slope and y-intercept determine the orientation and position of the straight line on the xy-plot, therefore our task is to find their best values. By best values we mean the ones for which each of the data points is as close to the line as possible. To quantify this statement we first note that the value of y at any x_i is y_i and the value of the straight line at that point is $mx_i + c$. The difference between the two at each data point, which can be called *residual*, is then given by

$$R = mx_i + c - y_i. \tag{9.7.2}$$

This residual should be minimum for each data point. In other words the sum of the residuals should be a minimum. However there is a problem with this scheme, namely the summed residuals would turn out to be zero since the positive residuals would cancel out the negative ones. To overcome this problem one can minimize the sum of the *squared* residuals instead. That is, we can demand that

$$\chi^2 = \sum (mx_i + c - y_i)^2 \tag{9.7.3}$$

is minimum. Here we have represented the sum of the squared residuals by χ^2 as it is the most commonly used notation for this quantity. Now we need to minimize this function with respect to c and m. Since the differential of a function vanishes at its minimum, therefore we have the following two conditions.

$$\frac{\partial \chi^2}{\partial c} = 0 = \frac{\partial}{\partial c}\left[\sum (mx_i + c - y_i)^2\right] \tag{9.7.4}$$

$$\frac{\partial \chi^2}{\partial m} = 0 = \frac{\partial}{\partial m}\left[\sum (mx_i + c - y_i)^2\right] \tag{9.7.5}$$

Performing these differentiations gives

$$\sum (mx_i + c - y_i) = 0 \tag{9.7.6}$$

$$\text{and} \quad \sum (mx_i + c - y_i)\, x_i = 0. \tag{9.7.7}$$

Now, we have two equations, which we can solve to determine the required c and m. A few algebraic manipulations finally yield

$$m = \frac{\sum x_i y_i - \sum x_i \sum y_i}{\sum x_i^2 - (x_i)^2} \tag{9.7.8}$$

$$\text{and} \quad c = \frac{\sum y_i \sum x_i^2 - \sum x_i \sum x_i y_i}{\sum x_i^2 - (x_i)^2}. \tag{9.7.9}$$

Note that these are the values for which the sum of the squared residuals is minimum. In other words, these values represent a line that is the best fit to the data.

It is evident that for a large dataset the computations involved to determine the best fit are enormous. Therefore normally one uses computer codes to perform the regression analysis. Luckily enough now most standard statistical analysis packages have built-in routines that can handle linear as well as more complicated regressions.

9.7.B   Nonlinear Regression

By nonlinear regression we mean fitting any nonlinear function to the data. This could be a polynomial of the order 2 or more, an exponential, a logarithmic, a combination of these, or some other function. There are different techniques available to handle the nonlinear regression problems but the two most practical and common ones are least squares regression and maximum likelihood regression. The maximum likelihood function has already been discussed earlier in the chapter and therefore will not be repeated here. One point that is worth noting here is that, although maximum likelihood regression is a very fine technique but in practice it is not very commonly used since it is computationally more involved than the least squares regression technique.

The basic idea behind least squares regression for nonlinear fitting is the same as we discussed for the simple linear regression. That is, one tries to minimize the sum of the squared residuals of the dataset. The problem in this case, however, is that due, to the nonlinear nature of the function, analytic forms for the coefficients can not be generally found. One then resorts to numerical techniques to solve the equations. These techniques normally solve the equations recursively, a process that may or may not converge to an acceptable solution. Even with this shortcoming, the least squares method is sill the most widely used technique for nonlinear as well as linear regression.

Since exact form of nonlinear regression equations depend on the type of function one is trying to fit, therefore it is not worthwhile to go into specific function details. However, to give the reader a general overview of the technique, we will have a look at its functional form. Suppose we want to fit a nonlinear function of the form $f(x, \alpha)$ to the data. Here x are the independent variables and α represents the coefficients that need to be determined. For example, the function may be a simple second order polynomial given by

$$f(x, \alpha) = \alpha_1 + \alpha_2 x + \alpha_3 x^2. \tag{9.7.10}$$

Following the procedure described for the case of linear regression, we define the sum of the squared residuals as

$$\chi^2 = \sum [f(x_i, \alpha) - y_i]^2, \tag{9.7.11}$$

where y_i is the data at x_i and $f(x_i, \alpha)$ represents the value of the function at x_i. To minimize χ^2 we differentiate it with respect to all the α's and equate the result to zero, that is

$$\frac{\partial \chi^2}{\partial \alpha} = 0 = \frac{\partial}{\partial \alpha} \left[\sum \{f(x_i, \alpha) - y_i\}^2 \right]. \tag{9.7.12}$$

It is obvious that the function $f(x, \alpha)$ can be of any type, which in fact is the strongest point of the least squares method. One can essentially fit any function provided numerical techniques can be developed to solve the resulting equations.

Example:
Determine the least squares equations to determine the coefficients of a second order polynomial.

Solution:
A second order polynomial can be written as

$$f(x, \alpha) = \alpha_1 + \alpha_2 x + \alpha_3 x^2.$$

We are required to obtain equations that can be solved to determine the three coefficients α_1, α_2, and α_3, This can be done by first constructing the expression for the sum of the squared residuals, which according to equation 9.7.11 can be written as

$$\chi^2 = \sum \left[\alpha_1 + \alpha_2 x_i + \alpha_3 x_i^2 - y_i\right]^2.$$

Now, according to equation 9.7.12, the first equation is

$$0 = \frac{\partial}{\partial \alpha_1}\left[\sum \left\{\alpha_1 + \alpha_2 x_i + \alpha_3 x_i^2 - y_i\right\}^2\right]$$
$$\Rightarrow 0 = 2\sum \left\{\alpha_1 + \alpha_2 x_i + \alpha_3 x_i^2 - y_i\right\}.$$

Similarly the second equation is

$$0 = \frac{\partial}{\partial \alpha_2}\left[\sum \left\{\alpha_1 + \alpha_2 x_i + \alpha_3 x_i^2 - y_i\right\}^2\right]$$
$$\Rightarrow 0 = 2\sum \left\{\alpha_1 + \alpha_2 x_i + \alpha_3 x_i^2 - y_i\right\} x_i.$$

And the third equation is

$$0 = \frac{\partial}{\partial \alpha_3}\left[\sum \left\{\alpha_1 + \alpha_2 x_i + \alpha_3 x_i^2 - y_i\right\}^2\right]$$
$$\Rightarrow 0 = 2\sum \left\{\alpha_1 + \alpha_2 x_i + \alpha_3 x_i^2 - y_i\right\} x_i^2.$$

9.8   Correlation

There are different techniques in statistics that can be used to determine how one dataset is associated with another one. The specific term used to determine such association is the *correlation analysis*. An example where such an analysis would be useful is to see how the change in the leakage current of a silicon detector is correlated with increase in the absorbed radiation.

The measure of the correlation, no matter what technique is used, always lies between -1 and +1. A correlation coefficient of +1 signifies perfect correlation while a value of -1 shows that the data are negatively correlated. Note that negative correlation does not mean no correlation, rather strong correlation but in an opposite sense. It would mean that if one variable is increasing the other is decreasing but in a perfectly correlated manner. A correlation coefficient of 0 represents no correlation.

Even though there are several techniques used to determine correlation, however the most commonly used technique is the so called *Pearson r* or *simple linear* correlation. We will therefore restrict ourselves to this correlation technique.

9.8.A Pearson r or Simple Linear Correlation

In essence, the simple linear correlation determines the extent of proportionality between two variables. The proportionality is quantified through the coefficient of correlation, which is related to the regression fit of the data. The correlation coefficient has several equivalent forms but is most conveniently determined from the relation

$$r = \frac{N \sum x_i y_i - \sum x_i \sum y_i}{\sqrt{\left[N \sum x_i^2 - (\sum x_i)^2\right]\left[N \sum y_i^2 - (\sum y_i)^2\right]}}, \tag{9.8.1}$$

where the summation ($\sum$) is over the whole dataset belonging to variables x and y.

The way correlation coefficient is interpreted has already been discussed. Fig. 9.8.1 shows a few examples of regression fits to different datasets and the corresponding correlation coefficients.

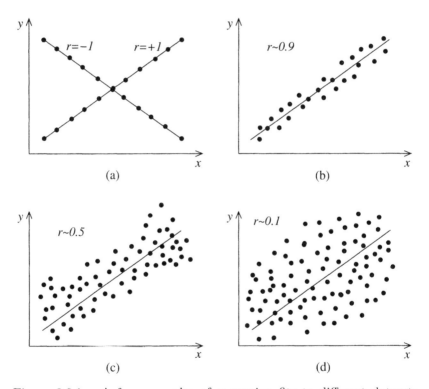

Figure 9.8.1: A few examples of regression fits to different datasets and their corresponding correlation coefficients. It should be noted that $r = -1$ in (a) represents anti-correlation, which in fact signifies perfect correlation just like $r = 1$. As the data get dispersed the correlation coefficient approaches no-correlation value of 0.

Example:

The following data are obtained in an experiment.

$x = \{102, 154, 200, 220, 267, 263, 352, 361, 423, 449, 512, 598, 601, 701, 711\}$
$y = \{28, 98, 132, 98, 129, 202, 265, 243, 291, 324, 376, 412, 524, 511, 560\}$

Determine if the data are well correlated.

Solution:

A plot of the data together with a regression fit is shown in Fig.9.8.2. It is evident that the data are very well correlated. To quantify our confidence we calculate the correlation coefficient. To do that we can use equation 9.8.1 to compute the correlation coefficient. To simplify the computations, we create a table with the needed terms computed individually. The correlation coefficient is then given by

$$
r = \frac{(15)(2113232) - (5914)(4193)}{\sqrt{\left[(15)(2875604) - (5914)^2\right]\left[(15)(1574209) - (4193)^2\right]}}
$$

$$
= 0.98.
$$

Hence we can say with high confidence that the variables are very well correlated.

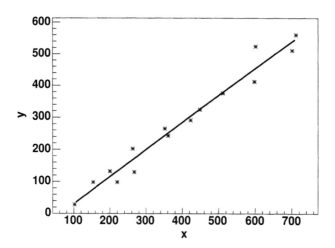

Figure 9.8.2: Plot of the data given in the example above together with a straight line fit.

Table 9.8.1: Parameters to compute the correlation coefficient for the example above. The last row is the sum of each column.

x	y	xy	x^2	y^2
102	28	10404	784	2856
154	98	23716	9604	15092
200	132	40000	17424	26400
220	98	48400	9604	21560
267	129	71289	16641	34443
263	202	69169	40804	53126
352	265	123904	70225	93280
361	243	130321	59049	87723
423	291	178929	84681	123093
449	324	201601	104976	145476
512	376	262144	141376	192512
598	412	357604	169744	246376
601	524	361201	274576	314924
701	511	491401	261121	358211
711	560	505521	313600	398160
5914	4193	2875604	1574209	2113232

 ## 9.9 Time Series Analysis

Most experiments involve taking data at regular intervals of time. These time series data are then analyzed using different techniques. The analysis depends on the application and the inferences to be drawn from the data. For example, one might be interested in simply determining the expectation value of the variables involved or a much complicated task of identifying the hidden structures in the time series. In the following sections we will discuss some of the important techniques used in time series analysis.

9.9.A Smoothing

As the name suggests, smoothing refers to getting rid of small fluctuations in data. It is generally done to smooth out the small local fluctuations in the data. In time series analysis, however, smoothing can cause information to be lost, which of course is not desirable. For example, data from detectors generally have low level fluctuations embedded on the signal output. Since these fluctuations can give insight into the noise sources therefore it is not a good idea to smooth out the data. However, in some instances one is interested in determining the baseline of the output as shown in Fig.9.9.1. Baselines are generally determined by taking the average of all the data points. This gives a constant value that can be used in the computer code as a reference to determine the parameters related to the pulse. Another smoothing technique is the so called *moving average*. This involves taking averages of smaller subsets of the data in succession. For example, let us suppose we want to compute the moving averages of the data: 3,4,2,6,4,3,2,5,....

To determine the moving average we first take the first three points and calculate their average, which turns out to be

$$\frac{3+4+2}{3} = 3.$$

Next we move one step forward and take the average of three data points excluding the first one, that is

$$\frac{4+2+6}{3} = 4.$$

In the next step we exclude one more point and compute the average of the next three numbers. Note that one can choose essentially any number of data points in one set and the choice depends on how the data are varying.

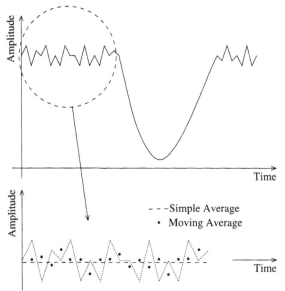

Figure 9.9.1: Simple and moving average smoothing of the base line of a typical detector pulse. The moving averages were calculated by taking the simple average of 3 point sets in succession such that at each subsequent step the set is moved forward by one data point.

Fig.9.9.1 also shows the result of moving averages for 3-point subsets of data. It is evident that such a technique is not suitable for determining the baseline. However there are other instances where this technique gives better results than the simple all-data averaging. Actually moving average is most suitable for situations where small scale fluctuations are superimposed on a large scale fluctuation. In other words if the data contain high frequency components on top of low frequency components then smoothing by simple average would smooth out the low frequency components as well. On the other hand, moving average smoothing would retain the low frequency components. In this respect we can say that moving average acts like a low pass filter with a pass band that depends on the number of data points selected in each set.

Simple and moving averaging are not the only smoothing techniques available. In fact, with the availability of fast computers, the trend now is to use the more involved techniques, such as *exponential smoothing*. The basic idea behind exponential smoothing is to perform weighted averages instead of the simple averages. The weights are assigned in an exponentially deceasing fashion, hence the name exponential smoothing.

9.10   Frequency Domain Analysis

Sometimes it is desired that the time series data are analyzed in the frequency domain rather than the time domain. This desire is generally driven by the need to obtain frequency and phase content of the signal. For example, one may want to see if the time series data from a detector has a 50 or 60 Hz components, which would indicate coupling of the system with a power line. Similarly one might be interested in determining the time periodicities in the signal, which stand out as frequency peaks in the frequency domain. Another application of frequency domain analysis is the filtration of digitized pulse, which allows the frequencies to be selectively filtered out.

The frequency domain analysis is also sometimes referred to as *spectral analysis*. However spectral analysis is a much broader term that involves more involved types of analysis as well, such as fractal dimensional analysis. Here by frequency domain analysis we mean transformation of the signal into frequency domain through Fourier transformation and then analyzing the resulting power spectrum. A power spectrum is the variation of the square of the Fourier transform amplitude with respect to frequency.

The basic idea behind Fourier transformation is that any signal can be *decomposed* into a sum of simple sinusoidal functions with coefficients that represent amplitudes. In other words, no matter how complex a signal is, it can be represented by a sum of sinusoids. The resulting function is known as a Fourier series. We will not go into the details of this series but will concentrate on how a time varying function can be transformed into frequency space. Any function $g(t)$ can be Fourier transformed according to

$$F(f) = \int_{-\infty}^{\infty} g(t)e^{-i2\pi ft}dt, \qquad (9.10.1)$$

where f represents frequency in Hz. This equation is good for continuous function that can be evaluated analytically and therefore if we want to transform experimental data, which generally can not be represented in a functional form, it is not very useful. For such a situation the above equation can also be written in discrete form, that is

$$F_n = \sum_{j=0}^{N-1} g_j e^{-i2\pi nj/N},\qquad\qquad(9.10.2)$$

where N is the number of data points.

Discrete Fourier transformation is an extensively used technique in data analysis. There are different algorithms exist to solve the above equations but the most common one is the so called *fast Fourier transform* or FFT. As the name suggests, this algorithm performs the transformation faster than any other technique, which is actually due to the lesser number of computations required by it. FFT performs $2N \log_2 N$ ($\log_2$ is the base 2 log) computations as compared to the usual algorithm that requires $2N^2$ computations.

Fig.9.10.1 shows an example of the utility of Fourier transformation. The figure shows a sinusoid function and its power spectrum obtained by taking the square of the amplitude of the Fourier transformed data. Suppose the sinusoid represents the output of an electronics chain in response to a perfect sinusoidal input. The broadness of the peak in the power spectrum tells us that the system is behaving as a damped harmonic oscillator. But we know that the damping is electronic systems is characterized by charge injection. Hence, if the peak is too broad we can say that there is significant charge injection in the circuitry and if it was used to read out a detector output the signal to noise ratio may not be acceptable.

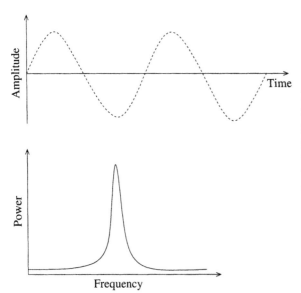

Figure 9.10.1: Fourier transformation of an imperfect sinusoidal function and its power spectrum. The power is obtained by taking the square of the amplitude of the Fourier transform. The broadness of the peak determines the *quality* of the sinusoid.

9.11   Counting Statistics

Detection of single events is perhaps the most widely used investigation method in particle physics, nuclear chemistry, nuclear medicine, radiology, and other related disciplines. In such applications one is interested in determining whether a particle interacted in the detector or not. This is not an enormously difficult task since here one is not really interested in the shape of the pulse, rather its presence or absence. In other words, one simply counts the *interesting* pulses. Which pulses are interesting, depends on a whole lot of factors including type of particles to be detected, signal to noise ratio of the system, and precision required. The *uninteresting* events or pulses are then either blocked from reaching the analysis software or are eliminated by the analysis software. For example, with a neutron detector one might want to eliminate the α-particle events. This can be accomplished easily since the energy deposited by an α-particle is generally much higher than deposited by a neutron and all one needs to do is to set an upper limit on the pulse amplitude.

   In counting experiments, statistics plays a crucial role. The reason is that most of the underlying phenomena are random in nature. Take the example of measuring activity of a radioactive sample. The emission of particles from the sample is a random phenomenon. To detect these particles one must allow them to interact with some detection medium. If the medium is a scintillator, it will produce light as a result of particle interaction. This process is governed by quantum mechanics, which is essentially a statistical theory. The light thus produced can be detected by a photomultiplier tube by first converting it into electrons and then by multiplying the electron population. All these steps are characterized by statistics.

   Hence in essence, we can say that statistics is of primary importance in counting experiments. The physical and detection limits should therefore be deduced by statistical methods. This is the subject of our next section.

9.11.A   Measurement Precision and Detection Limits

Most of the counting experiments follow Poisson statistics. Accordingly, the standard deviation of a measurement is equal to the square root of its magnitude, that is

$$\sigma = \sqrt{N}, \tag{9.11.1}$$

where N represents the observed counts. Note that here we have not yet taken into account any detector related issues. The standard error in the measurement obtained in this way reflects only the physics of the underlying processes. In other words, it gives us the *physical limit* of the measurement error. One can not have a dataset that is spread out less than what is suggested by σ. The physical limit of the error in measurement is then given by

$$\delta_N = \frac{\sigma}{N} \tag{9.11.2}$$

$$= \frac{\sqrt{N}}{N} = \frac{1}{\sqrt{N}}. \tag{9.11.3}$$

This quantity after multiplying by 100 is sometimes referred to as the physical limit of the measurement precision (multiplying by 100 simply evaluates the precision as a percentage). That is, one can not achieve a precision better than this value.

This statement needs some clarification, though. What we are assuming here is that the measurement time is not changing. If we increase the measurement time, N would increase and so would the measurement precision. Hence the physical limit on measurement precision actually depends on the measurement time.

The discussion above applies to measurement at a point only. How can we determine the precision in measurement if measurements are made at several points? An obvious example is energy spectroscopy, where the goal is to obtain an energy spectrum of particles. Such measurements generally produce one or more peaks over a background. The main analysis tasks are to identify the peaks and measure their respective areas. Both of these tasks are tied to the identification and elimination of background. The difficulty lies in determining the true area of the peak, that is the counts that contributed to the peak and not the background. To simplify the matter, let us first assume that the peak is background-free as shown in Fig.9.11.1. This is a Gaussian-like peak with two tails. Now, the tails are not part of the background but could have arisen because of the errors induced by the measuring device[1]. That is, there is some uncertainty associated with them. Hence we would want to exclude them from the measurement of peak precision. This requires selection of a *region of interest* in the peak as shown in the figure. It should be noted that there is no universally accepted method of selecting the region of interest but most experimenters use the area above a line that cuts the peak at 10% of its maximum. The width of the peak at this line is called *full width at one tenth of the maximum*.

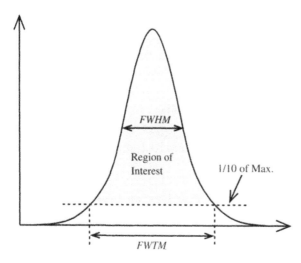

Figure 9.11.1: Distribution of counts in an experiment with no background. The region of interest (shaded portion), with an area of N_i, has been chosen such that the counts below 10% of the peak amplitude get discarded. The total area of the peak is N_t.

It is apparent that the best error estimate will be obtained if we take all the counts in the peak N_t, that is

$$\delta_{N_t} = \frac{1}{\sqrt{N_t}}. \tag{9.11.4}$$

---

[1]Some experimenters prefer to include tails in the background as well. However a better approach is to treat tails and background separately.

If we consider only the counts in the region of interest N_i, then the error in measurement will be given by

$$\delta_{N_i} = \frac{1}{\sqrt{N_i}}. \tag{9.11.5}$$

Of course, the value of δ_{N_i} will depend on our choice of the area of interest. As the area approaches N_t, the error in N_i approaches the error in N_t. This becomes obvious if we divide equation 9.11.5 by equation 9.11.4.

$$\frac{\delta_{N_i}}{\delta_{N_t}} = \sqrt{\frac{N_t}{N_i}}$$

$$\Rightarrow \delta_{N_i} = \sqrt{\frac{N_t}{N_i}} \delta_{N_t} \tag{9.11.6}$$

It was mentioned earlier that most peaks encountered in spectroscopic measurements are Gaussian-like and generally $FWTM$ is used to define the region of interest. For a Gaussian peak the $FWTM$ is given by[2]

$$FWTM \approx 4.29\sigma_t. \tag{9.11.8}$$

Dividing both sides by N_t gives

$$\frac{\sigma_t}{N_t} \equiv \delta_{N_t} = \frac{FWTM}{4.29N_t}. \tag{9.11.9}$$

We now substitute this expression in equation 9.11.6 and obtain

$$\delta_{N_i} = \sqrt{\frac{N_t}{N_i} \frac{FWTM}{4.29N_t}}$$

$$= \frac{FWTM}{4.29\sqrt{N_i N_t}}, \tag{9.11.10}$$

which can be written in a more convenient form as

$$\delta_{N_i}\sqrt{N_t} = \frac{FWTM}{4.29\sqrt{N_i}}. \tag{9.11.11}$$

If the quantity on the left hand side of the above equation approaches unity, the error in N_i will approach the least possible spread in N_i, that is $\sigma_t = \sqrt{N_t}$. This implies that for best measurement precision we should have

$$\frac{FWTM}{4.29\sqrt{N_i}} = 1$$

$$\Rightarrow FWTM = 4.29\sqrt{N_i}. \tag{9.11.12}$$

---

[2]This is obtained by using the definition of Gaussian distribution as follows:

$$-\frac{x^2}{2\sigma^2} = \frac{1}{10}$$

$$\Rightarrow x = \sigma\sqrt{2\ln(10)}$$

$$\Rightarrow FWTM = 2x = \sigma 2\sqrt{2\ln(10)} \approx 4.29\sigma$$

$$\tag{9.11.7}$$

The reader is encouraged to verify that, for a Gaussian peak, $FWHM \approx 2.35\sigma$.

Since it is not a requirement that $FWTM$ be used to define the region of interest, we can write the above equation in a general form as

$$w = \epsilon \sqrt{N_i}, \qquad (9.11.13)$$

where w is the width of the peak at the bottom of the region of interest and the factor ϵ depends on the definition of the region of interest. We now want to write equation 9.11.6 in terms of $FWHM$, which is a commonly measured quantity in spectroscopy. For that, we first note that

$$FWHM \approx 2.35\sigma_t. \qquad (9.11.14)$$

Dividing both sides by N_t gives

$$\frac{\sigma_t}{N_t} \equiv \delta_{N_t} = \frac{FWHM}{2.35N_t}. \qquad (9.11.15)$$

Substituting this expression in equation 9.11.6 gives

$$\begin{aligned}
\delta_{N_i} &= \sqrt{\frac{N_t}{N_i} \frac{FWHM}{2.35N_t}} \\
&= \frac{FWHM}{2.35\sqrt{N_i N_t}} \qquad (9.11.16) \\
\Rightarrow \delta_{N_i} \sqrt{N_t} &= \frac{FWHM}{2.35\sqrt{N_i}}. \qquad (9.11.17)
\end{aligned}$$

As in the case of $FWTM$, for the best precision in the measurement of N_i, the right hand side of the above equation should approach unity, that is

$$\begin{aligned}
\frac{FWHM}{2.35\sqrt{N_i}} &= 1 \\
\Rightarrow FWHM &= 2.35\sqrt{N_i}. \qquad (9.11.18)
\end{aligned}$$

Dividing equation 9.11.13 by equation 9.11.22 gives

$$\begin{aligned}
\frac{w}{FWHM} &= \frac{\epsilon}{2.35} \\
\Rightarrow w &= \frac{\epsilon}{2.35} FWHM. \qquad (9.11.19)
\end{aligned}$$

If we define the region of interest using full width at tenth maximum, then $\epsilon = 4.29$ and the above equation becomes

$$w \approx 1.82 FWHM. \qquad (9.11.20)$$

In general, it is recommended that the region of interest is chosen such that the width w is approximately twice the $FWHM$.

Up until now we have not taken the background into consideration, which of course is not a very realistic situation. In spectroscopic measurements, one generally finds peaks embedded on a background. Fig.9.11.2 shows such a peak. It is apparent that the background adds uncertainties to the measurements, which are not only due to channel-by-channel variations but also statistical fluctuations in background

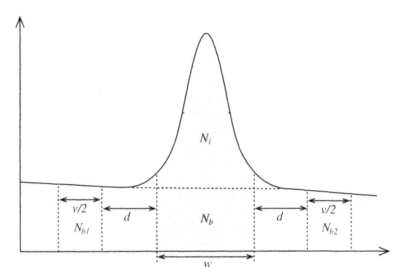

Figure 9.11.2: Subtraction of background for determination of peak area.

counts. In the example shown in the figure, the background has a linear trend, which is not really a representative of the majority of spectra. However, it would serve well to introduce the reader to a simple method of background elimination.

The first step in elimination of background is to define a region of interest as shown by shaded portion of the spectrum in Fig.9.11.2. This region has two parts as depicted by a horizontal line in the figure. The upper part is what we are interested in. The bottom part is the background that needs to be subtracted from the total area. The question is, how do we draw that horizontal line? One way to do it is to simply connect the two ends o f the peak. But this method may not work very well since the local fluctuations in the background may introduce too large an uncertainty in the result. Another simple method is to use two other regions to estimate the background counts as shown in Fig.9.11.2. We first choose two regions of equal widths $v/2$ on either side of the background area. The total area of these regions is $N_{b1} + N_{b2}$. We then assume that the ratio of this total area to the total width of these regions is equal to the ratio of the background area to the background area width, that is

$$\frac{N_b}{w} \approx \frac{N_{b1} + N_{b2}}{v} \tag{9.11.21}$$

$$\Rightarrow N_b \approx \frac{w}{v} [N_{b1} + N_{b2}]. \tag{9.11.22}$$

Having estimated N_b, we can now calculate the area of the peak from

$$N_i = N_t - N_b, \tag{9.11.23}$$

where N_t represents the total area of the region of interest (shaded portion in Fig.9.11.2).

We are now interested in determining the error in our measurement of the area. For that we can use the error propagation formula 9.5.5, according to which the standard deviation in measurement of N_i is given by

$$\sigma_i = \left[\sigma_t^2 + \sigma_b^2\right]^{1/2}, \tag{9.11.24}$$

where the subscripts i, t, and b refer to the peak, the total, and the background respectively. The spread in the total counts is given by $\sigma_t = \sqrt{N_t}$. To calculate the spread in the background counts we note from equation 9.11.22 that since v and w are constants, the spread in N_b is equal to the spread in $N_{b1} + N_{b2}$, that is

$$\sigma_b \approx \frac{w}{v}\left[N_{b1} + N_{b2}\right]^{1/2}. \tag{9.11.25}$$

This can be simplified by substituting the value of $N_{b1} + N_{b2}$ from equation 9.11.22. Hence we get

$$\sigma_b \approx \frac{w}{v}\left[\frac{v}{w}N_b\right]^{1/2}. \tag{9.11.26}$$

Substitution of σ_t and σ_b in equation 9.11.25 yields

$$\sigma_i = \left[N_t + \frac{w}{v}N_b\right]^{1/2}. \tag{9.11.27}$$

This equation can also be written in terms of N_i by substitution of $N_t = N_i + N_b$.

$$\begin{aligned}
\sigma_i &= \left[N_i + N_b + \frac{w}{v}N_b\right]^{1/2} \\
&= \left[N_i + \left(1 + \frac{w}{v}\right)N_b\right]^{1/2} \tag{9.11.28}
\end{aligned}$$

Dividing both sides of this equation by N_i gives us the error in measurement of N_i.

$$\frac{\sigma_i}{N_i} \equiv \delta_{N_i} = \frac{1}{\sqrt{N_i}}\left[1 + \left(1 + \frac{w}{v}\right)\frac{N_b}{N_i}\right]^{1/2} \tag{9.11.29}$$

Note that, in the absence of background ($N_b = 0$), this equation reduces to equation 9.11.5. This equation sets the limit to accuracy of the measurement of peak area using our simple background elimination technique. To get an idea of how it compares to the minimum possible error $\delta_{N_t} = 1/\sqrt{N_t}$, let us divide it on both sides by δ_{N_t}.

$$\frac{\delta_{N_i}}{\delta_{N_t}} = \sqrt{\frac{N_t}{N_i}}\left[1 + \left(1 + \frac{w}{v}\right)\frac{N_b}{N_i}\right]^{1/2} \tag{9.11.30}$$

One conclusion we can draw here is that the error in peak area depends on the total background counts. This is due to the statistical fluctuations in the background counts. So, the common notion that *the size of the background does not matter for as long as it is constant*, is not really correct. The only way background can be suppressed is by improving the experimental setup. For example, proper shielding can suppress background considerably.

Problems

1. Show that for large number of data points, the maximum likelihood function approaches the Gaussian distribution function.

2. Using maximum likelihood method show that the number of measurements needed to get a particular value of the parameter ξ with a certain accuracy can be obtained by

$$ N = \frac{1}{(\triangle\xi)^2} \int \frac{1}{L} \left(\frac{\partial L}{\partial \xi} \right)^2 dx, $$

where L is the likelihood function. The accuracy of the measurement is characterized by $\triangle\xi$. (Hint: Start by writing equation 9.3.49 for N number of measurements.)

3. In a radioactive decay experiment for measuring activity of a long lived sample, the following counts are observed:
   2012, 2154, 1993, 2009, 2028, 2129

   ▶ What would be your Maximum Likelihood function? (Hint: radioactive decay can be characterized as a Poisson process.)

   ▶ Estimate the value of the most probable activity (counts) using the maximum likelihood method.

4. For the function of the previous exercise, estimate how many measurements must be performed to get a value of $\mu = 0.5$ with 1% accuracy.

5. Given the following ADC counts observed in a radioactive decay counting experiment, perform the Student's t test to determine if the means of the two datasets differ significantly from each other.

   Measurement-1: 200, 220, 189, 204, 199, 201, 217
   Measurement-2: 190, 230, 179, 188, 218

6. In a research paper the following experimental results of two separate measurements of a quantity are given.

$$ 112.56 \ \pm \ 0.78 $$
$$ 104.23 \ \pm \ 0.34 $$

   If the number of data points for each measurement was 12, determine at 95% confidence level if the two means are significantly different from each other.

7. In two different experiments, the half life of a radioactive sample is found to be 15.5 ± 2.3 days and 16.2 ± 1.5 days. Determine the best estimate of the half life by combining the two results.

8. In a radiation hardness study of a silicon detector, the system is exposed to a constant flux of radiation over a long period of time. The damage caused by the integrated flux is measured by noting the leakage current at regular intervals of time. The following data are obtained (units are arbitrary)

Integrated flux: 0, 100, 200, 300, 400, 500, 600
Leakage current: 453, 638, 773, 958, 980, 1120, 1334

Determine the degree of correlation between the two datasets.

9. Suppose you want to fit an exponential function of the form $a\exp(-bx)$ to a dataset. Determine the least squares fit equations that you will need to solve to determine the coefficients a and b.

Bibliography

[1] Anderson, T.W., Finn, J.D., The New Statistical Analysis of Data, Springer, 1996.

[2] Barlow, R.J., Statistics : A Guide to the Use of Statistical Methods in the Physical Sciences, John Wiley & Sons, 1993.

[3] Bevington, P. et al., Data Reduction and Error Analysis for the Physical Sciences, McGraw-Hill, 2002.

[4] Caria, M., Measurement Analysis: An Introduction to the Statistical Analysis of Laboratory Data in Physics, Chemistry and the Life Sciences, World Scientific Publishing Company, 2001.

[6] Cowan, G., Data Analysis: Statistical and Computational Methods for Scientists and Engineers, Springer, 1998.

[6] Cowan, G., Statistical Data Analysis, Oxford University Press, 1998.

[7] Dey, I., Qualitative Data Analysis; A User-Friendly Guide, Routledge, 1993.

[8] Everitt, B.S., Dunn, G., Applied Multivariate Data Analysis, Hodder Arnold, 2001.

[9] Frodesen, A.G., Probability and Statistics in Particle Physics, Oxford Univ. Press, 1979.

[10] Fruehwirth, R. et al., Data Analysis Techniques for High-Energy Physics, Cambridge University Press, 2000.

[11] Gelman, A., Bayesian Data Analysis, Chapman & Hall/CRC, 2003.

[12] Jobson, J.D., Applied Multivariate Data Analysis: Volume 1: Regression and Experimental Design, Springer, 1999.

[13] Jobson, J.D., Applied Multivariate Data Analysis: Volume II: Categorical and Multivariate Methods, Springer, 1994.

[14] Lichten, W., Data and Error Analysis in the Introductory Physics Laboratory, Prentice Hall College Div., 1988.

[15] Montgomery, D.C., Runger, G.C., Applied Statistics and Probability for Engineers, John Wiley & Sons, 1998.

[16] Ramsey, F., Schafer, D., The Statistical Sleuth: A Course in Methods of Data Analysis, Duxbury Press, 2001.

[17] Roe, B.P., Probability and Statistics in Experimental Physics, Springer, 1992.

[18] Rice, J.A., Mathematical Statistics and Data Analysis, Duxbury Press, 1995.

[19] Sigworth, F.J., A **Maximum-Likelihood Approach to Single-Particle Image Refinement**, J. Struct. Bio., 122, 1998.

[20] Smith, D.L., **Probability, Statistics, and Data Uncertainties in Nuclear Science and Technology**, Amer. Nuclear Society, 1991.

Chapter 10

Software for Data Analysis

With the advent of electronic detectors and fast data acquisition systems, the amount of data available for analysis has grown tremendously. A natural consequence of these developments has been the application of sophisticated analysis techniques to the datasets. However this would not had been possible if there were not an exponential increase in the availability of affordable computing power and memory. Today's computers perform millions of floating point operations in a flash, something that was unthinkable before computers became available.

In earlier days of personal computing, the trend was to develop data analysis packages from scratch and according to the particular needs. Most of these packages were developed in the programming language *fortran*, which was and still is a very powerful language. The power of fortran comes from its compiler that has been specifically designed for scientific computing. Later on, as more and more program routines were made available by the community to the general user, the trend shifted towards implementing these small program snippets to develop the required software. This trend is still continuing, though with a bit of twist now. That is, now the developers use C++ *classes* instead of fortran *subroutines*. These classes are *reusable* programs, which can be modified by the programmer according to the specific needs without re-coding. This powerful concept is the heart of the *object-oriented* programming languages.

 ## 10.1   Standard Analysis Packages

In this section we will visit some of the commonly used data analysis package that are available to the general user. Before we start, one thing that should be pointed out is that the description of these software packages are not meant to be complete in any sense and no attempt has been made to extensively test or compare the packages. Each package has its own advantages and disadvantages and therefore we do not recommend any one over the others. To make an informed decision, the reader should go through the manuals of these packages as available on their respective web sites and elsewhere.

10.1.A   ROOT

ROOT is a very powerful data analysis and visualization package. It was originally developed for the high energy physics community by Rene Brun and Fons Rademakers at CERN in mid the 1990's. An important part of root is its C++ interpretor called CINT, developed independently by Masa Goto in Japan. CINT has been integrated into ROOT for its command line and script processing.

ROOT is a freely available software and is open source, meaning that its source code is also available for novel users to make modifications. This approach has enabled the users in the scientific community to modify the segments of the program independently and then make the modified code available to the general users. Hence from its inception, the software can be said to be in continuous development.

Although ROOT was originally developed to tackle analysis tasks encountered in high energy physics, however it is in no way limited to this field and is being successfully used in a variety of disciplines.

ROOT has an object oriented framework. This means that the utilities that are commonly needed for analysis have already been provided and the user can simply use them to develop a custom analysis code. One of the main advantages of this approach is that it shrinks considerably the amount of code that must be written. Other advantages include *customization* of the utilities and their repeated usage. Of course ROOT has inherited these aspects from the object oriented nature of C++.

A.1   Availability

ROOT can be downloaded free of charge from its dedicated server: http://root.cern.ch. At the time of writing this book, the developers claim it to have been successfully tested on the following platforms.

- ► Intel x86 Linux (g++, egcs and KAI/KCC)

- ► Intel Itanium Linux (g++)

- ► AMD x86 64 (Athlon64 and Opteron) (FC gcc)

- ► HP HP-UX 10.x (HP CC and aCC, egcs1.2 C++ compilers)

- ► IBM AI X 4.1 (x1.c compiler and egcs1.2)

- ► Sun Solaris for SPARC (SUN C++ compiler and egcs)

- ► Sun Solaris for x86 (SUN C++ compiler)

- ► Sun Solaris for x86 (KAI/KCC)

- ► Compaq Alpha OSF1 (egcs1.2 and DEC/CXX)

- ► Compaq Alpha Linux (egcs1.2)

- ► SGI Irix (g++, KAI/KCC and SGI C++ compiler)

- ► Windows NT and Windows95 (Visual C++ compiler)

- ► Mac Mk Linux and Linux PPC (g++)

- ► Hitachi HI-UX (egcs)

- ► Mac OS (Code Warrior, no graphics)

The installation of ROOT depends on the platform and therefore the user should read through the installation instructions available on the internet site mentioned above.

In the following sections we will briefly discuss the basic structure of ROOT and introduce the reader to its syntax using some examples. For further information the interested reader is encouraged to look at different manuals available on ROOT's internet site. Although all the examples that will follow have been tested only on an Intel x86 machine running under Fedora Core 1, the codes should work on any platform.

A.2   Data Handling, Organization, and Storage

ROOT can handle data in various formats. In fact, since it is based on C++, one can tailor the analysis routine to import/export data in any format.

For organizing and storage, ROOT authors have developed a specific method called *trees*. The tree structure is designed to optimize the access and storage of data. A tree consists of branches, buffers, and leaves. The structure is stored in a special binary format file with an extension *.root*. The ROOT file is composed of directories containing objects in several levels. Therefore it looks like a standard UNIX file directory. The best way to browse a ROOT file is to use the *TBrowser* method. This can be done by creating a *TBrowser* object on ROOT's command line by typing *TBrowser b;*, which creates a window similar to the one shown in Fig.10.1.1.

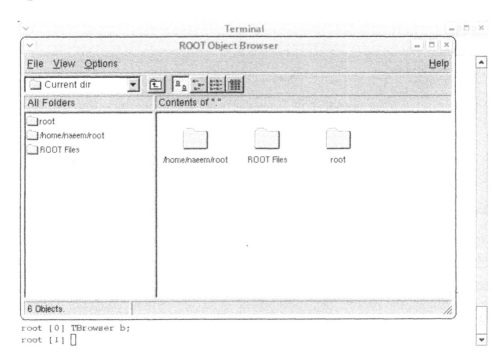

Figure 10.1.1:   TBrowser window of ROOT.

As the name suggests, the TBrowser window is a platform where the whole ROOT file can be browsed. The browser shows not only the current directory and the root directory but also a directory called *ROOT Files*, where the ROOT file being browsed exists. Note that ROOT does not create any new physical directory, rather loads the file to a location in the TBrowser's memory (see Fig.10.1.2). When one clicks twice on a ROOT file in any physical directory, the file is loaded into the ROOT Files location from where it can be browsed.

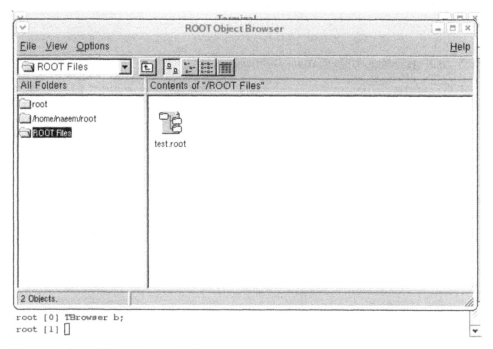

Figure 10.1.2:   When a ROOT file is clicked twice in any directory, the TBrowser loads it into the ROOT Files location.

Clicking twice the ROOT file in the ROOT Files location expands it into objects as shown in Fig.10.1.3.

The individual objects can then be clicked twice for examination. For example, Fig.10.1.4 shows the histogram that was obtained by double clicking the first icon in Fig.10.1.3. In the example shown the ROOT file contains only a few histograms. However the design of ROOT allows it to contain any allowed object. It can even contain *trees* as mentioned earlier.

The tree structure is a generalization of the *ntuple* structure, which contains only floating point numbers. A tree, on the other hand, may consist of many objects, such as histograms and arrays. The advantage of saving the data and analyses in a tree structure is that it makes further access and processing of the data very fast. The faster speed is a consequence of the way tree structure has been designed. It has a hierarchical structure consisting of leafs, branches, and buffers. These objects can be accessed independently of other similar objects in a tree and therefore reprocessing does not require the whole data to be accessed.

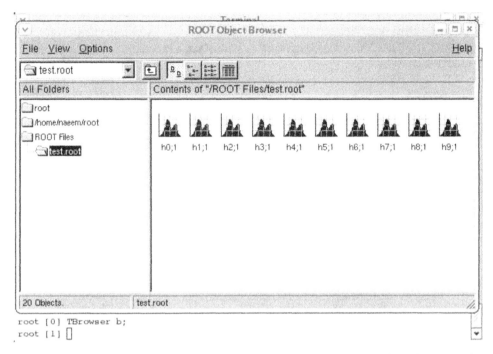

Figure 10.1.3: When a ROOT file in ROOT Files location is clicked twice, the TBrowser expands it into its constituents. The figure here shows a ROOT file consisting of a number of histogram objects.

A.3 Data Analysis Capabilities

Since ROOT has been specifically designed to address the needs of high energy physicists, therefore its libraries are geared towards the respective analyses and methods. However this does not in any way limits the utility of ROOT in other fields. The reason is that most of the methods adapted in high energy physics analyses can be applied to other fields as well. An obvious example is curve fitting, which is often required to model experimental data.

The main difference between ROOT and other analysis software is that ROOT provides the user with a framework on which a complete custom analysis package can be designed. This feature comes handy when designing a large scale analysis package that is intended to handle very large datasets. That is why ROOT is generally preferred in high energy physics experiments where immense amounts of data have to be analyzed.

At present ROOT has almost all the commonly used statistical analysis methods built into it. This includes correlation and curve fitting routines as well. ROOT also has a very efficient matrix manipulation library.

A.4 Graphics Capabilities

ROOT has very strong data visualization capabilities. Different types of one and two dimensional graphing functions are available, which produce publication quality

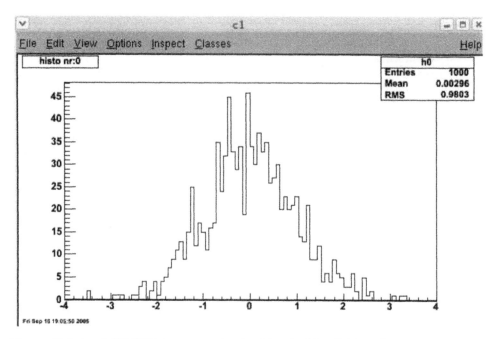

Figure 10.1.4: A ROOT histogram obtained by double clicking the respective icon in the *test.root* file.

pictures. The figures can be manipulated either through commands or through a graphical interface (see Fig.10.1.5).

The plotting methods available in ROOT have all the functionalities that are normally required. For example, one can draw error bars, choose type of data point display, manipulate labels and tiles, and save the figure in different formats.

One and two dimensional histograms can be filled, drawn, and displayed in ROOT with only a few lines of code. The histogram statistics can also be drawn on the same canvas. Modifications of the histogram are also permitted after it has been drawn.

A.5   Using ROOT

ROOT can be used in the following three different ways.

▶ **Interactive Mode:** In this mode the user invokes the CINT to interactively issue commands. If one types **root** at the command prompt, the CINT gets invoked with a screen dump similar to the one shown in Fig.10.1.6. At this point CINT is ready to invoke standard C++ or ROOT routines.

Running in interactive mode, however, does not mean that the user must issue all the commands successively one at a time. One also has the option to create a *macro* and then execute it from the command line. A macro is simply a collection of all the commands in the order in which they are supposed to be executed. A macro can be executed from command line by typing *root[ ] .x macrofile.C;*. Note that a macro does not have a *main function*.

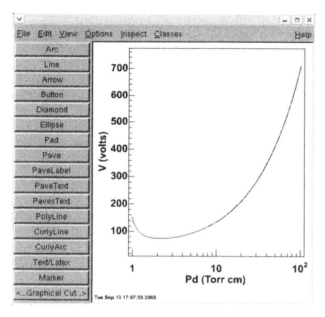

Figure 10.1.5: Figure editor of ROOT can be used to manipulate the figures, such as adding texts or graphics.

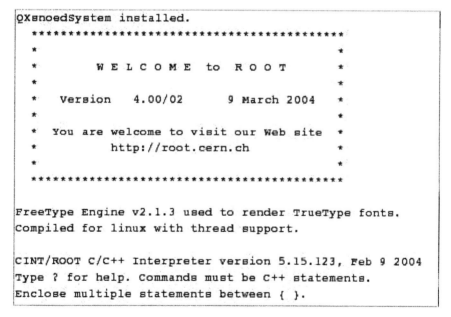

Figure 10.1.6: Typical screen dump at startup of ROOT's command line interpretor CINT.

▶ **Executable Mode:** A ROOT macro with all the required include files and a *main function* can be compiled and linked using a *Makefile*. The executable can then be run independent of the ROOT package.

▶ **Interface Mode:** This is perhaps the most common use of ROOT since in this mode the classes of ROOT are used to develop a custom analysis package according to the particular needs.

Since C++ is the programming language for ROOT therefore all the commands must be C++ commands. For some, this is a disadvantage since it requires getting familiar with C++. But for those, who already know C++, using ROOT is a simple matter. One thing that should be pointed out here is that for small to moderate analysis tasks, one only needs to know the basic C++ methods and therefore one can essentially start working with ROOT while learning C++.

A.6 Examples

Now that we have some idea of what ROOT is and how it can be used, let us have a look at some practical examples of its usage.

```
root [0] cout << "An alpha particle is simply a helium nucleus.\n";
An alpha particle is simply a helium nucleus.
root [1] printf("A proton is made up of three quarks.\n");
A proton is made up of three quarks.
root [2] for (int i=0;i<5;i++){printf("Learning ROOT is fun!\n");};
Learning ROOT is fun!
Learning ROOT is fun!
Learning ROOT is fun!
Learning ROOT is fun!
Learning ROOT is fun!
root [3] []
```

Figure 10.1.7: Screen dump of ROOT output to three text display commands.

We will start with the most basic of C and C++ commands in the interactive mode of ROOT. We want to see if CINT accepts standard output commands of C and C++. Fig.10.1.7 shows the screen dump of this exercise. It is apparent that the CINT syntax is exactly the same as that of C and C++. The first command is a C++ output stream command, which by default sends the output to the screen display. The second is a C command to display a text. In the third, a loop has been introduced that prints out a sentence five times on the screen.

As mentioned earlier, it is also possible to write all the commands in a file and then run it as a macro. To show how this is done, we will write all the above commands in a file named *example1.C* (see Fig.10.1.8) and then run it from ROOT. The macro can be executed with the command .x example1.C as shown in Fig.10.1.9.

Let us now go to a a little more involved example. Suppose we have some experimental data and we want not only to plot it but also to perform a regression fit. We will first write a macro to plot the data. For simplicity we will hardwire the data into the macro, which in fact is not a good practice. Usually one would read the data from a file using standard C or C++ commands. The macro together with comments explaining different commands is shown in Fig.10.1.10.

```
// example1.C
{
  cout << "An alpha particle is simply a helium nucleus.\n";
  printf ("A proton is made up of three quarks.\n");
  for (int i=0;i<5;i++){
    printf("Learning ROOT is fun!\n");
  }
}
```

Figure 10.1.8: Macro to run from the command line interface of ROOT.

```
root [0] .x example1.C
An alpha particle is simply a helium nucleus.
A proton is made up of three quarks.
Learning ROOT is fun!
Learning ROOT is fun!
Learning ROOT is fun!
Learning ROOT is fun!
Learning ROOT is fun!
root [1] []
```

Figure 10.1.9: Screen dump of ROOT output of the macro *example1.C* executed at the command line.

To run this macro we type *.x graph.C;* at the commandline interface of ROOT. This produces a graph of y versus x as shown in Fig.10.1.11. Our next step is to fit a regression line to the data.

As mentioned earlier, ROOT provides a number of libraries to fit standard and user defined functions to data. Here, since the data show an obvious linear trend, therefore we will try to fit a straight line. The easiest way to do this is to use ROOT's fit panel, which is a graphical user interface specifically designed for fitting. It can be invoked by right clicking on any data point in the graph of Fig.10.1.11 and then selecting the *FitPanel* method (see Fig.10.1.12).

Now, we want to use this panel to fit a straight line to our graph. Since we know that a straight line is a first order polynomial, we will select *pol1* from the panel. Subsequent clicking on *Fit* produces a screen dump on the command line interface as shown in Fig.10.1.13. Since we did not select the *"Do not draw function"* option from the fit panel, the corresponding fit line will also be drawn on the graph (see Fig.10.1.14). Fit panel provides the user with built in options of fitting an up to 9th order polynomial. Other built in functions include Guassian, Landau, and exponential. However one is not restricted to these functions only, since essentially

```
{
  /* File graph.C to produce a two dimensional graph */

  /***** Create a canvas to plot the data ******/
  TCanvas *c1 = new TCanvas("c1","Correlation Coefficient",0,0,400,300);

  int n = 15; // Number of data points

  /*** Create arrays containing data ***/
  Double_t x[n]={102,154,200,220,267,263,352,361,423,449,512,598,601,701,711};
  Double_t y[n]={28,98,132,98,129,202,265,243,291,324,376,412,524,511,560};

  /****** Create and custamize a TGraph object ******/
  tgr1 = new TGraph(n, x, y);
  tgr1->SetTitle("");
  TAxis *xaxis = tgr1.GetXaxis();
  TAxis *yaxis = tgr1.GetYaxis();
  xaxis->SetTitle("x");
  xaxis->SetTitleSize(0.05);
  xaxis->SetTitleOffset(1.1);
  xaxis->CenterTitle(kTRUE);
  xaxis->SetLabelSize(0.05);
  yaxis->SetTitle("y");
  yaxis->SetTitleSize(0.05);[]
  yaxis->SetTitleOffset(1.0);
  yaxis->CenterTitle(kTRUE);
  yaxis->SetLabelSize(0.05);

  /****** Draw graph ******/
  tgr1->Draw("A*");
}
```

Figure 10.1.10: Macro *graph.C* to produce a two dimensional graph.

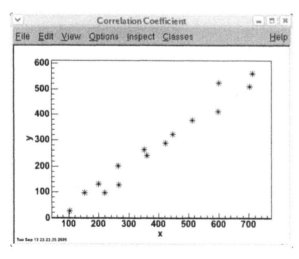

Figure 10.1.11: Result of running the macro *graph.C*.

any function can be fitted by clicking on the *user* tab on the fit panel. What this tab does is that it uses a function named *user* to perform the fit. This means that, before it could be used, it must have already been defined and loaded. The easiest way to do this is to write a short macro containing the definition of this function

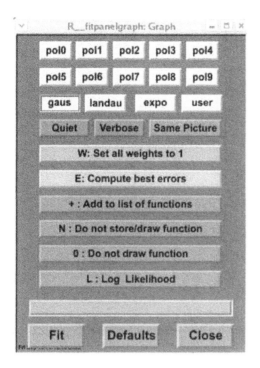

Figure 10.1.12: Fit panel of ROOT.

and run it before calling the *FitPanel*. The disadvantage with this scheme is that modifying the function would mean changing it in the macro and running it again.

```
MIGRAD MINIMIZATION HAS CONVERGED.
FCN=12969.9 FROM MIGRAD     STATUS=CONVERGED      24 CALLS           25 TOTAL
                    EDM=7.03718e-23     STRATEGY= 1   ERROR MATRIX UNCERTAINTY
1.9 per cent
  EXT PARAMETER                                   STEP         FIRST
  NO.   NAME       VALUE           ERROR          SIZE         DERIVATIVE
   1  p0          -5.39606e+01    5.72688e-01   -3.02111e-11  -6.33419e-11
   2  p1           8.45859e-01    1.39668e-03    7.57720e-14  -2.77338e-08
```

Figure 10.1.13: Fitting parameters produced by ROOT.

It is not absolutely necessary to use the fit panel for fitting. A few lines of code in a macro can also produce the same results. In fact performing a fit from a macro has the advantage of more flexibility in terms of choosing the fitting function. As we just saw, using a user defined function from the fit panel means defining it in a macro and running it. A better scheme would therefore be to simply use the fitting function in the main code and perform the fitting over there. Fit panel is most useful if one intends to fit the already defined functions.

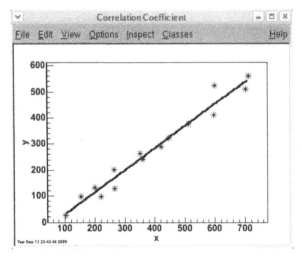

Figure 10.1.14: Straight line fit of the data.

10.1.B Origin®

Origin®[1] is an easy-to-use and a very versatile data analysis package developed by OriginLab® Corporation. One of its attractive features is its ability to produce production and presentation quality graphics. But perhaps its strongest point is its use of well tested numerical algorithms from Numerical Algorithms Group (NAG). These algorithms and their C libraries have been independently developed and tested by the NAG and are known to produce dependable results. Origin® provides access to these libraries from not only the simple menu driven point-and-click commands but also through its own C language called Origin® C. This slightly modified form of ANSI C language includes some features of C++ as well. Its compiler has been built to fully recognize ANSI C and therefore a user can choose to write code in simple C and still be able to call NAG routines.

Following are some examples of the types of analysis performed through NAG data analysis routines in Origin®.

▶ Descriptive statistics

▶ Correlation analysis

▶ Regression analysis

▶ Interpolation

▶ Curve fitting

▶ Surface fitting

▶ Nonparametric statistics

▶ Fourier transformation

---

[1]Origin® is a registered trademark of OriginLab Corporation.

Most of the tasks in Origin® can be performed through its menu driven commands. However to develop a complete customized data analysis package one can choose to use its programming environment. A typical screen shot of Origin® is shown in Fig.10.1.15. Note that here we have converted the image into gray scale from its original multicolor version.

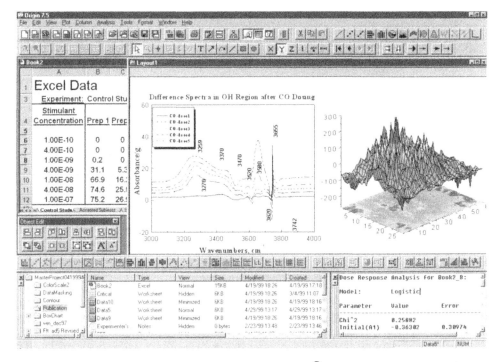

Figure 10.1.15: A typical screen shot of Origin® (courtesy of OriginLab)

B.1 Data Import Capabilities

In order to make Origin® compatible with most of the standard software packages, its authors have designed it such that it could handle very diverse data file types used by other standard softwares. Following are some of the data formats for which Origin® has built in import utilities.

► ASCII delimited text (DAT, CSV, TXT, etc.)

► Lotus

► Data Interchange Format

► Thermo Galactic SPC

► MATLAB®

► MiniTab

▶ LabTech

▶ SigmaPlot

▶ Sound (WAV)

▶ Mathematica

▶ Kaleidagraph

▶ pCLAMP (Axon binary and data files)

▶ Microsoft Excel spreadsheet

In addition to these formats, it can also access data from SQL[2] Server, MySQL[3], Microsoft Access and other databases using ODBC[4]. The user can also import data written in any other format as long as the details of the format are known. In this way Origin® can handle any ASCII or binary data type. The good thing about built in data import function is that the user can import data without worrying about the format.

The newer version of Origin® has a data import wizard that takes the user stepwise through the process. The user then has the option to select different options including the format of items to be imported. Several data filtration options are also available, which can save the user from performing these function later in the code.

B.2   Graphics Capabilities

The graphics capabilities of Origin® are one of the finest available in commercial data analysis packages. With Origin® one can generate essentially any type of two or three dimensional graph. The graphs can be generated either using Origin® C language or through menu driven commands. The graphs can also be edited using user interfaces.

B.3   Data Analysis Capabilities

Origin®'s data analysis libraries are based on NAG[5] routines, which are not only reliable but are also known for fast algorithms. This assures the user that the analysis is being performed by one of the best codes available.

Origin® provides the user with a very broad range of data analysis libraries and methods. Following are some examples of the types of analysis that can be performed in Origin®. This list is not complete and the interested reader should consult Origin® manual for an up to date list.

▶ Linear and Polynomial Regression

▶ Nonlinear Curve Fitting

---

[2]Structured Query Language or SQL is a computer language used for data handling from relational database management systems.

[3]MySQL is a multithreaded, multiuser SQL database management system.

[4]ODBC is an acronym for Open Database Connectivity. It provides a standard software application programming interface method for using database management systems.

[5]NAG stands for Numerical Algorithms Group. The group has released a large number of numerical computation libraries for C and C++.

- ▶ Signal Analysis

- ▶ Baseline and Peak Finding

- ▶ Curve smoothing

- ▶ Descriptive Statistics

- ▶ One and Two Way ANOVA

- ▶ t-Test

- ▶ Survival Analysis

B.4 Programming Environment

Origin® has a powerful development environment *Code Builder* to write, compile, link, run, and debug code in its specifically designed language *Origin® C*. Origin® C is fully compatible with ANSI C but also includes some features of C++. A strong base of NAG C libraries are also available to facilitate designing of custom analysis packages.

B.5 Examples

Let us now have a look at a few examples of Origin® usage. It should be pointed out that here we will be performing very simple analyses, which are not meant to analyze the effectiveness of Origin® as a data analysis package. We simply want that the user gets a glimpse of how one uses Origin® and how easy it is to use as an analysis platform.

Let us first assume that we have some data that we have gathered from a detector at equal intervals of time. First we want to quickly look at the data and have a qualitative idea of its time evolution. For this first we must import the data into Origin®'s spreadsheet. The newer version of Origin® has an *import wizard*, which can be used to import any type of data. Import wizard is a very user friendly interface that guides the user step wise through the data import process. At the end, it dumps the data in the chosen spreadsheet. We use this wizard to import the data and then plot it by first selecting the columns and then invoking the *Line* command from the *Plot* pull down menu. This produces the desired graph as shown on the left of Fig.10.1.16. Now, by looking at this time series we realize that there are some periodicity in the data. The best way to determine the periodicities is to perform a fast Fourier transform or FFT of the data. Performing FFT in Origin® is simply a matter of a few mouse clicks since a number of FFT algorithms are already built into its library. First we select the data columns and then from the *Analysis* pull down menu we invoke the *FFT* method. A pop up menu appears, where one could change the box method for FFT, choose between *Amplitude* and *Power*, and select some other functions. The result of the FFT appears in a new table that contains frequencies, real and imaginary components, phase, and amplitude or power. The amplitude or power density and phase are also automatically plotted in a publication quality template as shown on the right side of Fig.10.1.16. It is obvious that the frequencies in the data get clearly marked in the power density spectrum. This method of finding periodicities in the data is very common. It, however, has a

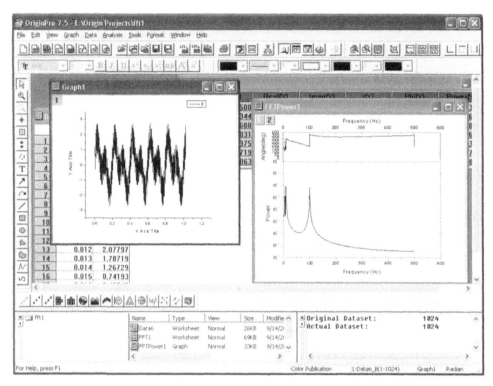

Figure 10.1.16: Plotting a time series and performing FFT in Origin®.

limitation that the data must be regularly spaced and must have been sampled at or above the Nyquist frequency (see chapter on data analysis and statistics). For irregularly spaced data one can use *Lomb* method instead of Fourier transform.

Let us now move to another example. What we now want is to fit a nonlinear function to some data. We follow the same steps as outlined in the previous example to import the data and to plot it. Then from the *Tools* pull down menu we select *Nonlinear fit wizard* method. A pop up window appears where one could choose the fitting function or define a new one. Origin® has a very large number of built in functions and one rarely needs to define another one. For our example, we chose to fit a 5th order polynomial to the data. The result is shown in Fig.10.1.17. The statistics related to the fit are also displayed along with the plot. The fitted curve itself is superimposed on the actual data plot, giving the user a qualitative way to judge the fit. Another very useful feature of fitting wizard is that it also shows a visual of the selected fitting function, which gives the user some idea of whether the function would be a good fit to the data or not.

All of the fitting functions available in the fitting wizard are also available in Origin® C. Hence, if one decides to write a code for data analysis, all these methods can be called by using built in Origin® C functions. This route is generally chosen for analysis of large datasets that have to be analyzed repeatedly through the same method. In such a case one is better off writing a code once and running it each time instead of invoking the menu driven functions time and again.

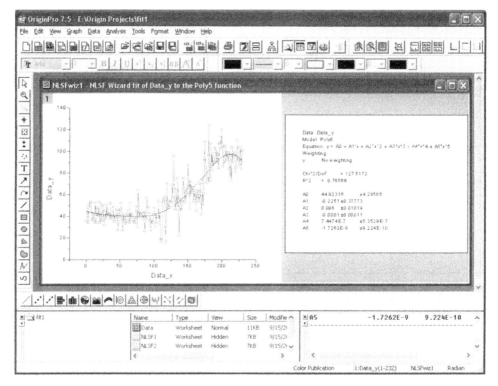

Figure 10.1.17:   Result of fitting a 5th order polynomial in Origin®.

We now go to another example. Suppose we have a dataset of which we want draw a histogram. Recall that a histogram is a plot of the frequencies at which the parameter assumes different values. A histogram is a very important tool since it characterizes the distribution function of the parameter under consideration. And if we know the distribution function of a parameter we can determine the expectation value of the parameter under any confidence interval (see chapter on data analysis). So, we have some data that we import using import wizard of Origin®. Then from the *Statistics* pull down menu we choose the *frequency* function. A pop up menu appear with minimum and maximum values of the parameter along with a bin size. All these values can be changed if needed. The result of the frequency count appears in a table, which can then be plotted by using the *Plot* pull down menu. Fig.10.1.18 shows the histogram plotted as a bar chart. As the second step to the process we tried to fit a Gaussian function to it using the fitting wizard. The result has been plotted on the same histogram.

10.1.C   MATLAB®

MATLAB®[6] is a very powerful high level computing language and interactive development environment developed by Mathworks. It is capable of handling extremely

---

[6]MATLAB® is a registered trademark of The MathWorks, Inc.

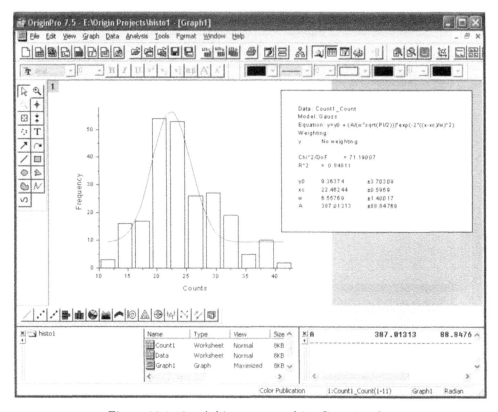

Figure 10.1.18:   A histogram and its Gaussian fit.

complicated data analysis and visualization tasks. The MATLAB® programming language is an easy-to-learn and easy-to-use high level language, which, together with its interactive environment, provides the user with the capability of integrating the code with other languages, such as C, C++, and Fortran. A typical screen shot of MATLAB®'s development environment is shown in Fig.10.1.19. It should be noted that here the original multicolor image has been transformed into gray scale.

C.1   Toolboxes

A very strong feature of MATLAB® is that it has a number of very powerful toolboxes to perform specific tasks. These toolboxes are add ons to the standard package and must therefore be installed before their respective functions could be used. Some of these toolboxes are listed below.

- ▶ Statistics toolbox

- ▶ Curve fitting toolbox

- ▶ Optimization toolbox

- ▶ Control system toolbox

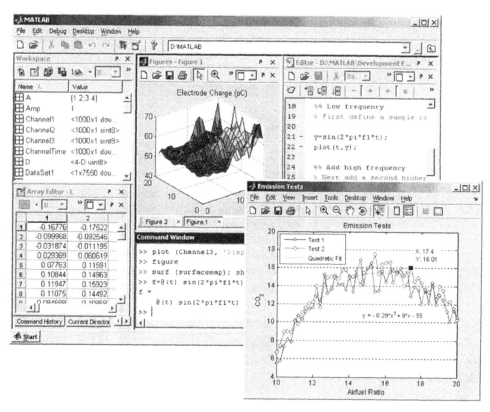

Figure 10.1.19:   A typical screen shot of MATLAB®'s interactive development environment (courtesy of Mathworks).

▶ Fuzzy logic toolbox

▶ Signal processing toolbox

▶ Filter design toolbox

▶ Image processing toolbox

▶ Image acquisition toolbox

▶ Data acquisition toolbox

▶ Instrument control toolbox

▶ Database toolbox

Of course here we have listed only a few of the full range of toolboxes available in MATLAB®. For the complete and up to date list the reader should refer to the web site of Mathworks. A quick look at this list reveals that MATLAB® is more than a simple data analysis package. Data analysis is merely a part of the extremely diverse tasks it can perform.

C.2 Data Acquisition and Import Capabilities

MATLAB® allows the user to import data from different sources in a variety of formats, such as

▶ ASCII text file,

▶ Microsoft Excel file,

▶ Image file in any standard graphics format,

▶ Sound file, and

▶ Video file.

The user can also access other types of binary files, files from web pages, and XML files through low-level functions available in MATLAB®.

MATLAB® also has a data acquisition tool box, which can be used to acquire data from hardware devices. Examples of such devices are computer's standard communication ports and custom designed computer interface cards. The online data acquisition coupled with real time analysis libraries and database support make MATLAB® a complete data acquisition, analysis, and storage system.

C.3 Data Analysis Capabilities

As mentioned before MATLAB® has a number of toolboxes to handle essentially any data analysis task. However the standard package also has most of the commonly used data analysis functions, such as

▶ Descriptive Statistics,

▶ Curve Fitting,

▶ Correlation,

▶ Filtering,

▶ Fourier transformation,

▶ Matrix analysis,

▶ Interpolation, and

▶ Smoothing.

C.4 Visualization Capabilities

MATLAB® has very strong and user friendly plotting libraries to produce publication level plots. It also has an interactive environment where users can modify and create graphics. This environment can be used to change the shape and properties of any object on the figure. One can also drag-and-drop new data sets onto an existing figure. Furthermore, operations like zooming in/out, rotating, and panning are also available.

Any type of two and three dimensional graph can be produced in MATLAB® through a few mouse clicks or by writing a few lines of code. The figures generated are of very high quality and ready for publication. The two dimensional plotting functions available in MATLAB® include the following.

► Scatter, line, bar, and pie charts

► Histograms

► Animations

► Polygons and surfaces

MATLAB® also has three dimensional plotting capabilities. For example, one can create surface, contour, or mesh plots from data and also create 3-D surfaces interactively.

C.5 Programming Environment

As stated earlier, MATLAB® is a high level programming language with very strong library support. There are very large number of mathematical and statistical functions available in MATLAB® that can be accessed through high level commands of its language. MATLAB® language is different from traditional programming languages in that it has a high level and very easy-to-use syntax. For example, the user does not have to worry about declaring variables or initializing them before use. A new user can learn the basic syntax very quickly and can even start writing the code in the first learning session.

C.6 Examples

Let us now have a look at a few examples of MATLAB® usage. What we want is to first generate a time series consisting of three sinusoidal frequencies and white noise. We then want to plot the series. After that we want to perform its Fourier transform and plot the power spectral density. Although we can perform these tasks on the command line interface of MATLAB® but it is always better to write all the commands in a macro and run it. The program written in the code editor of MATLAB® is shown in Fig.10.1.20. The different lines in the code are explained below.

► **Line 1-3:** Comments that are ignored by the compiler.

► **Line 5:** Generates the time axis from 0 to 5 seconds at steps of 0.001 second. This is equivalent to sampling the data at $1/0.001 = 1\ kHz$. Hence, according to the Nyquist theorem, we should be able to identify frequencies of up to 500 Hz in the dataset.

► **Line 6:** Generates the data (which could be ADC counts) consisting of four sinusoids with frequencies of 50 Hz, 100 Hz, 150 Hz, and 100 Hz. We have arbitrarily assigned weights to these sinusoids as well. The factor 300 is also arbitrary and is meant to keep the data values positive at all times.

► **Line 7:** Adds white (random) noise to the values generated in the previous step.

► **Line 8:** Creates first canvas for the figure. This is not needed if one intends to generate only one figure.

► **Line 9:** Plots the time series (see Fig.10.1.21).

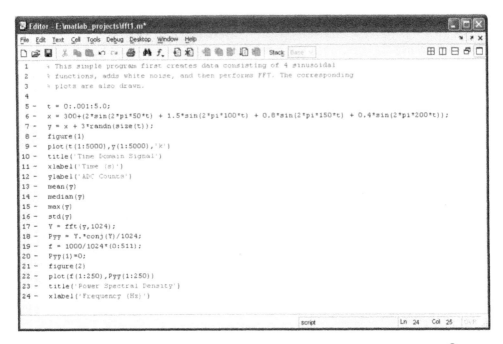

Figure 10.1.20:   A simple code written in the code editor of MATLAB®.

▶ **Line 10-12:** Set title and labels of the plot just generated.

▶ **Line 13-16:** Calculate mean, median, maximum value, and standard deviation of the data. The values are output on the standard output, which for MATLAB® is its command line interface. We did not need to compute these values for this example and they are meant to demonstrate this capability of MATLAB®.

▶ **Line 17:** Compute the FFT of the data.

▶ **Line 18:** Compute the power spectral density from the real and imaginary components computed in the previous step.

▶ **Line 19:** Normalize the frequency and select one half of the transform since the other half is the image copy of the first one.

▶ **Line 20:** Set the first value to be zero since it is simply the sum of all values.

▶ **Line 21:** Create another canvas to plot the second figure.

▶ **Line 22:** Plot the power spectral density (see Fig.10.1.22).

▶ **Line 23-24:** Set the title and x-axis label.

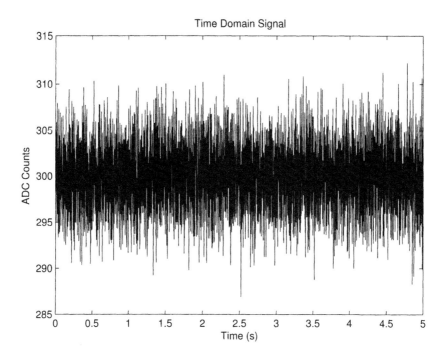

Figure 10.1.21:   A time series plot generated in MATLAB®.

10.2   Custom-Made Data Analysis Packages

Sometimes it is not feasible to use an existing software analysis package due to availability, cost, or customization difficulties. In such a situation one must develop a custom analysis code. This, in fact, is not as complicated a task as it sounds, thanks to the availability of a whole line of high level programming languages and graphics libraries. However, before we proceed any further, one point is worth noting. One should keep in mind that very large number of data handling and analysis routines have already been written and are available free of cost. There is no point in diluting one's efforts by reinventing the wheel. For example, if one wishes to incorporate an integration algorithm into the code, it is not necessary to write the code from scratch since there are a host of Fortran and C language library available that can be simply be added to the code. Now, this does not mean that writing a data analysis package means merely adding some available bits and pieces together. In fact, a whole package has many more components and not just the core analysis routines. In general, any data analysis package has the following four basic parts.

▶ Data import/export routines.

▶ Data analysis routines.

▶ Data visualization and storage routines.

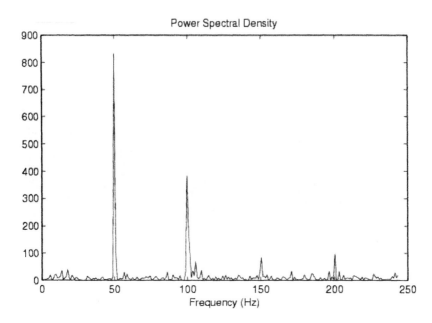

Figure 10.1.22: Power spectral density of the data shown in Fig.10.1.21.

► User interface.

Of course the most important amongst these are the data analysis routines, which depend on the particular methods and algorithms involved. Therefore during the development of any data analysis package most attention should be paid to development and implementation of the analysis methods, something that is unfortunately sometimes overlooked. As explained above, the smart thing to do is to incorporate the well tested routines into the code. But if the situation warrants design and development from scratch, then utmost care should be taken in not only designing the algorithms but also in implementing it.

In the following sections we will spend some time discussing the different parts of a typical data analysis package.

10.2.A Data Import/Export Routines

Such routines can be fairly easily developed in any modern high level programming language. Different languages provide the developer with different ways to handle input/output. For example, C++ has input/output streams that can import and export data in any format.

An important aspect of any data acquisition and storage system is the format of the data file it generates. Since the data analysis program has to read data from such files, one should be careful in choosing their format. The choice depends on many factors but perhaps the most important is the size of the dataset. For small data files, of the order of a few megabytes, one is better off writing data in delimited ASCII test format. The reason is that it becomes fairly easy for the user to visually inspect the entries in the datafile if needed. One can do this for a binary file as well but only with the help of a computer code or a special program. For very large data file, of the order of gigabytes or more, one should use binary format. The format of such a file depends on several factors, such as user accessibility and data handling routines. There are standard binary formats, such as *ZEBRA*, for which extensive read/write routines are available in different languages. Using such a standardized format is advantageous since it simplifies the task of the developer as well as the user.

A good thing about binary format is that its access is much more efficient and faster than the ASCII text format. The efficiency of reading ASCII data files decreases with file size and therefore for moderate size datasets one should weigh the pros and cons of saving the data in both formats before designing the package.

Coming back to the data analysis package, the developer should first understand the format of the data file and devise efficient and faster means of accessing the data. For example, extensive use of *rewinding* ASCII text files decreases the speed and therefore one is generally better off reading all the entries, even if they will be used in a later part in the code, and saving them in arrays.

10.2.B Data Analysis Routines

We have mentioned a few times before that it is always advantageous to use available data analysis routines that are based on tested algorithms. In some cases, however, one does not have an option but to resort to self design and development.

The prerequisite to writing data analysis routines is the algorithm. Algorithm is the heart of any data analysis system. If it has not been properly developed, the code can produce false results. The program in this case might not produce any errors, and if the values are not far off from the expectations, the user would never know that the results were actually wrong. Hence development and implementation of algorithm requires extreme care and attention to details.

The first two parts of algorithm development process are to understand the task and choose a method to handle it. For example, for a certain analysis package one might be interested in determining the correlation of two datasets belonging to two variables. Since there are different ways to determine the correlation therefore one should determine which method would suit best in that particular situation. In general there are different ways to solve a particular problem and the choice depends on several factors including

- ▶ the application,

- ▶ the required accuracy of the results,

- ▶ the available computing power, and

- ▶ the amount of available processing time.

After deciding on the method to solve the problem, the step by step computational tasks are defined in the so called *flow diagram*. A flow diagram is simply a graphical display of the computations that are to be performed. It helps the code developer in dividing the program into smaller segments, which makes the program not only more efficient but also helps to modify and debug it. Even though development of a flow diagram is a standard and very useful procedure but it is not universally practiced specially for smaller programs.

10.2.C   Code Generation

After the algorithm has been developed, the next step is to implement it in a programming language. At this point the developer has to decide on a particular language for coding the algorithm. Practically speaking, most of the times this decision is based on the personal liking and experience of the developer for a particular language. However, one should take into account at least the following factors while deciding on a language.

▶ Amount of computations

▶ Available computing power

▶ Display requirements

▶ Code re-usability requirement

▶ Code expandability requirement

The first two of these requirements are not of much significance due to the availability of today's computing power. It is arguable whether the languages specially designed for scientific computing, such as Fortran, are actually more efficient than other languages, such as C or C++, or not. Therefore the earlier trend of writing scientific codes in Fortran has now shifted to C++, which provides the user with more flexibility and better designing power as compared to Fortran. The codes written in C++ comprise of small segments called *Classes*, which can actually be reused. This strong feature of *object oriented* languages has revolutionized the program design and development.

10.2.D   Result Display

Displaying the results is perhaps the most difficult of the tasks as it requires special programming skills. The difficulty also comes from the fact that the display methods depend on the platform. One method that works well on an machine running under Microsoft Windows may not work at all on a Unix based system. This difficulty can be overcome by using a language that is platform independent, such as Java.

Bibliography

[1] Brun, R. et al., An Interactive Object Oriented Framework and its application to NA49 data analysis, Proc. of Computing in High Energy Physics, 1997.

[2] Gilat, A., MATLAB®: An Introduction with Applications, John Wiley & Sons, 2004.

[3] Hahn, B., Essential MATLAB® for Scientists and Engineers, Butterworth-Heinemann, 2002.

[4] Hanselman, D.C., Littlefield, B.L., Mastering MATLAB® 7, Prentice Hall, 2004.

[5] Noggle, J.H., Practical Curve Fitting and Data Analysis: Software and Self-Instruction for Scientists and Engineers, Prentice Hall, 1992.

[6] OriginLab, Origin Documentations, Available at the web site of OriginLab: http://www.originlab.com.

[7] Palm III, W.J., Palm, W., Introduction to MATLAB® 7 for Engineers, McGraw-Hill, 2003.

[8] Pratap, R., Getting Started with MATLAB® 7, Oxford University Press, 2005.

[9] ROOT Development Team, ROOT User's Guides, Available at ROOT's web site: http://root.cern.ch.

[10] Snow, J. et al., Use of ROOT in the D0 Online Event Monitoring System, Presented at CHEP2000, Padova, Italy, 2000.

[11] Weitzman, E., Miles, M.B., Computer Programs for Qualitative Data Analysis: A Software Sourcebook, SAGE Publications, 1995.

Chapter 11

Dosimetry and Radiation Protection

The extensive use of radiation in many fields has prompted the development of the field of radiation dosimetry. Originally the emphasis was on determining the integrated dose received by a person working in a radiation intensive environment with the intention to limit the exposure for personal safety. However, the discovery that radiation may also affect materials to the point that they become unusable, has stretched the applicability of this field to industrial and research purposes. We have already seen in the chapter on semiconductor detectors that radiation damage poses a major problem specially in high radiation fields. This has lead the researchers, such as physicists working at particle accelerators, to establish continuous dosimetry programs for their detectors. Other areas where dosimetry plays a central role are medical diagnostics and radiation therapy.

Most of the detectors used for dosimetry are based on the designs we have already discussed in earlier chapters. In this chapter we will look at the most commonly used dosimetry techniques.

11.1 Importance of Dosimetry

It is a universally accepted fact that radiation causes damage, which can range from a subtle cell mutation in a living organism to the bulk damage in a silicon detector. The type of damage depends mainly on the type and energy of radiation and the type of material. This damaging mechanism of radiation is sometimes exploited for the benefit of mankind. An obvious example is the radiation therapy of cancer, where cancer cells are targeted and destroyed by radiation.

Unfortunately the damage caused by radiation can not always be easily quantified. Some types of cell mutations caused by radiation take years to develop into detectable cancer. The same is true for electronic components that are in a hostile radiation field. The damage is so slow that often it is hard to notice the small degradation in performance. The question is, how do we then find out if a particular individual or equipment has received a high enough dose. The answer lies in the statistics. There have been extensive studies to determine the *safe* radiation levels for individuals, radiation workers, and equipments using statistical inferences of the damage data collected over a long period of time. Based on these studies, standards have been set for maximum allowable dose to humans. For materials, in most of the cases, such as silicon detectors in a particle accelerator, the issue is operational degradation and not safety and therefore no universally accepted standards exist.

11.1.A   Dose and Dose Rate

Radiation causes damage to the medium it passes through by depositing energy to the medium. This damage can be *acute* if a high level of dose is delivered in a short period of time. In such a case the dose delivered per unit time is so high that the material does not get enough time to repair the damage. In literature, this kind of dose is sometimes referred to as the *instantaneous dose*. However, this notion is somewhat misleading. The acute damage always depends on the time integrated energy deposited by the radiation. If the energy deposition time is shorter than the repair mechanisms, acute damage can occur. Hence the safe practice is to use *dose* and *dose rate* to refer to the integrated energy deposited and the energy deposited per unit time respectively. Note that *integrated* dose rate simply refers to dose.

Having warned the reader, in this book wherever the term high instantaneous dose is mentioned it will refer to high dose delivered in a short period of time. The following three cases give examples of the damage caused by various amounts of doses.

▶ Very high dose rate, such as 500 *rem/s* received due to a neutron burst near a reactor core.

▶ Low dose received over a long period of time. For example, 20 *mrem/s* received by a radiation worker for 5 days would exceed the 5000 *rem* limit set by most organizations.

▶ Very high dose rate received for an extended period of time. This serious scenario can lead to fatality.

The term *dosimetry* has traditionally been mostly used to refer to the measurement of dose. However modern dosimetry includes the measurement of dose rate as well. The mathematical definitions of these terms will be presented later in the chapter.

11.2   Quantities Related to Dosimetry

11.2.A   Radiation Exposure and Dose

Radiation is capable of causing damage to both living and nonliving things through different processes. It is natural to think that the severity of such a damage would depend on the amount of radiation absorbed by the material. This is true for the case of acute radiation effects. The severity of the stochastic effects, on the other hand, does not depend on the intensity of the radiation. One such effect is the mutation of cells, where even low levels of radiation can cause permanent damage. In this case the amount of radiation can only increase the probability of initiation of the damage, such as cancer, but not its severity. Many organizations keep track of the *fatal cancer risk factor* of their radiation workers by computing the integrated dose they have received during the time period. Apart from cancer, exposure to high radiation can also cause severe skin burns and tissue damage. In medicine, radiation dose calculations are routinely performed to minimize the harmful effects of radiation.

Because the amount of radiation is an important factor in quantifying the possible damage to materials, therefore the terms *radiation exposure* and *radiation dose* have been devised. It has been found that both living and nonliving things get affected by not only high levels of instantaneous doses but also by low levels of sustained doses. With the availability and use of high radiation environments, such as particle accelerators, the question of radiation damage to electronics circuitry and detectors is getting more and more attention. For example, a lot of work is being done to produce the so called radiation hard silicon detectors, which could be used for an extended period of time in high energy colliders. The present form of silicon detectors become unusable after being exposed to high radiation fields for some time (a couple of years) and must be replaced to continue the experiments. This is a major drawback of these otherwise highly sensitive and dependable detectors.

A.1   Roentgen (R)

Roentgen is a measure of exposure due to photons only (x-rays or gamma rays) and is based on the amount of ionizations they produce in air. 1 R equals the dose needed to ionize and produce 2.58×10^{-4} coulombs of positive and negative charges in one kilogram of air. Since it is not applicable to all tissues, for example, bones, and represents dose only due to photons, it is not used in dosimetry any more.

A.2   Absorbed Dose

The amount of energy deposited in a medium per unit mass of the medium by ionizing radiation is called *absorbed dose*. It is measured in units of J/kg. For absorbed dose the particular name of this unit has been chosen to be *gray* (Gy) with

$$1 \, Gy \; = \; 1 \, J/kg.$$

Apart from Gy, there is another unit of absorbed dose called *rad*. Even though *rad* has mostly been replaced by Gy it is still found in some modern literature. *Rad* was introduced in 1953 to replace Roentgen, which was the unit of exposure due to x-rays or γ-rays only. It is defined as the dose equivalent to the absorption of 0.01 joule of energy per kilogram of tissue.

A.3   Equivalent Dose

It is often quoted that absorbed dose is not capable of characterizing the biological effect of radiation. That fact of the matter is, absorbed dose is not capable of characterizing damage to any medium. All it tells us is how much energy has been absorbed by the medium and not what this deposited energy has done to the medium. In this respect absorbed dose treats all types of radiation equally. In other words, for the case of absorbed dose, there is no difference between a photon and an α-particle if they deposit the same amount of energy.

Hence when it comes to the effects of radiation, one can not use absorbed dose as the relevant quantity. Since dosimetry is primarily concerned with the safety of personnel, therefore a quantity called *equivalent dose* has been defined that characterizes the damaging effect of radiation on tissues. The basic idea is fairly simple: multiply the absorbed dose by a factor suited to the biological effectiveness of the particular type of radiation and the location of its source, that is

$$H_{T,R} = w_R \cdot D_{T,R}, \tag{11.2.1}$$

where $H_{T,R}$ is the equivalent dose due to radiation type R, $D_{T,R}$ is the mean absorbed dose delivered by radiation R, and w_R is the radiation weighting factor. The radiation weighting factor is given by

$$w_R = Q_R \cdot N_R, \tag{11.2.2}$$

where Q_R and N_R are the *quality* and *modified* factors for the radiation type R respectively. For external sources of radiation, N_R is taken to be unity. For internal sources, N_R is assigned a value by some relevant authority. In most cases a value of $N_R = 1$ can be taken and the knowledge of the quality factor is sufficient to calculate the weighting factor. The quality factors for different radiation types are listed in Table 11.2.1[1].

Note that the above formula is valid for only one type of radiation. In case of mixed field, the total equivalent dose can be obtained by simply summing the contribution due to individual types of radiation, that is

$$H_T = \sum_R w_R \cdot D_{T,R}. \tag{11.2.3}$$

Since the weighting factor is a dimensionless quantity, the equivalent dose is also measured in units of J/kg. However now the specific name given to the unit is sievert represented by the symbol Sv. There is also an older unit called *rad equivalent for man* or *rem*, which is still in limited use. The conversion from Sv to rm is fairly simple with

$$1 \, Sv \equiv 100 \, rem.$$

Example:
In a mixed radiation environment, a person receives 20 mGy of γ-ray dose and 2 mGy of slow neutron dose. Calculate the total equivalent dose received by the person.

Solution:
As the source is external, we can take $N_R = 1$ and the weighting factors for the two radiation types as given in Table 11.2.1 are

$$w_\gamma = 1 \quad \text{and} \quad w_n = 5.$$

---

[1]In table-11.2.1 as well as in earlier sections the particle energies have been given in *electron volt* units (keV, MeV) instead of the conventional SI units of *Joules*. This convenient unit of energy is defined as the energy attained by an electron when it is made to accelerate by an electric potential difference of 1 volt.

$$E = qV = (1.6 \times 10^{-19}C)(1\frac{J}{C})$$

This gives

$$1eV = 1.6 \times 10^{-19} J.$$

Table 11.2.1: Quality Factors for Different Types of Particles.

Type of Radiation		Quality Factor (Q_R)
x-rays, γ-rays		1
Electrons, Muons		1
α-Particles		20
Fission Fragments		20
Heavy Nuclei		20
Protons		5-10
Neutrons with E	$< 10\ keV$	5
	10-100 keV	10
	$>100\ keV$ up to 2 MeV	20
	2-20 MeV	10
	$> 20\ MeV$	5

The equivalent doses due to γ-rays and neutrons are given by

$$
\begin{aligned}
H_{T,\gamma} &= w_\gamma \cdot D_{T,\gamma} \\
&= (1)(20) \\
&= 20\ mSv \\
H_{T,n} &= w_n \cdot D_{T,n} \\
&= (5)(2) \\
&= 10\ mSv
\end{aligned}
$$

The total dose received by the person is then sum of these individual doses, that is

$$
\begin{aligned}
H_T &= H_{T,\gamma} + H_{T,n} \\
&= 20 + 10 \\
&= 30\ mSv
\end{aligned}
$$

A.4 Effective Dose

The equivalent dose as described above can be used for one tissue type only as it does not address the sensitiveness of tissue types to the same type of radiation. The question is, how we can determine the whole body equivalent dose to estimate the risk associated with a certain type of radiation environment. Or how one can estimate the whole body dose if the dose received by a particular organ is known. This is done by using the quantity *effective dose*, defined through the relation

$$E = w_T \cdot H_T, \tag{11.2.4}$$

where w_T is the tissue weighting factor and H_T is the equivalent dose in the tissue T. The rational behind weighting the equivalent dose with w_T is that, as explained earlier, each tissue or organ responds differently to radiation. As with equivalent dose, the effective dose is also measured in units of sievert. The w_T for different tissues and organs are given in Table 11.2.2. The reader should note that these numbers are based on the ICRP recommendations at the time of writing this book. For the most up to date factors the reader is encouraged to refer to the most recent ICRP publications.

Table 11.2.2: Tissue weighting factors w_T according to 1990 recommendations of ICRP(26).

Tissue or Organ	w_T
Gonads	0.20
Bone marrow (red), Colon, Lung, Stomach	0.12
Bladder, Breast, Liver, Oesophagus, Thyroid	0.05
Skin, Bone surface	0.01
Remainder	0.05

The effective dose as computed from the above relation is good for any single tissue or organ and only one type of radiation. In case of mixed radiation exposure to more than one tissues and organs, the total effective dose can simply by obtained by adding the respective contributions together, that is

$$E_{total} = \sum_i w_{T,i} H_{T,i}, \qquad (11.2.5)$$

where the subscript i refers to each organ. $H_{T,i}$ is the total equivalent dose for each organ i as calculated from equation 11.2.3.

Example:
During a CT scan of the stomach, that had to be repeated several times, a patient receives a total absorbed dose of 0.3 Gy. Compute the total effective dose received by the patient.

Solution:
Since CT scan is performed with x-rays therefore the radiation weighting factor $w_R = 1$. The equivalent dose received by the patient's stomach is

$$\begin{aligned} H_{T,R} &= w_R \cdot D_{T,R} \\ &= (1)(0.3) \\ &= 0.3 \; Sv. \end{aligned}$$

The tissue weighting factor for stomach is $w_T = 0.12$ as given in Table 11.2.2. The effective dose is then give by

$$
\begin{aligned}
E &= w_T \cdot H_{T,R} \\
&= (0.12)(0.3) \\
&= 0.036 \ Sv = 36 \ mSv.
\end{aligned}
$$

The usual effective dose received during a typical CT scan of abdomen is around 10 mSv, which means that this patient received more than three times the usual dose.

11.2.B Flux or Fluence Rate

We know that the ionizing power of radiation is directly related to the energy it carries. This can be understood by noting that in most gases the number of charge pairs produced per unit of absorbed energy is almost independent of the type of radiation (see chapter on gas filled detectors). In case of solids, although this independence is not guaranteed, still the ionization caused by radiation depends to a large extent to the energy it delivers. In dosimetry therefore one is interested in determining the amount of radiation carried by the radiation. *Energy flux* is a measure that can be used to characterize this quantity. Another quantity closely related to energy flux is the *particle flux*.

In order to define particle and energy fluxes, Let us first assume that we have a mono-energetic beam of particles incident on a material having a cross sectional area da. If we count the number of particles dN incident on this area in a time dt, then the particle flux can be calculated from

$$
\Phi_r = \frac{dN}{dtda}. \tag{11.2.6}
$$

This shows that the particle flux simply represents the number of particles incident on a surface per unit area per unit time. In the field of dosimetry this quantity is also known as *particle fluence rate*. Now, since we have a mono-energetic beam of particles therefore we can use the above relation to compute the energy flux as

$$
\Psi_r = \frac{E_p dN}{dtda}, \tag{11.2.7}
$$

where E_p is the energy carried by a single particle. The energy flux is a measure of the total energy incident on a surface per unit area per unit time. Another term used for this quantity, specially in dosimetry, is the *energy fluence rate*. Energy fluence rate is actually the preferred terminology adopted widely in the field of dosimetry. In fact, the usage of flux is discouraged in this field. It is evident that the energy fluence rate and particle fluence rate are directly related through the relation

$$
\Psi_r = E_p \Phi. \tag{11.2.8}
$$

Up until now we have considered the radiation beam to be absolutely mono-energetic, that is, all the particles had same energy. This is certainly not true since an absolute mono-energetic beam of particles is impossible to create. A realistic beam has an energy spectrum with a range that could be very narrow or very wide. If the energy spectrum is well defined, in most instances we can assign an average energy $\bar{E}_p$ to all the particles and still use the above formulas to compute the energy flux. However a better thing to do would be to compute the *energy flux spectrum* through the relation

$$d\Psi_r = \frac{E_{sp}dN}{dtda},\tag{11.2.9}$$

where $d\Psi$ represents the energy flux at the energy E_{sp}. A practical but very basic way to do this would be to increment the discriminator windows in a single channel analyzer in small steps and counting the particles for a certain time. This procedure when carried out for the whole range of particle energies would yield the energy flux spectrum. Of course a better way would be to use a multi channel analyzer, which generates the spectrum without the need to change the discriminator level repeatedly.

11.2.C   Integrated Flux or Fluence

Sometimes it is desired to compute the flux and fluence rates integrated over a certain time period. This could, for example, be required to determine the radiation dose received by a patient undergoing radiation therapy.

The flux integrated over a period of time is called *integrated flux* or *fluence*. Mathematically, it is given by

$$\Phi = \frac{dN}{da},\tag{11.2.10}$$

where, dN represents the number of particles passing through the area da. Since this relation does not explicitly contain time, it can also be interpreted to represent the number of particles incident on a surface area da *at any instant*. However this definition is somewhat misleading, since practically speaking, it is unmeasurable. We have repeatedly seen in the previous chapters that performing a counting experiment always involves time during which the counting is performed. This time can be made very small but its width is still limited by the timing resolution of the readout circuitry. It is therefore preferable to see the particle fluence as representing the particle flux integrated over a time period.

Just like the particle fluence we can also define the *energy fluence* or *integrated energy flux* as the amount of energy incident on a surface area within a certain time period. Mathematically speaking, this can be written as

$$\Psi = \frac{\bar{E}_p dN}{da},\tag{11.2.11}$$

where, as before, $\bar{E}_p$ represents the average particle energy.

Example:
1.5×10^4 photons having an average wavelength of 0.12 *nm* pass through a surface of area 1.8 *cm*^{2} per second. Determine the energy fluence rate and the integrated particle flux for 1 second of irradiation.

Solution:
The energy fluence rate for photons is given by

$$\Psi_r = \bar{E} \frac{dN}{dt\,da}$$
$$= \frac{hc}{\bar{\lambda}} \frac{dN}{dt\,da},$$

where $\bar{\lambda}$ is the average wavelength of the photons. Substituting the given values in this equation gives

$$\Psi_r = \frac{\left(6.63 \times 10^{-34}\right)\left(2.99 \times 10^8\right)}{0.12 \times 10^{-9}} \frac{1.5 \times 10^4}{(1)\left(1.8 \times 10^{-4}\right)}$$
$$= 1.37 \times 10^{-7} \; Jm^{-2}s^{-1}$$

The integrated flux can be computed as follows.

$$\Phi = \frac{dN}{da}$$
$$= \frac{1.5 \times 10^4}{1.8 \times 10^{-4}}$$
$$= 8.33 \times 10^7 \; m^{-2}$$
$$= 8.33 \times 10^3 \; cm^{-2}$$

11.2.D Exposure and Absorbed Dose - Mathematical Definitions

We have already defined the terms exposure and absorbed dose. Here we will look at their proper mathematical definitions.

Exposure X is defined as the charge dQ produced by heavy charged particles (ions) per unit mass dM of dry air when all the electrons liberated by the incident *photons* are completely stopped. Mathematically we can write this as

$$X = \frac{dQ}{dM}. \tag{11.2.12}$$

This definition of exposure has a fundamental problem that it is valid only for photons in dry air. Of course the intention for defining it in this way was to standardize the quantity and to make available a yardstick for comparison. However since the ways photons interact with target materials are very different from other particles, such as neutrons or α-particles, therefore exposure has not been found to be very useful in characterizing damage due to other types of radiation.

Example:
Estimate how much energy deposition of photons per unit mass of dry air is equivalent to 1 R.

Solution:
We know that
$$1\,R = 2.58 \times 10^{-4}\,C/kg.$$

To determine the number of charge pairs per kg corresponding to $1\,R$, we simply divide the right hand side of the above equivalence by the unit electrical charge, that is

$$
\begin{aligned}
N &= \frac{Q}{e} \\
&= \frac{2.58 \times 10^{-4}}{1.602 \times 10^{-19}} \\
&= 1.61 \times 10^{15} \text{ charge pairs per } kg. \qquad (11.2.13)
\end{aligned}
$$

The production of charge pairs is related directly to the energy deposited E and the energy needed to produce a charge pair, through the relation

$$N = \frac{E}{W}.$$

The W-value for most gases including air is approximately $34\ eV$. Using this and the value of N as calculated above we can determine the required energy equivalent to $1\,R$ as follows.

$$
\begin{aligned}
E &= NW \\
&= \left(1.61 \times 10^{15}\right)(34) \\
&= 5.47 \times 10^{16}\ eV\,kg^{-1} \\
&= 5.47 \times 10^{7}\ MeV\,g^{-1}
\end{aligned}
$$

In laboratory environments it is often desired to estimate the exposure expected from a known radioactive source. Intuitively thinking we can say that the exposure from a radioisotope is proportional to the following quantities.

▶ **Source Activity:** The more the activity the larger number of decays and hence higher exposure. Source activity is given by $\lambda_d N$ with λ_d and N being the sample's decay constant and the number of radioactive atoms in the sample respectively.

▶ **Inverse of Distance Squared:** Exposure decreases with increasing distance from the source. According to the inverse square law, the flux of radiation varies by inverse of the square of the distance from a point source. Since exposure is directly related to flux, it will also be inversely proportional to r^2, the distance between the source and the point of measurement. A point worth mentioning here is that a source can be considered a point source if the point of measurement is much larger than its mean radius.

▶ **Exposure Time:** Radiation exposure increases with increasing exposure time.

Combining all of the above, we can reach at the following relation for the exposure from a radiation source.

$$X \ \propto \ \frac{\lambda_d N t}{r^2}$$

$$= \ \Gamma \frac{\lambda_d N t}{r^2}, \tag{11.2.14}$$

where t is the exposure time and Γ is generally known as the *gamma constant*. Gamma constant is specific to the type of radioisotope and its values for most isotopes are available in literature (see Table 11.2.3). The usual quoted units of Γ are $Rcm^2/MBqh$ and $Rcm^2/mCih$.

Table 11.2.3: Gamma constants and predominant decay modes other than γ-decays of some radioisotopes (4). All values are given in $Rcm^2mCi^{-1}h^{-1}$.

Isotope (mode)	Γ	Isotope (mode)	Γ	Isotope (mode)	Γ
$^{227}_{89}Ac$ (β)	2.2	$^{198}_{79}Au$ (β)	2.3	$^{228}_{88}Ra$ (β)	5.1
$^{124}_{51}Sb$ (β)	9.8	$^{181}_{72}Hf$ (β)	3.1	$^{106}_{44}Ru$ (β)	1.7
$^{72}_{33}As$ (γ, e^+)	10.1	$^{124}_{53}I$ (EC)	7.2	$^{46}_{21}Sc$ $(\beta))$	10.9
$^{140}_{56}Ba$ (β)	12.4	$^{130}_{53}I$ (β)	12.2	$^{75}_{34}Se$ (EC)	2.0
$^{7}_{4}Be$ (EC)	0.3	$^{132}_{53}I$ (β)	11.8	$^{110}_{47}Ag$ (EC, β)	14.3
$^{47}_{20}Ca$ (β)	5.7	$^{192}_{77}Ir$ (β)	4.8	$^{22}_{11}Na$ (EC)	12.0
$^{11}_{6}C$ (EC)	5.9	$^{59}_{26}Fe$ (β)	6.4	$^{24}_{11}Na$ (β)	18.4
$^{134}_{55}Cs$ (EC, β)	8.7	$^{140}_{57}La$ (β)	11.3	$^{85}_{38}Sr$ (EC)	3.0
$^{137}_{55}Cs$ (β)	3.3	$^{28}_{12}Mg$ (β)	15.7	$^{182}_{73}Ta$ (β)	6.8
$^{38}_{17}Cl$ (β)	8.8	$^{52}_{25}Mn$ (EC)	18.6	$^{121}_{52}Te$ (EC)	3.3
$^{57}_{27}Co$ (EC)	0.9	$^{65}_{28}Ni$ (β)	3.1	$^{187}_{74}W$ (β)	3.0
$^{60}_{27}Co$ (β)	13.2	$^{95}_{41}Nb$ (β)	4.2	$^{88}_{39}Y$ (EC)	14.1
$^{154}_{63}Eu$ (EC, β)	6.2	$^{42}_{19}K$ (β)	1.4	$^{65}_{30}Zn$ (EC)	2.7
$^{72}_{31}Ga$ (β)	11.6	$^{226}_{88}Ra$ (α)	8.25	$^{95}_{40}Zr$ (β)	4.1

Example:
Calculate the exposure received in 2 hours at a distance of 3 cm from a 10 mCi cobalt-60 source.

Solution:
The gamma constant for cobalt-60 is 13.2 $Rcm^2mCi^{-1}h^{-1}$. To calculate the exposure we substitute this and the given values in equation 11.2.14.

$$X = \Gamma\frac{\lambda_d Nt}{r^2}$$

$$= 13.2\frac{(10)(2)}{3^2}$$

$$= 29.3\ R \tag{11.2.15}$$

Note that since $\lambda_d N$ corresponds to the activity of the material therefore we did not need individual values of λ_d and N.

Since exposure can not be used for particles other than photons therefore another quantity called *absorbed dose* has been introduced. It is defined as the average energy $d\bar{E}$ absorbed in an infinitesimal volume element dV per unit density ρ of the medium.

$$D = \frac{1}{\rho} \lim_{V \to 0} \frac{d\bar{E}}{dV} \tag{11.2.16}$$

Note that this definition has less parameters than the definition of exposure and therefore is much easier to measure and compare. The parameter that needs a bit of attention here is the volume element, which according to the definition should be *infinitesimally small.* The question is how small is infinitesimally small. The answer lies in the physical limit of signal to noise ratio set by the statistics of charge pair production. We know that the production of charge pairs is a statistical process with the standard deviation given by $\sqrt{N}$ for the production of N charge pairs. Now, the production of charge pairs is, among other things, dependent on the availability of molecules and therefore to the volume element itself. Hence the volume element should not be too small to cause high fluctuations in the charge pair production. Of course the fluctuations also depend on the energy flux of the radiation and therefore one can not generally say how big the volume element should be for all cases.

Whatever the volume is, one can in general express the absorbed dose in terms of energy absorbed per unit mass, that is

$$D = \frac{d\bar{E}}{dM}, \tag{11.2.17}$$

where $dM = V\rho$ is the mass of the sample. From this definition it is evident that the SI units for absorbed dose are J/kg. This is generally known as a *gray* and represented by the symbol Gy.

Radiation quantities and conversion factors relating to dosimetry are listed in Table 11.2.4.

Table 11.2.4: New and old units of quantities related to radiation dose and their conversion factors.

Old Unit		New Unit		Conversion
Name	Symbol	Name	Symbol	Factor
Radiation Absorbed Dose	rad	Gray	Gy	$1Gy = 100rad$
Rad Equivalent for Man	rem	Sievert	Sv	$1Sv = 100rem$
Curie	Ci	Bacquerel	Bq	$1Bq = 2.7 \times 10^{-11}Ci$

11.2.E   Kerma, Cema, and Terma

In this section we will describe three quantities, Kerma, Cema, and Terma, that are in common use in the field of dosimetry due to their effectiveness in characterizing the macroscopic effects of radiation.

E.1   Kerma

Kerma is an acronym of kinetic energy released in a medium per unit mass. It is defined as the total kinetic energy of all the charged particles liberated by uncharged particles per unit mass of the target material. Mathematically, it is written as the quotient of the charged particle's kinetic energy E_{kin} and the mass of the material dm, that is

$$K = \frac{dE_{kin}}{dm}. \tag{11.2.18}$$

Kerma is generally measured in the same units that are used for absorbed dose, that is, J/kg or $Gray$. Note that here dE_{kin} is the kinetic energy of the charges produces as a result of radiation interaction. This energy is not necessarily equal to the energy transferred by the incident radiation since some of the energy can also go into other processes, such as radiative and radiationless transitions. The other point to note here is that the mass element dm should be very small.

   The above definition of Kerma is actually a simplified form of the actual microscopic definition, which takes the form

$$K = \lim_{\triangle V \to 0} \frac{\triangle E_{kin}}{\rho \triangle V}. \tag{11.2.19}$$

A point worth noting here is that the condition of the volume element shrinking to zero ensures that the deposition of energy by radiative transfers gets ignored. In this way Kerma is actually an approximation of absorbed dose.

   Kerma is *not* independent of the type of the target material and therefore must always be defined with respect to the medium. For example, the energy released in air per unit mass of air should be written as *air Kerma*.

   The energy lost by the incident radiation has two main components: collisional loss and radiative loss. K can therefore be divided into K_{col} and K_{rad} corresponding

to collisional and radiative losses respectively. The total Kerma would then be equal
to their sum, that is

$$K = K_{col} + K_{rad}. \tag{11.2.20}$$

It is customary to denote the energy lost through radiative processes as a fraction of
the radiative to total Kerma. This fraction is generally known as *radiative fraction*
and is given by

$$g = \frac{K_{rad}}{K}. \tag{11.2.21}$$

Note that this ratio g is mainly a function of the energy loss through Bremsstrahlung,
which in most situations can be neglected, that is $g \approx 0$. The above equation implies
that the total Kerma can be evaluated from

$$K = \frac{K_{col}}{1 - g}. \tag{11.2.22}$$

Earlier in the chapter we introduced the term energy fluence, which has the dimensions of energy per unit area. If we multiply this quantity with the mass energy
transfer coefficient $\mu_{m,tr}$, the resultant will have the dimensions of Kerma. In fact,
this is how total Kerma is often computed, that is

$$K = \mu_{m,tr}\Psi, \tag{11.2.23}$$

where Ψ is the energy fluence. Note that here the use of mass energy transfer
coefficients ensures that we obtain total Kerma. If one wishes to evaluate only
collision Kerma, the relevant parameter to use would be mass absorption coefficient
$\mu_{m.col}$.

$$K_{col} = \mu_{m,col}\Psi \tag{11.2.24}$$

There also exists a direct relationship between air Kerma and the exposure we
introduced earlier in the chapter. The reader would recall that the exposure determines the amount of charge (either electrons or ions) produced per unit mass of air
while collision Kerma represents the kinetic energy released per unit mass due to
collisions. Now, since this energy goes into creation of charge pairs therefore the two
terms should be directly related. To derive this relationship we first note that the
total charge produced in air can be written as

$$dQ = eN = e\frac{dE_{col}}{W_{air}}, \tag{11.2.25}$$

where N is the total number of charge pairs created, dE_{col} is the collision energy
loss, and W_{air} is the energy needed to create an electron ion pair in air. Dividing
both sides of the above equation by the mass element dm, we get

$$\frac{dQ}{dm} = \frac{e}{W_{air}}\frac{dE_{col}}{dm} \tag{11.2.26}$$

$$\text{or} \quad X = \frac{e}{W_{air}}K_{col,air}, \tag{11.2.27}$$

which is our required relation between exposure and collision Kerma in air. Now,
since the factor $W_{air}/e \approx 33.85$ is constant for air, therefore we can also write

$$K_{col,air} \approx 33.85X. \tag{11.2.28}$$

This collision Kerma can also be used to determine the total air Kerma. To do this we combine equations 11.2.22 and 11.2.28 together and obtain

$$K_{air} = \frac{33.85X}{1 - g}. \tag{11.2.29}$$

In most instances one is interested in collision Kerma as opposed to total Kerma. The reason is that most of the photons produced as a result of radiative energy transfer escape the medium without depositing any energy. Such photons do not contribute to the dose and can therefore be ignored. However if the medium is large enough or if the distribution of energy transferred is very wide, some radiative losses may lead to absorption of energy.

Example:
In an earlier example, a 2 hour exposure at a distance of 3 cm from a 10 mCi cobalt-60 source was found to be 29.3 R. Compute the air Kerma if the quality is approximately given by 0.01.

Solution:
The air Kerma can be computed from equation 11.2.29. But first we need to convert the exposure in units of Ckg^{-1}. Using the conversion $1R = 2.58 \times 10^{-4}$ Ckg^{-1} we get

$$X = (29.3)(2.58 \times 10^{-4}) = 7.56 \times 10^{-3} \ Ckg^{-1}.$$

Substituting this and the given value of g into equation 11.2.29 gives us the required air Kerma.

$$
\begin{aligned}
K_{air} &= \frac{33.85X}{1 - g} \\
&= \frac{(33.85)\left(7.56 \times 10^{-3}\right)}{1 - 0.01} \\
&= 0.258 \ Jkg^{-1}.
\end{aligned}
$$

Another parameter commonly used to quantify the distribution of energy between the absorbed dose D and radiative losses is given by

$$\eta = \frac{D}{K_{col}}. \tag{11.2.30}$$

The value of η changes as the beam traverses into the medium as shown in Fig. 11.2.1. The three regions in the graph based on the value of η are described below.

▶ $\eta < 1$: This refers to the case when absorbed dose is less than the collisional energy loss of radiation. It occurs at short depths of the target. As the radiation penetrates the material, more and more collisional energy gets absorbed and contribute to the total dose.

▶ $\eta = 0$: At some depth the collisional loss becomes just equal to the absorbed dose. That is, all the energy lost by the radiation through collisional processes

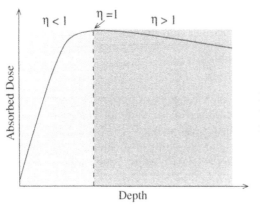

Figure 11.2.1: Plot of absorbed dose with respect to the depth of the target material.

contributes to the dose and all the radiative energy transfers result in photons escaping from the target material.

▶ $\eta > 1$: As the radiation penetrates deeper, the probability of radiative loss photons being absorbed inside the medium increases, mainly due to their broader energy spectrum. Hence the dose can actually become higher than the collision Kerma.

E.2   Cema

One limitation of Kerma is that it is defined only for uncharged particles, such as photons. For charged particles, another quantity called Cema has been introduced. It is analogous to Kerma in definition and is mathematically written as

$$C = \frac{dE_{con}}{dM},\tag{11.2.31}$$

where dE_{conv} is the energy lost by charged particles in a material of mass dM. The SI units of Cema are J/kg or $Gray$.

E.3   Terma

In chapter 2 we discussed different mechanisms of energy transfer when radiation passes through matter. We saw that the absorption of radiation may lead to several different types of excitations as well. Some of these excitations my not lead to the release of kinetic energy and therefore would not be included in the definition of Kerma. To overcome this shortcoming, another quantity called Terma has been defined. Terma is the acronym of total energy released in a medium per unit mass and can be evaluated from

$$T = \frac{dE_{total}}{dM}.\tag{11.2.32}$$

where dE_{total} is the total energy transferred by the incident radiation to the medium having mass dM. Just like Kerma and Cema, the usual units for Terma are also J/kg of $Gray$.

11.2.F Measuring Kerma and Exposure

We saw earlier that air Kerma is directly related to the exposure through the relation (see equations 11.2.27 and 11.2.29)

$$K_{air} = \frac{W_{air}}{e(1-g)} X. \qquad (11.2.33)$$

where g is a factor that characterizes the loss of energy through radiative processes. This equation implies that measuring Kerma is essentially equivalent to measuring exposure. This is one of the reasons that even though exposure is termed as an obsolete quantity, it is still widely mentioned in the literature and used in measurements.

Exposure and Kerma can be measured by different kinds of detectors including semiconductor devices. However ionization chambers have been, and still are, extensively used for the purpose. Later on in the chapter when we discuss the ionization chamber dosimetry we will look at two kinds of ionization chambers, namely *free-in-air* and *cavity*, which are most commonly used to determine exposure and related quantities.

11.2.G Cavity Theories

Absorbed dose is a very useful quantity in terms of determining the strength of radiation interactions in a material. For example, one might be interested in determining the dose received by a patient during radiation therapy to evaluate the effectiveness of the method. Measuring the dose, however, is not as easy as it may sound. The reason is that to measure dose one has to use a detector and every detector has a detection medium, which might not be the same as the surrounding material. Hence the dose measured from the instrument would be different from what it should be. For example, if one uses a carbon dioxide filled ionization chamber to measure the dose, the results will not be directly applicable to the air surrounding the chamber. This is where cavity theories come into play since they relate the dose measured by the detector to the dose in the surrounding medium. In the following we will discuss two of the most important cavity theories.

G.1 Bragg-Gray Cavity Theory

A Bragg-Gray cavity refers to such a small detection volume that it does not influence the particle fluence when used in a medium. Theoretically this would mean constructing a detector having a volume that shrinks to zero. Of course this is practically not possible but in most cases one can say that it holds up to a good approximation.

According to Bragg-Gray cavity theory, the ratio of the dose absorbed in a cavity (that is, a detector) to the dose absorbed in its surrounding medium D_{med} is given by

$$\frac{D_{med}}{D_{cav}} = \frac{\left(\bar{L}/\rho\right)_{med}}{\left(\bar{L}/\rho\right)_{cav}}, \qquad (11.2.34)$$

where the factor $\bar{L}/\rho$ represents the spectrum averaged unrestricted mass collision stopping powers of the cavity and the medium. The term unrestricted implies that

that the δ-electrons (that is, secondary electrons) are not considered at all. This, as we will shortly see, is one of the problems with this theory. The reader might recall that the term mass stopping power was introduced in chapter 2 as well. Although there we used a different notation but in effect both represent the same parameter, that is

$$\frac{\bar{L}}{\rho} \equiv \frac{1}{\rho}\frac{dE}{dx}. \tag{11.2.35}$$

Since the right side of equation 11.2.34 represents the ratio of stopping powers therefore it is conventionally represented by

$$\left(\frac{\bar{L}}{\rho}\right)_{cav}^{med} \equiv \frac{(\bar{L}/\rho)_{med}}{(\bar{L}/\rho)_{cav}}. \tag{11.2.36}$$

The unrestricted mass collisional stopping power can be evaluated fairly accurately for most materials and radiation types using energy spectrum that ignores the δ-electrons. Hence if we measure the dose using a detector, which to a good approximation can be considered a Bragg-Gray cavity, we can determine the absorbed dose expected in the medium under same conditions. Let us suppose we use a gas filled ionization chamber to determine the absorbed dose in another material, say water. We first need the ratio of the spectrum averaged mass collision stopping powers for water and gas. This can be evaluated from the known spectrum of electron fluence. The second quantity that we need is the dose measured by the ionization chamber. We know that an ionization chamber does not directly measure the dose but it can be used to determine the total charge produced by the radiation. If the total charge is dQ_{gas} then the dose is given by

$$
\begin{aligned}
D_{gas} &= \frac{dE_{gas}}{dm_{gas}} \\
&= \frac{dQ_{gas}}{dm_{gas}}\frac{W_{gas}}{e},
\end{aligned}
\tag{11.2.37}
$$

where dE_{gas} refers to the total energy deposited in the mass element dm_{gas} and W_{gas} is the energy needed to produce a charge pair. If the gas is air then $W_{air}/e \approx 33.9$ J/C and the above relation reduced so

$$D_{air} = 33.9\frac{dQ_{gas}}{dm_{gas}}. \tag{11.2.38}$$

At first sight it might seem that determination of the absorbed dose is fairly straightforward. Unfortunately the situation is not that simple. The main problem lies in the determination of mass of the gas. Since the electric field inside an ionization chamber is not uniform, the charge collection is not the same at all locations. The easiest way to account for these nonlinearities is to use an effective mass instead. This effective mass is calculated by calibrating the detector.

Bragg-Gray cavity theory has many limitations. For example, it assumes that the cavity is infinitesimally small, which certainly is not true. Another problem is that it assumes that the walls of the cavity are of the same material as the medium. Since most ionization chamber dosimeters are constructed with graphite walls therefore they do not quite meet this condition. Uniform irradiation of the cavity is another condition, which is not always possible to satisfy. However the good news is that even if one does not account for these anomalies the computed dose values are generally correct by more than 95% for well designed dosimeters. A well designed dosimeter usually refers to a chamber with walls made of a low Z material such as graphite. If the walls are made of high Z materials, the Bragg-Gray theory gives false results. The reason for this anomaly lies in the way the right hand side of equation 11.2.34 is computed. In Bragg-Gray theory this quantity, that is the stopping power ratio, is calculated by using the spectrum of the electron fluence, which has been computed in the continuous slowing down approximation. In this approximation the production of δ-electrons during the slowing down process of primary electrons is completely neglected. Now, if these δ-electrons are also taken into account then the total electron fluence is much larger, especially at the low end of the energy spectrum. This effect is more pronounced if the walls of the cavity are made of high Z materials, which favor the production of δ-electrons.

G.2 Spencer-Attix Cavity Theory

The main shortcoming of the Bragg-Cavity theory, the fact that it does not take into account the δ-electrons, was overcome by the Spencer-Attix Cavity theory. In this theory the δ-electrons are divided into two distinct groups based on the energy they possess. The first group contains the δ-electrons that are locally absorbed and thus do not contribute to the electron fluence. The other group represents the rest of the δ-electrons that have energy high enough to escape the local absorption. These electrons do contribute to the total electron fluence. The energy level used to distinguish between these two groups is traditionally represented by the symbol $\triangle$. Although there are different ways in which this energy can be defined, the original definition proposed by Spencer and Attix is still the most widely used. This definition is based on the idea that $\triangle$ should correspond to the energy needed by an electron to cross the cavity. This implies that $\triangle$ depends on the cavity dimensions and can therefore be easily calculated.

The result of including δ-electrons in the fluence spectrum is that the total electron fluence assumes a higher value. Note that the basic formula to compute the dose in the Spencer-Attix cavity theory is the same as in the Bragg-Gray cavity theory, that is

$$\frac{D_{med}}{D_{cav}} = \left(\frac{\bar{L}}{\rho}\right)^{med,\triangle}_{cav,\triangle}. \tag{11.2.39}$$

where, as in the case of Bragg-Gray equation, the term on the right hand side represents the ratio of the stopping powers in the medium to that in the cavity, that is

$$\left(\frac{\bar{L}}{\rho}\right)^{med,\triangle}_{cav,\triangle} \equiv \frac{\left(\bar{L}/\rho\right)_{med,\triangle}}{\left(\bar{L}/\rho\right)_{cav,\triangle}}. \tag{11.2.40}$$

The stopping power as defined here is different from the one calculated in the Bragg-Gray cavity theory. Here it is evaluated considering δ-electrons as well. In terms of conventional notation for stopping power, we can write the equivalence

$$\left(\frac{\bar{L}}{\rho}\right)_{\triangle} \equiv \left(\frac{1}{\rho}\frac{dE}{dx}\right)_{unres}, \tag{11.2.41}$$

where the subscript *unres* stands for unrestricted and implies that the δ-electrons are to be considered while evaluating the quantity.

11.2.H   *LET* and *RBE*

Earlier in the chapter we used the term linear energy transfer or *LET* on a few occasions. Due to its importance in dosimetry it is worth spending a little time describing this quantity. *LET* is a measure of the energy absorbed in a medium when charged particles pass through it and is closely related to stopping power $(-dE/dx)$ of charged particles as presented in chapter 2. The difference between the two is that *LET* is supposed to exclude delta rays that carry away energy from the dosimeter. The equivalence of *LET* and stopping power is therefore guaranteed for very large targets in which the secondary electrons also get fully absorbed. For smaller targets, such as biological cells, the target volume is quite small and therefore a significant fraction of the total energy lost by charged particles manages to leave the target without getting absorbed. This escaped energy does not cause any damage to the target and should therefore be excluded from the dose calculations. That is why, instead of the stopping power, generally *LET* is used in dosimetry. Later on, during the discussion on microdosimetry, we will see that *LET* is also replaced by a more useful statistical quantity called lineal energy.

Now, let us turn our attention to a very important parameter extensively used in dosimetry, that is the relative biological effectiveness or *RBE*. To understand this quantity we first note that the main purpose of dosimetry is to quantify the effectiveness of radiation on biological organisms. The problem, however, is that the quantity *absorbed dose* does not faithfully represent the damage a particular radiation field is capable of causing. The difference actually lies in the fact that different kinds of radiation can cause different types and amounts of damage. To account for such difference, the quantity relative biological effectiveness or *RBE* has been defined as

$$RBE = \frac{D_{ref}}{D_{test}}, \tag{11.2.42}$$

where D_{ref} is the x-ray dose and D_{test} is the dose from the test radiation that produces the same biological effect. Unfortunately it is not at all trivial to determine *RBE* analytically as can be appreciated from the following list of parameters on which it depends.

► Dose

► Dose rate

► Type of biological entity

► Linear energy transfer

Before 1990, the average *RBE* for different biological effects were approximated by an average quality factor. Since the quality factor has now been replaced by weighting factor, we will not discuss it here. The weighting factors will be described in the next subsection.

11.2.1 Beam Size

An important parameter, that must be known while making dose calculations for radiation beams, is the size of the beam. The reason is that the absolute exposure or dose can not be calculated without knowing the area of the beam. To clarify further, let us take the example of a radiotherapy setup. The radiation beam, produced by a source, such as an accelerator, is made to pass through a set of collimators before the subject is exposed to it. The collimators are the ones that define the shape of the beam. If one knows the beam flux (number of particles per unit time per unit area), one can multiply it with the area of the beam to find the total number of particles per unit time. The situation for dose calculations is not that simple, though. In most of the formulae derived for dosimetry, one generally assumes a circular or a square beam. However, in practice, a beam can have different shapes: square, rectangular, circular, or even irregular. The problem is that one can not simply calculate the area of a beam having arbitrary shape and substitute the result in these equations. One needs to know the *effective* area that would have the same impact as a beam having circular or square area.

The case of an irregularly shaped beam is fairly complex but, to a very good approximation, any irregular shape can be assumed to be circular, square, or rectangular. A square beam can be equated to a circular beam having the same area, that is

$$A_{sq} = A_{cir}$$
$$\Rightarrow x^2 = \pi r^2$$
$$\Rightarrow x = \sqrt{\pi} r, \tag{11.2.43}$$

where x represents a side of the square and r is the radius of the equivalent circle.

Now, the area of a rectangular beam can not be simply equated to the area of a circular or a square beam. However, one can *translate* a rectangular beam into a square beam such that the ratio of their areas to their perimeters remain the same, that is

$$\frac{A_{sq}}{P_{sq}} = \frac{A_{rect}}{P_{rect}}, \tag{11.2.44}$$

where A and P represent area and perimeter, and the subscripts $rect$ and sq refer to rectangular and square beams respectively. In terms of sides, we can then write

$$\frac{x^2}{4x} = \frac{yz}{2(y+z)}$$
$$\Rightarrow x = \frac{2yz}{y+z}, \tag{11.2.45}$$

where x represents a side of the square, and y and z represent the adjacent sides of the rectangle.

What these equations imply is that one can translate essentially any beam shape into either a square or a circular shape (see example below). This greatly simplifies computations of the parameters related to dosimetry, specially in the field of radiation therapy.

Example:
Calculate the equivalent radius of a rectangular beam of sides 0.5 *mm* and 1.2 *mm*.

Solution:
The radius of the equivalent circle can be found by equating equations 11.2.43 and 11.2.45.

$$\sqrt{\pi} r = \frac{2yz}{y+z}$$
$$\Rightarrow r = \frac{2yz}{\sqrt{\pi}(y+z)}$$
$$= \frac{(2)(0.5)(1.2)}{\sqrt{\pi}(0.5+1.2)}$$
$$= 0.4 \ mm.$$

11.2.J Internal Dose

Now that we are on the topic of dose and dose calculations, we will discuss a very important class of dosimetry, namely calculation and measurement of internal doses. By internal dose we mean the dose received by internal organs of a living thing (actually we are almost always concerned with doses received by humans) by a source or sources inside the body. These sources may or may not be uniformly distributed throughout the organ under consideration. However, for most practical purposes we can assume their uniform distribution. The reason is that the sources digested or inhaled follow the usual metabolic path and then reside in an organ, such as liver. The slow intake of the source assumes uniform distribution throughout the organ tissues. Now, the question is how we know which source ends up where. For example, one would assume that an inhaled source may reside inside the lung for an extended period of time. This is certainly true but the source can also get distributed to other organs through blood and diffusion. In fact, this is something that has been discovered through experimentation and the medical physicists can fairly accurately tell where in body to expect the most significant amount of any particular source. For example, the radium and strontium almost always end up in bones. Cesium and tritium get distributed throughout the body, while iodine finds its way into the thyroid glands.

With advancement in microdosimetry, it has become possible to measure dose inside the human body due either to external or internal sources. In certain situations, it is not really required that the dose be actually measured. For example, in medical diagnostics where radiations sources are introduced into the body, one is interested in determining the *expected* dose. It is certainly impractical and also unnecessary to actually measure the dose with equipment.

In the next two subsections we will discuss the doses received by tissues due to charged particles and thermal neutrons. A number of diagnostics and therapeutic radioactive sources emit such particles and therefore estimation of their respective doses is routine practice in nuclear medicine.

J.1   Internal Dose from Charged Particles

Both α and β particles can inflict damage to tissues. There is a large number of naturally occurring and man made radionuclides that emit these particles. If one knows the activity concentration of the radionuclide in the tissue and the energy of the particles, the dose can be calculated from

$$D = A_m E \triangle t, \tag{11.2.46}$$

where A_m is the activity per unit mass of the tissue, E is the average particle energy, and $\triangle t$ is the exposure time (or the time for which the integrated dose is to be calculated). Usually A_g is given in the units of Bq/g and E in MeV/disintegration. If $\triangle t$ is in seconds, the above equation with these units can be written as

$$D[Gy] = 1.6 \times 10^{-10} A_m E \triangle t \tag{11.2.47}$$

J.2   Internal Dose from Thermal Neutrons

Thermal neutrons are known to be extremely hazardous due to their ability to penetrate deep into the atom and get captured by the nucleus. Recall that a thermal neutron has an energy in the vicinity of 0.025 eV. A tissue is mostly composed of light elements: hydrogen, carbon, oxygen, and nitrogen. The cross sections of thermal neutrons for these elements, specially hydrogen and nitrogen, are fairly high. For example, a thermal neutron reacts with hydrogen according to

$$^1_1H + n \rightarrow {}^2_1H + \gamma(2.224~MeV), \tag{11.2.48}$$

with a cross section of 0.33 barns. The 2.2 MeV γ-rays thus emitted may deposit all of their energy in the tissue. Same is true for other elements in the tissue. The cross section for nitrogen (1.7 barns) is even higher than that for hydrogen and in this case a proton is emitted. The proton, being a charged particle, quickly looses its energy along its track. In such a case the probability that the released energy gets deposited within the tissue under consideration is much higher than in case of capture by hydrogen. Therefore even though nitrogen density is more than an order of magnitude lower than that of hydrogen in a typical tissue, its effect is significant and can not be ignored.

The dose due to neutron capture by a single element can be estimated from

$$D = \frac{\Phi}{\rho_m} \sigma_c E \rho_a, \tag{11.2.49}$$

where Φ is the incident photon flux, ρ_m is the mass density of the tissue, σ_c is the capture cross section for thermal neutrons in the tissue, E is the energy released, and ρ_a is the atom density of the element in the tissue.

The above formula is valid for one element in the tissue. The total dose can be obtained by simply summing the contributions from all elements, that is

$$D = \frac{\Phi}{\rho_m} \sum_i \left(\sigma_{c,i} E_i \rho_{a,i} \right), \tag{11.2.50}$$

where now the subscript i refers to the ith element in the tissue. The atom densities and capture cross sections of different elements in a typical soft tissue are listed in table 11.2.5. To compute the energy released, one must first write the equation for the capture reaction and then calculate the Q-value using energy conservation (see example below).

Table 11.2.5:   Percentage by mass, atom densities and capture cross sections of elements in a typical soft tissue.

Element	Percentage (by mass)	ρ_a (cm^{-3})	σ_c $(barn)$
Oxygen	76.2%	2.45×10^{22}	1.9×10^{-4}
Carbon	11.1%	9.03×10^{21}	3.5×10^{-3}
Hydrogen	10.1%	5.98×10^{22}	0.33
Nitrogen	2.6%	1.29×10^{21}	1.70

Example:
Compute the energy released after the capture of a thermal neutron by nitrogen-14 at rest.

Solution:
The capture reaction can be written as

$$^{14}_{7}N + n \rightarrow ^{14}_{6}C + p.$$

Conservation of energy implies that

$$m_{0,1}c^2 + m_{0,n}c^2 = m_{0,2}c^2 + m_{p,2}c^2 + E_p,$$

where $m_{0,1}$ and $m_{0,2}$ represent the *rest* masses of nitrogen and carbon respectively. $m_{0,n}$ and $m_{0,p}$ are the rest masses of the neutron and the proton respectively. E_p represents the kinetic energy of the released proton. Note that here we have ignored the kinetic energies of all particle, except of course of the released proton. The reason is that the nitrogen and carbon are bound in the bulk mass of the tissue by chemical bonds and therefore their kinetic energies (mostly vibrational) are very small compared to their rest energies. The thermal neutron on the other hand, though moving, has a kinetic energy insignificantly smaller as compared to its rest energy.
The energy released is therefore given by

$$E_p = (m_{0,1} + m_{0,n} - m_{0,2} - m_{0,p})c^2 \ J$$
$$\Rightarrow E_p = (m_{0,1} + m_{0,n} - m_{0,2} - m_{0,p}) \, 931.48 \ MeV,$$

where the masses in the first equation are in kg while in the second are in amu. The second equation is much more convenient to use since it gives the result in units of MeV. Substituting the values of rest masses from atomic data tables in the above equation gives

$$E_p = (14.003074 + 1008665 - 14003242 - 1.007825)\,931.48$$
$$= 0.626\,MeV.$$

11.3 Passive Dosimetry

Passive dosimetry involves the use of a material to *record* dose and then take the material out of the radiation environment to read the recorded value. This method is not suitable for measuring instantaneous dose rates but is highly successful and convenient for measuring integrated dose. Some of the commonly used passive dosimetry techniques are described below.

11.3.A Thermoluminescent Dosimetry

In the chapter on solid state detectors we briefly mentioned that there is a class of solids which can store the energy they are exposed to. Upon heating they release the stored energy in the form of photons. Since the released energy is proportional to the stored energy therefore these materials can be used as dosimeters. Such materials are known as thermoluminescent materials and the detectors based on them are known as thermoluminescent dosimeters or somply TLDs.

TLDs can measure only integrated doses since they must be allowed to absorb and store energy. Even though other more efficient detectors are now available, still due to their simplicity of operation, TLDs are being used extensively in different fields. Generally the intention is to determine the dose received by a person or an equipment, such as a semiconductor detector, over a long period of time. For example, radiation workers are required to wear badges made of some TL material whenever they are working in an environment where radiation is higher than the nominal background level. After a predefined period of time, usually several months, the badge is read out to determine the integrated dose received by the person. The dose level is then compared with the limit set by the organization to ensure that the person has not received more than what she or he is supposed to.

TL dosimeters are also being routinely used in experiments at particle accelerators, where the aim is to determine the integrated dose received by the radiation vulnerable devices, such as silicon detectors. This approach has been highly successful since these devices do not need any electronic circuitry for operation and can be easily installed and retrieved. The drawback is that they can not be used to measure instantaneous doses and are also not as accurate as electronic detectors.

Another area in which TL dosimeters are gaining interest is the clinical dosimetry, which has traditionally been occupied by ionization chambers and semiconductor detectors.

A.1 Working Principle and Glow Curve

The way TL materials save information, that is dose, is by trapping the electrons
and holes in the lattice defect sites. When these charges are released from those sites,
they give off energy in the form of light photons. The intensity of the emitted light is
proportional to the energy transferred by the incident radiation to the material. The
rate of release can be enhanced by providing uniform heat energy to the material.
It has been found that the probability of release increases with temperature. If we
increase the temperature of the material at a constant rate, the charges in different
defect levels will escape at different temperatures. Hence a plot of output light
intensity versus temperature (or time as it is proportional to the temperature for
linear temperature increase) should show different peaks corresponding to charges
stored at different levels.

The main building blocks of a TLD readout system are shown in Fig.11.3.1. The
material is slowly heated by a heater supply. The emitted light is filtered and then
detected by a photomultiplier tube. The temperature of the material is recorded
through the thermocouple.

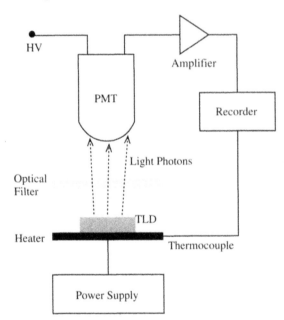

Figure 11.3.1: Block diagram of a
simple TLD readout system.

A typical plot of the output light intensity versus temperature is shown in Fig.11.3.2.
The same curve can also be drawn with time on the x-axis as explained above. Such
a plot is known as the *glow curve*. Since peaks in the curve correspond to the defect energy levels, therefore each TLD material has a specific shape of glow curve.
The height of each peak, however, can be different for different dose levels, types of
materials, and heat transfer rates.

The question as to whether the number of charges stored at each level is proportional to the absorbed dose or not, depends on the radiation field. In case of single
radiation field, such as only γ-rays or only neutrons, each peak is approximately
proportional to the dose absorbed at the corresponding energy level. In a mixed

Photopeak

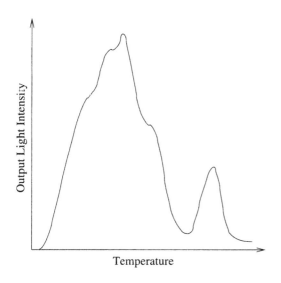

Figure 11.3.2: Typical glow curve obtained by constant rate heating of a TLD exposed to radiation. Different peaks in the curve correspond to the defect energy levels of the material. The total area under the curve is proportional to the total absorbed dose.

field, however, this is generally not true since the probability of populating a certain level is dependent on the type of incident radiation. This problem is generally solved by using coupled TLDs, which we will describe later.

A.2 Common TL Materials

There are quite a few TL dosimeters available commercially, some of which are listed in Table.11.3.1. Each of them has its own characteristics, which should be taken into consideration while deciding on a particular type. In some situations, choosing a TLD can be hard, though. This can happen, for example, in a situation where one type of TLD has good performance in high radiation fields but has very poor neutron detection efficiency, while another one is good for neutrons but can not handle the high radiation field. In such a situation one might even be tempted to use both types of TLDs at the same time. There are no universally accepted guidelines for choosing a TLD but the decision should normally be based on the following criteria.

▶ **Minimum Dose Rate and Total Expected Dose:** It is always good to have an idea about the expected dose rate since then it will be possible to estimate the integrated dose over the course of deployment. Sometimes the length of time the TLD has to stay in the area is determined by other considerations. For example, in a particle accelerator, the TLDs can only be removed during a shutdown period. On the other hand in a medical application the time may be very limited. With the expected dose rate and the time as uncontrollable parameters, one can calculate the expected integrated dose. This total dose can then be used to decide on the best possible TL material based on signal-to-noise ratio considerations.

▶ **Energy Response:** Not all TL materials respond equally to the energy deposited by the incident radiation. It is therefore recommended to match the expected energy spectrum of the incident radiation with the energy response parameters supplied by the TLD manufacturer.

▶ **Fade Characteristic:** Every TL material fades with time, i.e., the trapped electrons in the defect levels escape even at room temperatures. However most of the materials are designed such that this fading is not significant. Still it is a good practice to compare the fade characteristics of the material to the total length of time from its deployment to the eventual read out.

▶ **Type of Incident Radiation:** As stated earlier, not every TL material is suitable for all types of radiation. One should therefore be careful in making a choice for a particular application. In particular, thermal and epithermal neutron dosimetry can be done only with special TLDs, most of which have either lithium-6 or boron-10 as the neutron-active medium. The neutrons react with these materials according to

$$n + {}_3^6 Li \quad \rightarrow \quad {}_1^3 H + \alpha \qquad (11.3.1)$$
$$\text{and} \quad n + {}_5^{10} B \quad \rightarrow \quad {}_3^7 Li + \alpha. \qquad (11.3.2)$$

The α-particles produced then deposit all their energy into the material. Although these materials can be and are used for neutron dosimetry, there are two problems associated with them. First of all, these materials have low cross sections for fast neutrons and therefore the TLDs made with them have low fast-neutron-efficiencies. Secondly it has been observed that the response of these materials to γ-rays decreases with neutron exposure. Hence they can yield inaccurate results if used in situations where neutrons are also accompanied by γ-rays, which actually is the case for fast neutron fields.

A.3   Advantages and Disadvantages of TL Dosimeters

Following are some of the reasons for preferring TL dosimeters over other types of detectors.

▶ **No Electronics Circuitry:** A TL dosimeter does not require any high voltage or readout circuitry. It essentially acts as a memory device from which the information can be retrieved *offline* using a TL reading device. Hence its deployment in a radiation environment does not require any cable connections.

▶ **Small Size:** A typical TL dosimeter is less than $5mm$ long and $2mm$ wide. This makes it perfect for installation in very narrow spaces where other detectors would be hard to fit in.

▶ **Wide Dynamic Range:** General TL dosimeters can record doses in a wide range.

TL dosimeters also has some undesirable characteristics that do not allow them to be used in certain situations.

Table 11.3.1: Common TL dosimeters.

Material	Commercial Name
$LiF : Mg, Ti$	TLD-100
$LiF : Mg, Cu, P$	TLD-100H
$LiF : Mg, Ti$	TLD-600
$LiF : Mg, Cu, P$	TLD-600H
$LiF : Mg, Ti$	TLD-700
$LiF : Mg, Cu, P$	TLD-700H
$CaF_2 : Dy$	TLD-200
$CaF_2 : Mn$	TLD-400
$AlO_2 : C$	TLD-500
$Li_2 B_4 O_7 : Mn$	TLD-800
$CaSO_4 : Dy$	TLD-900

▶ **Dose Rate:** TL dosimeters can not be used to measure dose rate. However, the *average* dose rate can be estimated by dividing the absorbed dose by the exposure time. For accurate dose rate measurements, an electronic detector is more suitable.

▶ **Types of Radiation:** There is no universal TL material that can be used to measure dose from all types of radiation. For example, there are some very good materials that measure dose from γ-rays but they are not good for neutrons. A good strategy in this case is to use a combination of materials in the environment where different types of radiation are expected to be present.

11.3.B   Optically Stimulated Luminescence Dosimetry

Thermal stimulation is not the only means of retrieving stored energy from materials. There are also materials, which emit light when stimulated by light photons. These materials, called optically stimulated luminescence or simply OSL materials, form a new class of dosimeters with qualities much superior than the conventional TL dosimeters. The OSL materials can store energy in the same way as TL materials but they give off light when stimulated optically instead of thermally as in the case of TLDs. This is a relatively new technology and was mostly unheard of before 1992.

Some of the advantages of OSL dosimetry are listed below. We have made comparison with TL dosimetry only since functionally it is closest to the OSL dosimetry.

▶ **Faster Processing:** OSL dosimetry requires light to retrieve the dose information. This process is significantly faster and more efficient as compared to heating the TLDs.

▶ **Higher Precision:** Since the light used to retrieve the dose information can be controlled much more precisely than the heat, therefore the dose measurements fro OSL dosimeters are much more accurate than their TLD counterparts.

▶ **High Dynamic Range:** OSL dosimeters are much more sensitive to very low and very high doses as compared to TLDs. Typical crystals can be used from about 1 *mrad* up to several thousand *rads*.

▶ **Multiple Readouts:** Most of the energy stored in a TLD material gets released upon heating, leaving no possibility of re-reading the material. This is not the case with OSL dosimeters since they release only a fraction of the absorbed energy when optically stimulated. Hence an OSL dosimeter can be read out a number of times, which reduces the uncertainty in the results.

▶ **Shapes and Forms:** OSL crystals can be formed in a variety of forms and shapes. They can even be produced in the form of fibers and powders.

▶ **Mechanical Stability:** OSL materials are highly stable with respect to large variations in temperature and humidity.

Looking at the list above, it is quite evident that OSL materials have many desirable characteristics. Therefore since the realization of their potential, OSL dosimeters are becoming more and more popular.

B.1   Working Principle and OSL Curve

It has been observed that most TL materials, at least to some extent, possess OSL characteristics. This points to the possibility that the underlying energy retrieval mechanism for the two types of stimulation should be similar. In other words, in both cases the energy storage mechanism is the same. However there have been some experiments, which have pointed to the fact that a fairly intact TL glow curve can be obtained from the same material after optical stimulation (see Fig.11.3.3). On this finding it has been argued that the two storage mechanisms should be intrinsically different. It should, however, be pointed out that, as pointed out earlier, the energy released by optical luminescence is very small as compared to thermal luminescence. In fact, it is only a fraction of the energy stored by the material. Therefore the argument that they are intrinsically different does not hold.

Just like glow curve for TL materials, one can also obtain OSL curve from OSL materials. A typical OSL curve is shown in Fig.11.3.3. It is apparent that the OSL curve has a much faster decay time than glow curve making the reading process much faster.

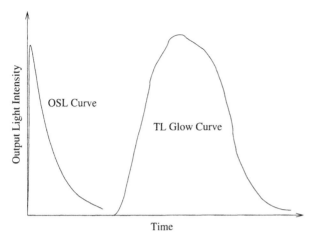

Figure 11.3.3: Typical OSL and glow curves obtained in succession from a material that exhibits both OSL and TL properties.

B.2 Common OSL Materials

Although a number of materials have been identified with good OSL characteristics but the most commonly used one is the carbon doped aluminum oxide or $Al_2O_3 : O$. This material is good for γ-ray and electron dosimetry but can not be very effectively used for neutrons.

A big advantage of OSL materials is that they do not have to be heated and hence do not have to possess high temperature stability. This one-less constraint on the choice of material has been shown to be most advantageous for fast neutron dosimetry specially in the field of medicine. In medical dosimetry of fast neutrons one strives to use a material that is as close in content to the tissue as possible. Since most of the fast neutrons are absorbed in the hydrogen of the tissue, the most suitable dosimeter would have high hydrogen content. However the TL materials, having high hydrogen content, have very low temperature stability and can not be used as dosimeters. On the other hand, one can use an OSL material with a high percentage of hydrogen without worrying about its temperature stability since the material does not have to be heated. An example of such a material is $NH_4Br : Tl$, which has been shown to possess good dosimetry properties for fast neutrons.

11.3.C Film Dosimetry

Film dosimetry is based on the so called *radiochromatic* materials, which change color when exposed to radiation. Since the amount of this coloration is proportional to the delivered dose, therefore radiochromatic films provide a direct means of dose measurement.

The coloration of the film becomes evident when white light is made to pass through it after irradiation. Since white light is composed of photons of different wavelengths, its transmission through the film is dependent on the absorption coefficient at each wavelength. Hence the film appears colored. The degree to which the light is blocked by the film depends on its *optical density*, which is defined by

$$OD = \log(I_0/I), \qquad (11.3.3)$$

where I_0 and I are the incident and transmitted radiation intensities respectively. The reason for defining optical density in this way is that the ratio I_0/I has an exponential dependence on dose D. Hence we can write

$$D \propto OD. \tag{11.3.4}$$

This relation forms the basis for radiochromatic film dosimetry. With proper calibration, the above relation guarantees a linear relationship between the measured optical density and the absorbed dose.

C.1   Advantages and Disadvantages of Film Dosimeters

Following are the main advantages of using radiochromatic materials for dosimetry.

▶ **No Post Irradiation Processing:** The radiochromatic films do not have to be processed after irradiation as in the case of TL dosimeters.

▶ **High Spatial Resolution:** The cost to benefit ratio of film dosimetry is much smaller than other types of position sensitive dosimetry techniques, such as TLD arrays or electronic detectors. Spatial resolution of a fraction of a millimeter is not uncommon for typical films.

▶ **Good Spatial Uniformity:** Radiochromatic films can be manufactured with high degree (typically better than 95%) of spatial uniformity.

Film dosimetry has some disadvantages as well, some of which are mentioned below.

▶ **Fading:** As with TL materials, radiochromatic films also show post irradiation fading. This fading is highest immediately after irradiation and then slows down after some time. It is therefore recommended to delay the optical density for about two days to allow the film to get somewhat stabilized. However one must ensure that this delay is the same as used during calibrations.

▶ **Temperature Dependence:** The optical density of radiochromatic films has strong temperature dependence. Therefore the temperature must be kept as uniform as possible during irradiation and optical scan.

▶ **UV Sensitivity:** Most of the radiochromatic materials are sensitive to ultra-violet light.

▶ **Film Orientation:** Radiochromatic materials are non-isotropic crystals and therefore their orientation matters. The optical density should be measured at the same orientation at which the film was exposed to radiation.

C.2   Common Radiochromatic Materials

The most commonly used radiochromatic film is called *Gafchromic*. These films are made of sub-micron sized crystals of a monomer, which get polymerized after irradiation. Since this polymerization is highly localized and does not spread therefore the spatial resolution from these films can be achieved up to the size of the crystal.

11.3.D   Track Etch Dosimetry

Up until now we have been talking about the problems associated with radiation damage to the materials. There is one instance where this damage can be exploited to determine the dose delivered by the radiation. The process is fairly simple and starts with letting the material absorb radiation for an extended period of time. After the exposure the material is etched with a suitable solution, which reveals the damaged zones or *tracks* of particles. The number of these tracks and their geometry are a measure of the particle fluence and can therefore be used to estimate the dose.

Track etch detectors are usually made of plastics in the form of polymer foils. Since these materials are fairly inexpensive, they are widely used for personal and environmental radiation monitoring.

Estimation of dose from track etch detectors is based on the principle that the damage caused by radiation is directly related to the linear energy transfer or *LET*. Therefore the dimensions of the track bear direct relationship to the stopping power of the radiation. And since stopping power is related to the atomic number of the incident particles, the track dimensions are proportional to the atomic number as well. A particle entering the material deposits its energy as it traverses the material as shown in Fig.11.3.4. The damaged zone is generally a deformed cone but can also have sharp boundaries and other shapes. Depending on the thickness of the material and the energy and type of radiation the incident particle may or may not get fully stopped.

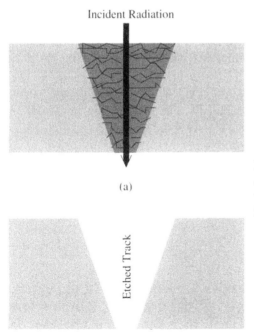

(a)

(b)

Figure 11.3.4: (a) Damage caused by incident radiation to a track etch dosimeter. (b) The damaged track after chemical etching of the material.

After irradiation the material is etched with a suitable chemical, such as hot sodium hydroxide solution. In earlier days of track etch dosimetry the tracks were counted by visual inspection, a process that was prone to human errors. Modern systems are equipped with automated microscope readers and are therefore much more precise and accurate.

D.1   Advantages and Disadvantages of Track Etch Dosimeters

The advantages of track etch detectors include the following.

▶ **Cost Effectiveness:** Due to their low cost, these dosimeters have found wide applications.

▶ **Good Sensitivity:** Track etch polymers are sensitive to almost all types of ionizing radiation, with the exception of photons.

▶ **Operationally Safe:** Since these materials are non-toxic, they can be handled without any extra precautions. This has made them highly suitable for use in households by non-specialists for radon dosimetry surveys.

Track etch detectors also suffer from several disadvantages, some of which are listed below.

▶ **Track Fading:** Since the tracks left by the radiation fade away with time therefore the readout can not be delayed.

▶ **Insensitivity to Photons:** Track etch dosimeters are not sensitive to photons and can therefore not be used to properly assess the dose in a mixed radiation environment.

▶ **Dependence on Etching Rate:** The results depend to some extent on the etching process. In particular the etching rate has been seen to affect the dose estimation.

The most commonly used track etch material is the so called *CR39* plastic, which has the composition of polyallyl diglycol carbonate. The material can be bought in the form of large sheets, which can then be cut into desired sizes.

11.4   Active Dosimetry

Active dosimetry involves use of an electronic detector and is suitable for measuring both instantaneous and integrated doses. The most commonly employed active dosimeters are described below.

11.4.A   Ion Chamber Dosimetry

Ionization chambers have long been used in all types of dosimetry because of their simplicity in design and low operating cost. In clinical practice, they have have become standard dosimetry tools, though the trend is now shifting towards other types of detectors such as semiconductor devices.

In the following sections we will look at two methods commonly used in ion chamber dosimetry and discuss their pros and cons.

A.1   Free in Air Ion Chamber Dosimetry

As the name suggests, this type of ionization chamber measures exposure or Kerma in free air. Such chambers are commonly used to determine air Kerma for x-rays. The main idea behind this method is to ensure secondary electron equilibrium or SEE. The SEE is said to exist if the energy taken away by the secondary electrons escaping from a small mass element is compensated by the energy brought into the element by the secondary electrons produced outside the element. This concept is graphically depicted in Fig.11.4.1. The primary electrons, shown by small open circles, loose all their energies within the measurement volume (large circle) but the secondary electrons (solid circles), having larger range, deliver their energies outside the volume element. This amounts to a loss of energy and consequent underestimation of the dose. But if the irradiation is constant outside the volume element as well then the secondary electrons generated there may deposit their energies inside the measurement volume. Now, if the volume element is very small then we can assume that the secondary electrons escaping the element will be compensated by the secondary electrons generated outside but absorbed inside the measurement volume.

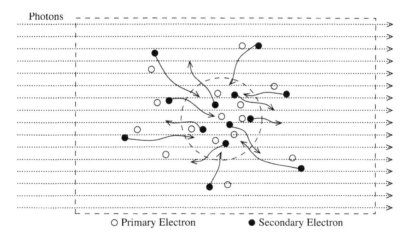

Figure 11.4.1:   Concept of secondary electron equilibrium for free air ion chamber dosimetry. See text for explanations.

The requirement of secondary electron equilibrium can, to a good approximation, be realized in a parallel plate ionization chamber. A simplified diagram of such a dosimeter is shown in Fig.11.4.2. The small shaded volume element subtended by the narrow beam and the electric lines of force between anode and cathode form the measurement volume. The energy absorbed in this volume is actually measured and constitutes the absorbed dose or air Kerma. The secondary electrons escaping from this volume are compensated by the secondary electrons produced in the *compensating* volume elements on either side. The guard electrodes are mainly used to *smooth out* the electric field non-linearities at the edges of the anode and also provide safety against high voltage on the anode.

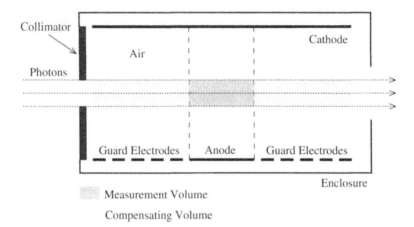

Figure 11.4.2: Simplified diagram of a free in air ionization chamber designed for air Kerma or dose measurement.

Let us now see how we can measure air Kerma using this chamber. Earlier in the chapter we saw that air Kerma can be computed from (see equation 11.4.2)

$$K_{air} = \frac{W_{air}}{e(1-g)} X.$$

The exposure X here was defined earlier by

$$
\begin{aligned}
X &= \frac{dQ}{dm} \\
&= \frac{dQ}{\rho dV},
\end{aligned}
$$

where dm is the mass of air having density ρ in the volume element dV. dQ is the total charge of either sign created by the radiation in the volume element dV. For the case of the ionization chamber shown in Fig.11.4.2 this volume element is the darker shaded measurement area. Substituting this expression for exposure in the above expression for air Kerma we get

$$K_{air} = \frac{W_{air}}{e(1-g)} \frac{dQ}{\rho dV}. \tag{11.4.1}$$

Note that this equation can only be used for an ideal detector since it does not take into account the errors introduced by different parts of the system. The easiest way to do this is to multiply the above equation by a *correction factor* k_t, which gives

$$K_{air} = \frac{W_{air}}{e(1-g)} \frac{dQ}{\rho dV} k_t. \tag{11.4.2}$$

This correction factor k_t is composed of a number of correction factors related to individual sources of error, that is

$$k_t = k_d k_h k_r k_s k_f k_p k_i k_a k_o, \tag{11.4.3}$$

where

k_d is   the correction due to variation in density of air.

k_h is   the correction for the dependence of humidity on W_{air},

k_r is   the correction for the loss of charges due to recombination.

k_s is   the correction for scattering of electrons outside the region of interest,

k_f is   the correction for the non-uniformity in the electric field,

k_p is   the correction for penetration of radiation through the collimator material,

k_e is   the correction for loss of secondary electrons on electrodes,

k_a is   the correction for attenuation before the measurement volume, and

k_o is   the correction for other design specific error sources.

The uncertainties due to these factors in well designed chambers lies well below 1%. For example, k_d for 10-100 keV x-rays is only 0.03%. In fact the uncertainties associated with other factors in equation 11.4.2 are much larger. The error introduced by W-value, for example, can be as high as 0.25%. The same is true for the g factor.

A.2   Cavity Ion Chamber Dosimetry

The free air ionization chamber technique we just studied is only good for moderate photon energies up to about 400 keV. As the mean energy of the incident photons increases, the average energy of the secondary electrons also increases. This results in secondary electrons traveling farther in the chamber before being absorbed and even escaping from the active volume. One could argue that a solution would be to increase the volume of the chamber. However, this has associated engineering difficulties. For example, as the chamber size becomes larger the electrode distance also widens and requires higher voltage to achieve ionization chamber plateau. Recall that ionization chamber plateau is the flat region in the voltage-pulse height curve that corresponds to the minimum recombination and collection of almost all charge pairs created by the incident radiation. Now, the higher the voltage the more probable it becomes that a electrical discharge between one of the electrodes and a nearby metal occurs. Hence one can not indefinitely increase the bias voltage.

The solution to this problem is to use a Bragg-Gray cavity, which has been described earlier in the chapter. Since we are now not required to assure secondary electron equilibrium, there is no need to construct a large chamber. On the contrary, now the Bragg-Cavity theory requires the chamber to be as small as permissible by the following two conditions.

1. The fluence of the primary, secondary, and all subsequent generations of electrons should be uniform throughout the detector's active volume.

2. The total energy delivered by the radiation to the air molecules should be much larger than the energy delivered to the secondary electrons.

A simple ion chamber dosimeter that fulfills the Bragg-Gray cavity conditions is shown in Fig.11.4.3. The chamber consists of a cylindrical cathode with a thin anode wire stretched across its axis. The chamber is filled with air under standard atmospheric conditions. The main problem with this chamber is that it has walls with a material that is very different from air. Therefore, the dose measured from

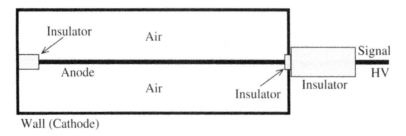

Figure 11.4.3: A simple air filled ion chamber dosimeter.

the chamber inevitably has errors due to interaction of charges and radiation in the wall material. We will later see how these errors are taken into account but first let us see how the dose can be calculated from the chamber.

The radiation passing through the chamber produces electron-ion pairs that drift in opposite directions under the influence of the externally applied electric potential. The number of charge pairs created by the radiation and the total charge carried by them can be calculated if one knows the energy deposited by the radiation and the W-value of air through the following relations.

$$N_{air} = \frac{E_{dep}}{W_{air}} \tag{11.4.4}$$

$$\Rightarrow Q_{air} = e\frac{E_{dep}}{W_{air}} \tag{11.4.5}$$

Here E_{dep} is the total energy deposited by the radiation inside the active volume of the detector. The charge Q_{air} is a measurable quantity since it is proportional to the current flowing through the chamber, which can be measured using suitable electronics readout circuitry. If Q_{air} is known, the deposited energy E_{dep} can be calculated from the above relation. This energy can then be used to calculate the absorbed dose according to

$$\begin{aligned} D_{air} &= \frac{E_{dep}}{M_{air}} \\ &= \frac{W_{air}Q_{air}}{eM_{air}}, \end{aligned} \tag{11.4.6}$$

where M_{air} is the mass of the air in the chamber. Up until now we have not considered any sources of error. In other words, the dose calculated from the above expression is good only for an ideal detector. A practical system, no matter how perfectly it is built, introduces some uncertainties in the measurements. To take these uncertainties into account we multiply the above expression by a correction factor k_t, that is

$$D_{air} = \frac{W_{air}Q_{air}}{eM_{air}}k_t, \tag{11.4.7}$$

where k_t, to a large extent, depends on the design of the chamber as well as environmental conditions. It can be factorized into individual correction factors exactly as in equation 11.4.3.

This is all good as long as we wanted to measure the dose *inside the chamber*. How we relate this dose to the material surrounding the chamber is something that can be dealt with by the Bragg-Gray cavity theory. Note that even if the detection medium and the material surrounding the chamber are under exact same conditions, the dose calculated from the above expression does not represent the expected dose in the medium. To apply the Bragg-Gray cavity theory we make use of equation 11.2.34. Substituting the expression for D_{air} in this equation gives

$$
\begin{aligned}
D_{med} &= D_{air} \left(\frac{\bar{L}}{\rho} \right)_{air}^{med} \\
&= \frac{Q_{air} W_{air}}{e M_{air}} \left(\frac{\bar{L}}{\rho} \right)_{air}^{med} k_t.
\end{aligned}
\tag{11.4.8}
$$

The above equation, though very simple, gives a fairly accurate measure of the dose, provided the chamber walls are not made of high Z elements and the chamber size is small. As mentioned earlier in the chapter, the dose computed from the Bragg-Gray theory has a few sources of error, the most important of which are listed below.

▶ Measuring mass of air in the chamber is not trivial since it has dependence on temperature and pressure. This is further complicated by the fact that, due to non-uniformity of the electric field inside the chamber, the charge collection efficiency deviates from the ideal case. This is compensated in the above equation by taking an *effective mass* of air instead of the absolute mass.

▶ The Bragg-Gray theory assumes continuity between the two media. Since ionization chambers have walls therefore this assumption is not strictly valid specially for the walls that are made of high Z elements.

▶ The irradiation might not be uniform as assumed by the Bragg-Gray theory.

The uncertainties due to these errors are included in the correction factor k_t in the above equation with the exception of wall correction, which is generally done separately. The wall correction is actually a two step process. The first step involves estimating the dose in the wall material using Bragg-Gray equation 11.4.8. that is

$$
D_{wall} = \frac{Q_{air} W_{air}}{e M_{air}} \left(\frac{\bar{L}}{\rho} \right)_{air}^{wall} k_t.
\tag{11.4.9}
$$

The dose in the surrounding medium can then be obtained by simply multiplying the above equation by the ratio of the mass energy absorption coefficients of the medium and the wall. This gives

$$
D_{med} = \frac{Q_{air} W_{air}}{e M_{air}} \left(\frac{\bar{L}}{\rho} \right)_{air}^{wall} (\mu_{m,en})_{wall}^{med} k_t,
\tag{11.4.10}
$$

where

$$
(\mu_{m,en})_{wall}^{med} \equiv \frac{(\mu_{m,en})_{med}}{(\mu_{m,en})_{wall}},
\tag{11.4.11}
$$

with $\mu_{m,,en}$ being the mass energy absorption coefficient. This coefficient should be computed from

$$\mu_{m,en} = \frac{\mu_{m,tr}}{1-g},\tag{11.4.12}$$

where $\mu_{m,tr}$ is the mass energy transfer coefficient and g is the fraction of loss of energy of the secondary charged particles through radiative processes, such as Bremsstrahlung.

So far we have explicitly used Bragg-Gray cavity theory to derive expressions for dose equivalence. However, as we saw earlier in the chapter, even with all the corrections made, this theory does not give very accurate results specially in situations where the wall is thick and is made of high Z elements. Since Spencer-Attix theory gives better results than the Bragg-Gray theory, general practice is to start with the Spencer-Attix equation and then apply some correction factors to it. The modified form of this equation for an ion chamber dosimeter can be written as

$$\frac{D_{med}}{D_{cav}} = \left(\frac{\bar{L}}{\rho}\right)^{med,\triangle}_{cav,\triangle} k_t,\tag{11.4.13}$$

where, as before, k_t represents the total correction factor and

$$\left(\frac{\bar{L}}{\rho}\right)^{med,\triangle}_{cav,\triangle} \equiv \frac{\left(\bar{L}/\rho\right)_{med,\triangle}}{\left(\bar{L}/\rho\right)_{cav,\triangle}}$$

Using the expression for $D_{cav} = D_{air}$ as derived earlier for the Bragg-Gray cavity, the computational form of the Spencer-Attix equation becomes

$$D_{med} = \frac{Q_{air} W_{air}}{e M_{air}} \left(\frac{\bar{L}}{\rho}\right)^{med,\triangle}_{cav,\triangle} k_t.\tag{11.4.14}$$

As noted earlier, there are uncertainties associated with various quantities in the expressions we have derived so far. Fortunately, in well designed chambers the combined uncertainty is typically less than 1%. Still, it is a good practice to carefully analyze all sources of uncertainties before inferences from the measurements are drawn. Table.11.4.1 lists some of the parameters and their typical uncertainties. The reader should note that these values can not be used for any type of chamber, and uncertainties for specific dosimeters should either be experimentally determined or, in case of a commercial product, evaluated from the data provided by the manufacturer.

11.4.B   Solid State Dosimetry

We have already seen a few examples of solid state dosimeters, namely the TL, OSL, and film dosimeters. Apart from these so called passive dosimeters, there are also active solid state dosimeters that are being extensively used specially in the field of radiation therapy. Perhaps the most commonly used active solid state dosimeters are based on silicon diodes. The silicon detectors used in dosimetry are similar to the usual silicon diode detectors we studied in the chapter on solid state detectors. These mostly single channel devices have many advantages over their gas

Table 11.4.1: Typical uncertainties in air filled ionization chambers used in dosimetry.

Source	Uncertainty (%)
Measurement of sensitive volume (each dimension)	0.05
Humidity	0.05
High voltage ground shifts	0.02
Inhomogeneity of electric field	0.02
Pulse height measurement	0.02
Ion collection	0.10
Charge measurement (capacitance uncertainty)	0.05
W-value for dry air	0.20
Stopping power ratio	0.40
Perturbatioin correction factor	0.15

filled counterparts. Most importantly, due to their small sizes they can be used in *in vivo*[2] dosimetry, something that is highly desired in radiation therapy.

We will not discuss here the construction details and working principles of common semiconductor dosimeters based on junction diodes since similar devices have already been discussed in the chapter on solid state detectors. However, a relatively newer form of semiconductor dosimeter, namely the MOSFET device, will be discussed in some detail in the next section. Later on we will also discuss diamond dosimeter, which is now gaining popularity due to its radiation hardness.

B.1 MOSFET Dosimeter

MOSFET is an acronym of Metal Oxide Semiconductor Field Effect Transistor. This technology has existed for a number of years but its use has been limited to electronic devices. Its possible role as a radiation dosimeter was investigated a few years ago. It was found that it could not only work as an efficient dosimeter but also provided many advantages over conventional dosimeters. Ever since, its use has considerably increased in the field of radiation therapy.

The simplified diagram of a p-channel type MOSFET dosimeter is shown in Fig.11.4.4. Such a device is fabricated on an n-type silicon substrate having a typical thickness of 500 μm.

---

[2]*In vivo* means *in live*.

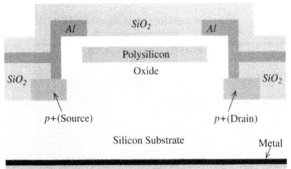

Figure 11.4.4: Simplified schematic of a p-channel MOSFET dosimeter.

The working principle of the device shown in Fig.11.4.4 is fairly simple. Application of a high voltage to the polysilicon forces a large number of holes to move from the surrounding regions into the oxide layer and the adjacent silicon substrate. If a large number of holes are gathered in that area, they form a current channel between the source and the drain SiO_2 regions. A small voltage, called the *threshold voltage*, can then initiate the current flow. The radiation passing through the oxide region produces electron hole pairs. The holes move towards the interface of silicon and SiO_2, where they get trapped. This excess positive charge induces current in the channel between the source and the drain. Consequently the threshold voltage shifts to an extent that is proportional to the positive charge buildup. And since this charge buildup is proportional to the energy deposited by the incident radiation, the change in the threshold voltage is a measure of the radiation dose delivered to the material.

Note that, in principle, it is possible to operate a MOSFET dosimeter without any applied bias. However in this situation the recombination of electrons and holes is significantly large, resulting in a nonlinear signal loss. Hence, in most instances MOSFET dosimeters are operated with a negative bias applied to the polysilicon.

MOSFET dosimetry has many advantages over conventional ion chamber and even silicon diode dosimetry. Some of its advantages are listed below.

▶ **Small Size:** MOSFET dosimeters are very small in size with typical dimension of less than a millimeter. This makes them highly suitable for *in vivo* dosimetry.

▶ **Good Spatial Resolution:** Due to their small size, MOSFET dosimeters offer excellent spatial resolution.

▶ **Good Isotropy:** The axial anisotropy exhibited by typical MOSFET dosimeters is ±2% for 4π, which is acceptable for most applications.

▶ **Large Dynamic Range:** Because of their large dynamic range they can be used in very low to very high radiation environments.

▶ **Radiation Type Sensitivity:** Since MOSFET dosimeters work on the principle of electron hole production by the incident radiation, therefore they can be used for dosimetry of any type of radiation. Their use in photon, electron, and proton dosimetry is very common.

MOSFET dosimeters also have some disadvantages, some of which are listed below.

▶ **Sensitivity to Bias Instability:** MOSFET devices are very sensitive to drifts in bias voltage and therefore require highly stable power supplies.

▶ **Radiation Damage:** Being semiconductor devices, MOSFET dosimeters are highly susceptible to the damage caused by the radiation. Therefore if they are used in high radiation fields, their response may change after irradiation.

▶ **Temperature Dependence:** Like all semiconductor devices, their response is vulnerable to small drifts in temperature. However, specially designed *double* MOSFET devices have shown good temperature stability.

B.2   Diamond Dosimeter

Diamond is a radiation hard substance, which has made it a material of choice in for building detectors used in hostile radiation environments. The detectors based on CVD diamonds have been discussed in the chapter on solid state detectors. The dosimeters based on CVD diamond have the same type of structure but they are smaller in size. Therefore we will not discuss the construction details of such detectors but it is certainly worthwhile to discuss their advantages.

▶ **Radiation Hardness:** Diamond has high radiation hardness, which makes it suitable for use in very high radiation environments. The particle beams generally used in radiation dosimetry are fairly intense and therefore the use of diamond based dosimeters is gaining interest in this field.

▶ **Stability against Water:** Diamond does not absorb water, which is a added advantage for their use in *in vivo* dosimetry.

▶ **Good Temperature Stability:** Diamond dosimeters have shown very good temperature stability, with a typical value of less than 0.1% per degree Celsius. This makes them superior to conventional MOSFET dosimeters and comparable to double MOSFET dosimeters.

▶ **Good Isotropy:** Their behavior is fairly isotropic, making them easy to use in practical applications.

▶ **Good Energy Response:** The energy response of diamond dosimeters is fairly linear and therefore they can be used over a large dynamic range without the need of correction factors.

▶ **Small Size:** Typical diamond dosimeters are fairly small, of the order of a few mm^3. This allows them to measure dose without appreciable affecting the radiation field.

▶ **Good Spatial Resolution:** Due to their small size, diamond dosimeters show very good spatial resolution.

▶ **Tissue Equivalence:** Since diamonds are based on carbon, they have good tissue equivalence. This makes them advantageous over semiconductor dosimeters, which require energy corrections due to their tissue nonequivalence. Of course this only holds for medical dosimetry applications.

The disadvantages of diamond dosimeters include the following.

▶ **Dose Rate Dependence:** Diamond dosimeters show some dependence on dose rate.

▶ **Polarization:** If diamond dosimeters are not in a radiation field they tend to become polarized. Therefore to minimize this effect they must be irradiated before being deployed.

11.4.C   Plastic Scintillator Dosimeter

A scintillator is a material that produces light when irradiated. The detectors based on such materials are called scintillation detectors. We have discussed these materials and the different types of detectors based on them in the chapter on scintillation detectors. The interested reader is encouraged to read through the relevant sections of that chapter to understand how such devices work.

Scintillators are available in many shapes and forms but for dosimetry purposes the most popular are plastic scintillators. A typical plastic scintillator dosimeter consists of a plastic scintillator, an optical fiber to carry the light photons, and a PMT to count the photons. Sometimes another fiber and a PMT is added to the setup to measure and subtract the background light. In the following we list some advantages of plastic scintillators.

▶ **Small Size:** Typical dimensions of plastic scintillator dosimeters are less than 1 mm^3, which makes them suitable for medical dosimetry.

▶ **Temperature Stability:** Plastic scintillators show very good temperature stability and therefore the response of dosimeters does not change significantly with drifts in temperature.

▶ **Good Spatial Resolution:** Due to their small size, plastic scintillator dosimeters show excellent spatial resolution.

▶ **Good Energy Response:** The linear energy response of these devices makes them highly suitable for use in radiation therapy applications.

▶ **Radiation Hardness:** The response of plastic scintillators does not deteriorate significantly with irradiation.

▶ **Good Isotropy:** The isotropy of plastic scintillator dosimeters is comparable to that of diamond detectors.

▶ **Large Dynamic Range:** The large dynamic range makes them suitable for use in diverse radiation environments.

▶ **Water Equivalence:** The composition of plastic scintillators is approximately equivalent to water. This makes them suitable for phantom measurements in clinical dosimetry.

Perhaps the biggest disadvantage of plastic scintillator dosimeters is that they are highly susceptible to background Cherenkov light. This can be overcome by using a background subtraction counter, which requires another set of fibers and PMT. However the development of highly efficient light guides and PMTs has made the task simpler than before. Consequently these dosimeters are now gaining popularity.

11.4.D Quartz Fiber Electroscope

Quartz fiber electroscope is a pocket dosimeter that has the capability of instantaneous dose readout. It is therefore generally called a Self-Reading Pocket-Dosimeter or SRPD. It consists of the following main components.

▶ **Ionization Chamber:** This consists of a very small volume gas filled chamber where the incident radiation produces electron ion pairs.

▶ **Quartz Fiber Electrometer:** This is used to measure the charge produced by the radiation in the ionization chamber. It consists of a movable quartz fiber and a metal frame forming its two electrodes. The separation of these electrodes depends on the amount of charge on the metal electrode.

▶ **Microscope:** The purpose of the microscope is to facilitate the reading of the fiber image.

The working principle of the electroscope is quite simple. The incident radiation produces charge pairs in the chamber volume, which creates a current flow. Consequently the charge on the electrode decreases. Due to this decrease, the quartz fiber moves and its deflection can be projected by a light source to a calibrated scale through an objective lens. To do this, the user points the dosimeter to any external light source while looking into the microscope eyepiece. Since the deflection of the fiber is proportional to the current flowing through the chamber, therefore it is a measure of the dose delivered by the radiation. The electroscope is pre-calibrated and therefore no calibration is required by the user. Upon full discharge, the electroscope can no longer be used before being charged again. This is done by applying high voltage through a charging pin.

Fig.11.4.5 shows the working principle of a typical quartz fiber electroscope. The electrometer of an electroscope that has not seen any radiation is fully charged and the reading pointer is at 0 position. As the incident radiation produces electron ion pairs in the ionization chamber, the effective voltage on the fixed electrode decreases. Consequently the quartz fiber moves towards the fixed electrode and the reading pointer move forward.

D.1 Advantages and Disadvantages of Quartz Fiber Electroscope

The advantages of this instrument include the following.

▶ **Visual Readout:** Even though the quartz fiber electroscope is an active dosimeter, it does not require any electronic circuitry to read the dose. The readout is done visually at any time.

▶ **Good Accuracy:** Although the accuracy of the electrometer can not be compared with that of an electronic dosimeter, it gives a modestly accurate measure of the integrated dose. In most organizations the electroscope is used in conjunction with some other dosimeter, such as a TLD.

▶ **Repeated Usage:** The electroscope can be used repeatedly after being recharged, which is fairly simple and is done through a special unit.

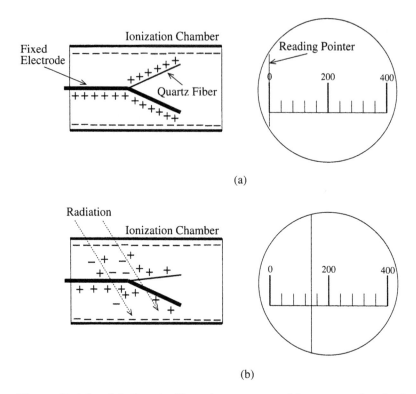

Figure 11.4.5: (a) Quartz fiber electroscope without any absorbed dose. (b) Incident radiation produces charge pairs in the ionization chamber that decreases the charge on the electrometer pushing the electrodes close together. The eyepiece of the microscope is shown to read a dose of 140 units.

▶ **Easy Handling:** This device is of the size of a pen and can be carried in a pocket. Its outer metallic surface makes it sturdy against mechanical shocks and therefore it does not require any special handling requirements.

Quartz fiber electroscope also suffers from some disadvantages, some of which are listed below.

▶ **Readout Errors:** The readout scale of the electrometer is fairly narrow and can not be very accurately read out. The readout process is also prone to human errors.

▶ **Low Dynamic Range:** The very small volume of the ionization chamber does not allow accumulation of large doses and, in hostile radiation environments, the instrument gets saturated fairly quickly.

▶ **Low Accuracy:** The accuracy of the instrument is lower than other types of dosimeters, which does not make it suitable for accurate dose measurements.

11.5 Microdosimetry

Microdosimetry is a rapidly evolving technique that is being used to measure the stochastic distribution of energy deposited by radiation in microscopic sites. It has been found to be an extremely valuable tool in assessing energy transfer to very small volumes, such as cellular and subcellular structures. For example, in radiation therapy one is often interested in determining the distribution of energy transfer inside single cancer cells so that the cells could be effectively targeted. Conventional dosimetry does not provide such information since it deals with macroscopic description of energy transfer in large volumes. Microdosimetry has, in fact, been mostly developed for dealing with such issues in radiation therapy.

The biggest issue faced in microdosimetry is the statistical fluctuation involved in the energy transfer process. In conventional dosimetry this is not of much problems since there the number of particles interacting is so large that the statistical fluctuations are much smaller than the systematics of the measurement process. To understand this further the reader is pointed to the fact that the energy transfer can be considered a Poisson process. The statistical fluctuations of a Poisson process are equal to the square root of the number of interacting particles or $\sqrt{N}$. If N is small the fluctuations are large and so is the measurement uncertainty.

11.5.A Microdosimetric Quantities

The quantities used in microdosimetry are somewhat different than the ones used in conventional dosimetry. This section is therefore devoted to the discussion of the most commonly used microdosimetric quantities.

A.1 Linear Energy Transfer and Dose

In microdosimetry one is interested in measuring or estimating the distribution of energy transfer along a particle's track. Earlier in this chapter and also in chapter 2 we introduced a term, the linear energy transfer or LET, to characterize how the particles loose energy along their track. This quantity can be used in microdosimetry, but one should be cautioned that it actually is a macroscopic quantity. Let us first understand how LET can be used to characterize the absorbed dose.

The absorbed dose is related directly to the linear energy transfer through the relation

$$D = \frac{1}{\rho} L_t \bar{\Phi}, \tag{11.5.1}$$

where $\bar{\Phi}$ represents the average fluence of particles passing through the medium of density ρ. L_t represents the track average LET, which can be computed from

$$L_t = \int L f(L) dL, \tag{11.5.2}$$

where $f(L)$ is the normalized probability distribution function for LET and the integration is carried out over the whole track length. Since we are essentially taking the weighted average of LET at each point along the track, the L_t is referred to as the track average LET. One can similarly define the *dose average LET* through

$$L_d = \int L D(L) dL, \tag{11.5.3}$$

where $D(L)$ represents the normalized dose distribution as a function of LET. It is given by

$$D(L) = \frac{1}{\rho}L\Phi(L), \tag{11.5.4}$$

with $\Phi(L)$ being the fluence distribution as a function of LET.

One problem with the use of LET in microdosimetry is that it does not properly characterize the differences in the biological effectiveness of different radiation types. In other words, the biological effectiveness of two types of radiation could be different even though they might have same LET. This implies that one should be careful in using this quantity to derive dosimetric inferences. Because of this particular reason, the statistical quantities, such as specific energy and lineal energy are more commonly used.

A.2 Specific Energy

The specific energy is defined as the ratio of the energy imparted by radiation to the matter in a volume to the mass of the matter, that is

$$z = \frac{E}{m}, \tag{11.5.5}$$

where m represents the mass of the matter and E is the energy imparted. Since the energy impartation is a statistical process and suffers from random fluctuations, the specific energy has similar uncertainties associated with it.

A.3 Lineal Energy

The lineal energy is defined as the ratio of the energy imparted to the medium in a single event to the average chord length, that is

$$y = \frac{E}{\bar{l}}. \tag{11.5.6}$$

where E is the energy imparted in a single event and $\bar{l}$ is the average chord length. The lineal energy is generally measured and computed in units of $keV\,\mu m^{-1}$. To understand the average chord length we first note that any particle track in a medium leaves a number of randomly oriented chords behind. The average length of these chords is what is represented by the parameter $\bar{l}$. In general, both the chord lengths and their orientations are randomly distributed.

Note that, as with specific energy, here too the imparted energy is a random variable. Therefore the lineal energy is also a statistical quantity and is described by its own frequency distribution function $f(y)$. Now, just like LET, we can use this distribution to determine the average track lineal energy, that is

$$\bar{y}_t = \int yf(y)dy, \tag{11.5.7}$$

where we have assumed that the distribution $f(y)$ has been normalized. Similarly we can also define a dose average lineal energy by

$$\bar{y}_d = \int yD(y)dy, \tag{11.5.8}$$

where $D(y)$ represents the normalized dose distribution as a function of lineal energy. It can be evaluated from

$$D(y) = \frac{1}{\bar{y}_t} y f(y). \tag{11.5.9}$$

Substituting this into the expression for $\bar{y}_d$ reveals that the dose average lineal energy is simply the second moment of the distribution $f(y)$, that is

$$\bar{y}_d = \frac{1}{y_t} \int y^2 f(y) dy. \tag{11.5.10}$$

The prime objective of microdosimetry is the determination of lineal energy distribution along particle's track. Since the dimensions involved in such a study are extremely small, typically ranging from a nanometer to a few micrometers, the experimental techniques are also somewhat modified versions of the conventional dosimetry.

It was mentioned earlier that statistical fluctuations constitute the most significant source of errors in microdosimetry, provided the dosimeter has been carefully designed. The quantities we just described for microdosimetry are all statistical in nature and therefore suffer from uncertainties because of statistical fluctuations. These fluctuations can be due to a number of factors, such as

▶ delta rays, which are the electrons emitted as a result of radiation interaction in the medium,

▶ lineal energy variation with particle track,

▶ energy straggling,

▶ range straggling, and

▶ chord length variations.

11.5.B   Experimental Techniques

Microdosimetry techniques are somewhat different from the conventional dosimetry. The reason is that in conventional dosimetry one is only concerned with macroscopic deposition of energy while this macroscopic picture is not much useful to derive microdosimetric quantities. In this section we will survey some of the commonly used micrdosimeters..

B.1   Tissue Equivalent Proportional Counter (TEPC)

We have already discussed the working principle of proportional counters at various places in the book. In essence, these counters work on the principle of charge multiplication in the presence of externally applied high electric field. The tissue equivalent proportional counter is a modified version of the conventional proportional counter. The basic difference lies in constructing the chamber such that it is a close in physical properties to a tissue as possible. A dosimeter that is not *equivalent* to the tissue can not be measure the true dose absorbed by the tissue.

The most common approach to building such a chamber is based on simulating the tissue volume by a relatively larger filling gas volume. The requirement of equivalence can be approached by assuring that the energy losses of charged particles in the gas is the same as in the tissue, that is

$$\bar{E}_g = \bar{E}_t, \tag{11.5.11}$$

where $\bar{E}_g$ and $\bar{E}_t$ represent the mean energy losses in the active gas volume of the chamber and the tissue. Now, let us suppose the mean distance traveled by a charged particle in the gas is d_g. For a small spherical chamber, this could be the diameter of the sphere. If the mass stopping power for the particles in the gas volume is $S_{\rho,g}$ (see chapter 2 for discussion on mass stopping power) and the density of the gas is ρ_g, the mean energy deposited will be given by

$$\bar{E}_g = S_{\rho,g}\rho_g d_g. \tag{11.5.12}$$

Similarly the mean energy deposited in the tissue can be evaluated from

$$\bar{E}_t = S_{\rho,t}\rho_t d_t, \tag{11.5.13}$$

where $S_{\rho,t}$ is the stopping power of same particles in the tissue having density ρ_t. The equivalence of absorbed energy then requires that

$$S_{\rho,g}\rho_g d_g = S_{\rho,t}\rho_t d_t. \tag{11.5.14}$$

Now, suppose that the atomic composition of the gas and the tissue are identical. In this case the mass stopping powers in the two median will be the same, that is $S_{\rho,g} = S_{\rho,t}$. With this condition the above equation becomes

$$\rho_g d_g = \rho_t d_t. \tag{11.5.15}$$

This condition for TEPCs is generally written as

$$\rho_g = k\rho_g, \tag{11.5.16}$$

where $k \equiv d_t/d_g$ is equal to the ratio of the tissue diameter to the chamber diameter for spherical chambers. Looking closely at the above relation we find that, for a fixed chamber dimension, the only parameter that can be varied to assure the equivalence of energy deposition is the gas density. This is the basic working principle of a TECP, that is, the density of the gas is adjusted such that the above relation is satisfied. In such a chamber, the absorbed dose is equivalent to the dose absorbed in the tissue.

Adjusting the gas density is generally done through adjusting the gas pressure. This simple principle gives fairly accurate measure of the dose in tissues. The low pressure TEPCs are therefore the foremost microdosimeters constructed and are still widely used. A typical TEPC has a spherical or cylindrical geometry. Except for the lower gas density, its working principle is the same as the conventional proportional counters. Such detectors have already been discussed at length in the chapter on gas filled detectors.

Design and construction of a TEPC requires special attention to the uncertainties that can be caused by the walls of the chamber. These uncertainties can be due to several effects. One such effect is due to the so called *re-entry* mechanism. What happens is that an electron produced by the incident radiation can move out of the chamber's active volume and then re-enter it at a later time. In a real tissue, the probability of this happening is small due to smaller volume (see Fig.11.5.1(a)). The dose measured in this case will then be higher than the dose expected in the tissue. Another effect that can add uncertainty to the measurements is the scattering of electrons. An incident particle can scatter off another particle before entering the active volume of the chamber. If the scattered particle deposits its energy withing the chamber volume, it will lead to a larger pulse or an extra event, thus overestimating the dose (see Fig.11.5.1(b)). Fig.11.5.1(c) shows an effect that is said to produce *V-tracks*. A neutron interacting with a nucleon in the wall material may break it up into two heavy charged particles. These particles traverse the gas volume in different directions forming a V, hence the name V-tracks. A relative smaller volume tissue will not see both the particles. This effect is significant only at high neutron energies, of the order of 20 *MeV* and above. Another class of events that can cause error in measurements are related to delta rays. The delta rays produced outside the chamber's active volume can also lead to double events as shown in Fig.11.5.1(d). A read tissue, on the other hand will miss delta rays at a higher probability.

The discussion in the preceding paragraph leads us to the conclusion that chamber walls is not a good idea. A good approach would then be to build a *wall-less* chamber. The reader might recall that during discussion on conventional dosimetry we introduced a wall-less ionization chamber, in which the volume was subtended by the electric lines of force and the beam of incident particles. That is, there was no physical wall defining the active volume of the chamber. The same approach has been adopted in building TEPCs. However such dosimeters suffer from small variations in the electric field intensity, which lead to uncertainties in the definition of active volume.

Another approach to building a TEPC is to use gridded walls. Using a gridded wall minimizes the wall material and hence reduces the uncertainties associated with the wall effects. The main advantage of this approach is that the active volume of the chamber remains well defined even with slight variations in the electric field intensity. The disadvantage, of course, is that the counter is not really wall-less and therefore suffers from associated uncertainties, though at a smaller scale than a walled counter.

B.2 Solid State Nuclear Track Detector ((SSNTD)

Earlier in the chapter we discussed track etch dosimeters. Such solid state nuclear track detectors or SSNTDs can also be used as microdosimeters. To remind the reader, a particle passing through a track etch detector leaves behind a damaged zone. This zone can be revealed through the process of chemical or electrochemical etching. The reason these detectors can be used as microdosimeters is that the track parameters are directly related to the linear energy transfer and the type of incident particle. In modern systems these parameters are determined through an automated process, which is less prone to human errors. The apparatus generally consists of an optical image analyzer with a built-in microscope.

Determination of the *LET* is done through a parameter called etch rate ratio given by

$$V = \frac{V_t}{V_b}, \tag{11.5.17}$$

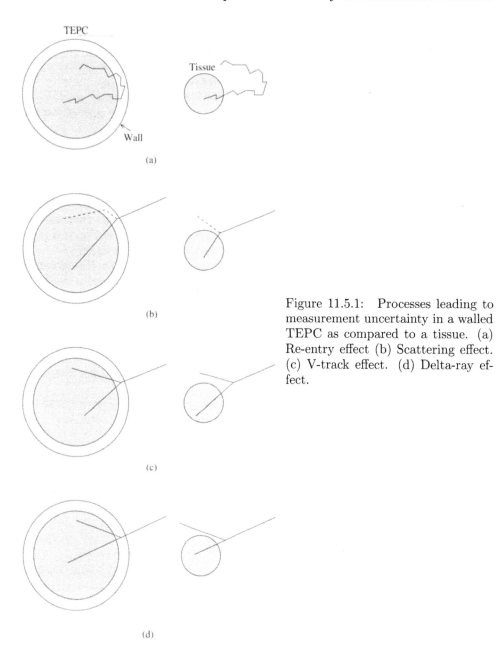

Figure 11.5.1: Processes leading to measurement uncertainty in a walled TEPC as compared to a tissue. (a) Re-entry effect (b) Scattering effect. (c) V-track effect. (d) Delta-ray effect.

where V_t and V_b are respectively the track and bulk etch rates. Since the track ratio V is proportional to the linear energy transfer, therefore with proper calibration it can be used to determine the absolute LET of the radiation. The calibration can be performed by irradiating a similar SSNTD with known radiation.

B.3   Silicon Microdosimeter

Microdosimeters based on silicon are a relatively newer development. As we will see they offer some advantages over conventional TEPCs and therefore they are gaining popularity in the field of hadron therapy.

A silicon microdosimeter is similar to a conventional pn diode detector. However it has some features that are specific to the requirements of microdosimetry. These requirements are listed below.

▶ **Tissue Equivalence:** Tissue equivalence is of course the basic condition in microdosimetry for radiation therapy. A silicon detector should be constructed such that it satisfies this condition as closely as possible.

▶ **Small Size:** The active volume of the detector should be so small that it could simulated a biological cell.

▶ **Defined Collection Volume:** We saw earlier that TEPCs suffer from disadvantage of either not having a well defined volume or suffering from wall effects. The condition of well defined volume is important in the case of silicon microdosimeters as well.

▶ **High Sensitivity:** The detector should be able to measure very low lineal energies, of the order of 1 $keV/\mu m$.

The condition of well defined collection volume is not satisfied in conventional pn junction diode detectors. The main cause of the uncertainty in active volume is the diffusion of charges from the bulk silicon. To overcome this problem, a new type of silicon device, called silicon-on-insulator or SOI, has been introduced. The main idea is to subtend the depletion region between insulators such that the possibility of charge transfer from the bulk silicon is minimized. A simplified diagram of such a device is shown in Fig.11.5.2.

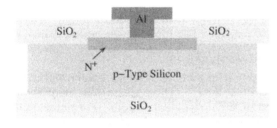

Figure 11.5.2: Structure of a typical silicon-on-insulator microdosimeter. The potential to the substrate is applied through a P^+ layer (not shown) similar to the N^+ layer. The connection to the supply is made by feed through aluminum implant. A typical mirodosimeter system has a number of similar structures implanted on a single substrate.

There are many advantages of the silicon-on-insulator technique, some of which are listed below.

▶ **Well Defined Volume:** The active volume of the dosimeter is defined by the depletion region. Since the region is surrounded by insulator therefore the charge diffusion from outside is minimum and hence the volume is well defined.

▶ **Good Spatial Resolution:** A typical silicon microdosimeter system is based on a number of closely spaced detectors as the one shown in Fig.11.5.2. With small cell and pitch dimensions, of the order of 10-40 μm, high spatial resolution can be achieved.

▶ **Low Voltage Operation:** The active layer of p-type silicon of a SOI has a width of about 2 μm. This enables the operation of the detector at very low voltages as opposed to TEPC that requires fairly high voltage to allow charge multiplication.

Another advantage of silicon microdosimeter with respect to TEPC is that its overall dimensions can be made fairly small by integrating the front-end electronic components on the same silicon substrate. This makes the handling of the system easy by avoiding the need to install external electronic components near the detector.

SOI is not the only technology available to build silicon microdosimeters. In fact the earliest silicon dosimeters were simple pn diode detectors. The SOI technology has attracted more interest due to its advantages as described above. However it should not be considered that the microdosimeters based on this technology are perfect devices. In general, silicon microdosimeters (or for that matter, microdosimeter based on any semiconductor material) suffer from some disadvantages as well, some of which are listed below.

▶ **Tissue Equivalence:** Semiconductor materials can not be considered equivalent to tissues. Hence the dose measured by them can not be said to represent dose in tissue. To overcome this problem, some semiconductor microdosimeters are equipped with a converter material.

▶ **Geometry:** The geometry of a silicon microdosimeters is parallelepiped as compared to a TEPC that can have spherical geometry. This implies that the chord length distribution in a silicon microdosimeter is different than in a TEPC.

▶ **Radiation Damage:** Semiconductors are not very radiation tolerant, which makes them vulnerable to radiation damage. The lifespan of a silicon detector is therefore shorter than a TEPC.

▶ **Field Funneling:** The high *LET* particles passing through a semiconductor produce local distortions in the electric field . These distortions may favor collection of charge pairs produced in the undepleted regions. For newer SOI technology, this is not of much concern because of the insulator layers.

11.6   Biological Effects of Radiation

All living organisms are made of cells, which themselves are made of atoms. The radiation can interact with these atoms in many ways, as we have been discussing throughout this book. At the most basic level, it can either ionize the atom or interact with its nucleus. In terms of biological damage, both of these interactions can have serious consequences if they result in weakening the bonds between atoms. A weekend bond may eventually break up and cause the cell to malfunction.

Now, a cell is composed of a very large number of atoms. Not all of them have the same importance in terms of cell functioning. The most important part of the cell are the chromosomes, which carry the DNA. A damage to the chromosome has the potential of causing cell mutations leading to genetic effects and cancer development. Unfortunately these effects occur very slowly and take a long time, on the order of years, to become noticeable. All types of ionizing radiation, no matter how intense, are capable of causing such damages. However the probability of their occurrence is seen to decrease with decreasing radiation flux or intensity.

Cell mutation is not the only type of damage caused by radiation. A high level of dose can, in fact, cause so much ionization that the cell dies. If a large number of cells in an organ die, they can cause the organ to malfunction or even lead to total failure. The gravity of the situation can be appreciated by noting that, according to different studies, it is believed that receiving a whole body dose of approximately 500 *rem* for a few minutes can cause death to the 50% of the population. However such a high dose is expected to be delivered only in extreme radiation environments, such as atomic explosions or accidents of nuclear power reactors.

Fig.11.6.1 shows the possible effects of radiation on biological cells. As shown in the figure, one possibility is that the cell simply dies after irradiation. This is of not much concern unless a large number of cells in an organ die, causing it to malfunction or fail. The other possibility is that the cell does not die but gets damaged. By damage we mean that the bonds between atoms get weakened and as a result new chemicals are formed. The natural repair mechanisms of the cells, for most part, take care of such localized damages. The probability that the repair is successful depends on the extent and locality of the damage. However it should be mentioned that any part of cell can get repaired including the chromosomes. In fact, the cell damage and repair occur constantly in biological organisms even without irradiation. The damages caused by low level radiation are therefore not much different than the damages caused by other sources, such as foreign chemicals and autoimmune disorders.

Unfortunately not all repair mechanisms always function perfectly. For example, if at the time of irradiation a cell is in the middle of performing a certain task, the repair may not go through successfully. This might cause the cell to start operating abnormally. Such a cell is the one that can eventually cease to reproduce or start reproducing uncontrollably. The latter is the source of the development and spread of cancer or genetic disorders. If the cell does not reproduce itself, it can still cause operational problems and genetic effects. However such effects will be localized unless a very large number of cells start behaving in like.

A point to note is that all cells are not equally prone to radiation damage. For example, the cells that produce blood have been found to be highly susceptible to radiation damage. In general, the more reproductively active a cell is, the more sensitive it is to the damage.

Now that we understand the basic radiation damage mechanisms to individual cells, let us discuss the macroscopic effects of radiation dose.

11.6.A   Acute and Chronic Radiation Exposure

It has been found that both high instantaneous and high integrated doses are harmful to health. A high instantaneous dose leads to the so called *acute exposure*, while a low

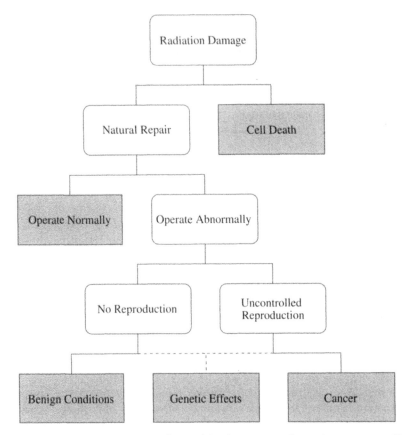

Figure 11.6.1: Diagram of possible damage and repair steps a cell may follow after irradiation.

instantaneous dose received for a long period of time is termed as *chronic exposure*. We will discuss the effects of such exposures in the following two subsections.

A.1 Acute Exposure

High instantaneous dose received in a short period of time can cause severe damage to skin and internal organs. It is generally accepted that a radiation dose of greater than 0.1 *Gy* received within a short period of time (hours to a few days) can be classified as a source of acute exposure. The severity of the damage caused by acute dose depends on the dose level while the recovery depends on the general health of the individual.

In the following the damage to specific organs and the resulting symptoms have been listed according to the dose level. Note that the higher the dose the more probable it is to cause multiple kinds of damages.

▶ **Dose > 50** *Gy*: This extremely high dose can cause permanent damage to cells of the nervous system. The damage can result in internal bleeding and fluid build-up in the brain ultimately leading to death. The exposure can cause severe shock and even coma. At the minimum, the exposed person can suffer from confusion, speech difficulty, and coordination loss.

▶ **Dose > 25** Gy**:** This much exposure can cause epidermal and deep skin necrosis.

▶ **Dose > 10** Gy**:** A dose higher than 10 Gy can cause damage to the lining of intestines and stomach leading to gastrointestinal complications. The exposed person initially suffers from nausea, vomiting, and diarrhea, which can lead to dehydration. Severe effects include formation of ulcers and inability to digest food. Other probable damage is the atrophy of the skin with complications.

▶ **Dose > 5** Gy**:** It can lead to visual impairment and cataract.

▶ **Dose > 3.5** Gy**:** This much dose can lead to permanent sterilization if the reproductive organs are directly exposed.

▶ **Dose > 2** Gy**:** A dose higher than 2 Gy can cause hair loss and erythema or reddening of the skin.

▶ **Dose > 1** Gy**:** A number of organs such as bone marrow, spleen, and lymphatic tissues are highly susceptible to radiation damage and therefore even the moderately high dose of 1 Gy can cause severe damage to their cells. The damage can cause internal bleeding leading to bacterial infections and possibly death.

▶ **Dose > 500** mGy**:** Benign tumor growths in thyroid glands are common after receiving a dose higher than 500 mGy. Other effects include damage to bone marrow and opacities in the ocular lens.

A.2 Chronic Exposure

In the preceding section we saw that an acute dose can bring serious health hazards. The same is true for a low level dose received in a long period of time. Such an exposure is said to be chronic. The distinguishing aspect of the chronic exposure is that it is very difficult to assign deterministic effects to it.

The effects of chronic exposure become obvious after a long time, of the order of several years. For example, some kinds of cancers can take up to 25 years to produce symptoms and get diagnosed. At that time it becomes difficult to determine whether the cause of the cancer was radiation exposure or something else. Such a long *incubation* period of the disease makes it difficult to determine safe levels of dose for workers and general population. One way to derive such statistical inferences is to study the disease rates in populations that are exposed to different levels of radiation. Such studies have been undertaken by many researchers but the results have not been found to be statistically significant to make predictions. Due to this difficulty the so called *safe* levels have been devised by extrapolating acute doses with known deterministic effects.

11.6.B Effects and Symptoms of Exposure

In the preceding sections we discussed how radiation affects living organisms and also looked at its effects on human body. In general, the effects on humans can be divided into two categories: somatic effects and genetic effects.

B.1 Somatic Effects of Radiation

The somatic effects are the ones that appear in the person that is exposed to radiation. As discussed earlier, the effect can become obvious immediately after the exposure or can take a long time to cause noticeable changes. For example, high dose to skin can cause immediate reddening of the exposed area. Such a prompt exposure is typical of high levels of radiation delivered in a short period of time. On the other hand, most cancers take many years to develop and cause symptoms that could lead to diagnosis. Such delayed effects are caused by cell mutations and subsequent uncontrolled reproduction. Whether the effect is prompt or delayed, if it appears in the person being exposed, it is called a somatic effect.

B.2 Genetic Effects of Radiation

We saw earlier that if the chromosome of a cell gets permanently damaged, it can lead to genetic changes. If the exposed cell is related to reproduction, the damage to its DNA can lead to developmental problems in the offspring of the person. Such cell changes are generally termed as *germline mutations* and do not affect the exposed person. The hazards associated with germline mutations range from premature death and miscarriage to cancer in later life.

A number of studies have been performed to determine the probability of damage due to germline mutations. Since it is difficult to determine whether a germline mutation in an individual has occurred or not, these studies have largely been based on finding the correlation of dose received by radiation workers and the genetic disorders in their children. Fortunately up until now, no conclusive evidence of strong correlation between the exposure and childbirth defects has been found. However a few studies have indicated that the occurrence of leukemia on the children whose fathers received occupational exposures was about 2 to 2.5 times higher than in general population. Unfortunately these studies have suffered from low statistics, that is, the number of individuals available for analysis were fairly small. Because of this, many researchers do not regard this inference as conclusive evidence of leukemia occurrence in children. It should, however, be mentioned that the type of exposure to father has also been found to be important in this regard. The internal exposure to radionuclides is considered as a high risk cause of childhood cancer as opposed to whole body exposure.

11.6.C Exposure Limits

We just saw that radiation can not only induce acute damage but also cause subtle cell mutations. Unfortunately the effects of cell mutations become obvious only after an extended period of time. For example, it has been observed that people, who are continuously exposed to low level of radiation, have a high probability of developing some kind of cancer. The question then arises as to what are the safe limits of dose and when should one start worrying about the exposure. Unfortunately there is no simple answer to this since most of the effects of radiation are statistical in nature. One could, of course, argue that for extremely high dose the damaging effects are almost certain. This is definitely true but the effect is still statistical since in such a case due to very high intensity of radiation the probability of cells being damaged becomes very high. Having said that, there have been several extensive studies to determine the safe levels for radiation workers and general public. Based on those studies the limits for radiation workers and general public have been suggested by international commission on radiological protection. These limits, as given in Table 11.6.1, are followed by most organizations.

Table 11.6.1:   Exposure limits of different body parts for radiation workers and general public (26).

Exposure Type	Maximum Permissible Equivalent Dose (mSv)	
	Occupational	General Public
Skin	500/year	50/year
Eye lens	150/year	15/year
Hands	500/year	

The effective whole body dose limit set by ICRP is 20 mSv per year for radiation workers and 1 mSv per year for general public (26). This limit and the ones given in Table 11.6.1 are taken as the absolute maximum and therefore no radiation worker should ever be allowed to keep working in radiation environments after receiving doses higher than these limits. For additional safety, another principle that has now been almost universally adopted, is the so called ALARA principle. ALARA stands for As Low As Reasonably Achievable. What this means is that one should strive to aggressively limit the exposure to individuals within the limitations of the working environment and the cost.

Table 11.6.2:   A typical 3-level ALARA system for radiation workers of age 19 and older.

Maximum Permissible	ALARA (mSv)		
Whole Body Dose	Level-1	Level-2	Level-3
20/year	0.2/month	0.3/2 months	0.4/3 months

The ALARA principle does not relax the maximum permissible doses as those are still the ones given in Table 11.6.1. On the contrary, it actually provides a

framework for the radiation safety personnel to monitor the workers more closely
and take safety precautions accordingly. The usability of ALARA principle becomes
evident when one considers the possibility that a worker gets a high dose within a
short period of time, say a month, such that the yearly dose does not get exceeded.
This can be problematic since continuing working in such an environment may prove
to be harmful due to the increased probability of damage. For such situations, most
organizations define different ALARA levels. Two level systems are fairly common
in which if the dose received by the worker exceeds the first level the environment
is surveyed for the conditions and the worker is put on an aggressive monitoring
routine. In case the worker exceeds the level 2 she or he may then be barred from
working in any radiation environment for an extended period of time. Most of the
time ALARA levels are defined according to the dosimeter readout periods. The
dosimeter badges are generally read out every one or two months, at which time the
dose is monitored according to the ALARA guidelines specific to the organization.
Table 11.6.2 gives an example of a three level ALARA guideline for the whole body
dose for radiation workers.

 ## 11.7   Radiation Protection

Though radiation has been found to be extremely useful in different fields, its use
has extreme radiation hazards associated with it. The protection of personnel and
equipment from harmful effects of radiation is of primary concern in any radiation
environment. In this section we will concentrate on discussion of radiation protection
for personnel. The issue of radiation damage to materials has already been discussed
in various chapters in this book.

11.7.A   Exposure Reduction

In many situations it is extremely difficult, if not impossible, to completely eliminate the possibility of radiation exposure to personnel. However, one can work on
minimizing the exposure using some simple techniques mentioned in the following
sections.

A.1   Time

Low level dose integrated over a long period of time can cause significant harm to
personnel. The dose can therefore be minimized by carefully planning the activities
such that the exposure time is as short as possible. The workers in radiation environments are required to wear a dosimeter to record the dose. After some time the
dosimeter is read out through a reading device and the exposure level is compared
with the limit set by the corresponding authority. In case the limit has been approached the worker is not allowed to enter any other radiation environment for a
specific period of time.

A.2   Distance

Radiation follows inverse square law unless there is significant scattering and absorption in the medium. This law states that the radiation flux is proportional to

the square of the distance from the point source, that is

$$\phi \propto \frac{1}{r^2}, \qquad\qquad (11.7.1)$$

where r is the distance of the measurement point from the *point source*[3].

A point that is worth mentioning here is that for inverse square law to be valid the radiation should not suffer from significant scattering and absorption in the medium. In other words this law holds for all types of radiation only in vacuum. This does not mean that the law is not valid in laboratory environment where the medium is generally air, since the photons traveling in air suffer very little scattering and absorption. Therefore x-rays and γ-rays traveling in air do follow inverse square law to a very good approximation. On the other hand the massive particles, such as electrons and α-particles, the law does not hold at all. These particles quickly loose their energy in the medium and removed from the beam after traveling very short distances. For example, the range of even highly energetic α-particles in air is only a few centimeters.

Now assume that we have radiation in an environment that does not significantly absorb or scatter radiation. The question is how we can minimize the dose received by a person working in that environment. Of course this can be done by maximizing the distance of the person from the source. Since dose is directly related to the radiation flux therefore we can also write the inverse square law for dose as

$$D \propto \frac{1}{r^2}. \qquad\qquad (11.7.2)$$

Certainly we can write a similar expression for the exposure as well, that is

$$X \propto \frac{1}{r^2}. \qquad\qquad (11.7.3)$$

Note that this principle of maximizing the distance also holds for radiation that do not follow inverse square law. However the safe distance in this case is much smaller than photon beams.

Example:
A person is expected to receive an exposure of 500 mR if he stays at a distance of 12 cm from the source for 8 hours. What would be the total exposure if he is asked to stay at a distance of 1.2 m from the source for the same 8 hours?

Solution:
According to equation 11.7.3 the ratio of exposure at two positions is given by

$$\frac{X_1}{X_2} = \frac{r_2^2}{r_1^2}.$$

---

[3]As a reminder to the reader, a point source is the one whose dimensions are much smaller than the distance from the measurement point.

If subscripts 1 and 2 represent distances from the source of 12 cm and 1.2 m then the required exposure is given by

$$X_2 = \left(\frac{r_1}{r_2}\right)^2 X_1$$

$$= \left(\frac{12}{120}\right)^2 500$$

$$= 5 \, mR.$$

This example clearly demonstrates the advantage of increasing distance from the source. Increasing the distance by 10 times decreases the exposure by 100 times.

A.3   Shielding

Shielding is an integral part of any radiation facility. In most situations it is the most effective way of reducing radiation exposure to personnel. In chapter 2 we saw that the intensity of a photon beam decreases exponentially with distance, that is

$$I = I_0 e^{-\mu x}, \tag{11.7.4}$$

where I is the photon intensity at a depth x, I_0 is the initial intensity of the beam, and μ is the attenuation coefficient of the material. Recall that μ is a function of the material as well as the energy of the radiation. At any specific energy its value depends strongly on the atomic number of the material. The high Z elements, such as lead, are therefore the most effective means of shielding. Use of lead as a shielding material is very common in laboratories working with radiation sources. In radiation hostile environments, a combination of high Z elements and concrete are used to construct very thick shields.

Since the exposure and absorbed dose are directly related to the radiation intensity, the above equation can also be written as

$$X = X_0 e^{-\mu x} \tag{11.7.5}$$

$$\text{and} \quad D = D_0 e^{-\mu x}. \tag{11.7.6}$$

Of course, these expressions hold for the respective rates as well. One might argue that, since exposure is defined for air only, the first of these expressions can not be used for materials other than air. This is certainly not true since one can always assume that the exposure is being measured outside the material. For example, it can be used to determine the thickness of a shielding material such that the exposure falls to a certain level, provided the exposure in air without shield is known (see example below).

Example:
A ^{241}Am source emits 60 keV γ-rays. If the exposure rate in air at a certain distance from the source is 500 $mR/hour$, estimate the thickness of lead needed to decrease the exposure rate to 10 $mR/hour$ at the same location. Assume the attenuation coefficient to be 50 cm^{-1}.

Solution:
According to equation 11.7.5, the required thickness is given by

$$x = \frac{1}{\mu} \ln\left(\frac{X_0}{X}\right)$$

$$= \frac{1}{50} \ln\left(\frac{500}{10}\right)$$

$$= 0.078 \; cm.$$

Two terminologies that are commonly used in dosimetry applications. These are *half-value layer* or *HVL* and *tenth value layer* or *TVL*. The *HVL* is defined as the thickness of the shield that reduces the intensity by a factor of two. Similarly a *TVL* decreases the thickness by a factor of 10. The values of *HVL* and *TVL* depend on the value of the attenuation coefficient. This can be seen by substituting $I = I_0/2$ (decrease in intensity by a factor of 2) in the above expression of intensity variation as

$$\frac{I_0}{2} = I_0 e^{-\mu HVL}$$

$$\Rightarrow HVL = \frac{\ln(2)}{\mu}. \tag{11.7.7}$$

Similary the expression for *TVL* can be obtained by substituting $I = I_0/10$ in equation 11.7.4, that is

$$\frac{I_0}{10} = I_0 e^{-\mu TVL}$$

$$\Rightarrow TVL = \frac{\ln(10)}{\mu}. \tag{11.7.8}$$

The relationship between *HVL* and *TVL* can be obtained by taking the ratio of equations 11.7.7 and 11.7.8, that is

$$\frac{HVL}{TVL} = \frac{\ln(2)}{\ln(10)}$$

$$\Rightarrow HVL = 0.3 \times TVL.$$

Example:
^{137}Cs emits 662 keV γ-rays. Determine the *HVL* and *TVL* for lead having attenuation coefficient of 1.15 cm^{-1}.

Solution:
The *HVL* and *TVL* can be computed from equations 11.7.7 and 11.7.8 as

follows.

$$HVL = \frac{\ln(2)}{\mu}$$
$$= \frac{\ln(2)}{1.15}$$
$$= 0.6 \; cm$$
$$TVL = \frac{\ln(10)}{\mu}$$
$$= \frac{\ln(10)}{1.15}$$
$$= 1.0 \; cm$$

A point to note here is that when very high energy photons pass through a material, such as lead, they produce high energy electrons. These electrons quickly loose their energy and as a result give rise to Bremsstrahlung radiation. This radiation is also highly penetrating and therefore should be considered while estimating the thickness of shielding material. Such computations are fairly complex and are generally done by standard computer codes that simulate the passage of radiation through matter. For safety reasons it is always a good practice to use shields much thicker than required by simple calculations of exponential attenuation.

Problems

1. *In laboratory environments, radioactive samples are tagged with their initial activities. If considerable time has elapsed as compared to its half life, the current activity of a sample should be computed using the decay rate equation. This and the next example would refresh the reader's mind as to how simply it can be done.* A radioactive sample having a half life of 20.6 days has an initial activity of 80 μCi. What would be its activity after 200 days?

2. The half life of a radioactive sample is found to be 21.5 days. How long would it take for its activity to decrease by 70%?

3. A radiation worker receives an exposure of 25 R after working for 3 hours at a distance of 5.5 cm from a cobalt-60 source. Compute the activity of the source.

4. If the worker of the previous exercise had to work at the same distance for the same amount of time in front of a cesium-137 source, estimate the activity of the source if in the end he received the same amount of exposure as in the case of cobalt-60 source.

5. In a radiation therapy setup it was assumed that the incident radiation beam was circular with a radius of 0.3 mm. However the actual beam was rectangular. If one side of the rectangle was 0.05 mm, compute the other side. What would be the area of an equivalent square beam?

6. Compute the energy released after the capture of thermal neutrons by oxygen-12 and carbon-14 nuclei.

7. Estimate the dose absorbed by a soft tissue if an internal source of thermal neutrons remains in a soft tissue for 24 hours. Assume that the source emits, on the average, 10^3 neutrons/$cm^2 s^{-1}$.

8. A patient is exposed to 0.02 nm wavelength photons for a period of 2.5 seconds. If 6.7×10^6 photons strike the patient per second, compute the particle and energy fluence rates.

9. Calculate the exposure rates from a 30 mCi cobalt-60 source at distances of 10 cm and 100 cm.

10. For the exposure computed in the previous exercise, determine the air Kerma. Assume the quality factor to be 0.01.

11. In a laboratory experiment, a 20 mCi radioactive source emitting γ-rays is to be shielded. Estimate the width of lead blocks needed to shield the source such that the exposure rate falls to 5% of its value at 10 cm in air.

12. If the mass attenuation coefficient of lead for γ-rays emitted by ^{241}Am (60 keV) is 50 cm^{-1}, calculate the half- and tenth-value layers.

13. Suppose a GM counter having an efficiency of 15% measures a count rate of 3500 counts per minute in a radiation environment. Determine the activity it would correspond to if the measurement is made at a distance of 1.6 m from the source.

14. A radiation worker works for 4 hours a day for 5 days in the radiation environment depicted in the previous exercise. If, on the average, he remained at a distance of 2 m from the source during the whole time, determine the total dose he would have received. Should this worker be allowed to keep working in radiation environments?

15. Estimate the thickness of lead required to decrease the exposure rate from a cobalt-60 source by 99%. Assume the attenuation coefficient to be 0.6 cm^{-1}.

▟ Bibliography

[1] Alberts, W.G. et al., **Advanced Methods of Active Neutron Dosimetry for Individual Monitoring and Radiation Field Analysis (ANDO)**, PTB-Bericht PTB-N-39, 1999.

[2] ASTM, **Standards on Dosimetry for Radiation Processing**, ASTM International, 2004.

[3] Attix, F.H., **Introduction to Radiological Physics and Radiation Dosimetry**, Wiley-Interscience, 1986.

[4] Auburn University, **Radiological Safety Reference Handbook** Auburn University, AL, USA, 2004.

[5] Boetter-Jensen, L., McKeever, S.W.S., **Optically Stimulated Luminescence Dosimetry**, Elsevier Science, 2003.

[6] Butson, M.J. et al., **Radiochromic Film for Medical Radiation Dosimetry**, Materials Sci. and Engg. R, 41, 2003.

[7] Cameron, J.R., **Thermoluminescent Dosimetry**, University of Wisconsin Press, 1968.

[8] Dietze, G. et al., **Determination of Dose Equivalent with Tissue-Equivalent Proportional Counters**, Radiation Prot. Dosimetry, 28, 1989.

[9] Forshier, S., **Essentials of Radiation Biology and Protection**, Thomson Delmar Learning, 2001.

[10] Genthon, J.P., **Reactor Dosimetry**, Springer, 1987.

[11] Gerweck, L.E., Kozin, S.V., **Relative Biological Effectiveness of Proton Beams in Clinical Therapy**, Radiother. Oncol., 50, 1999.

[12] Grandolfo, M., **Biological Effects and Dosimetry of Nonionizing Radiation:Radiofrequency and Microwave Energies**, Springer, 1983.

[13] ICRU, **Linear Energy Transfer, Report 16**, Intern. Comm. on Rad. Units and Meas., 1970.

[14] ICRU, **Microdosimetry, Report 36**, Intern. Comm. on Rad. Units and Meas., 1983.

[15] ICRU, **Stopping Powers and Ranges for Protons and Alpha Particles, Report 49**, Intern. Comm. on Rad. Units and Meas., 1993.

[16] ICRU, **Fundamental Quantities and Units for Ionizing Radiation, Report 60**, Intern. Comm. on Rad. Units and Meas., 1998.

[17] Jani, S.K., **Handbook of Dosimetry Data for Radiotherapy**, CRC, 1993.

[18] Jayaraman, S., Lawrence, H., **Clinical Radiotherapy Physics: Basic Physics and Radiation Dosimetry**, CRC-Press, 1996.

[19] Kraft, G., **Tumor Therapy with Heavy Charged Particles**, Prog. Part. Nucl. Phys., 45 (Suppl. 2), 2000.

[20] Ranogajec-Komor, M., **Thermoluminescence Dosimetry - Application in Environmental Monitoring**, Rad. Safety. Mgmt., Vol.2, No.1, 2002.

[21] Lim, M., **Medical Radiation Dosimetry: Principles and Practice**, Mosby-Year Book, 1992.

[22] Lindskoug, B.A., **Clinical Applications of Thermoluminescent Dosimetry**, Strahlentherapie, 161, 1985.

[23] Martin, J.E., **Physics for Radiation Protection**, Wiley-Interscience, 2000.

[24] McKinlay, A.F., **Thermoluminescence Dosimetry**, Taylor & Francis, 1981.

[25] Orton, C.G., **Radiation Dosimetry**, Springer, 1986.

[26] ICRP, **Summary of Recommendations of Publication 60**, International Commission on Radiological Protection, 1990.

[27] Paganetti, H. et al., **Relative Biological Effectiveness (RBE) Values for Proton Beam Therapy**, Int. J. Radiat. Oncol., Biol., Phys., 53, 2002.

[28] Paganetti, H., **Nuclear Interactions in Proton Therapy: Dose and Relative Biological Effect Distributions Originating from Primary and Secondary Particles**, Phys. Med. Biol., 47, 2002.

[29] Paic, G., **Ionizing RadiationProtection and Dosimetry**, CRC Press, 1988.

[30] Profio, A.E., **Radiation Shielding and Dosimetry**, John Wiley & Sons Inc., 1978.

[31] Rajan, G.K.N., **Advanced Medical Radiation Dosimetry**, Prentice-Hall, 2004.

[32] Sabol, J., Weng, P.-S., **Introduction to Radiation Protection Dosimetry**, World Scientific Pub. Co. Inc., 1995.

[33] Skarsgard, L.D., **Radiobiology with Heavy Charged Particles: a Historical Review**, Phys. Medica, 14 (Suppl. I), 1998.

[34] Shani, G., **Radiation Dosimetry: Instrumentation and Methods**, CRC Press, 2000.

[35] Shapiro, J., **Radiation Protection: A Guide for Scientists, Regulators and Physicians**, Harvard University Press, 2002.

[36] Simmons, J.A., Watt, D.E., **Radiation Protection Dosimetry: A Radical Reappraisal**, Medical Physics Publishing Corporation, 1999.

[37] Turner, J.E., **Atoms, Radiation, and Radiation Protection**, Wiley-Interscience, 1995.

[38] Wagermans, J. et al., **Reactor Dosimetry in the 21st Century: Proceedings of the 11th International Symposium on Reactor Dosimetry Brussels, Belgium 18 - 23 August 2002**, World Scientific Publishing Company, 2003.

[39] Wambersie, A., **RBE, Reference RBE and Clinical RBE: Applications of these Concepts in Hadron Therapy**, Strahlenther. Onkol., 175 (Suppl. II), 1999.

[40] Watt, D.E., **Quantities for Generalized Dosimetry of Ionizing Radiations in Liquid Water**, CRC, 1996.

[41] Wilkens, J.J., Oelfke, U., **Analytical Linear Energy Transfer Calculations for Proton Therapy**, Med. Phys., 30(5), 2003.

Chapter 12

Radiation Spectroscopy

Radiation spectroscopy refers to a class of techniques that utilize different properties of radiation to study materials and particles. Its usefulness in research and industrial environments is well established. Perhaps the biggest advantage of radiation spectroscopy is that it is essentially a non-destructive technique.

Because of its direct industrial applications and research value, radiation spectroscopy has seen immense developments not only in techniques but also in related instrumentation. Even though this entire chapter has been devoted to the discussion of this field, it is not intended to be a comprehensive guide. However every attempt has been made to give an overall picture of different techniques and spectrometers that are in common use.

Most of the discussions in this chapter have been organized with respect to the types of radiation. That is, the spectroscopy of different radiation types have been discussed separately. This was necessary due to the differences in the techniques and instruments that are used for different kinds of radiation. At the end of the chapter, the technical aspects of *mass spectroscopy* and *time spectroscopy* are also discussed.

12.1   Spectroscopy of Photons

12.1.A γ-Ray Spectroscopy

Because of their deep penetrating power, γ-rays are routinely used for nuclear spectroscopy. The lower energy photons that originate from electronic transitions (such as x-rays) are not suitable for this purpose mainly due to their lower penetration power as compared to γ-rays.

The standard system used for γ-ray spectroscopy consists of a $NaI(Tl)$ scintillator and a photomultiplier tube. The scintillator converts the γ-ray photons into visible light photons, which are detected by the photomultiplier tube. A photomultiplier tube is not absolutely necessary since, in principle, any photodetector with appropriate efficiency and sensitivity can be used. However, the scintillator and photomultiplier combination is more popular. The reader is encouraged to go through the relevant sections in the chapter on scintillators and photodetectors for more details. Here we will concentrate on their use in γ-ray spectroscopy.

A typical γ-ray spectroscopy setup using NaI scintillator is shown in Fig.12.1.1. The shield shown in the figure has two purposes. One is to protect the personnel from radiation exposure and the other is to minimize the background in the detector. The source and the detector are positioned such that most of the γ-ray photons from the source reach the scintillator, where they get converted into light photons. To guide the light photons towards the photomultiplier tube the scintillator is surrounded by a light reflector assembly. The light photons are then absorbed by the photocathode of the PMT, which emits photoelectrons. These photoelectrons get multiplied through the dynode assembly and ultimately lead to a measurable pulse.

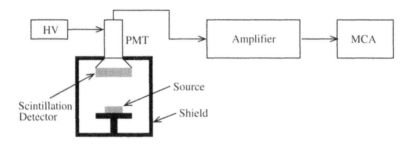

Figure 12.1.1: Block diagram of a typical γ-ray spectroscopy setup with a scintillation detector (such as NaI(Tl)) and a PMT.

In the above description we have not considered the γ-ray photons that hit the shielding material. Since these photons can produce sizable effect on the output, they must be taken into account. Fig.12.1.2 shows the interaction mechanisms in the detector as well as in the surrounding shield. In chapter 2 we discussed different ways in which a photon can interact with material. We saw that the three most important interaction mechanisms are photoelectric effect, Compton scattering, and pair production. The γ-ray photons, as well as secondary photons, interact with the detector material and the shield through all of these modes, provided they carry enough energy.

The usual method adopted in spectroscopy is the pulse height analysis. This technique is based on the fact that the height of the output pulse is proportional to the energy deposited in detector's active medium. Therefore if one plots the number of pules obtained with respect to height of the pulses, it would correspond to the energy spectrum of the deposited energy. Now, since the deposited energy is directly related to the energy carried by the incident radiation, the spectrum obtained actually corresponds to that of the spectrum of the incident radiation. Pulse height analysis can be done using a single channel analyzer, though the process in that case is fairly tedious and labor intensive. The best and the most commonly used technique is the use of a multi channel analyzer or MCA. An MCA records the number of pulses in each *pulse bin*. The size of the pulse bin can normally be defined by the user and is usually selected according to the particular constraints of the experiment, such as required resolution and available time. The MCA spans its full dynamic range in equal width pulse bins and generally displays the output on a screen. The data can be saved, printed, or transferred. Modern MCAs can also be directly interfaced to a computer for further analysis, display, and storage of the data.

As shown in Fig.12.1.1, the photons from a radioactive source are emitted in all directions. The ones moving directly to the scintillator deposit the most energy and produce highest energy peak. The $FWHM$ of this peak determines the resolution

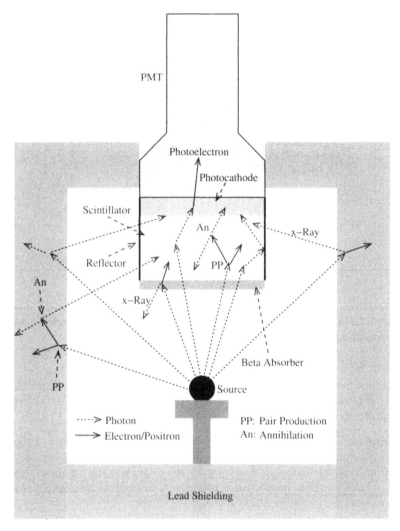

Figure 12.1.2: Photon interaction mechanisms in a γ-ray spectroscopy setup.

of the system at that particular energy. In a typical γ-ray source spectrum, other peaks are also visible. In chapter 2, during the discussion on Compton scattering, we saw that the two other peaks generally found are Compton edge and the backscatter peak. The Compton edge appears at approximately $E_\gamma - 255 \; keV$ and corresponds to the Compton scattering of electrons at an angle such that the electrons carry away the maximum possible energy. This scattering can actually be visualized as a head-on collision of an incoming photon with an electron at rest such that the electron starts moving forward and the photon scatters back at 180^0. The reader is encouraged to review the Compton scattering section in chapter 2 for detailed discussion of the Compton edge.

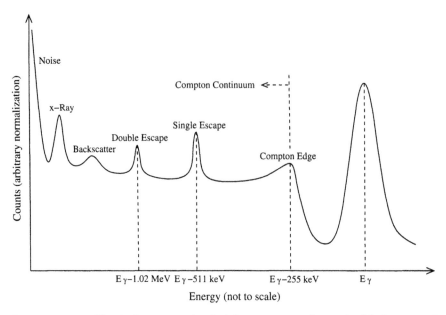

Figure 12.1.3: Typical γ-ray pulse height spectrum for a shielded setup of source and detector.

The highest energy peak and the Compton edge are sketched in Fig.12.1.3, which represents a typical energy spectrum of a radioactive source. The other notable features of the spectrum are:

▶ **Compton Continuum:** The curve below the Compton edge is called Compton continuum. This corresponds to the distribution of energy between incident photons and the scattered electrons during Compton scattering at different angles. Since Compton scattering is possible at all angles therefore this continuum can extend up to the beginning of the spectrum.

▶ **Noise Peak:** Most MCAs allow the user to gate off the noise counts from the spectrum. This is done be selecting a threshold pulse height below which no events are recorded and displayed. However in some situations one might be interested in keeping this information in the data stream. The noise peak shown in Fig.12.1.3 refers to this case.

▶ **Escape Peaks:** If the energy of the incident photon is greater than the threshold for pair production, that is $E_{gamma} > 1.02\ MeV$, it can produce an electron-positron pair in the detector as well as in the shield. A pair production event in the detector is shown in Fig.12.1.4. The positron thus produced has very short half life and quickly combines with a nearby electron to produce photons, a process known as annihilation. Most of the annihilation events produce two photons traveling in opposite directions. Now, as shown in Fig.12.1.4, it can happen that one or both of these photons escape the detector volume without depositing any appreciable energy. If one photon escapes and the other deposits most of its energy, it leads to a *single escape peak* in the

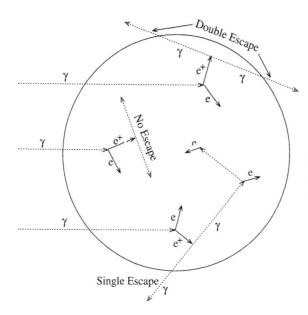

Figure 12.1.4: Energy deposition of high energy photons above pair production threshold in a detector.

spectrum (see Fig.12.1.3). The single escape peak can be easily identified since it appears at an energy of $E_\gamma - 0.511\ MeV$. Just like single escape peak, a double escape peak is also sometimes visible in the spectrum. This corresponds to the escape of both annihilation photons from the detector. The double escape peak appears at $E_\gamma - 0.511\ MeV$. It could happen that none of the annihilation photons escapes the detector. In this case the full energy carried by the incident photon gets deposited in the detector. The height of the output pulse then corresponds to the full energy peak of the spectrum (E_γ in Fig.12.1.3).

▶ **Shield Scattering Peaks:** These are the peaks that appear when photons hit the shielding material and are scattered into to the detector. Since these photons lose most of their energy during the collision, these peaks appear at the lower end of the spectrum. A very commonly seen shield scattering peak is the so called *backscatter peak*. This peak corresponds to the deposition of energy by photons that are backscattered from the shield into the detector. The backscatter peak is generally fairly broad due to scattering of photons at different angles that are close to 180^0.

▶ **X-Ray Peak:** When γ-ray photons interact with the high Z material of the shield, they can produce x-ray photons. The energy of these photons is substantially lower than that of the original γ-ray photons. Therefore, as shown in Fig.12.1.3, the energy deposited by them in the detector produces a peak at the lower end of the energy spectrum.

12.1.B Calibration

Calibrating the apparatus is the first step for any spectroscopic measurement. Here by calibration we mean determination of the relationship between MCA channels and the energy absorbed in the detector. A good multi channel analyzer has a very linear response in its full dynamic range. That is, the bins or channels of the MCA are directly proportional to the energy absorbed in the detector. This implies that if one plots the absorbed energy with respect to the MCA channel, the result should be a straight line. Without this linearity it would become very difficult, if not impossible, to calibrate the system.

In essence, calibration is a fairly straightforward process and can be performed using one or more known radioactive sources. We saw in Fig.12.1.3 that the highest peak corresponds to the energy of the γ-ray. Therefore one can determine the bin interval of the MCA corresponding to this peak. Repeating the experiment with another source would produce another bin-interval to energy relation. These two points can be used to determine the calibration curve. Once could also use other peaks in the spectrum to obtain more statistics and hence better calibration parameters.

One might be tempted to calibrate the system using only one peak from a source. This can be done provided the null points of the MCA bin and the energy scale coincide. In general, this is not true due to the presence of an offset on the MCA bin axis. That is, the zero energy does not correspond to the zero pulse height. This offset is due purely to electronics and can be determined by using an external pulser. However if one uses several known peaks in the spectrum and fits a straight line to the data, the offset can be determined by extrapolating the line to the x-axis. This point will become clearer when we discuss the calibration procedure for the α-particle spectroscopy systems.

12.1.C   X-ray Spectroscopy

X-ray, due to its high penetration power, is an excellent material probe. But this is not all, since spectroscopic techniques based on x-rays can be used to determine intricate details about the material they pass through. With the advent of synchrotron radiation sources, the field of x-ray spectroscopy has seen immense research and development. In the following sections we will look at the most commonly used x-ray spectroscopic techniques.

C.1   X-ray Absorption Spectroscopy

We know that the absorption probability of x-rays depends on the type of material they interact with. Hence if somehow we could determine the extent of the absorption, we could determine the details about the material. This is the main purpose of x-ray absorption spectroscopy. Specifically, one strives to determine the x-ray absorption fine structure or XAFS of the material by measuring the photon absorption at different x-ray wavelengths. The concept is fairly simple but since it requires very intense source of x-rays, the experiment can only be done at specialized laboratories. In fact, most XAFS measurements are done at synchrotron radiation facilities. Synchrotron radiation was introduced in the first chapter as a highly intense beam of x-rays. At all accelerators where charged particles are accelerated in curved paths, synchrotron radiation is produced as a byproduct. Since the realization of its potential as an excellent non-destructive tool for material research, it has found more prominence in the scientific and industrial communities. So much that now dedicated synchrotron radiation facilities exist in different parts of the world where experimental time is made available to researchers.

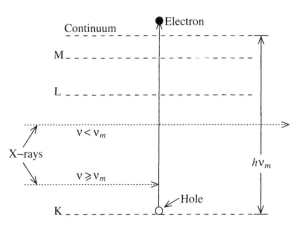

Figure 12.1.5: Graphical depiction of the process of photoelectric absorption of x-ray photons by an atom. An atom in the ground state is transparent to the photons having energy less than the binding energy of the core electron. X-rays having energy greater than the binding energy are absorbed by an inner core electron. The electron then gets elevated to the continuum and becomes free to move around. The vacancy left by the electron is called a *hole* since it acts as an attractor for electrons from outer shells.

The principle aim of x-ray absorption spectroscopy is to determine the XAFS of the material. To understand how it is done we should first revisit the mechanisms of photon interaction with matter. Fortunately for x-rays we should not be concerned with all the interaction mechanisms since x-ray photons interact with atoms predominantly through the process of photoelectric absorption. This interaction is graphically depicted in Fig.12.1.5. Whenever a photon having energy equal to or greater than the binding energy of the core electron interacts with the atom, it sets the electron free. The vacancy created in the K-shell acts as an effective positive charge and is sometimes referred to as a *hole*. We will shortly discuss its role in absorption spectroscopy. First let us see how the elevation of the electron can lead to the determination of XAFS. The process of photoelectric absorption completely eliminates the photon, resulting in the reduction in beam intensity. Measurement of the beam intensity after the material can therefore lead to the estimation of absorption by the material. And since the absorption is related to the binding energy of the material, it can lead to determination of electronic states of the material and possibly identification of the elements composing the material. This is the basic principle of *transmission* XAFS. The reason why transmission XAFS works so well can be understood by recalling that the absorption of photons in a material follows an exponential of the form

$$I_x = I_0 e^{-\mu(E)d}, \tag{12.1.1}$$

where I_0 and I_x are the incident and transmitted intensities. d is the thickness of the material and μ is the absorption coefficient of the material at that energy. This relation can also be written as

$$\mu(E) = \frac{1}{d} \ln \left(\frac{I_0}{I_x} \right). \tag{12.1.2}$$

Note that the absorption coefficient μ is energy dependent. If the energy of the photons is increased slowly, there comes a point when the absorption coefficient suddenly increases (see Fig.12.1.6). This happens when the photon energy equals the binding energy of the core electron. The sudden increase in the absorption coefficient results in a corresponding decrease in the measured intensity. This change in absorption coefficient is known as an *edge*. A material can have a number of edges corresponding to electronic transitions from different shells. For x-ray energies, one generally deals with K-edges. The important point to note here is that an edge is a unique signature of an element and can therefore be used to identify it from a mixture. This technique is generally known as *transmission* spectroscopy since here one simply measures the intensity of transmitted x-rays at different energies. The simplicity of this technique should not undermine its usefulness, though. As a matter of fact, as we will shortly see, besides being simple it is also the most efficient technique available for absorption spectroscopy.

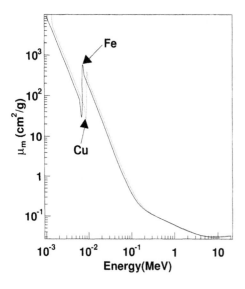

Figure 12.1.6: Variation of mass absorption coefficients of iron and copper for photons. The edges shown correspond to K-shell transitions.

Fig.12.1.7 shows a typical absorption spectrum obtained through transmission spectroscopy. As shown in the figure, such spectra are generally divided into two regions. The one near the absorption edge contains the near edge x-ray fine structure or NEXAFS. The spectroscopy done in this region is then referred to as NEXAFS or XANES (x-ray absorption near edge structure) spectroscopy. The second region on the side in Fig.12.1.7 contains the extended x-ray absorption fine structure or EXAFS. These structures are explored through the EXAFS spectroscopy.

The attenuation coefficient shown in Fig.12.1.7 has been plotted against the x-ray photon energy. However the general practice is to plot it with respect to the photo-electron wavenumber instead. The rationale behind this is that the fine structures are generated due to the interference effect and are therefore more closely related to the wave nature of the photoelectron. The transition from the photon energy to the photoelectron wavenumber is fairly straightforward. We start with the basic

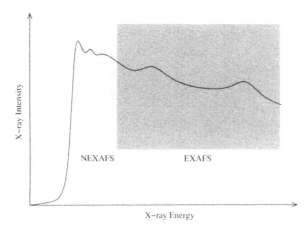

Figure 12.1.7: Typical x-ray absorption spectrum. The spectrum is divided into two regions of near-edge and extended fine structures.

relation between the wavelength of an electron λ_e and its momentuam p_e, that is

$$\lambda_e = \frac{h}{p_e},\tag{12.1.3}$$

where h is the Planck's constant. Using non-relativistic consideration, we can write the momentum of the electron as

$$\begin{aligned} p_e &= m_e v_e \\ &= \sqrt{2 m_e E_e},\end{aligned}\tag{12.1.4}$$

where m_e is the mass of the electron, v_e is its velocity, and E_e is its kinetic energy given by $E_e = m_e v_e^2/2$. Substituting $p_e = \sqrt{2 m_e E_e}$ in the above equation gives

$$\lambda_e = \frac{h}{\sqrt{2 m_e E_e}}.\tag{12.1.5}$$

The kinetic energy E_e of a photoelectron is relation to the energy of the incident photon E_γ through

$$E_e = E_\gamma - E_b,\tag{12.1.6}$$

where E_b is the binding energy o f the atom. Hence the wavelength of a photoelectron can be evaluated from

$$\lambda_e = \frac{h}{\sqrt{2 m_e \left(E_\gamma - E_b \right)}}.\tag{12.1.7}$$

Now, the wavenumber of a particle having wavelength λ is given by $k = 2\pi/\lambda$. The above equation can then also be written as

$$\begin{aligned} \frac{2\pi}{k} &= \frac{h}{\sqrt{2 m_e \left(E_\gamma - E_b \right)}} \\ \Rightarrow k &= \left[\frac{2 m_e \left(E_\gamma - E_b \right)}{\hbar^2} \right]^{1/2},\end{aligned}\tag{12.1.8}$$

where $\hbar \equiv= h/2\pi$. Using this expression the attenuation coefficient of the material for x-rays can be scaled to k. A typical plot of $\mu(k)$ with respect to k is shown in Fig.12.1.8. Note that the scaling of μ with respect to k does not affect its shape.

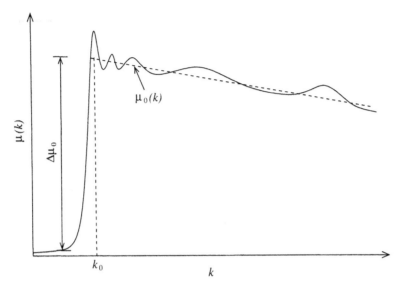

Figure 12.1.8: Typical variation of x-ray absorption coefficient with respect to photoelectron wavenumber as obtained by XAFS spectroscopy.

The variation of the absorption coefficient with respect to energy or wavenumber are generally known as *oscillations*. These oscillations are believed to be due to interference effect. This implies that in the absence of any interference, that is when there is only an isolated atom, one should have a nearly flat response. Note that this does not mean that μ should not vary with respect to k. In Fig. reffig:spectxray6 such a case is depicted by the function $\mu_0(k)$. This function is very useful in XAFS since it can be used to quantify the oscillations according to the following equation.

$$\chi(k) = \frac{\mu(k) - \mu_0(k)}{\triangle\mu_0(k_0)} \tag{12.1.9}$$

Here $\triangle\mu_0(k_0)$ is simply a normalization factor, called the *edge step*. A plot of $chi(k)$ with respect to k depicts the oscillations. The structure of these oscillations gives insight into the structure of the material and its constituents. A typical XAFS oscillation curve is shown in Fig.12.1.9.

Let us now discuss how XAFS spectra can be obtained using the transmission technique. The basic building blocks of transmission experiment are shown in Fig.12.1.12 and described below.

▶ **X-ray Source:** One of the basic requirements of an XAFS measurement setup is the intensity of x-rays, which should be as high as possible. The reason is that the number of photons surviving the absorption should be large enough to obtain a well defined spectrum showing the fine structure details. Another important requirement is that the x-rays should be tunable. That is, one should be able to tune the laser to particular frequencies. Synchrotron radiation sources fulfill both of these requirements since they produce highly intense

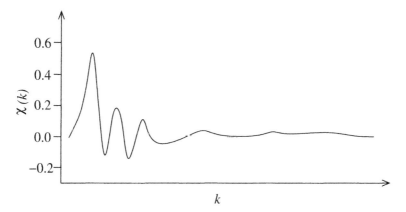

Figure 12.1.9: Typical XAFS oscillation curve normalized to the edge step.

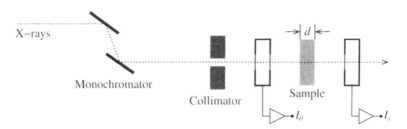

Figure 12.1.10: A basic setup for XAFS spectroscopy using transmission technique.

beams of x-rays of continuous spectra. Hence most XAFS experiments are performed using synchrotron radiation. In fact, modern synchrotron facilities are equipped with dedicated XAFS beam lines, which are designed to provide highly stable beams of x-rays.

▶ **Monochromator:** The job of a monochromator is to tune the synchrotron beam to well defined frequencies. Most monochromators in use today are made of silicon. They exploit the process of Bragg diffraction to tune the x-rays. It is important that the monochromator provides good energy resolution in order to reveal all the fine structures. In general one strives to have an energy resolution of better than 10^{-4}, which is achievable through silicon devices.

▶ **Collimator:** A collimator is used to align the beam to minimize dispersion.

▶ **Detectors:** In general one needs two detectors to correctly identify the fine structures. The detector before the sample ensures that any structure in the beam itself does not get wrongly identified as a fine structure. However there are stringent requirements on the working of this detector. For example, it should not attenuate the beam too much, say more than 10%. At the same time its resolution should be high enough to identify fine details in the spectrum. A resolution of better than 10^{-4} is generally required but is difficult to achieve. The main reason is the conflicting requirements of the resolution and the attenuation. For higher resolution the detector should absorb large number of photons. Recall that the physical limit on resolution is set by the statistical noise of photon absorption. The larger the number of photons absorbed, the lower the relative uncertainty or *noise*. Since the detector after the sample does not have the attenuation constraint, it can absorb as many photons as required to achieve good signal to noise ratio. In most XAFS applications, use to ionization chambers as transmission detectors is fairly common. Such a detector is based on parallel plate geometry and generally has several chambers.

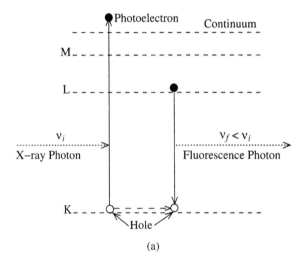

(a)

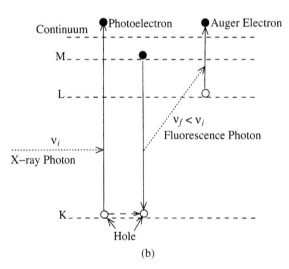

(b)

Figure 12.1.11: Generation of (a) fluorescence photon and (b) Auger electron following x-ray photon absorption.

Even though it is the most widely used one, transmission measurement is not the only technique available for absorption spectroscopy. The other two techniques used are the so called *fluorescence* spectroscopy and *Auger electron* spectroscopy. To understand these methods, the reader is referred to Fig.12.1.11. The sketches drawn here depict the possible relaxation modes of an atom after absorbing an x-ray photon. Note that absorption of an x-ray photon, resulting in the emission of a photoelectron, makes the atom unstable. To regain the stability one or more electronic transitions take place quickly after the photoelectric absorption. Let us first concentrate on Fig.12.1.11(a). Here the vacancy left behind by the photoelectron is shown to have been filled by an electron from the L-shell. Since this electron was originally at a higher energy level therefore, while making the transition to the lower level, it releases the excess energy in the form of a fluorescence photon. Obviously the energy of this photon is less than the energy of the original x-ray photon. The fluorescence spectroscopy involves detecting these photons and deriving the absorption coefficient from the measured fluorescence intensity. However now the absorption coefficient for x-rays does not follow the relation 12.1.2. In fact the actual dependence is fairly complicated due to the self absorption of these photons in the material. The reader should note that now we are dealing with two attenuation coefficients, one for the x-rays and the other for the fluorescence photons. The fluorescence photons, having longer wavelengths as compared to x-ray photons, generally get absorbed more quickly in the material. Nevertheless a very simple relationship between the absorption coefficient for x-rays and the intensity of fluorescence photons exists and is given by

$$\mu(E) \propto \frac{I_f}{I_0}. \tag{12.1.10}$$

Note that here I_f refers to the intensity of the fluorescence photons while I_0 is the intensity of incident x-rays. This relationship assumes little or no self absorption of fluorescence photons and therefore should be used carefully while interpreting the data. The most damaging effect of self absorption is the damping of the fine structures making the technique useless as far as spectroscopy is concerned. Fig.?? shows a basic setup designed for fluorescence and transmission spectroscopy of x-rays. Note that, in terms of detection setup, fluorescence spectroscopy is more demanding than its transmission counterpart. The main reason is that the fluorescence photons are essentially emitted in all directions.

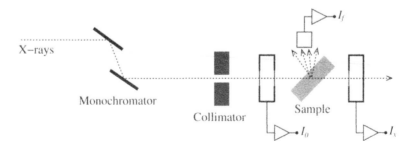

Figure 12.1.12: A basic setup for XAFS spectroscopy using transmission and fluorescence techniques.

As mentioned earlier, another type of absorption spectroscopy is the Auger spectroscopy. It involves detection of the Auger electrons that are produced following photoelectric absorptions. The mechanism is graphically depicted in Fig.12.1.11(b). Here, the vacancy created by the photoelectron transition to the continuum is shown to have filled by an electron in the M shell. The excess energy carried by this electron is given off in the form of a fluorescence photon. This photon has the ability to knock off an electron from the L or M shells. The figure shows transition of an L shell electron to continuum after absorbing the fluorescent photon. Such an electron is called Auger electron. The energy of these electrons is fairly low and therefore they can get absorbed near their generation site. Only the ones produced near the surface of the material manage to escape and get detected. Auger electron spectroscopy is therefore quite a challenging task. In a very simplified case, for the Auger electrons produced on and near the surface of the material, one can assume that the intensity of these electrons is related to the absorption coefficient of x-rays through the relation

$$\mu(E) \propto \frac{I_A}{I_0}. \tag{12.1.11}$$

where I_A and I_0 represent the intensities of Auger electrons and the x-ray photons respectively. μ is the absorption coefficient of the material for the x-ray photons. As with fluorescence spectroscopy, here also this relation does not take into account the damping of fine structures due to self absorption of Auger electrons.

As a final note, it should be mentioned that that here we are dealing with absorption of radiation and not its scattering. In the techniques that use scattering as the basis for spectroscopy, the material should have crystalline structure. This limits the applicability of such methods to a narrow band of materials that can be crystallized. As opposed to that, x-ray absorption spectroscopy can be used for virtually any material in any form, thus making it a highly desirable materials research tool.

C.2 X-ray Photoelectron Spectroscopy (XPS)

The XAFS measurements we studied in the previous section mostly involve detection of transmitted x-ray photons or fluorescence photons. Only in a very limited number o f cases one resorts to the technique involving detection of Auger electrons. No matter what technique we use, main process of x-ray photon absorption is always photoelectric effect. During our discussion on XAFS we did not talk about these electrons at all. However, for spectroscopic purposes these electrons can play an important role. To explain this, the reader is referred to the basic definition of the photoelectric effect, which states that this process involves total absorption of a photon by a bound electron such that the electron assumes the energy given by

$$E_e = E_\gamma - \Phi, \tag{12.1.12}$$

where E_γ is the incident photon energy and Φ is the work function. This work function refers to the binding energy of the most loosely bound electron in the atom. Since there are other electrons as well, which can absorb photons, we can write a generalized photoelectric equation as

$$E_{e,i} = E_\gamma - E_{b,i}, \tag{12.1.13}$$

where $E_{e,i}$ refers to the energy of the photoelectron having a binding energy of $E_{b,i}$. This implies that if one measures the energies of the emitted electrons, the corresponding spectrum would show peaks at the binding energies of the atom. Now, since binding energies are specific to the element, the spectrum can be used to identify the constituents of the sample. This technique is called x-ray photoelectron spectroscopy or XPS. Note that, as opposed to x-ray absorption spectroscopy, here one does not need to perform measurements at different x-ray energies. An x-ray source producing nearly single-wavelength photons is sufficient for photoelectron spectroscopy.

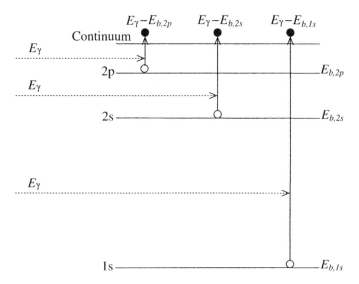

Figure 12.1.13: Principle of x-ray photoelectron spectroscopy.

The principle of XPS is graphically depicted in Fig.12.1.13. The most tightly bound electron is the 1s electron, which assumes the lowest energy after absorbing an x-ray photon. The electrons in the higher energy levels are emitted with higher energies. Detection of these photons and measurement of their energies then leads to a spectrum as shown in Fig.12.1.14. The peaks correspond to the transitions shown in Fig.12.1.13.

Let us now discuss the practical aspects of x-ray photoelectron spectroscopy. The experimental setup required for XPS is more complicated than for XAFS spectroscopy. The reason is that here one is required to not only detect electrons but also to measure their energies. Since electrons have very short range even in air therefore the sample and the detector must be kept in a vacuumed chamber (see Fig.12.1.15). The setup also requires an electron energy analyzer to measure the photoelectron energy.

Different types of energy analyzers exist but probably the most common is the so called hemispherical deflection analyzer or HDA. HDA is also known as concentric hemispherical analyzer or CHA since it mainly consists of two concentric hemispheres. The purpose of these hemispheres is to set a passband of electron energies. That is, to block all electrons that are outside a defined energy range. A sketch of the HDA is shown in Fig.12.1.16. The first part of the apparatus consists

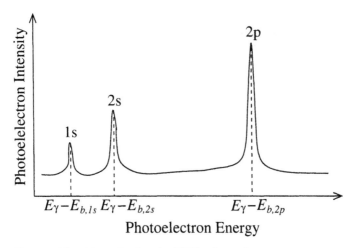

Figure 12.1.14:   A simple XPS photoelectron spectrum. Each peak corresponds to emission of photoelectrons from an atomic level. A typical spectrum contains a number of peaks corresponding to different elements in the sample.

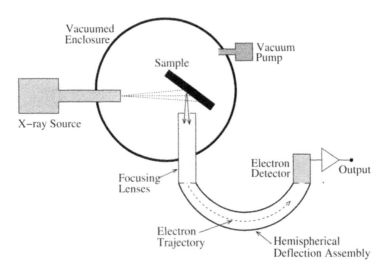

Figure 12.1.15:   Conceptual design of an electron energy analyzer.

of a series of focusing lenses. These lenses perform two tasks. The first one is to focus the electrons to the center of the slit to minimize their loss. The other task is to slow down the electrons such that they fall within the required energy passband of the analyzer. This slowing down is however not arbitrary since then the energy information will be lost. The second main component of an HDA consists of two metallic hemispheres. Application of electric potentials on these hemispheres creates

an electric field in the volume subtended by them. All the electrons coming from
the focusing lenses experience this electric force. The higher energy electrons get
deflected toward the outer hemisphere while the lower energy ones go toward the inner hemisphere. Hence only electrons having energy within a narrow band will pass
through the hemispheres and get detected by the electron detector (see Fig.12.1.16).
The energy and intensity measurements are done by the lens parameters and the
electron detector counts.

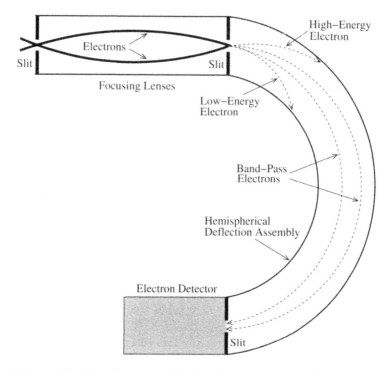

Figure 12.1.16: Conceptual design of a hemispherical deflection
analyzer.

C.3   X-ray Diffraction Spectroscopy (XDS)

The XAFS spectroscopy and the XPS we discussed earlier do not give us insight into
the internal physical structure of the material. The reason is that these techniques
exploit the processes leading to the absorption of x-ray photons and then detection
of particles emitted as a result. Since there is no correlation between the emitted
particles and the bonding structure of the material, therefore one can not deduce
the physical structure from the measurements.

    X-ray diffraction spectroscopy or XDS is a technique that can actually be used to
determine the internal structure of crystalline materials. As the name suggests, this
technique is based on the process of diffraction of x-ray photons. To understand the
process, the reader is referred to Fig.12.1.17, which depicts a simplified picture of
three atomic planes of a crystalline lattice. Parallel beams of x-rays are shown to be

reflected from these planes. The reflected rays can interfere constructively as well as destructively. The condition for constructive interference of two parallel x-rays, according to Bragg's law, is given by

$$2d \sin \theta = n\lambda, \tag{12.1.14}$$

where λ is the wavelength of x-rays, d is the distance between the two atomic planes, θ is the angle of incidence of x-rays with respect to atomic planes, and n is an integer.

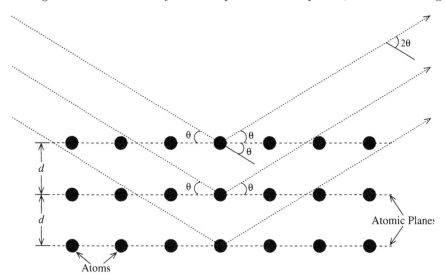

Figure 12.1.17: Reflection of parallel x-rays from atomic planes of a crystalline lattice.

Bragg's relation implies that if we scan the intensity of diffracted x-rays of known wavelength at all angles, we can determine the spacing of atomic planes. Since a typical lattice has planes with different spacings therefore the distribution of scattering angles can give information about the shape of the lattice as well. This is the principle of x-ray diffraction spectroscopy.

Let us now move on to the practical aspects of XDS. What we need is an x-ray source and a detector that can measure the diffracted x-ray intensities at different angles. The choice of x-ray source depends on the type of sample, desired accuracy, and cost. Although synchrotron radiation is the best choice in most instances but its use is tagged with fairly high beam time cost. If cost is a limiting factor, one can resort to conventional x-ray beams capable of delivering x-rays in the required energy range. The other apparatus we need for the XDS setup is a photon detector to detect diffracted x-rays. Certainly the best choice would be to use a position sensitive detector covering the full range of the angles of diffraction. However this may not be possible due to cost and complexity. A more practical approach is to use a single channel detector mounted on an assembly such that it could be moved around the sample. Such a setup is shown in Fig.12.1.18. Generally the mounting assembly is designed such that it has the capability to rotate the sample as well. This enables one to study the diffraction patterns with respect to different lattice planes.

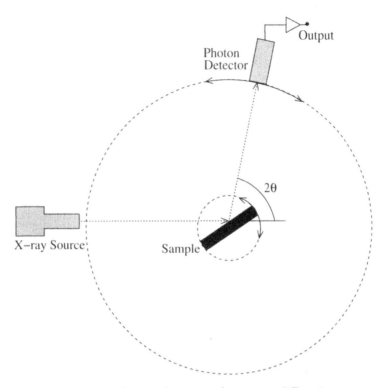

Figure 12.1.18: A simple setup for x-ray diffraction spectroscopy. The diffracted x-rays make an angle of 2θ with respect to the direction of motion of incident x-rays (see also Fig.12.1.17).

A typical distribution of photon intensity with respect to the angle of reflection θ is shown in Fig.12.1.19. Each of the peaks shown in the figure corresponds to constructive interference of diffracted x-rays. Since these peaks satisfy Bragg's criteria, therefore the angle information can be used to determine the atomic plane spacing through equation 12.1.14.

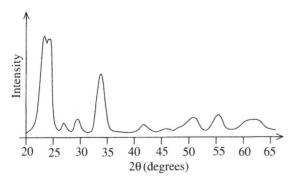

Figure 12.1.19: XDS pattern of WO_3 powder obtained after heating in air for 8 hours at $90\ ^0C$ (redrawn from (19)).

Example:
In an XDS experiment a first order diffraction peak was found to be at $2\theta = 37^0$. If the wavelength of the x-rays used in the study was 1.3 Å, find the atomic plane spacing corresponding to the observed peak.

Solution:
A diffraction peak occurs when the Bragg condition 12.1.14 is satisfied, that is

$$2d \sin \theta = n\lambda.$$

This can be rearranged to give

$$d = \frac{n\lambda}{2 \sin \theta}.$$

Substituting the given values of $n = 1$, $\lambda = 1.3 \times 10^{-10}$ m, and $\theta = 37^0/2 = 18.5^0$, in the above equation gives

$$
\begin{aligned}
d &= \frac{(1)\left(1.3 \times 10^{-10}\right)}{(2)\left[\sin\left(18.5^0\right)\right]} \\
&= 2.048 \times 10^{-10} \ m \\
&= 2.048 \ \text{Å}.
\end{aligned}
$$

12.2   Spectroscopy of Charged Particles

12.2.A α-Particle Spectroscopy

A large number of radionuclides emit α-particles of well defined energy. Since these particles are emitted from the nucleus of the atom, their spectroscopic parameters allow one to deduce information about the nuclear structure. This makes α-particle spectroscopy a very valuable tool for nuclear physics. The usefulness of the spectroscopy has fueled a lot of research and development into not only the techniques and methods but also the associated equipment.

A typical α-particle spectroscopy setup is shown in Fig.12.2.1. It consists of a vacuumed enclosure for source and the detector, a pulse amplification system, and a pulse analyzer. In the following we will look at these units individually.

A vacuumed enclosure for the source and the detector is absolutely necessary to avoid parasitic absorption of α-particles in the space between the source and the detector. The reader might recall that the range of α-particles emitted by radioactive sources in air is very small, of the order of a centimeter. Even if the source and the detector are placed very near to each other, there would still be some air between them. Since the cross section of air molecules for α-particles is very high, there would be significant absorption leading to loss of information and lowering of the signal to noise ratio. The latter effect is due to the statistical nature of α-particle interactions leading to addition of a statistical noise term in the expression for the signal to noise ratio. There are two ways to avoid the parasitic absorption of α-particles. One is to place the source inside the active volume of the detector, which would ensure total

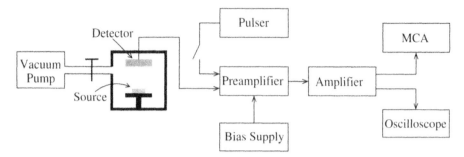

Figure 12.2.1: Block diagram of a typical setup for α-particle spectroscopy.

absorption of energy. This can only be done with gas filled detectors and carries handling and operational difficulties. The other technique, as shown in Fig.12.2.1, is to place the source and the detector in a vacuumed enclosure.

Let us now move on to the types of detectors suitable for α-spectroscopy. In principle, any detector capable of detecting α-particles can be used. However the most commonly used ones are the semiconductor detectors. In the early years of α-spectroscopy, mostly gas filled detectors were used as the semiconductor detector technology was in its infancy at that time. There are a few reasons behind preference of semiconductor detectors over their gas filled counterparts, as described below.

▶ **Resolution:** Resolution is a very important parameter as far as spectroscopy is concerned. By resolution we mean how well the detector can differentiate between two particles depositing unequal energies into its active volume. Note that we are not talking about the electronics resolution yet. The resolution of the detector is a separate issue and is concerned with how well the deposited energy corresponds to the pulse height.

▶ **Linearity:** It is an established fact that semiconductor detectors show better linearity for particles of low to moderate energies.

▶ **Parasitic Absorption:** A big issue with gas filled detectors is that they need an enclosure to contain the filling gas. The enclosure is generally made of a metal with an entrance window for the radiation. One can make such a window very thin, of the order of a few microns, provided the gas is kept at approximately normal temperature. However even in such thin windows there is a high probability of α absorption. Of course this parasitic absorption leads to signal loss and non-linearity in the response. A way around this problem is to place the source inside the active volume of the detector, which is associated with handling and operational difficulties. Semiconductor detectors do not share this problem since they do not have any entrance windows. The incident α-particles directly interact with the material in the active volume and deposit their energies.

The analog signal from the detector has to be amplified and shaped before being processed by the pulse height analyzer. These analog units are generally installed very close to the detector output. The output of the amplifier/shaper is a well shaped pulse with height proportional to the energy deposited by the incident α-particle.

The shaped pulse is then input into a pulse height analysis system. Earlier in the book we discussed one such instrument called the multichannel analyzer or MCA. In brief, an MCA first digitizes the analog pulses and then increments the counts of its channels according to the height of the pulse. The channels of an MCA, also called *bins*, span its whole dynamic range. These bins are divided into intervals or *bin widths*. In general all the bin widths are of equal size but some modern MCAs allow the user to manually set different bin widths as well. The bin width used in the analysis and the overall MCA resolution set the limit on peak resolution capability of the system. This concept is graphically depicted in Fig.12.2.2. As shown in the figure, if the peaks are too close together, they can not be resolved and hence appear as a single peak in the MCA spectrum.

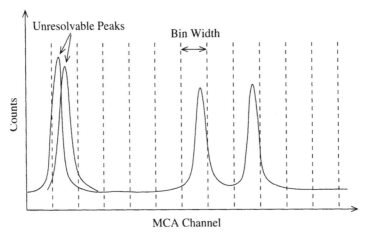

Figure 12.2.2:   Effect of bin width of an MCA on its energy resolution.

Before the spectrum of an unknown source is obtained, the detection system must be calibrated. This is usually done with the help of a pulser and a known radioactive source. The commonly used sources for calibration are listed in Table.12.2.1. It should be noted that except for americium, none of the sources listed here have an energy spectrum with two or more *resolvable* peaks.

Calibration of the instrument is extremely important for absolute measurements. We will therefore spend some time discussing how it is practically done. Suppose we try to calibrate the setup using a source that emits mono-energetic *alpha*-particles. The spectrum captured by the MCA will show a single peak occupying some of its channels. We can try to calibrate the whole dynamic range of the MCA using the center of this peak. However since this method uses a single point, it will be prone to large uncertainties. A better technique would be to use two sources separately or their mixture. This would ensure that two calibration points corresponding to the two peaks are clearly identifiable. An important point that must be considered during calibration is the so called *zero point energy*. In a perfect system one would expect that the zeroth bin of the MCA would correspond to the zero energy. That is, the straight line between the energy and the MCA channel would cross the axes at the origin. However this in general is not true due to the

Table 12.2.1:   Common α-particle sources used for calibrations.

Source	Isotope	Energy MeV	Branching Ratio
Polonium	$^{210}_{84}Po$	5.304	100%
Plutonium	$^{239}_{94}Pu$	5.157	73.3%
		5.144	15.1%
		5.106	11.5%
Americium	$^{241}_{95}Am$	5.486	84.5%
		5.443	13%
		53.882	16%
Curium	$^{244}_{96}Cm$	5.805	76.4%
		5.763	23.6%

unavoidable system offsets. It is therefore very important that the zero point energy is included in the calibration curve. Some of the modern MCAs also enable the user to reset the zeroth channel to correspond to the zero energy. Generally this is done before the calibrations, which ensures that the zero point is also available to determine the calibration curve.

We just mentioned that at least two sources are required for the calibrating the instruments. However, it is also possible to calibrate the equipment with the help of a source and a pulse generator. We will shortly see how it can be done.

A typical calibration setup is shown in Fig12.2.1. The pulser is shown connected to the preamplifier by a dotted line because this is an optional instrument and not all preamplifiers are equipped with a external charge injection circuitry. Since the range of α-particles in air is very small therefore the source must be kept in a vacuumed container. The oscilloscope is used to look at individual α-particle pulses and helps in adjusting the gain of the preamplifier. The following steps are typically taken to perform the calibration using a source and a pulser.

▶ **Step-1:** The source is placed in the enclosure, which is then pumped for high vacuum.

▶ **Step-2:** The bias supply is turned on and the high voltage to the detector is slowly increased until the recommended value is reached.

▶ **Step-3:** The gain of the preamplifier is varied until a good enough peak is observable on the oscilloscope. The amplitude of the pulse is noted down.

▶ **Step-4:** The MCA is reset and data acquisition started.

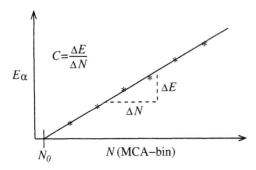

Figure 12.2.3: Typical calibration curve obtained from a known source and an pulser. The pulser helps in determining the overall system linearity and the offset.

▶ **Step-5:** The MCA channel corresponding to the peak is identified. The peak is generally symmetric but has a finite width due to statistical fluctuations of various processes involved in the detection. The centroid of the peak is taken to represent the expected energy of the particles.

▶ **Step-6:** Next, the pulser is calibrated against this peak to determine the linearity of the system and the offset. If the system is linear then the energy of the α-particles will be related to the MCA channel number N by

$$E_\alpha = C(N - N_0), \tag{12.2.1}$$

where C is the calibration factor in units of energy per channel (such as $MeV/$MCA-bin). N_0 is the channel offset. Ideally there should not by any offset, that is, zero energy should correspond to the zeroth channel. In practice, however, this is rarely seen and therefore it is always recommended to use a pulser to determine the offset.

Before connecting the pulser to the preamplifier, the source should be taken out of the container. The pulses are injected into the preamplifier and the signal amplitude is varied until the oscilloscope shows the same pulse height as in the case of the source. MCA spectrum of the pulser is then accumulated and it is verified that the pulser peak is at the same channel number as the source peak. This signal amplitude of the pulser is noted down.

▶ **Step-7:** The signal amplitude of the pulser is varied in steps such that it spans the whole expected range of particle energy. For example for a polonium-210 source it could be from 0 to $6MeV$. At each step the MCA spectrum is obtained and the channel number corresponding to the peak is recorded.

▶ **Step-8:** The channel number is plotted against the energy corresponding to the pulser's signal amplitude.

▶ **Step-9:** A straight line fit on these data points is performed. If there is no offset, the line should end at the origin of the plot. If not, the offset is one of the calibration parameters and must be included in any further calculations. The slope of the straight line gives the required calibration constant C in terms of energy per channel (see Fig.12.2.3).

We mentioned earlier that a pulse generator is not always required to calibrate the equipment. One can also use two or more sources instead to acquire more data points. The procedure would then be simply to first adjust the zero point of the MCA, obtain the spectra of the two sources, and then draw the calibration curve.

A.1   Energy of an Unknown α-Source

In the previous section we saw how a detection system can be calibrated. After the calibration has been performed, one could essentially place any source in the vacuumed container, obtain the MCA spectrum, and determine the energies corresponding to the various peaks observed. The peaks can then be used to identify the isotopes present in the sample.

A.2   Range and Stopping Power of α-particles in a Gas

Using the setup shown in Fig.12.2.1, one can determine the range and stopping power of α-particles in air provided the vacuum system is equipped with a precision pressure gauge.

A.3   Activity of an α Source

A very commonly performed experiment in radiation laboratories is to find the activity of an α source. An α spectroscopy system can be used for the purpose. The experiment is fairly straightforward and involves capturing the α peak of the source. The source is placed at a known distance away from the detector as shown in Fig.12.2.4 and the energy spectrum is captured for a known period of time. The spectrum shows a peak (see Fig.12.2.5), which characterizes the activity due to the α particles.

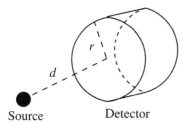

Figure 12.2.4: Source-detector geometry for determination of α activity of a source.

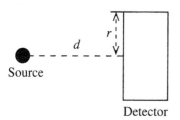

The α activity can then be calculated from the relation

$$A = \frac{C_\alpha}{T} \frac{4\pi d^2}{\pi r^2}.$$

(12.2.2)

Here C_α represents the area under the α peak, T is the measurement time, d is the perpendicular distance between the source and the detector, and r is the radius of the detector. The second term on the right side of this equation corrects for the solid angle subtended by the source at the detector. This is necessary since we know that a radioactive source emits particles in all directions with equal probability.

A point worth mentioning here is that determination of the area under the α-peak requires subtraction of the background. The reason is that the peak is actually embedded on top of the background activity (see Fig.**??**). The simplest procedure is to discard the area below a line stretched between two baseline points of the peak. The baseline points at the two ends can be determined by taking averages of a few points on the two tails of the peak. Other methods also exist but require fitting the peak with a Gaussian function and then determining the end points using some criterion based on the properties of the function. Such methods, however, do not offer much improvement over the simple method mentioned above and are therefore not generally used.

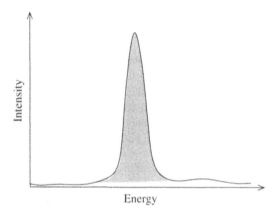

Figure 12.2.5: Typical α spectrum of a source. The area under the peak is proportional to the source activity.

12.2.B   Electron Spectroscopy

Just like α-particle spectroscopy is used for nuclear structure analysis, electron spectroscopy can be used for characterization of atomic structure. The experimental setups and procedures for electron spectroscopy with radionuclides are similar to the α-particle spectroscopy we have already discussed.

There are other kinds of electron spectroscopy as well. In fact, we have already discussed two such techniques, namely XPS or x-ray photoelectron spectroscopy and AES or Auger electron spectroscopy. We saw that the spectra of the electrons, which are emitted as a result of photoelectric absorption of x-ray photons, can reveal the intricate details of the atomic orbitals and the bonding between atoms. That is why XPS is sometimes referred to as *electron spectroscopy for chemical analysis* or ESCA. In Auger electron spectroscopy, instead of photoelectrons, Auger electrons are used for spectroscopic purposes.

ESCA and Auger electron spectroscopy are very useful techniques to analyze the surfaces of solids. The reason is that if such electrons are produced deep inside the solids, they can not escape to the surface and be detected. In most materials the escape depth of these electrons is not more than a few nanometers. This is a good thing in the sense that one can then analyze the surfaces without worrying about the interactions deep inside the material.

It is also possible to use electron beams for spectroscopic analysis. The *electron energy loss spectroscopy* or EELS is one such technique. Here, the material under investigation is bombarded with electrons. Some of these electrons suffer inelastic scatterings with the atoms and cause electronic transitions. The resulting spectrum can then be used to determine the chemical structure of the sample.

12.3 Neutron Spectroscopy

Spectroscopy of neutrons is an emerging and a very active area of research. Its usefulness in materials research and chemistry has been well established. The biggest advantage of neutrons as a material probe is their high penetration power, owing to their neutral electrical charge character. In this section we will look at different techniques and instruments used in spectroscopy with neutrons. But before we do that, let us quantitatively see why neutrons are useful for spectroscopy.

12.3.A Neutrons as Matter Probes

In the first chapter of this book we discussed the idea of wave-particle duality. We saw that sometimes particles behave as waves and can actually be assigned a wavelength. This wavelength is related to particle's momentum p through the relation

$$\lambda = \frac{h}{p}, \tag{12.3.1}$$

where h is the usual Planck's constant. This relation can also be written in terms of particle's energy by noting that

$$E = \frac{p^2}{2m}, \tag{12.3.2}$$

where m is the particle's mass. Substituting p from equation 12.3.5 in this relation gives

$$E = \frac{h^2}{2m\lambda^2}. \tag{12.3.3}$$

For a neutron this relation reduces to

$$\begin{aligned} E_n &= \frac{\left(6.626 \times 10^{-34}\right)^2}{(2)\left(1.6749 \times 10^{-27}\right)\lambda^2} \\ &= \frac{1.3106 \times 10^{-40}}{\lambda_n^2} \ J. \end{aligned} \tag{12.3.4}$$

where λ_n is in units of meter. It is more convenient to transform this relation such that λ_n can be input in units of angstroms and the energy is obtained in units of electron volts. For this, we multiply the above relation with joules to electron volts and meters to angstroms conversion factors. This gives

$$
\begin{aligned}
E_n &= \frac{1.3106 \times 10^{-40}}{\lambda^2} \frac{1}{(1.602 \times 10^{-19})(10^{-20})} \\
&= \frac{81.8128}{\lambda_n^2} \; meV,
\end{aligned}
\tag{12.3.5}
$$

where meV stands for *milli* electron volts and λ_n is in $\mathring{A}$. Now that we have a relation between neutron energy and wavelength we can estimate the wavelength of neutrons of any kinetic energy. Let us see what we get for thermal neutrons, which typically have an energy of 25 meV. Using above equation we get

$$
\begin{aligned}
\lambda_n &= \left[\frac{81.8128}{E_n}\right]^{1/2} \\
&= \left[\frac{81.8128}{25}\right]^{1/2} \\
&= 1.81 \; \mathring{A},
\end{aligned}
\tag{12.3.6}
$$

which corresponds to the typical interatomic distance in most materials. This makes slow neutrons an excellent tool for studying structure of materials. Note that 25 meV approximately corresponds to the thermal excitations at room temperature (see example below).

Example:
Determine the thermal excitation energy corresponding to the room temperature of 300 K.

Solution:
The energy of thermal excitations is given by

$$
E = k_B T,
$$

where k_B is the Boltzmann's constant and T is the absolute temperature. For $T = 300 \; K$ this equation gives

$$
\begin{aligned}
E &= (1.3806 \times 10^{-23})(300) \\
&= 4.1418 \times 10^{-21} \; J \\
&= \frac{4.1418 \times 10^{-21}}{1.602 \times 10^{-19}} = 0.0258 \; eV \\
&= 25.8 \; meV.
\end{aligned}
$$

In neutron spectroscopic analyses, most experimenters prefer to use the term wave vector instead of wavelength. Wave vector was previously defined for x-rays as $k = 2\pi/\lambda$. Same definition applies to neutron wavelength as well. Hence the expression 12.3.5 can also be written as

$$
E_n = 2.0723 k_n^2,
\tag{12.3.7}
$$

where k_n has units of $\overset{\circ}{A}{}^{-1}$. This equation can be used to estimate the magnitude of the wave vector for any given neutron energy (see example below). Note that here the energy is in units of meV.

Example:
Determine the magnitude of the wave vector for a neutron having energy equivalent to the thermal excitation energy at room temperature of $300\ K$.

Solution:
In the previous example we saw that the energy of thermal excitations at room temperature of $300\ K$ is $25.8\ meV$. Substituting this value in equation 12.3.7 gives the required value of the wave vector.

$$k_n = \left[\frac{E_n}{2.0723}\right]^{1/2}$$

$$= \left[\frac{25.8}{2.0723}\right]^{1/2}$$

$$= 3.5284\ \overset{\circ}{A}{}^{-1}$$

As mentioned earlier, the biggest advantage of neutrons as scattering probes is that their wavelengths correspond to the thermal excitations. The other advantage is their low cross section for most materials, which allows them to penetrate deep into the materials. Now, of course, the wavelength and cross section also have energy dependence but for spectroscopic purposes one mostly uses either thermal neutron sources or cold neutron sources. The former have a energy range of up to about $100\ meV$ while the latter can assume an energy of up to about $10\ meV$. At these energies, a diffraction pattern can reveal atomic structure of the material, which is also evident from the computations performed earlier.

12.3.B Neutron Spectrometry Techniques

In this section we will look at some of the common neutron spectrometry techniques and discuss the related instrumentation. But before we do that let us have a quick look at he process of neutron scattering. Suppose a beam of neutrons interacts with a sample and gets reflected as shown in Fig.12.3.1. The neutrons can undergo elastic as well as inelastic scattering with the nuclei. These processes are described below.

▶ **Elastic Scattering:** In an elastic scattering event the energy of the scattered electrons is equal to that of the incident neutrons. That is

$$E_f = E_i$$
$$\text{and} \quad k_f = k_i,$$

where the subscripts i and f refer to the states before and after scattering. In such as case we will have

$$E_i - E_f = \hbar\omega = 0 \qquad (12.3.8)$$
$$\Rightarrow \omega = 0. \qquad (12.3.9)$$

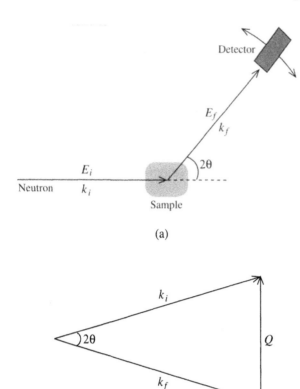

Figure 12.3.1: (a) A simple setup to study neutron scattering. The movable detector can be positioned to detect diffraction maxima. (b) Addition of initial and final wave vectors to obtain the momentum transferred to the sample.

▶ **Inelastic Scattering:** During an inelastic scattering process a neutron can loose or gain energy, that is

$$E_f \neq E_i$$
$$\text{and} \quad k_f \neq k_i.$$

This implies that

$$\hbar\omega = E_i - E_f \neq 0.$$

The quantity on the left hand side of the above equation represents the energy transferred to the sample. It can be obtained in terms of wave vector by using the relations $p = h/\lambda = \hbar k$ and $E = p^2/2m$, that is

$$E = \frac{p^2}{2m} = \frac{\hbar^2 k^2}{2m}$$

$$\Rightarrow \hbar\omega = \frac{\hbar^2}{2m}\left[k_i^2 - k_f^2\right]$$

$$\Rightarrow \omega = \frac{\hbar}{2m}\left[k_i^2 - k_f^2\right]. \tag{12.3.10}$$

Another important quantity is the momentum transfer, which is generally represented by Q. It can be obtained in terms of the wave vector $\bar{k}$ by noting that the magnitude of the wave vector is directly related to the momentum through $p = \hbar k$. The momentum transfer is generally represented only as a difference of the two wave vectors, that is

$$\bar{Q} = \bar{k}_i - \bar{k}_f. \tag{12.3.11}$$

Note that here all the quantities are vectors and should therefore be treated accordingly (see also Fig.12.3.1(b)). The quantity Q is a function of the angle of reflection and its value depends solely on the structure of the sample. This implies that it can be used to deduce information about the structure. This is true but the usual practice is to derive a *scattering function* S based not only on Q but also on ω. The information contained in this function is richer than the parameter Q alone and is therefore able to provide deeper insight into the structure of the material. To be more specific, the parameter ω gives information about the type of scattering, which is also a function of the material and its structure. A typical plot of S with respect to the energy transfer to the sample is shown in Fig.12.3.2.

▶ **Quasielastic Scattering:** The elastic peak shown in Fig.12.3.2 is broadened at higher values of energy transfer. This broadening is due to the so called *quasielastic* scattering process. Note that in an ideal case the elastic peak should be a delta function, that is, should not have any width. However, due to crystal imperfections and other effects related to measurements, the peak assumes a finite width. Further broadening of this peak due to the quasielastic process is separate from this, though. Quasielastic scattering is a physical process characterized by small energy transfers of the order of 2 meV.

A very important point to note here is that in the process of quasielastic and inelastic scatterings the scattered neutron energy can be lower or higher than the incident neutron energy. In other words, the incident neutron can also gain energy during the scattering process.

Now that we understand the basics of neutron scattering, we can move on to the discussion of the practical aspects of neutron spectroscopy. As mentioned earlier, there are two parameters, namely Q and ω, which can reveal information about the sample's internal structure. Therefore the neutron spectroscopy is concerned with determining the neutron intensity as a function of these two parameters. Since these parameters depend on the wave vectors and the scattering angle, the spectrometer should be able to determine the neutron intensity at different angles with respect to initial direction of neutrons. The wave vector can be determined by a number of instruments that employ techniques that are conceptually different from one another. In the following we list the three most commonly used techniques together with their respective instruments.

▶ Bragg Diffraction

    – Triple-Axis Spectrometer

    – High Flux Backscattering Spectrometer

    – Filter Analyzer Spectrometer

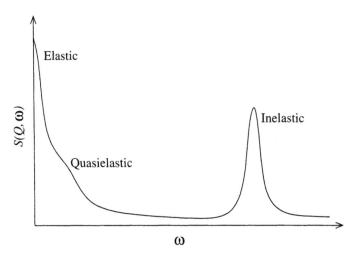

Figure 12.3.2: Typical variation of neutron scattering function with respect to the energy transfer. The contributions of elastic and inelastic scatterings are clearly visible as distinguishable peaks while that of quaielastic scattering appears as broadening of the elastic peak.

▶ Time-of-Flight

  − Disk Chopper Spectrometer

  − Fermi Chopper Spectrometer

▶ Larmor Precession

  − Spin Echo Spectrometer

A comprehensive discussion on these spectrometers is out of the scope of this book. However it is worthwhile to briefly introduce the reader to the principles of their designs. The interested reader can then refer to advanced texts on neutron spectroscopy to obtain further details.

B.1 Triple-Axis Spectrometry (TAS)

As the name suggests, this instrument is based on three axes to determine the wave vectors. The principle of TAS is shown in Fig.12.3.3. The spectrometer consists of a monochromator, a sample holder, an anlyzer, and a detector. All these components can be rotated along their axes.

The neutrons emitted by the source are elastically scattered by the monochromator, which works on the principle of Bragg diffraction. We have already studied that the Bragg condition of constructive interference is given by

$$n\lambda_i = 2d_m \sin\theta_m, \qquad (12.3.12)$$

where d_m is the atomic plane spacing of the monochromator, λ_i is the wavelength of the neutrons scattered off the monochromator, and θ_m is the monochromator

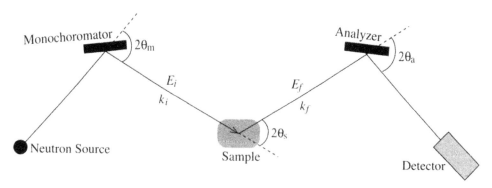

Figure 12.3.3: Principle of triple-axis neutron spectrometry.

scattering angle. The above equation can also be written in terms of wave vector by using the identity $k_i = 2\pi/\lambda_i$.

$$\frac{2\pi n}{k_i} = 2d_m \sin\theta_m \tag{12.3.13}$$

For first order diffraction ($n = 1$), this equation reduces to

$$k_i = \frac{\pi}{d_m \sin\theta_m}. \tag{12.3.14}$$

Since d_m for the monochromator is known a priori and θ_m is measured, the wave vector can be determined from this relation. The second wave vector, that is k_f, can be determined in a similar manner using the analyzer (see Fig.12.3.3). The defining equation for k_f is

$$k_f = \frac{\pi}{d_a \sin\theta_a}. \tag{12.3.15}$$

where d_a is the atomic plane spacing of the analyzer and θ_a is the analyzer scattering angle. Once we know k_i and k_f we can determine ω. Additionally, using the knowledge of sample's scattering angle θ_s we can calculate the momentum transfer $\bar{Q}$ through the relation 12.3.11.

Let us now turn our attention to the energy resolution of the system. We first note that it depends on the resolution of the neutron wavelength. This can be seen by differentiating equation 12.3.3 on both sides, which gives

$$dE = -\frac{h^2}{m\lambda^3} d\lambda \tag{12.3.16}$$

$$\Rightarrow \delta E \propto \lambda^{-3} \delta\lambda. \tag{12.3.17}$$

Next we differentiate equation 12.3.12 (with $n = 1$) on both sides to get

$$\delta\lambda = 2d\cos\theta\delta\theta. \tag{12.3.18}$$

Substituting $d\lambda$ into equation 12.3.17 gives

$$\delta E \propto \frac{d}{\lambda^3}\cos\theta\delta\theta, \tag{12.3.19}$$

which implies that the energy resolution has dependence on angular resolution of the system as well as wavelength of neutrons and the atomic plane spacing of the crystal. Triple axis spectrometers can typically provide up to about 1 meV of resolution, which is good for most applications. Further enhancement in resolution is possible by using lower energy neutrons, the so called cold neutrons.

B.2   High Flux Backscattering Spectrometer (HFBS)

As we saw in the previous section, the energy resolution of a triple axis spectrometer has a dependence on the spread in the wavelength of neutrons (see equations 12.3.18 and 12.3.18). It is apparent from equation 12.3.18 that one could in principle obtain perfect resolution if $\theta = \pi/2$, which will give $\delta\lambda = 0$ and $\delta E = 0$. This is the case when neutrons are backscattered. Of course the ideal case of $\theta = 0$ for all neutrons is not practically achievable but one can design an instrument such that most of the neutrons are backscattered. In this case we will have

$$\delta E \to 0 \quad \text{if} \quad \theta \to \frac{\pi}{2}.$$

Practically speaking, this condition requires higher flux of incident neutrons as compared to the case of triple axis spectrometer. That is why the spectrometer based on this condition is called *high flux backscattering spectrometer* of HFBS. Such instruments are able to offer energy resolutions on the order of μeV.

B.3   Filter Analyzer Spectrometer (FAS)

Filter analyzer spectrometer is similar to the triple axis spectrometer with on exception: the energy analyzer part of the instrument is made of a *filter analyzer* instead of a crystal analyzer. The analyzer is still made of a crystalline structure but is used such that it filters out all the Bragg scatterings. In other words, it is assured that very minimal Bragg scatterings occur in the crystal. For this, the neutron energy is chosen such that the wavelength does not correspond to any of the atomic lattice plane spacings. Generally one chooses very low energy neutrons that have very long wavelengths, far away from the Bragg wavelengths of the crystal. With possibility of elastic scattering minimized, the other two processes that could lead to neutron loss are absorption and inelastic scattering. To avoid absorption of neutrons, one uses a material composed of atoms having very low neutron absorption cross sections. And to minimize the possibility of inelastic scattering, the whole instrument is cooled to liquid nitrogen temperature. The operation of FAS consists of detecting filtered neutrons at a fixed final energy. The experiment is performed at different incident neutron energies.

A sketch of the principle of FAS design in shown in Fig.12.3.4. Note its similarity to the triple axis spectrometer. The filter assembly is made of several layers of filters. Common materials used for filter are bismuth, graphite, and beryllium. The filter and detector assembly is generally constructed such that it can be moved around the sample. However, use of two fixed wedge shaped filter-detector assemblies removes the need for additional movement and alignment structures.

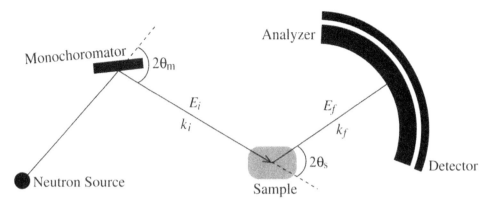

Figure 12.3.4: Principle of filter analyzer spectrometer. The whole filter assembly is kept at liquid nitrogen temperature to avoid inelastic scattering of neutrons. The filter is generally made of layers of beryllium, graphite, and bismuth.

B.4   Disk Chopper Spectrometer

Disk chopper spectrometer is an instrument that uses the time-of-flight information of neutrons to determine the inelastic scattering spectrum of the sample. Its working principle is fairly simple: mono-energetic neutrons are allowed to pass through the sample and the scattered neutrons are detected through a large array of detectors. The times of arrival of the neutrons is used to determine the type of scatterings they have gone through. The basic idea is that the neutrons that gain energy during the scattering process arrive earliest. They are followed by the elastically scattered neutrons. The neutrons that lose energy during collisions arrive latest. This implies that the timing information can give insight into the dynamics of scattering.

The principle design of a disk chopper spectrometer is shown in Fig.12.3.5. Such an instrument is generally designed to work on a nuclear reactor where high intensity beam of neutrons is available. The experiment is performed in neutron bursts. The reason is that the precise time or arrival of neutrons at the sample is required for proper time-of-flight measurements. Since the neutron sources produce neutrons continuously therefore DCS uses a chopper assembly to produce burst of neutrons, hence the name disk chopper spectrometer.

A typical disc chopper spectrometer has about 1000 neutron detectors for time-of-flight measurements. Since the detectors are used for timing measurements, their timing resolution is of high importance.

An important practical consideration of disk chopper and similar instruments is that they should have proper beam stop assemblies. The reason is that a typical neutron source used in such an experiment produces an intense beam of neutrons. Not all of these neutrons get reflected from the sample. In fact, a good fraction of these neutrons simply pass through the sample without undergoing any interaction. These neutrons have to be stopped for protection of personnel and safety. In Fig.12.3.5 such a beam stopping assembly is symbolically represented by a black circle.

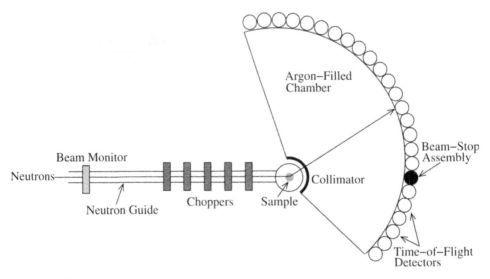

Figure 12.3.5: Principle design of a disk chopper spectrometer.

B.5 Fermi Chopper Spectrometer (FCS)

A Fermi chopper spectrometer is similar to the disk chopper spectrometer except that it uses a suitable crystal to produce monochromatic beam of neutrons. The monochromator is followed by a chopper assembly to create neutron bursts as in teh case of DCS. The conceptual design of a Fermi chopper spectrometer is shown in Fig.12.3.6.

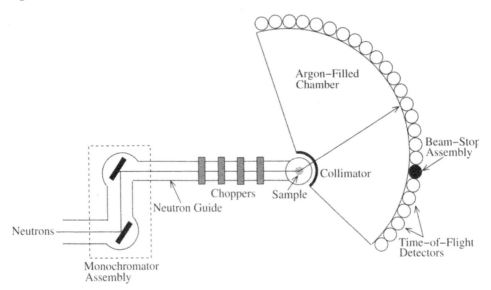

Figure 12.3.6: Principle design of a Fermi chopper spectrometer.

B.6 Spin Echo Spectrometer

All of the neutron spectroscopy techniques we have discussed so far demand use of single-wavelength neutrons at a time. A typical reactor source of neutrons emits neutrons having a broad energy spectrum. One then needs to use a monochromator to select the neutrons of the right energy needed for the analysis. This is not very convenient as the process greatly reduces the neutron flux, which is not a very desirable effect. Spin echo spectrometry is a technique that does not require the neutrons to be monochromatic for ensuring good energy resolution.

The idea behind spin echo spectrometry is graphically depicted in Fig.12.3.7. The neutrons from the source are first selected by a *velocity selector*. The velocity selector is not required to have a high resolution, as a wavelength spread of 10-15% can be tolerated. The selected neutrons are then made to pass through a polarizer, which aligns their spins in one direction. The next step is to flip the neutron spin by $\pi/2$ through a *flipper* using externally applied magnetic field. This orients the neutron spin perpendicular to the magnetic field. The neutrons then travel through the magnetic field and as a result precess with an angular frequency that is given by

$$\omega = \gamma B, \tag{12.3.20}$$

where γ is the neutron's gyromagnetic ratio and B is the applied magnetic field. ω is generally known as Larmor frequency. The neutrons keep on precessing until they reach the sample. We can determine the total precession angle by simply multiplying the frequency by the time it takes them to reach the sample, that is

$$\theta = \omega t. \tag{12.3.21}$$

The time t can be determined from

$$t = \frac{d}{v}, \tag{12.3.22}$$

where d is the distance traveled by the neutrons having velocity t. The total precession angle is then given by

$$\theta = \frac{\gamma B d}{v}. \tag{12.3.23}$$

Before the neutrons could interact with the sample, their spin is flipped by 180^0 by a π-flipper. These neutrons then interact with the sample and undergo elastic, quasi-elastic, and inelastic scatterings. The presence of the magnetic field after the sample ensures that the neutron keeps on precessing. However now since their polarization has been changed by 180^0, they precess in opposite direction as compared to before the sample. The total angle of precession swept by the neutrons after traveling a distance d' in a magnetic field B' is given by

$$\theta' = -\frac{\gamma B' d'}{v'}. \tag{12.3.24}$$

where v' is the velocity of the neutron after leaving the sample. Here the negative sign simply signifies the fact that the precession is in the opposite direction. Now, for simplicity assume that the magnetic field and the distance traveled before and after the sample are equal, that is, $B' = B$ and $d' = d$. In this case the above equation becomes

$$\theta' = -\frac{\gamma B d}{v'} \tag{12.3.25}$$

This equation implies that in case of elastic scattering in which there is no change in neutron's velocity, that is $v' = v$, we will have

$$\theta' = -\theta, \tag{12.3.26}$$

which is the so called *echo condition*, hence the name *spin echo spectroscopy*. If the neutron undergoes inelastic scattering, it can either gain or loose energy. Let us assume that the neutron looses some energy. In this case $v' < v$ and consequently $|\theta'| > |\theta|$. This shift in the angle of precession can be measured by detecting the neutron polarization. However an easier way to do this is to vary the magnetic field such that the echo condition gets satisfied. The magnitude of the magnetic field will then be a measure of the change in neutron velocity. This echo condition according to equations 12.3.23 and 12.3.24 can be written as

$$\frac{B'd'}{v'} = \frac{Bd}{v}. \tag{12.3.27}$$

And if the distances traveled by the neutrons before and after the sample are equal then the magnetic field needed for echo condition will be give by

$$B' = B\frac{v'}{v}. \tag{12.3.28}$$

The values of B and v are known a priori while that of B' is determined when then the echo condition is satisfied. The final velocity of the neutrons can then be determined from the above equation. Fig.12.3.7 shows the basic components of a spin echo spectrometer.

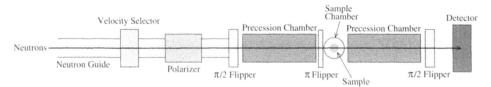

Figure 12.3.7: Principle design of a spin echo neutron spectrometer.

12.4 Mass Spectroscopy

Mass spectroscopy is a methodology that is extensively used to determine properties of charged particles. The basic principle of this technique is to dissociate the sample into smaller fragments and then let them pass through a high magnetic field. The charged particles get deflected in the magnetic field with the degree of deflection proportional to their mass to charge ratio (m/e). These deflected particles are then collected by a position sensitive detector, such as a microchannel PMT.

The incident particle of choice for mass spectroscopy is the electron due to several reasons. First of all, it is a fundamental particle and does not fragment itself during the collision. Also its small mass makes it easier to be accelerated in a small scale accelerator. A highly intense beam of electrons is therefore easy to produce as compared to other heavier particles, which require big accelerators.

A typical mass spectrometer is shown in Fig.12.4.1. The sample is first vaporized by a heater and then bombarded by a beam of electrons. Impact of electrons dissociates the sample into small fragments. The positively charged fragments are accelerated towards a powerful magnet through electrodes that are kept at high electrical potentials. Upon their passage through the magnet, the fragments flay apart from each other. The amount of deflection is proportional to their mass to charge ratio. However since most of the fragments have unit positive charge, their separations are proportional to their masses. The result is the production of several beams of like-mass particles, which are detected by a position sensitive detector. In the earlier days of mass spectroscopy, photographic films were used to detect the particles. With the advent of easily available large area position sensitive electronic detectors, the use of photographic films has faded away.

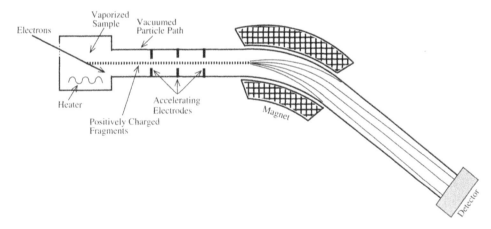

Figure 12.4.1: Working principle of a mass spectrometer.

12.5 Time Spectroscopy

In time spectroscopy one is interested in determining the relationship between the arrival times of particles emitted as a result of some event. Note that here we are not talking about coincidence timing, rather difference in the arrival time of pulses. However a time spectroscopy system can as well be used for coincidence spectroscopy.

Any detector having sufficient time resolution and response time can be used to build time spectrometer. In most cases, a combination of scintillator and photomultiplier tube (PMT) is used. A typical time spectrometry setup is shown in Fig.12.5.1. The two detectors individually produce pulses as a result of passage of radiation. These pulses are made to pass through two so called *time pickoff* circuits. A pickoff circuit produces a sharp logic pulse as soon as the input pulse amplitude crosses a set threshold. In other words, it simply puts a stamp on the arrival time of the pulse. The logic pulses thus produced are fed into a time-to-amplitude converter (TAC). The TAC produces a pulse with an amplitude proportional to the difference in the arrival times of the logic pulses. The way it is accomplished is fairly simple. One of the logic pulses is delayed by a set amount. The direct logic pulse, as it enters into the TAC circuitry, starts charging (or discharging) a capacitor through some internal or external source. The capacitor keeps on charging (or discharging) until the second delayed logic pulse arrives. The output of the TAC is therefore a voltage pulse with an amplitude proportional to the difference in arrival times of the two logic pulses.

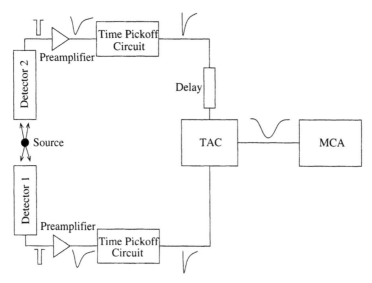

Figure 12.5.1:   A simplified setup for time spectroscopy.

Each event recorded by the two detectors gives rise to one TAC pulse. The distribution of the amplitudes of these pulses is called the timing spectrum. In most cases one simply hooks up a multichannel analyzer to the TAC to obtain the timing spectrum. However, for more demanding applications, one could digitize the pulse and then use a software to perform analysis tasks on the digitized data.

As mentioned earlier, the setup shown in Fig.12.5.1 can also be used for coincidence spectroscopy. Here one is interested in determining whether the two detector pulses were coincident or not. This is done by looking at the FWHM o f the distribution (that is, the timing spectrum). A typical timing spectrum for coincidence events is Gaussian-like as shown in Fig.12.5.2.

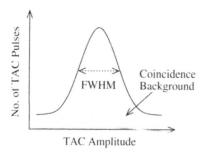

Figure 12.5.2: A typical time spectrum for coincidence spectroscopy.

Problems

1. Compute the atomic plane spacing of a crystal if an x-ray beam of 1.542 $\mathring{A}$ produces a first order diffraction peak at $2\theta = 43.5^0$.

2. Compute the wavelength of thermal neutrons.

3. Compare the wave vector and wavelength of 1.5 MeV neutrons with photons having same energy.

4. A triple-axis spectrometer is modified such that its angular resolution improves by 10% for neutrons of a certain energy. Assuming that d remains the same, estimate the change in the energy resolution.

Bibliography

[1] Alfano, R.R., **Semiconductors Probed by Ultrafast Laser Spectroscopy**, Academic Press, 1985.

[2] Barr, J.D. et al., **Photoelectron Spectroscopy of Reactive Intermediates using Synchrotron Radiation**, J. Elect. Spectrosc. Rel. Phen., 108, 2000.

[3] Benfatto, M. et al., **Theory and Computation for Synchrotron Radiation Spectroscopy (Aip Conference Proceedings)**, Amer. Inst. Phys., 2000.

[4] Briat, B. et al., **Spectroscopic Characterization of Photorefractive CdTe:Ge**, J. Cryst. Growth, 197, 1999.

[5] Brundle, C.R., Baker, A.D., **Electron Spectroscopy: Theory, Technique and Application**, Academic Press, 1981.

[6] Busch, O. et al., **Transition Radiation Spectroscopy with Prototypes of the ALICE TRD**, Nucl. Instr. Meth. A, 522, 2004.

[7] Campagna, M., Rosei, R., **Photoemission and Absorption Spectroscopy of Solids and Interfaces With Synchrotron Radiation: Proceedings of the International School of Physics "Enrico Fermi"**, North-Holland, 1991.

[8] Engel, T., Reid, P., **Quantum Chemistry and Spectroscopy**, Benjamin Cummings, 2005.

[9] Gabrielse, G. et al., **Precision Mass Spectroscopy of the Antiproton and Proton Using Simultaneously Trapped Particles**, Phys. Rev. Lett. 82, 1999.

[10] Hammes, G.G., **Spectroscopy for the Biological Sciences**, Wiley-Interscience, 2005.

[11] Hollas, J.M., **High Resolution Spectroscopy**, John Wiley & Sons, 1998.

[12] Hollas, J.M., **Modern Spectroscopy**, John Wiley & Sons, 2004.

[13] Huefner, S., **Photoelectron Spectroscopy**, Springer, 2003.

[14] Iizuka, T. et al., **Synchrotron Radiation in the Biosciences**, Oxford University Press, 1994.

[15] Iwasawa, Y., **X-Ray Absorption Fine Structure (Xafs for Catalysts and Surfaces)**, World Scientific Publishing Company, 1996.

[16] Koningsberger, D.C., Prins, R., **X-Ray Absorption: Principles, Applications, Techniques of EXAFS, SEXAFS and XANES**, Wiley-Interscience, 1988.

[17] Knoll, G.F., **Radiation Detection and Measurement**, Wiley, 2000.

[18] Kuzmany, H., **Solid-State Spectroscopy: An Introduction**, Springer, 2002.

[19] Kuzmin, A. et al. **X-ray Diffraction, Extended X-ray Absorption Fine Structure and Raman Spectroscopy Studies of WO_3 Powders and $(1 - x)WO_{3-y}.xReO_2$ Mixtures.**, J. App. Phys., Vol.84, No.10, 1998.

[20] Lakowicz, J.R., **Principles of Fluorescence Spectroscopy**, Springer, 1999.

[21] Lakshami, R.J. et al., **Tissue Raman Spectroscopy for the Study of Radiation Damage: Brain Irradiation of Mice**, Radiation Research, Vol.157, No.2, 2002.

[22] MacFarlane, J.J. et al., **X-ray absorption spectroscopy measurements of thin foil heating by Z-pinch radiation**, Phys. Rev. E, 66, 2002.

[23] Matsui, S. et al., **X-Ray Diffraction Gratings for Synchrotron Radiation Spectroscopy: a New Fabrication Method**, Appl. Opt., 21, 1982.

[24] McHale, J.L., **Molecular Spectroscopy**, Prentice Hall, 1998.

[25] Misra, P., **Ultraviolet Spectroscopy and UV Lasers**, CRC, 2002.

[26] Pavia, D. et al., **Introduction to Spectroscopy**, Brooks Cole, 2000.

[27] Penent, F. et al., **Multielectron Spectroscopy: The Xenon 4d Hole Double Auger Decay**, Phys. Rev. Lett. 95, 2005.

[28] Peterson, R.S., **Experimental Gamma Ray Spectroscopy and Investigations of Environmental Radioactivity**, Spectrum Techniques Inc., 1996.

[29] Ramsey, N.F., **Spectroscopy With Coherent Radiation: Selected Papers of Norman F. Ramsey With Commentary**, World Scientific Publishing Company, 1997.

[30] Saisho, H., Gohshi, Y., **Applications of Synchrotron Radiation to Materials Analysis**, Elsevier Science Pub. Co., 1996.

[31] Sham, T.-K., **Chemical Applications of Synchrotron Radiation**, World Scientific Pub. Co. Inc., 2002.

[32] Schmidt, V. et al., **Electron Spectrometry of Atoms using Synchrotron Radiation**, Cambridge University Press, 1997.

[33] Shafroth, S.M., **Scintillation spectroscopy of gamma radiation**, Gordon and Breach Science Publishers, 1967.

[34] Smith, E., Dent, G., **Modern Raman Spectroscopy: A Practical Approach**, John Wiley & Sons, 2005.

[35] Steinfeld, J.I., **Molecules and Radiation: An Introduction to Modern Molecular Spectroscopy**, Dover Publications, 2005.

[36] Stevenson, J.R., **Reflectance spectroscopy and auger spectroscopy of metals and alloys**, School of Physics, Georgia Institute of Technology, 1974.

[37] Stoehr, J., **NEXAFS Spectroscopy**, Springer, 2003.

[38] Stuart, B.H., **Infrared Spectroscopy : Fundamentals and Applications**, John Wiley & Sons, 2004.

[39] Svanberg, S., **Atomic and Molecular Spectroscopy : Basic Aspects and Practical Applications**, Springer, 2004.

[40] Tanaka, S., Mukamel, S., **Coherent X-Ray Raman Spectroscopy: A Nonlinear Local Probe for Electronic Excitations**, Phys. Rev. Lett., 89, 2002.

[41] Teo, B.K., **EXAFS: Basic Principles and Data Analysis**, Springer, 1986.

[42] Thurn-Albrecht, T. et al., **Photon correlation spectroscopy with high-energy coherent x rays**, Phys. Rev. E, 68, 2003.

[43] Turri, G. et al., **Probing the Molecular Environment Using Spin-Resolved Photoelectron Spectroscopy**, Phys. Rev. Lett., 92, 2004.

[44] Valeur, B., **Molecular Fluorescence: Principles and Applications**, Wiley-VCH, 2001.

Chapter 13

Data Acquisition Systems

Data acquisition is an extremely important part of any radiation detection system. The development of highly sensitive and fast detectors has stretched the requirements on data acquisition systems so much that it has lead to advancements in the technology of high end electronic devices. A typical modern data acquisition system is composed of analog and digital processing modules, computer interface, computer, software, and storage device. There are essentially three ways in which a complete data acquisition system can be designed. The first one is to use the so called modular instruments, which consist of pre-built modules. These modules are designed such that they can be joined together to form a complete system. This approach was very popular in early days of electric detectors due to ease in design and implementation. Their main downsides are the use of the modules as black boxes and be restricted to the vendor's specifications. Even with these disadvantages, such systems are still popular in small scale systems where building a complete application specific system is cost prohibitive. Another approach is to design a completely application specific system. This requires large investment of capital, manpower, and time but is highly advantageous for large scale systems. The third approach is to build a hybrid system, that is, a system based on modular instruments and some application specific devices. This chapter to reviews of modular and application specific devices.

We will start this chapter with discussions on complete data acquisition systems without going into the details of their individual building blocks. Later on we will look at the standards of modular instruments. The last part of the chapter is devoted to the PC based data acquisition systems.

13.1 Data Acquisition Chain

In this section we will look at the broader picture of how specific tasks related to radiation measurements can be performed. Design of a data acquisition system, to a large extent, depends on the application. For example, a system designed for slow pulse counting can not handle high data rates. Therefore it is instructive to give examples of systems for specific applications that are commonly encountered in radiation measurements.

13.1.A Pulse Counting

Pulse counting is perhaps the most widely used operation mode of radiation detectors. The basic idea is to count the number of *good* pulses generated by individual particles as they deposit energy into the active volume of the detector. A good pulse is generally understood to represent a pulse that has been generated by the incident radiation. But the definition is somewhat arbitrary since the experimenter might want to look at the noise pulses only or pulses generated by particles of certain energy. Whatever the definition of the good pulse is, the objective is to discriminate the good pulses from all the pulses reaching the electronics. That is, all the unwanted pulses must be blocked. To accomplish this task, a typical pulse counting system has a device called *discriminator*. A typical discriminator discriminates the pulses based on their amplitudes. To do this it compares the pulse amplitude to a preset *good pulse amplitude window*. The design of this discriminator and other devices depend mainly on the rate requirements as discussed in the following two subsections.

A.1   Slow Pulse Counting

Most of the radiation measurements in laboratory and industrial environments fall into this category. The reason is that the rate of radiation emitted by typically used radioactive sources is not very high. However in case of particle accelerators the rate is too high and might not be handled by a slow counting system.

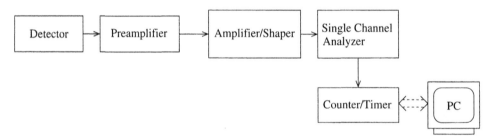

Figure 13.1.1:   Block diagram of a simple slow pulse counting system.

The building blocks of a slow pulse counting system is sketched in Fig. 13.1.1. The pulse from the detector is first preamplified. The preamplifier is usually installed near the detector output. The preamplified pulse is then transported to the main amplifier/shaper, where the pulse is properly amplified and shaped according to the input requirements of the next stage. The amplified pulse is fed into a single channel analyzer (SCA), which generates a digital pulse whenever the input pulse amplitude is within the preset window. In the next stage this digital pulse is counted. A counter module simply increments the count by one each time an SCA pulse arrives. The output of the counter can then be fed into a computer for further analysis and storage.

A.2   Fast Pulse Counting

The response time of a single channel analyzer can be prohibitive for its use in fast pulse counting applications. In such a case, a simple fast pulse discriminator can be used. Such a device would block all the pulses whose amplitude lies outside a preset window. The pulses can then be counted by a specially designed fast counter (see Fig.13.1.2). Note that a fast pulse counting application also requires a fast preamplifier. The counter output can be read out by a computer for further processing and storage.

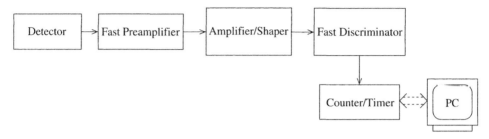

Figure 13.1.2: Block diagram of a simple fast pulse counting system.

13.1.B   Energy Spectroscopy

Energy spectroscopy has been thoroughly discussed in the chapter on spectroscopy. We saw that the best method for this kind of analysis is to use a multichannel analyzer (MCA). A multichannel analyzer can be thought to consist of an array of single channel analyzers with adjacent windows. It has a number of channels, each of which corresponds to a specific range of pulse heights. A pulse with height corresponding to one of these channels gets counted. In this way the whole spectrum of pulse heights gets recorded by the analyzer. Since the pulse height is proportional to the energy deposited by the incident radiation, the spectrum obtained corresponds to the energy spectrum of the radiation. Such a spectrum is also sometimes referred to as pulse height spectrum. The block diagram of a typical energy spectroscopy system is shown in Fig.13.1.3.

Figure 13.1.3: Block diagram of a simple energy spectroscopy system.

13.1.C   Time Spectroscopy

In some applications it is desired that the arrival time of detector pulses be precisely recorded. For example, measurement of lifetime of atomic states requires one to measure time of arrival of photons emitted as a result of those transitions. Such a measurement is generally done through the so called *time-to-amplitude* converter or TAC. A simple TAC device converts the time into a voltage pulse with a height proportional to the time. The idea is to start charging (or discharging) a capacitor at the arrival of a pulse and stop the process as soon as another pulse arrives. The charge remaining on the capacitor is then proportional to the time. Of course this requires proper calibration since the relationship between time and charge may not be perfectly linear. Apart from a TAC, one also needs a *timing discriminator* to build a time spectroscopy system. A timing discriminator can simply be a fast single channel analyzer. Recall that a single channel analyzer produces a logic pulse whenever the input pulse amplitude is between a preset amplitude window. The logic pulse thus produced also has a definite relationship to the arrival time of the linear pulse. That is the leading edge of the logic pulse can be used to determine the arrival time of the linear pulse, provided the internal electronic delay is precisely known. Most single channel analyzers also provide the user the with possibility of adding additional delay if required.

The time resolution required for time spectroscopy depends on the particular requirements of the experiment. However for typical applications it ranges from a few nanoseconds to a few picoseconds. Obtaining such a high resolution requires fairly fast electronic components with fast response and settling times. Fig.13.1.4 shows the block diagram of a typical time spectroscopy system. All the components in the chain must have fast timing to achieve the required overall timing resolution.

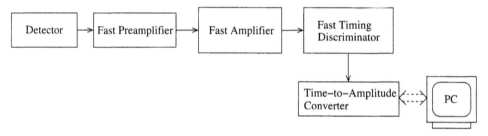

Figure 13.1.4: Block diagram of a simple time spectroscopy system.

Sometimes it is desired to perform both timing and energy spectroscopy at the same time. In such a case one can simply split the preamplifier output into two pulses such that one goes into the time spectroscopy system while the other passes through the energy spectroscopy system.

13.1.D Coincidence Spectroscopy

Time coincidence units are used to determine the coincidence between time of arrival of particles. For example, in a PET scanner two back-to-back γ-rays in time coincidence represent an electron-positron annihilation event. Additionally in some experiments one might also want to obtain the energy spectrum seen by one of the detectors.

Fig.13.1.5 shows a simple but complete time coincidence system. The particles are detected by two separate detectors that are usually of same type. The pulse is first amplified and shaped before being transported to the timing single channel analyzer. Whenever the height of a pulse is within the acceptable SCA window set by the user, a logic pulse gets sent to the coincidence module. If two such pulses arrive at the module within a time window set by the user, it generates a logic pulse. This logic pulse can be used by a multi channel analyzer to retain the pulse height information of one of the pulses related to that particular event. Note that the coincidence logic pulse must arrive at the MCA before the direct pulse from the detector reaches the MCA. To ensure this a delay amplifier is generally used to delay the arrival of the amplified detector pulse to the MCA.

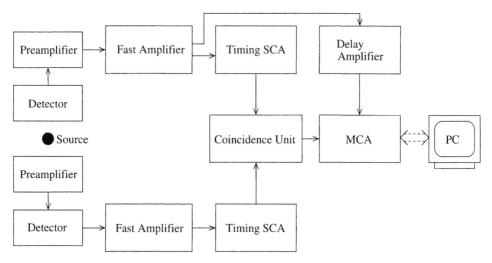

Figure 13.1.5:   Block diagram of a coincidence system capable of pulse height spectroscopy.

13.2   Modular Instruments

Modular instruments are very popular in research and industrial environments due to their off-the-shelf applicability and ease in use. The basic idea is to build a whole data acquisition system by connecting standalone modules. The obvious advantage of such systems is that they greatly reduce the cost and time involved in application specific designs and construction. The downside is that they are not suitable for large scale systems. Also the user is restricted in using the available I/O and controls as designed by the vendor instead of dictating these according to the system requirements.

In this section we will look at three most commonly used modular systems: NIM, CAMAC, and VME.

13.2.A   NIM Standard

Nuclear Instrumentation Methods (NIM) is an old standard that came into effect in 1964. The basic idea behind its establishment was to introduce standalone and replaceable modules, which could be combined together to form a complete data acquisition system. The convenience of using modules interchangeably without disturbing other parts of the system has been the main source of its widespread use by physicists and engineers. Even though new standards and modular systems have been introduced but due to its simplicity of integration and use, NIM is still used in many places.

A.1 NIM Layout

A complete NIM system consists of a *NIM crate* or *NIM bin* and *NIM modules*. A typical NIM crate has 12 slots for insertion of modules.

The communication and data transfer between the modules is realized through a built-in backplane. One of the deficiencies of the standard NIM backplane is that it does not have a digital data bus to allow computer based control. However some of the modern NIM modules have built-in communication ports, which allow them to be controlled and read out through a computer. Examples of typical communication ports are RS-232, Ethernet, USB, and IEEE-488.

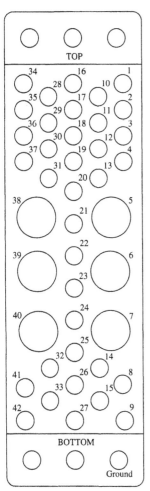

Figure 13.2.1: Rear view of a standard NIM bin. A module connector is simply a mirror image of this. The connector assignments can be found in Table 13.2.1.

A standard NIM crate has a built-in power supply that draws power from 110 V or 240 V AC outlets, though crates having other specific input voltage ratings are also available. The DC voltages are generated within the crate, that is, no extra DC supply module is needed. The voltages are distributed through the backplane and are also available in front of the crate for other user-specific requirements.

The modules are connected to the backplane through special NIM-connectors. Each slot in a NIM crate is equipped with one such socket. The connector configuration of a standard NIM bin is shown in Fig.13.2.1. The connector assignments can be found in Table 13.2.1.

Table 13.2.1: Connector assignments of a standard NIM bin (see Fig.13.2.1).

Pin	Function	Pin	Function	Pin	Function
1	Reserved	15	Reserved	29	-24 V
2	Reserved	16	+12 V	30	Spare
3	Bin Gate	17	-12 V	31	Spare
4	Reserved	18	Spare	32	Spare
5		19	Reserved	33	117 V_{DC}
6		20	Spare	34	Ground (Power Return)
7		21	Spare	35	Reset
8	+200 V_{DC}	22	Reserved	36	Gate
9	Spare	23	Reserved	37	Spare
10	+6 V	24	Reserved	38	
11	-6 V	25	Reserved	39	
12	Reserved	26	Spare	40	
13	Spare	27	Spare	41	117 V_{AC} (Neutral)
14	Spare	28	+24 V	42	Ground (High Quality)

A.2   NIM Modules

A standard NIM module has a height of 8.75 inches. The width of the module can be variable but it must be a multiple of 1.35 inches. The pin layout of a standard NIM module is a mirror image of the crate connector assignments shown in Fig.13.2.1. Both front and rear sides of a typical NIM module also contain a number of connectors to facilitate different I/O operations. Some of the modern NIM modules also contain communication ports (such as RS232) to allow control and readout through a computer. These ports are necessary due to unavailability of digital data bus in NIM crates.

As mentioned earlier, NIM modules are well suited for small scale systems that do not require complicated data handling and control. Most of the NIM modules are simply plug-n-play devices, that is, one can develop a whole data acquisition system by simply interconnecting off-the-shelf modules together in the right sequence.

A.3 NIM Logic

Standard NIM modules are designed to deliver and respond to slow-to-medium positive logic signals and fast negative logic signals. Since fast negative logic signals are more popular, they are generally said to constitute the standard *NIM logic*. These negative levels have been defined in terms of current ranges. However since the 50 W input/output impedances are also a part of the standard, the levels can be expressed in terms of voltages as well. The current standards are listed in Table 13.2.2. These fast negative pulses can have rise times as short as 1 ns. This makes them susceptible to reflections and must therefore be properly terminated. The general practice is to use 50Ω coaxial cables, such as RG-58 or RG-174, terminated with 50Ω.

Table 13.2.2:   Standard fast negative NIM logic levels.

Type	Input (mA)	Output (mA)
Logic 0	-4 to +20	-1 to +1
Logic 1	-12 to -36	-14 to -18

The levels corresponding to the slow-to-medium positive logic signals are listed in Table 13.2.3.

Table 13.2.3:   Standard slow-to-medium positive NIM logic levels.

Type	Input (V)	Output (V)
Logic 0	+1.5 to -2	+1 to -2
Logic 1	+3 to +12	+4 to +12

The reader would note that these standards are fairly different from other commonly used ones, such as TTL and ECL. This is a potential problem since it could limit the integration capability of NIM modules with systems based on other standards. To overcome this difficulty TTL-NIM and ECL-NIM translation modules have been made available. This added versatility has significantly broadened the scope of applicability of NIM based systems.

A.4   Signal Transport

Each NIM module has several signal and data I/O ports available on front and rear panels. The connection can be made through BNC, LEMO, or SMA connectors. The choice of cable is highly application dependent but for typical applications coaxial cables are generally used. For transporting digital signals, sometimes using a flat ribbon cable is more advantageous and space saving.

13.2.B   CAMAC Standard

CAMAC is an acronym of Computer Automated Measurement And Control. This standard was originally defined in 1969 by the ESONE Committee and was later on jointly standardized by the NIM and ESONE committees. With its built-in controllers and interface capabilities, CAMAC provides a more versatile architecture than NIM, though at the expense of much more complicated and difficult customization. The standard CAMAC backplane is called DATAWAY, which can be directly interfaced to a computer. This feature of CAMAC system makes it far more advantageous over its NIM counterpart where backplane does not have this functionality. In this way one can talk to any module in the crate through simple CAMAC commands without the need to connect it directly to a computer.

B.1   CAMAC Layout

CAMAC modules are housed in a CAMAC crate, which can accommodate up to 24 normal CAMAC modules. Each module slot is called a *station*. Besides these 24 stations there is another one reserved for the crate controller module. The controller module is an integral part of the system and can not be replaced by some other module. Also, some controller modules have double widths and therefore take up two slot positions. In such a case the crate can accommodate up to 23 normal modules. The pin allocation of one of these 23 normal stations is shown in Fig.13.2.2.

As mentioned earlier, the backplane of a CAMAC crate is called DATAWAY. DATAWAY consists of not only control, data, and bus lines but also module power lines. These lines are connected to the modules through sockets. There are standard guidelines for the current consumption at these sockets. The current should not exceed more than 3 A at any of the sockets. Furthermore the power dissipation per station should not exceed 8 W. However this rating can be relaxed in certain situations up to a maximum of 25 W. The power rating for the whole crate is 200 W.

B.2   CAMAC Controllers

A data acquisition system may consist of more than one CAMAC crate. The modules in each of these crates are controlled by individual crate controller modules. These modules are connected to a Branch Driver through a parallel Branch Highway. Each of the crate controllers can be identified on the Branch Highway through this driver. The Branch Driver is directly connected to a data acquisition computer, which can be programmed to issue the control commands.

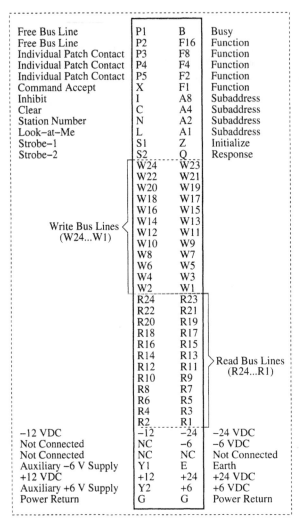

Free Bus Line	P1	B	Busy
Free Bus Line	P2	F16	Function
Individual Patch Contact	P3	F8	Function
Individual Patch Contact	P4	F4	Function
Individual Patch Contact	P5	F2	Function
Command Accept	X	F1	Function
Inhibit	I	A8	Subaddress
Clear	C	A4	Subaddress
Station Number	N	A2	Subaddress
Look-at-Me	L	A1	Subaddress
Strobe-1	S1	Z	Initialize
Strobe-2	S2	Q	Response
	W24	W23	
	W22	W21	
	W20	W19	
	W18	W17	
	W16	W15	
Write Bus Lines	W14	W13	
(W24...W1)	W12	W11	
	W10	W9	
	W8	W7	
	W6	W5	
	W4	W3	
	W2	W1	
	R24	R23	
	R22	R21	
	R20	R19	
	R18	R17	
	R16	R15	
	R14	R13	Read Bus Lines
	R12	R11	(R24...R1)
	R10	R9	
	R8	R7	
	R6	R5	
	R4	R3	
	R2	R1	
−12 VDC	−12	−24	−24 VDC
Not Connected	NC	−6	−6 VDC
Not Connected	NC	NC	Not Connected
Auxiliary −6 V Supply	Y1	E	Earth
+12 VDC	+12	+24	+24 VDC
Auxiliary +6 V Supply	Y2	+6	+6 VDC
Power Return	G	G	Power Return

Figure 13.2.2: Pin allocation of one of 23 normal stations in a CAMAC crate (as viewed from front of the crate).

B.3 CAMAC Logic

The CAMAC logic conforms to the standard TTL and DTL series logic with one exception that the signal convention is inverted such that the high state corresponds to logic 0 and low state to logic 1.

13.2.C VME Standard

VME stands for Versa Module Europa. It was jointly introduced in 1981 by Mostek, Motorola, Phillips, and Thompson. The original motivation of VME development and standardization was to introduce modular approach to highly intense computing tasks. However it was quickly adopted by researchers in other fields as well, so much that it now probably stands as the most used system in research laboratories. Due to its flexibility and expandability, it has received its fair share of popularity in industrial environments as well.

The original VME standard is now referred to as VME32 as it could handle up to 32 bits of data. The new standard is called VME64 and has the capability of 64 bit data transfers. The salient features of VME64 are listed below.

▶ 64 bit data transfer and addressing modes.

▶ Two distinct data-address transfer modes: data transfer with the address and address followed by the data.

▶ Multiplexed block transfer of data for faster delivery.

▶ 40 bit addressing and 32 bit data transfer modes for P1-only systems.

▶ Jumper selection of address is not required any more.

Another recent modification to the VME64 is the VME64x. This standard has provided more flexibility and ease in operation with faster data transfer rates. Following are the distinguishing features of VME64x.

▶ Addition of z and d rows to the P1 and P2 connectors bringing the total number of pins to 160 per connector.

▶ User defined connections available on the rear side of the crate (backplane).

▶ Better grounding scheme.

▶ More DC voltage points for P1 and P2 connectors.

▶ A number of spare pins for future additions.

Further modifications in these architectures have been done leading to faster data transfer rates. For example, the so called *2 edge VME* or 2eVME can essentially double the peak block data transfer rate of standard VME64 and VME64x up to 160 MB/second. In the subsequent discussions we will be mainly concerned with VME32 and VME64x.

C.1   VME Layout

VME was originally based on the Eurocard standard. A standard VME crate has a maximum of 21 slots, out of which 20 can be used for inserting modules. The first one is used by the *crate manager*, just like the *crate controller* of the CAMAC system. Typical VME crates are made with 20 slots so that they could fit into standard racks. Crates with lesser number of slots are also available.

The height of a VME crate is measured in units of U with $1U = 43.60\ mm$. Standard VME crate heights are $3U$, $6U$, and $9U$. The standard modules or *cards* are therefore available in these heights only. Similarly the two standard depths of a VME crate are 160 mm and 340 mm. The thickness of a VME card should not exceed 1.8 mm since then it might not slide in the rails.

C.2   VME Backplane

The backplane of a VME crate is generally referred to as *chassis*. The chassis contains connectors for the VME cards and has different buses for data, addresses, signals, and power. Following are the buses found in standard VME backplanes.

- ▶ **Utility Bus:** VME32 has +5 V and ±12 V supplies. VME64 has in addition a 3.3 V supply as well. These voltages are supplied to the modules through the utility bus. Other lines present on this bus are system clock, system reset, serial data, system failure, and AC failure.

- ▶ **Data Bus:** All the data transfer between modules takes place through this bus. The width of the bus depends on whether the system is VME32 or VME64. Data bus also contains data strobe lines. These lines contain the information about the availability and size of the data. The read and write operations are distinguished by the state of a dedicated WRITE line of the data bus. The data transfer error signal is carried by a bus error line and the data transfer completion is signaled by a transfer acknowledge line.

- ▶ **Address Bus:** Each module in a VME crate is identified by a unique address of its register. A module can therefore be accessed by invoking its register through the address bus line. The length of the address and the type of data cycle are carried by the address modifier lines.

- ▶ **Interrupt Bus:** The interrupt request lines allow initiation of a new data cycle.

C.3   VME Modules

A large number of standard VME modules are readily available in the market. With these off-the-shelf units one can essentially develop a whole data acquisition system. However many researchers prefer to develop customized modules according to their own requirements. This has also led to the development of non-standard VME backplanes. These backplanes are based on the basic VME principles but differ in bus topology and size.

The standard VME backplanes have specific connectors to which the modules must conform. The two most commonly used connector types are referred to as P1 and P2. The reader should however note that the pin assignments of these connectors for VME32 are different from those of VME64x. The VME32 P1 and P2 connectors have 3 rows of 32 pins each with a total of 96 pins. On the other hand the VME64x connectors have 5 rows of 32 pins each with a total of 160 pins. The pin assignments of these connectors can be found in Appendix C. Note that P1 connector is mandatory irrespective of the module size. Other connector types are also available, such as 95 pin P0 connector, which has 5 rows of 19 pins each.

C.4   VME Logic

VME is a TTL based system. That is, its internal functioning follows TTL logic. However modules accepting and producing other logic signals can be integrated into the system.

13.2.D   FASTBUS Standard

FASTBUS was originally developed in 1986 jointly by the U.S. NIM and the European ESONE committees. The main motivation of its development was to introduce capabilities that were not available in the NIM system. Just like the original VME architecture, the FASTBUS standard supports 32 bit data transfer and addressing. The bussed architectures allows it to be directly linked to a computer for control and data handling. The distinguishing feature of FASTBUS is its use of multiple processor buses linked together in parallel.

D.1   FASTBUS Layout

A standard FASTBUS crate has 26 slots. The user should compare this number with the standard VME crate that can have a maximum of 21 slots. The larger number of slots in a FASTBUS crate is due to its compactness and density. In fact this feature of FASTBUS makes it fairly cost effective as compared to other architectures.

One of the slots in a typical FASTBUS crate is occupied by the so called Geographical Address and Control card or GAC. This card is generally installed in the last slot but can also go into 25th slot. Another card that is an integral part of a complete FASTBUS system is the Arbitration Timing Control card or ATC. The ATC can occupy the first or the second slot in the crate.

Unlike most of the other architectures, power supply is not an integral part of a FASTBUS crate. The power requirements of FASTBUS modules therefore dictate the installation of the proper power supply. The FASTBUS standard specifies the use of $+5$ V, -5.2 V, -2 V, ±15 V, and ±28 V power supplies.

D.2   FASTBUS Backplane

A FASTBUS backplane consists of two distinct parts, namely *segment* and *auxiliary* units. The auxiliary backplane is generally not used for data transfer and is reserved for implementing custom features. The main data transfer and addressing is done over the segment backplane. The timing signals needed for addressing and arbitration are provided by the ATC. The GAC, installed on the other side of the crate, terminates the signal lines. Its main purpose is to establish communication between a slot's slave and a master. The GAC can address individual slots by broadcasting their hardwired geographic addresses.

13.3   PC Based Systems

A new trend in data acquisition is to employ PC based modules. These modules can be directly connected to a PC and generally provide a complete solution including analog processing of the input signals. In this section we will briefly discuss such solutions.

13.3.A   PCI Boards

Since most PCs have at least one PCI bus slot available, one can build a data acquisition system by installing a signal conditioning unit into this slot. Such modules provide the complete solution including signal conditioning, A/D conversion, and computer interfacing. The user has to simply connect the inputs and configure the software.

Since PCI modules are directly connected to the computer bus, they can handle high data transfer rates. Fast data polling can therefore be easily implemented in such systems. However one important point to note here is that the data transfer rate is actually limited by how the PCI board (or DAQ[1] module) communicates with the random access memory (RAM) of the computer. The microprocessor speed is not an issue here since modern microprocessors have clock frequencies in the gigahertz range. However since a microprocessor in a typical PC does multitasking therefore it can delay the data transfer to the RAM, thus damping the data acquisition frequency. This is the mode in which conventional DAQ boards work (Fig.13.3.1(a)). The data acquisition frequency can be increased if the DAQ module performs the so called *bus mastering*. What it does is that it bypasses the microprocessor and sends the data directly to the RAM. Microprocessor then acquires data from the RAM and performs the required operations (see Fig.13.3.1(b)). Since modern computers have big enough RAMs, data is generally not lost. It is, however, up to the programmer to ensure that the routines do not allow complete filling of the data buffers at any time.

The PCI based modules are good for building small scale data acquisition systems since typical modules can handle up to only 30 input channels. This is good enough for few channel detection devices but for large scale systems one should resort to other solutions.

13.3.B   PC Serial Port Modules

Most PCs contain one or more serial ports. The most commonly available port is the RS232. This port can be used to communicate with a data acquisition module. However such systems can only be used for slow data acquisition due to the serial nature of data transfer. The good thing is that the RS232 port can be used to transmit data and control signals to large distances. The RS232 standard specifies a maximum cable length of 50 feet but this does not mean that larger cables can not be used. In fact, the maximum length of the cable depends on the environment, baud rate, and the cable type. In an electromagnetically hostile environments even a few feet long unshielded cable might not work. On the other hand at low baud rate of about 110 one can use a shielded cable that is as long as 5,000 feet. Hence one should not feel constraint by the RS232 50-feet specification as good shielding and lower baud rates can assure perfect data transfer through very long cables. Another option is to use the so called RS485 port, which is specified to transmit data to about 5,000 feet.

According to the RS232 standard the devices should use 25-pin connectors. However only 9 of those are used for communication (plus one for ground). Therefore most computers have 9-pin connectors instead. The pin configurations of 25-pin and 9-pin male connectors are shown in Fig.13.3.2. The pin assignments can be found in table 13.3.1.

The most problematic things with serial (or parallel) communication is that it need direct connection of a PC with the module. This may become impractical if

---

[1]DAQ stands for data acquisition.

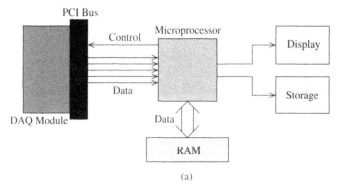

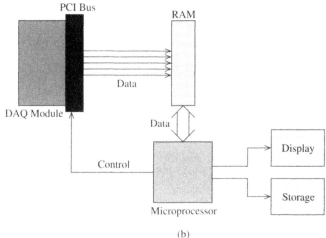

(a)

Figure 13.3.1: PCI based DAQ systems. (a) Conventional PCI module data flow. (b) Data flow with PCI module having bus mastering capability.

(b)

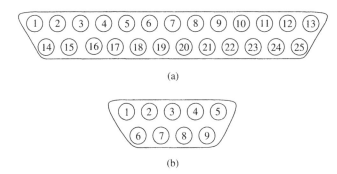

(a)

(b)

Figure 13.3.2: (a) 25-pin and (b) 9-pin RS232 male connectors. Pin assignments are given in table 13.3.1.

the module is to be installed far away from the PC. In such a situation one can use a serial to TCP/IP converter and connect it to the existing computer network. The data and control signals can then be transported to and from a remote PC anywhere on the network. We will discuss such solutions shortly.

Table 13.3.1: Pin assignments of standard 25-pin and 9-pin RS232 male connectors (see Fig.13.3.2).

		Pin Number	
Signal	Description	25-Pin Connector	9-Pin Connector
TD	Transmitted Data	2	3
RD	Received Data	3	2
RTS	Request To Send	4	7
CTS	Clear To Send	5	8
DSR	Data Set Ready	6	6
SGND	Signal Ground	7	5
CD	Carrier Detect	8	1
DTR	Data Terminal Ready	20	4
RI	Ring Indicator	22	9

13.3.C PC Parallel Port Modules

Just like the serial ports, the PC parallel ports can also be used to communicate with data acquisition modules. Parallel ports support much faster data transfer rates than the serial ports and are therefore more suitable for high rate applications. The biggest disadvantage of such systems is that the generally available parallel ports can not transmit signals beyond a few feet.

13.3.D USB Based Modules

This is a relatively new development in small scale data acquisition. Instead of installing a module into a PCI slot, the user simply connects the module to the computer through a USB board.

USB based modules have two advantages over PCI boards. One is their ease in operation since they are simply plug and play devices. The other advantage is that in most cases there is no need for an additional power supply since the USB port takes power from the PC. Of course the detector must be biased through a separate power supply. All the signal conditioning units can generally run on the USB power.

13.3.E TCP/IP Based Systems

Over the last few years a new technology, using the TCP/IP protocol, has emerged. The main idea is to simply connect the data acquisition modules to the existing network through Ethernet connectivity. This makes the data and control available to users on any point on the local network. If the local network is connected to the Internet, the user can access the system from any geographical location.

The most widely used and well established PC based data acquisition is through the RS232 communication port where the data acquisition hardware is directly connected to computer's serial port. The main problem with this scheme is the direct link between the two hardware, which becomes difficult to establish if there is substantial distance between them. The TCP/IP based systems can remedy this situation without taking away the excellent serial communication hardware. The trick is to connect the DAQ module with a serial-to-TCP/IP converter. The TCP/IP module is then connected to the local area network through Ethernet connectivity. The communication between the module and the PC is then carried through the network. If the local area network is connected to the Internet, the user can communicate with the module from any Internet access point.

There are many advantages of TCP/IP based system over conventional PCI based systems, some of which are listed below.

▶ The system is expandable at minimum cost.

▶ Eliminates the need to run long cables as the modules can be connected to any nearby network switch or hub.

▶ Each module gets its own IP address through which it can be accessed from anywhere.

▶ Each module works independently and therefore one malfunction does not bring down the whole system.

A newer class of TCP/IP based modules incorporate built-in signal conditioning units. That is, the whole system is in a small box, which can be simply connected to the sensor and the network. Some of these modules even have web servers installed in their hardware. The user can then simply point to the module's IP address through a web browser for complete monitoring and control of the module. For small scale systems with low polling frequency these modules provide an excellent alternative to the conventional serial or parallel port PC based systems. However for large scale systems one is better off employing a modular approach using standards such as NIM, CAMAC, and VME.

Bibliography

[1] Austerlitz, H., **Data Acquisition Techniques Using PCs (IDC Technology**, Academic Press, 2002.

[2] Beyon, J.Y., **Hands-on Exercise Manual for LabView Programming Data Acquisition and Analysis**, Prentice Hall PTR, 2000.

[3] Black, J., **The System Engineer's Handbook**, Morgan Kaufmann, 1992.

[4] Chiang, H.H., **Electronics for Nuclear Instrumentation: Theory and Application**, Krieger Pub. Co., 1985.

[5] DoE, **Standard NIM instrumentation system**, U.S. Dept. of Energy, Office of Energy Research, 1990.

[6] Gadre, D.V., **Programming the Parallel Port: Interfacing the PC for Data Acquisition & Process Control**, CMP Books, 1998.

[7] Himelblau, H., **Handbook for Dynamic Data Acquisition and Analysis**, Institute of Environmental Sciences and Technology, 1994.

[8] IEEE, **Camac Instrumentation and Interface Standards**, IEEE, 1982.

[9] IEEE, **IEEE Standard Modular Instrumentation and Digital Interface System**, IEEE, 1997.

[10] IEEE, **IEEE Standard Subroutines for Camac**, IEEE, 1997.

[11] IEEE, **IEEE Recommended Practice for Block Transfers in Camac Systems**, IEEE, 1997.

[12] IEEE, **IEEE Standard Parallel Highway Interface System (Camac)**, IEEE, 1997.

[13] IEEE, **IEEE Standard, Real-Time Basic for Camac**, IEEE, 1997.

[14] James, K., **PC Interfacing and Data Acquisition : Techniques for Measurement, Instrumentation and Control**, Newnes, 2000.

[15] Miner, G.F., Comer, D.J., **Physical Data Acquisition for Digital Processing: Components, Parameters, And Specifications**, Prentice Hall, 1998.

[16] Ozkul, **Data Acquisition and Process Control Using Personal Computers**, CRC, 1996.

[17] Park, J., Mackay, S., **Practical Data Acquisition for Instrumentation and Control Systems**, Newnes, 2003.

[18] Piegorsch, W.U., **A General Purpose CAMAC Interface for Large Scale Instrumentation Projects**, Kitt Peak National Observatory, 1978.

[19] Sprawls, P., **The Physics and Instrumentation of Nuclear Medicine**

[20] VITA, **American National Standard for VME64 Extensions** VMEbus International Trade Association, October 1998.

[21] Webster, J.G., **The Measurement, Instrumentation and Sensors Handbook**, CRC, 1998.

Appendices

Essential Electronic Measuring Devices

A.1 Multimeters

Multimeter is a general purpose device used to measure voltage, current, and impedance. Some multimeters are also equipped with capacitance measuring circuitry. There are essentially two kinds of multimeters: analog and digital. Before we go into their descriptions, it is worthwhile to discuss what is actually meant by measuring voltage and current.

A.1.A Measuring Voltage and Current

The information carried by a signal is contained in its power given by

$$P = VI$$
$$\Rightarrow P = \frac{V^2}{R} \quad \text{or} \quad P \propto V^2,$$
$$\text{and similarly} \quad P = I^2 R \quad \text{or} \quad P \propto I^2.$$

This implies that the relevant quantity for measurement is actually the average of the *squared* voltage or current and not the simple average. In fact, if we take an example of a simple pure sinusoidal wave and take its average, the result will be 0. The simple average is therefore not a good measure of the power information contained in the signal. Therefore the multimeters actually measure the average of the squared voltage and current. This quantity is known as the root mean square value and is defined for periodic and continuous voltage and current by

$$V_{rms} = \left[\frac{1}{T} \int_0^T (V(t))^2 \, dt \right]^{1/2} \tag{A.1.1}$$

$$I_{rms} = \left[\frac{1}{T} \int_0^T (I(t))^2 \, dt \right]^{1/2}, \tag{A.1.2}$$

where $T = 1/f$ is the period (peak-to-peak time) of the signal having frequency f.

To get a feeling of the *rms*-voltage with respect to the peak voltage, let us look at the example of a pure sinusoidal wave given by

$$V(t) = V_0 \sin(2\pi f t), \tag{A.1.3}$$

where V_0 is the peak voltage or amplitude of the signal. Using the above relation, the root mean square voltage is

$$V_{rms} = 0.707 V_0. \tag{A.1.4}$$

This shows that the root mean square value is only 70.7% of the peak value.

A.1.B   Analog Multimeter (AMM)

An analog multimeter has a needle pointer to indicate the measured parameter value on a printed scale. The scale is generally graduated and difficult to read because of small subdivisions. Reading is also prone to errors if the user's line of sight is not perpendicular to the pointer.

The input impedance of an analog multimeter is generally not very high. This is a potential problem for testing circuits having comparable or higher impedances. A typical analog meter has an impedance of around 200-300 $k\Omega$, which may not be high enough for most circuits. One should therefore be very careful in interpreting the test results. The input impedance of an analog meter is generally not stated by the manufacturers. Instead its sensitivity is quoted. The impedance can be obtained from the formula

$$R = S \times R_{max},$$

where S is the sensitivity of the meter (usually given in $k\Omega/V$) and R_{max} is the maximum range of the meter in volts.

Another problem with analog meters is that their pointers can get damaged if the input DC voltage polarity is opposite to what it should be. In such a case the pointer moves in the opposite direction and can get damaged by the force pushing it against the opposite end. The user should remember that conventionally the positive terminal of the meter is colored red while the negative terminal is colored black.

A.1.C   Digital Multimeter (DMM)

A digital multimeter uses a digital display to indicate the measured value of the parameter and the units. The display is usually made of LCD but LED based multimeters are also fairly common.

A good thing about digital multimeters is that they have a high constant input impedance. Most DMMs come with an input impedance of about 10 $M\Omega$, which is fairly high for typical electronic circuits. The impedance is generally written on the back of the DMM and the user should always confirm that it is much higher than the impedance of the circuit being tested. Care should be taken because some low cost DMMs can have input impedance in the range of 1 $M\Omega$.

A.1.D   Measuring Voltage

Measuring voltage is analogous to measuring potential difference. That is, the voltage is always measured between two points. The reference point is generally chosen to be the circuit ground. In this case, the common lead is connected to *any* ground point of the circuit and the other lead is connected to the point where voltage is to be measured.

It was mentioned earlier that the analog meters do not have very high input impedances. If the meter's impedance is not much greater than the circuit's impedance, it could lead the meter to draw some current from the circuit with the consequence of incorrect voltage reading. Therefore one should make sure that the meter's impedance is much greater than the circuit's impedance at the point where voltage is being measured. The rule of thumb is that the meter's impedance should be at least 10 times that of the highest resistance in the vicinity of the measurement point. Typical digital meters have high enough input impedance to allow them to be used in virtually any practical circuit. Preferring a digital meter over an analog one therefore has more to it than just a convenience of display.

A.1.E Measuring Current

As opposed to a voltmeter, an ammeter has a very low input impedance. The reason is that the current it is measuring must flow through it. To measure the current with an ammeter, one must allow the current to directly flow through it. That is, the circuit must be broken at the point of measurement. The ammeter should then be used to bridge the two leads. printed circuit boards it is generally not practical to measure the current through this scheme. It is most suitable in situations where the current is flowing through a lead. If the lead is not soldered to one of the ends, it can be disconnected from the circuit and then connected through the ammeter.

It is worth mentioning here that a class of instruments called *hall current sensors* are also capable of measuring the current flowing through a cable. A hall sensor generally has a donut shaped coil that is clipped over the cable. This allows a *non-invasive* measurement of current since it does not require breaking the circuit.

A.2 Oscilloscope

The biggest problem with a multimeter is that it measures only the rms value of the signal. The shape of the signal, which carries a wealth of information about the system, is totally ignored by such meters. Oscilloscope is a device that can be used to actually see the pulse and measure its properties, such as rise time, decay time, frequency, and amplitude. Moreover the amount of noise in the signal, and its AC/DC components can also be determined.

There are essentially two types of oscilloscopes: analog and digital. For typical measurement tasks one can use either of them but, as we will see later, the digital scope has a few advantages and is generally preferred.

A.2.A Analog Oscilloscope

An analog oscilloscope has two main components: display and trigger. The trigger system is coupled to the display system to create a visual effect based on the time profile of the signal. In effect two properties of the signal are displayed: the amplitude and the time variation. The vertical and horizontal scales of an oscilloscope's display represent amplitude and time respectively.

A typical display system of an oscilloscope consists of an electron gun, deflection plates, grid, focusing electrodes, and a fluorescent tube. All of these are collectively called *cathode ray tube* or simply CRT. This whole assembly is kept in a vacuumed glass enclosure to minimize parasitic electron absorption and scattering.

The basic components of an analog oscilloscope, as shown in Fig.A.2.1, will be discussed in the following subsections.

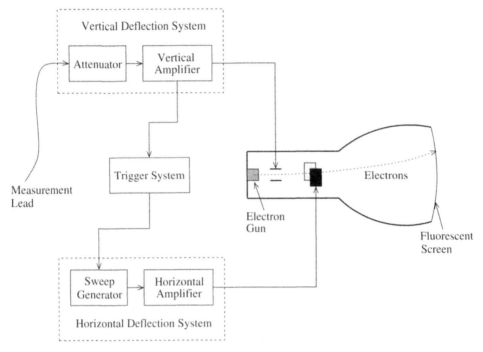

Figure A.2.1: Main components of a typical analog oscilloscope.

A.1 Attenuator

The role of attenuator is to reduce the signal amplitude according to a factor set by the user. This effectively allows the user to zoom-in the signal trace on the CRT display.

A.2 Electron Gun

The electron gun is the source of electrons that are used to create the image on the fluorescent screen. It is typically made of a filament that emits electrons when heated. In the absence of an electric field, these electrons would form a cloud around the filament. However the CRT is provided with a high potential gradient, which directs these electrons to the fluorescent screen. It should be noted that production of electrons is not initiated by the input signal. In fact, neither the production of electrons nor their intensity is dependent on the presence of signal at the oscilloscope input. Whenever the oscilliscope is turned on, the filament gets heated and starts emitting electrons.

A.3 Electron Beam Deflection Systems

The variation in the input voltage with time is displayed as a waveform in an oscilloscope. Generation of this waveform requires deflection of the electron beam in both vertical and horizontal directions. The vertical position at any instant in time is proportional to the input voltage at that time, while the horizontal position carries the time information. In this way, the value of the signal at any instant in time can be obtained.

Both vertical and horizontal deflections are carried out by letting the electron beam pass through electric potentials. The potential difference at the vertical plates depends on the input signal voltage and the range set by the user. The range is usually set by a knob provided on the front panel and is available in units of volts per division, where division is a large grid division on the screen. Typical available range set points or *deflection coefficients* are $10mV/div$, $20mV/div$, $50mV/div$, $0.1V/div$, $0.2V/div$, ... , $5.0V/div$. Some oscilloscopes also provide fine range settings between such coarse steps. The deflection coefficient set by the user can then be simply multiplied by the height of the signal trace to determine the absolute voltage at that time.

In almost all of the modern oscilloscopes there is also a possibility of letting the system adjust the deflection coefficient automatically. In this case it is not possible to determine the absolute voltage level. This setting is useful when determination of absolute voltage is not required and the signal amplitude has unexpected variations.

All oscilloscopes have at least two input channels with BNC connectors and are able to display two traces at the same time or alternately. These input channels of most oscilloscopes are terminated with $1M\Omega$ resistor in parallel with $20pF$ capacitor. This high input impedance ensures least distortion of the input signal for proper acquisition. However some of the high frequency oscilloscopes are equipped with 50Ω impedance inputs.

Most oscilloscopes provide different ways in which multiple channels can be displayed on the screen.

- ▶ **ALT Mode:** In this *alternate* mode the display is switched between channels after each sweep.

- ▶ **CHOP Mode:** In this mode the switching between channels is done in a small time step instead of the full sweep as in ALT mode. The alternating frequency can be very high ($500kHz$ or more), creating the illusion that both channels are being simultaneously displayed.

- ▶ **ADD Mode:** This mode can be used when the instantaneous sum of the two channels is required. The resulting waveform is simply the sum of the two waveforms at each instant in time.

So far we have only discussed how the amplitude of the signal is displayed. We mentioned waveform but didn't explain how the time component of the waveform is generated. This is done by the horizontal deflection system, which is used to deflect the electron beam in horizontal direction, thus producing a chart of the signal amplitude with respect to time. There are two main components of this system: a sweep generator and an amplifier. The purpose of the sweep generator is to start generating a sawtooth wave as soon as it receives a trigger initiated by the trigger system, which we will shortly discuss. The sawtooth wave is then amplified and connected to the horizontal deflection plates allowing the electron beam to sweep from left to right. The shape of the sawtooth waveform ensures that rate of this sweep remains constant till the beam reaches the end of the display. The beam is then switched off till the trace comes back to the start position on the left, as set by the user.

The time can be measured on the horizontal axis by multiplying the divisions by the current time per division setting. Usually the time per division is set through a knob on the front panel and is available in a wide range.

It is not always necessary to use the sweep generator. Sometimes it is desired to plot one signal with respect to another. Depending on the design of the oscilloscope, this can be done in more than one ways, the most common of which is to connect the two signal outputs to the two oscilloscope input channels and use the XY mode of operation.

A.4   Trigger System

We just saw that the horizontal sweep of the beam does not start until the sweep generator gets a trigger signal. This signal is generated by the trigger system, which is designed in such as way that, at every sweep, it initiates the trigger signal at the same height of the input signal. This ensures that each trace of a periodic signal synchronizes with the previous one and the display looks stable.

Most modern oscilloscopes provide the following three distinct ways to generate trigger.

- ▶ **NORM Mode:** In this *normal* mode, the user chooses a threshold level of the original signal such that whenever the signal crosses this level, a trigger signal is generated. This trigger generation can be set to be either with rising or falling edge of the signal, which are generally referred to as SIGNAL+ or SIGNAL- on the trigger panel of the oscilloscope.

- ▶ **AUTO Mode:** In this mode, the trigger system automatically generates a trigger if after a predefined time no trigger signal was generated. This of course depends on the preset triggering conditions. The advantage of this mode is that it always displays a signal even if the trigger conditions are not met.

- ▶ : **SINGLE Sweep Mode:** This triggering mode is very useful for studying single or non-periodic events, since here the sweep generator sweeps the beam only once. If the trigger conditions are properly set according to the expected signal shape, then as soon as a signal trace is found and displayed, the trigger system goes into an idle state. The user then has the opportunity to study the signal or save it in memory.

Oscilloscopes are also equipped with option of triggering on on an external signal, which can be connected to the EXT input. It is also possible to trigger on the power or line voltage, generally represented by LINE on the trigger panel.

A.2.B   Digital Oscilloscopes

In digital version of oscilloscopes, the input signal is first sampled and then converted to digital counts with an A/D converter before being displayed on the screen. The advantages are the ability to perform complex mathematical operations on the signal and storage of data in memory. The biggest downside is the dead time associated with the whole process. For example, it takes some finite time for the ADC to convert the analog signal, which then has to be converted back to analog signal

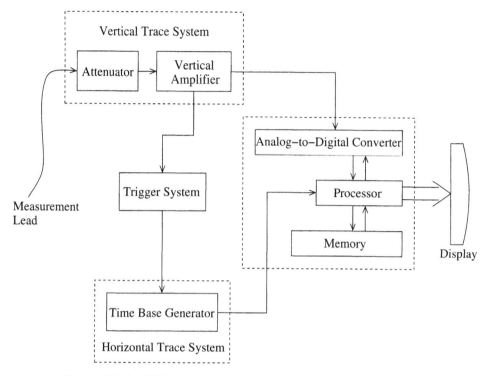

Figure A.2.2:   Main components of a typical digital oscilloscope.

using a digital to analog converter (DAC) so that it could be displayed on the fluorescent screen in much the same way as in an analog oscilloscope.

The bandwidth of a digital oscilloscope is limited by its sampling speed. Any signal frequency above the sampling frequency is ignored by the oscilloscope. This is a serious limitation of such oscilloscopes as compared to their analog counterparts, which can be built with very high bandwidths.

A good feature of digital oscilloscopes is that their operations, such as triggering, can be controlled by external logic units. Similarly the digital data can also be retrieved and transported through data bus to the processing and analysis units.

A.2.C   Signal Probes

A signal probe is used to carry the input signal to be measured to one of the oscilloscope inputs. There are different kinds of probes suitable for different kinds of input signals. But broadly speaking we can divide them into two categories: passive and active. In the following we will look at some commonly used probes belonging to both of these categories.

C.1   Passive Probes

A passive probe consists of passive electronic components and a cable to carry the signal to the oscilloscope.

A typical passive probe is shown in Fig.A.2.3. This is the so called high impedance compensated probe. Due to its high dynamic range, it can be used in a variety of circuits. In fact, due to its versatility, most oscilloscope manufactures include this in the standard oscilloscope package. The most problematic thing about this probe is its very low bandwidth, which does not allow it to be used for highly sensitive measurements. As shown in Fig.A.2.3, this probe has a variable capacitor. This allows the user to match the capacitance to the input capacitance of the oscilloscope. The compensation unit of the probe is provided in the termination box, which is usually at the other end of the probe, that is, at the oscilloscope side. The input impedance of the probe is very high, on the order of a few mega ohms. However probes having input impedances of about 1 $M\Omega$ are also not uncommon. For typical probes the input resistance is chosen such that the probe provides a 10:1 attenuation. Probes with other attenuation factors are also available.

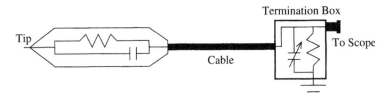

Figure A.2.3: A shigh impedance compensative passive probe.

The probe shown in Fig.A.2.3 does not have high bandwidth. This shortcoming does not allow it to be used in highly sensitive measurements, specially when the signal has a high or wide bandwidth. A much simpler probe, as the one shown in Fig.A.2.4 can solve this problem. Such a probe consists of a resistor in series with a 50 Ω signal carrying cable. Typically the value of the resistor is chosen such that it ensures a signal attenuation of 10:1. However higher attenuation probes are also available. This kind of probe is good for probing low level voltage signals specially when timing information is important. Such probes offer excellent time measurements due to their very wide bandwidths.

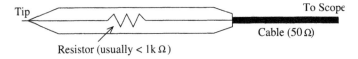

Figure A.2.4: A simple low impedance passive divider probe.

The two probes we just discussed are not the only passive probes available in the market. There is a variety of probes available and the user should carefully chose one according to the requirements.

C.2 Active Probes

A simple active probe is shown in Fig.A.2.5. As the name suggests this probe has an active electronic component, which is usually an operational amplifier. Such a probe offers excellent resistive and capacitive loading together with wide bandwidth. The dynamic range of such a probe is usually limited by the specifications of its active circuitry, which for typical probes is fairly low.

Figure A.2.5:   A simple passive probe.

All of the probes we have visited so far measure signals with ground as reference. If we wanted to measure a signal that has some other reference point, we can not use such probes. In such a situation we need the so called *differential probe*. A differential probe has two tips to probe the two points on the circuit. A simple active differential probe is shown in Fig.A.2.6.

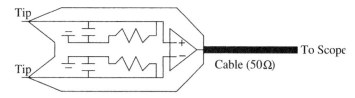

Figure A.2.6:   A simple differential probe.

Constants and Conversion Factors

 B.1   Constants

Constant	Symbol	Value	Units
Atomic mass constant	m_u	1.660×10^{-27}	kg
Atomic mass constant (energy equivalent)	$m_u c^2$	1.492×10^{-10}	J
		931.49	MeV
Avogadro's number	N_A	6.022×10^{23}	$mole^{-1}$
Boltzmann's constant	k_B	1.380×10^{-23}	$J K^{-1}$
		8.617×10^{-5}	$eV K^{-1}$
Elementary charge	e	1.602×10^{-19}	C
Plank's constant	h	6.626×10^{-34}	Js
		4.135×10^{-15}	$eV s$
Speed of light in vacuum	c	2.99×10^8	ms^{-1}
Thomson cross section	σ_e	6.652×10^{-29}	m^2

B.2   Masses and Electrical Charges of Particles

Particle	Symbol	Charge (multiples of e)	Mass
alpha	α, $^4_2He^{++}$	+2	6.644×10^{-27} kg 3727.379 MeV
deuteron	d, $^2_1H^+$	+1	3.343×10^{-27} kg 1875.612 MeV
electron, beta	e, β	-1	9.109×10^{-31} kg 0.511 MeV
neutron	n	0	1.675×10^{-27} kg 939.565 MeV
positron	e^+, β^+	+1	9.109×10^{-31} kg 0.511 MeV
proton	p, $^1_1H^+$	+1	1.672×10^{-27} kg 938.171 MeV

▌ B.3 Conversion Prefixes

Prefix	Symbol	Factor
peta	P	10^{15}
tera	T	10^{12}
giga	G	10^{9}
mega	M	10^{6}
kilo	k	10^{3}
hecto	h	10^{2}
deka	da	10
deci	d	10^{-1}
centi	c	10^{-2}
milli	m	10^{-3}
micro	μ	10^{-6}
nano	n	10^{-9}
pico	p	10^{-12}
femto	f	10^{-15}
atto	a	10^{-18}

VME Connector Pin Assignments

Table C.0.1: Pin assignments of standard VME32 P1 and P2 connectors. UD stands for user defined. The asterisk (*) specifies the opposite state of the logic: high is 0 and low is 1.

Pin	P1			P2		
	Row-A	Row-B	Row-C	Row-A	Row-B	Row-C
1	D00	BBSY*	D08	UD	+5 V	UD
2	D01	BCLR*	D09	UD	GND	UD
3	D02	ACFAIL*	D10	UD	RETRY	UD
4	D03	BG0IN*	D11	UD	A24	UD
5	D04	BG0OUT*	D12	UD	A25	UD
6	D05	BG1IN*	D13	UD	A26	UD
7	D06	BG1OUT*	D14	UD	A27	UD
8	D07	BG2IN*	D15	UD	A28	UD
9	GND	BG2OUT*	GND	UD	A29	UD
10	SYSCLK	BG3IN*	SYSFAIL*	UD	A30	UD
11	GND	BG3OUT*	BERR*	UD	A31	UD
12	DS1*	BR0*	SYSRESET*	UD	GND	UD
13	DS0*	BR1*	LWORD*	UD	+5 V	UD
14	WRITE*	BR2*	AM5	UD	D16	UD
15	GND	BR3*	A23	UD	D17	UD
16	DTACK*	AM0	A22	UD	D18	UD
17	GND	AM1	A21	UD	D19	UD
18	AS*	AM2	A20	UD	D20	UD
19	GND	AM3	A19	UD	D21	UD
20	IACK*	GND	A18	UD	D22	UD
21	IACKIN*	SERCLK	A17	UD	D23	UD
22	IACKOUT*	SERDAT*	A16	UD	GND	UD
23	AM4	GND	A15	UD	D24	UD
24	A07	IRQ7*	A14	UD	D25	UD
25	A06	IRQ6*	A13	UD	D26	UD
26	A05	IRQ5*	A12	UD	D27	UD
27	A04	IRQ4*	A11	UD	D28	UD
28	A03	IRQ3*	A10	UD	D29	UD
29	A02	IRQ2*	A09	UD	D30	UD
30	A01	IRQ1*	A08	UD	D31	UD
31	-12 V	+5 V SB	+12 V	UD	GND	UD
32	+5 V	+5 V	+5 V	UD	+5 V	UD

Table C.0.2: Pin assignments of standard VME64x P1 connector. The asterisk (*) specifies the opposite state of the logic: high is 0 and low is 1.

Pin	Row-z	Row-A	Row-B	Row-C	Row-d
1	MPR	D00	BBSY*	D08	VPC
2	GND	D01	BCLR*	D09	GND
3	MCLK	D02	ACFAIL*	D10	+V1
4	GND	D03	BG0IN*	D11	+V2
5	MSD	D04	BG0OUT*	D12	RsvU
6	GND	D05	BG1IN*	D13	-V1
7	MMD	D06	BG1OUT*	D14	-V2
8	GND	D07	BG2IN*	D15	RsvU
9	MCTL	GND	BG2OUT*	GND	GAP*
10	GND	SYSCLK	BG3IN*	SYSFAIL	GA0*
11	RESP*	GND	BG3OUT*	BERR	GA1*
12	GND	DS1*	BR0*	SYSREST	+3.3 V
13	RsvBus	DS0*	BR1*	LWORD	GA2*
14	GND	WRITE*	BR2*	AM5	+3.3 V
15	RsvBus	GND	BR3*	A23	GA3*
16	GND	DTACK*	AM0	A22	+3.3 V
17	RsvBus	GND	AM1	A21	GA4*
18	GND	AS*	AM2	A20	+3.3 V
19	RsvBus	GND	AM3	A19	RsvBus
20	GND	IACK*	GND	A18	+3.3 V
21	RsvBus	IACKIN*	SERCLK	A17	RsvBus
22	GND	IACKOUT*	SERDAT*	A16	+3.3 V
23	RsvBus	AM4	GND	A15	RsvBus
24	GND	A07	IRQ7*	A14	+3.3 V
25	RsvBus	A06	IRQ6*	A13	RsvBus
26	GND	A05	IRQ5*	A12	+3.3 V
27	RsvBus	A04	IRQ4*	A11	LI/I*
28	GND	A03	IRQ3*	A10	+3.3 V
29	RsvBus	A02	IRQ2*	A09	LI/O*
30	GND	A01	IRQ1*	A08	+3.3 V
31	RsvBus	-12 V	+5 V Standby	+12 V	GND
32	GND	+5 V	+5 V	+5 V	VPC

Table C.0.3: Pin assignments of standard VME64x P2 connector. UD stand for user defined.

Pin	Row-z	Row-A	Row-B	Row-C	Row-d
1	UD	UD	+5 V	UD	UD
2	GND	UD	GND	UD	UD
3	UD	UD	RETRY	UD	UD
4	GND	UD	A24	UD	UD
5	UD	UD	A25	UD	UD
6	GND	UD	A26	UD	UD
7	UD	UD	A27	UD	UD
8	GND	UD	A28	UD	UD
9	UD	UD	A29	UD	UD
10	GND	UD	A30	UD	UD
11	UD	UD	A31	UD	UD
12	GND	UD	GND	UD	UD
13	UD	UD	+5 V	UD	UD
14	GND	UD	D16	UD	UD
15	UD	UD	D17	UD	UD
16	GND	UD	D18	UD	UD
17	UD	UD	D19	UD	UD
18	GND	UD	D20	UD	UD
19	UD	UD	D21	UD	UD
20	GND	UD	D22	UD	UD
21	UD	UD	D23	UD	UD
22	GND	UD	GND	UD	UD
23	UD	UD	D24	UD	UD
24	GND	UD	D25	UD	UD
25	UD	UD	D26	UD	UD
26	GND	UD	D27	UD	UD
27	UD	UD	D28	UD	UD
28	GND	UD	D29	UD	UD
29	UD	UD	D30	UD	UD
30	GND	UD	D31	UD	UD
31	UD	UD	GND	UD	GND
32	GND	UD	+5 V	UD	VPC

Index

Printed and bound by CPI Group (UK) Ltd, Croydon, CR0 4YY

03/10/2024

01040418-0019